THE LOGIC
AND
METHODOLOGY
OF SCIENCE
AND
PSEUDOSCIENCE

THE LOGIC
AND
METHODOLOGY
OF SCIENCE
AND
PSEUDOSCIENCE

FRED WILSON
University of Toronto

Canadian Scholars' Press Inc.   Toronto   2000

The Logic and Methodology of Science and Pseudoscience
Fred Wilson

First published in 2000 by
Canadian Scholars' Press Inc.
180 Bloor Street West, Suite 1202
Toronto, Ontario
M5S 2V6

CSPI acknowledges the financial support of the Government of Canada through the Book Publishing Industry Development Programme for our publishing activities.

Canadian Cataloguing in Publication Data

Wilson, Fred, 1937–
    The logic and methodology of science and pseudoscience

Includes bibliographical references and index.
ISBN 1-55130-175-X

1. Science — Philosophy. 2. Science — Methodology. I. Title.

Q175.W66 2000   501   C00-932001-6

Managing Editor: Ruth Bradley-St-Cyr
Marketing Manager: Susan Cuk
Page layout: Brad Horning

00   01   02   03   04   05   06   6   5   4   3   2   1

Printed and bound in Canada by Mothersill

Dedication:
to May Brodbeck,
friend and teacher,
from whom we have all learned a lot

Felix qui potuit rerum cognoscere causas
Fortunate is he who can discern the causes of things
(Virgil, II *Georgics*)

ACKNOWLEDGEMENTS

Material in Chapter Four, section (iii) is derived from F. Wilson, *Hume's Defence of Causal Inference* (Toronto: University of Toronto Press, 1997), Ch. 3, sec. 9.

Material in Chapter Four, section (iv) is derived from F. Wilson, "Empiricism and the Epistemology of Instruments," *The Monist*, 78 (1995), pp. 207-229.

This material is used with permission.

The photograph of the author is by Kenneth Quinn.

PREFACE

During the late '30s, through the '40s and well beyond, Soviet biological research was organized around the theories of Trofim Lysenko. These theories had little to support them by way of evidence — only some crude uncontrolled experiments. Moreover, they contradicted accepted genetic theory that had a mass of evidence to support it. But because Lysenko had the ear first of Stalin and later of Khrushchev he was able quite literally to wipe out those who disagreed with him, and to impose his crackpot views on the scientific community in the Soviet Union.

Commenting on this affair, Michael Polanyi argues[1] that when the final authorities, the politicians and the bureaucrats, who control science are not scientists themselves, then sooner or later quacks and cranks will flourish and ultimately dominate the field. Even if we suppose what is not always the case with regard politicians and bureaucrats who constitute the state authority, that this authority has the best of intentions, those who are not themselves practising scientists cannot decide scientific questions and cannot, in the last analysis, choose between scientists and charlatans, between science and pseudoscience.

Polanyi reports how a general Soviet conference in 1932 "decided that genetics and plant-breeding should henceforth be conducted with a view to obtaining immediate practical results and on lines conforming to the official doctrine of dialectical materialism, research being directed by the state."[2]

He then continues:

No sooner had these blows been delivered against the authority of science than the inevitable consequences set in. Any person claiming a discovery in genetics and plant-breeding could henceforth appeal directly over the heads of scientists to gullible practitioners and fallacious theories advanced by dilettantes, cranks and imposters could now gain currency, unchecked by scientific criticism.[3]

His warning is just. There is no doubt that a triumph of the actions of those who defend "creation science" to get their pseudoscience into the school curriculum would have as devastating effect on biology in North America as Lysenko's activities had in the Soviet Union. The existence of a section on alternative medicine in the National Institutes of Health in the United States, the success of chiropractors in getting politicians to recognize them as a medical profession, and similar activities show that those who are unable to recognize pseudoscience can be persuaded to give it the status to which science alone, through its cognitive successes, can lay claim.

Polanyi's argument is that science ought to be autonomous, that it cannot continue to make progress, unless the politicians and bureaucrats are kept at an arm's distance and scientists alone make decisions relevant to the direction and progress of research. There is a certain point to this. It certainly is valid for the moving edge of science, where scientists alone, and often only a very few of these, are in a position to decide where and when it is worth committing resources.

Nonetheless, the blanket claim that non-scientists cannot learn to distinguish science from pseudoscience is, it seems to me, false. In order to distinguish science from pseudoscience what one has to do is grasp a few of the fundamental cognitive norms that define science, some of the elements of what is known as the philosophy of science such as a sense of the logic of the experimental method, and some of the basic concepts of science, in particular the science of Newton and the science of Darwin. If these are ready to hand, then it is easy enough to see why creation science is pseudoscience, why homeopathy is pseudoscience, and so on. Judge William Overton mastered the fundamental points when he rendered his judgment in Arkansas that ruled creation science was not science but religion disguised as science, and concluded that it therefore had no place in the classrooms in the United States. The politicians who insisted that resources of the National Institutes of Health in the United States be devoted to alternative medicine did not master the fundamental points. They should have done so. They should have undertaken the simple course of learning the basic standards of science, as Judge Overton did, before they made their decision. But the present point is the simple one that these norms and standards can, contrary to what Polanyi argues, be learned by any reasonably intelligent person.

This book has been written with that conviction in mind. It grows out of lectures that I have given over the years to a mixed group of undergraduates in a Philosophy course entitled "Science and Pseudoscience." On the whole the students have reacted positively and enthusiastically to the material. Their response has encouraged the present version of the same material. I hope that others than simply my own students will find it useful.

Endnotes

1   M. Polanyi, "The Autonomy of Science," *Scientific Monthly*, 60 (1945), pp. 141-150.

2   *Ibid.*, p. 147.
3   *Ibid.*

TABLE OF CONTENTS

INTRODUCTION

We all know if only somewhat vaguely what science is. It is a body of knowledge in terms of which we understand the world. Moreover, it provides us with a knowledge of causes of things in a way that enables us to predict and sometimes control things and events and processes in the natural world. With the knowledge of physics and astronomy we have been able to understand in a remarkable way the motions of the planets about the sun in our solar system. Physics has also provided us with knowledge of the parts of matter that are too small to see. Using this knowledge, we have been able to build all sorts of marvellous things, from telephones to television. Atmospheric physics has enabled us, if not to control the weather, at least to improve our ability to forecast it and thereby to take steps to help mitigate its effects. Biology has provided us with knowledge of the structure of the human DNA, which carries information from one generation to the next as organisms reproduce. This has provided us with important information about heritable diseases. Chemistry has provided us with medical marvels such as sulfa drugs and penicillin. It has enabled us to discover what it is about willow bark that helps relieve pain and has enabled us to synthesize it as ASA and to provide it in less risky pure forms in standard doses with predictable effects.

Science, then, is a body of knowledge about the order that is in the world, about its causal structure. Such knowledge of causes enables us to understand the world. It also enables us to predict changes in the world. The causal knowledge that it provides enables us to intervene effectively to achieve desirable ends.

Of course, it also enables us to do things that are undesirable. If we have penicillin, we also have atomic bombs. The knowledge of causes that science yields does not dictate whether that knowledge will be used for good or for ill.

But whether or not the knowledge that is science is used for good or for ill, it is still knowledge. In contrast, there is a lot of apparent nonsense that seems to lie outside science. There are still cranks who believe in a flat earth.

Here are some points about pseudoscience.

- Nancy Reagan at one time regularly advised her husband Ronald on affairs of state on the basis of the advice of her astrologer. Some people take Jean Dixon and other astrologers seriously, many reject their claims as pseudoscience. But it does seem profoundly wrong when not only the fate of empires but that of humankind rests of the questionable lore of astrologers.
- Some investment advisors base their recommended investment strategy on the deliverances of astrologers. Many would wonder whether this is the wisest of practices.
- Some people believe that one can give scientific evidence to support the claim that species of living things were specially created not too many years

ago by a supernatural Being; many reject this claim as pseudoscientific. The State Board of Education of Kansas in the United States has in effect banned the teaching of Darwin's theory of evolution by natural selection in state public schools. Others do not go so far, arguing only that teachers in schools ought to give equal time to "creation science" as to Darwin's theory. The objections to Darwin are often based on nothing more than a failure to distinguish evolution from natural selection. But in any case, it seems remarkably perverse to attempt to banish Darwin's theory or to suggest that it is somehow profoundly wrong when the whole of scientific biology, medical research included, proceeds everywhere on the basis of Darwin's theory and at no point on the notions of "creation science."

- Some people think that human civilization must have originated from a visit by extraterrestrial beings, many reject this claim as pseudoscientific: Einstein's theory of relativity renders it unlikely in the extreme that the planet Earth has had extraterrestrial visitors. Moreover, it certainly seems degrading to our ancestors, if not racist, to suggest that they are incapable of certain engineering feats, such as building pyramids in the desert, for example, and required the help of beings from elsewhere in the universe.

- Some parents reject medical treatments based on scientifically validated procedures and insist instead that their children be treated by practitioners of so-called "alternative medicine." Since the "alternative" treatments have not been tested, whether or not they have curative powers is in question. It is therefore also possible that these regimes will have dangerous effects. Is this fair to the children?

This is the problem with pseudoscience: it is dangerous — politically, cognitively, and socially.

The interesting question is this: how do we distinguish science from nonsense? To use the example just given, just what are the arguments that establish that the earth is, roughly, round, and that it is certainly not flat? Most people are quite confident that the earth is not flat, but

would be hard put to say exactly why the claim that the earth is flat is nonsense.

There are other similar views on the fringes of science that many scientists claim are nonsense. For example, there are various kinds of alternative medicine such as naturopathy or homeopathy that most physicians would dismiss as nonsense. But why are they nonsense? And what about UFOs? There are many who believe that apparent UFOs have in fact been identified and that they are vehicles transporting to our planet extraterrestrial intelligent beings. Most scientists dismiss these views as nonsense. But why? More importantly, why do they do it so peremptorily? In the case of things such as alternative medicines and UFOs, there are bodies of literature, various regular journals, societies of practitioners, and so on, all of which give legitimacy to these practices and views. And yet regular scientists dismiss these views without any serious and in-depth examination of the relevant literature. Have these regular scientists in fact lost the open-mindedness and sense of objectivity that are supposed to characterize the genuine searcher after truth?

The difficulty is that, if you do not have a clear idea of what science is then you will be hard put to say what it is about pseudoscience that makes it pseudo, what it is about nonsense that makes it non-sense. What is it about the theories of science that make it unreasonable to take seriously the theories of, say, homeopathy? What is it about scientific observation that makes it unreasonable to think that people have actually observed vehicles piloted by extraterrestrial intelligences? In order to answer questions such as these, it is necessary to know what science is, what are the patterns of reasoning and inference that are involved in science, and why these patterns are reasonable while those of pseudoscience and nonsense are not.

Of course, one of the best ways to come to understand what science is, and why it is reasonable, is by contrasting it to pseudoscience and nonsense. The sort of nonsense we are talking about has its own patterns of inference, its own systems of reasoning. Examining these can be illuminating: to find out what they are and to find out why they are unreasonable can help us understand what science is and why science, unlike pseudoscience, is reasonable.

Our aim is to discover the patterns of inference that we find in science, and the logical structure of the laws and theories to which such inferences lead. We shall try to do this by contrasting these patterns and structures to those of pseudoscientific theories of the sort that abound on the fringes of science. We shall also try to find out why science is reasonable in its inferences while pseudoscience is not. We shall, in other words, try to find out what makes science rational and pseudoscience pseudo.

We can, perhaps, begin to see what science is by noting what one informed observer, Thomas Kuhn, has said. Kuhn has delineated certain standards "which have held for scientists at all times," a "set of commitments without which no [person] is a scientist."[1]

> The scientist must … be concerned to understand the world and to extend the precision and scope with which it has been ordered. That commitment must, in turn, lead him [or her] to scrutinize, either for himself [or herself] or through colleagues, some aspect of nature in great empirical detail. And, if that scrutiny displays pockets of apparent disorder, then these must challenge him [or her] to a new refinement of his [or her] observational techniques or to a further articulation of his [or her] theories.[2]

Kuhn makes a number of points here which should be noted.

The first point to note is that the scientist is *concerned* to understand the world. This concern is a form of *interest*. It is an interest in understanding or coming to understand the world. It is an interest, in other words, that is satisfied by *knowledge*. It is thus a *cognitive interest*. So, scientists are moved by a cognitive interest in coming to understand the world.

What would the *motive* be for such a cognitive interest? We have already noted that the knowledge provided by science can be useful. It enables us to predict better than we could without it, e.g., in weather forecasting. It also enables us in certain cases at least to intervene effectively to achieve desirable ends, e.g., in the use of antibiotics in medicine. So one of the motives we might have for a cognitive interest in scientific knowledge is a *pragmatic interest* in the *application* of such knowledge. But one could also simply be interested in the knowledge for its own sake. One could, that is, be

interested in the knowledge for no other reason than that of *idle curiosity*.

We can therefore distinguish two sorts of motive behind the cognitive interest of science in understanding nature. One is the motive of idle curiosity. The other is some pragmatic interest.

But in what does such understanding of nature consist? What Kuhn submits is that the scientist aims to understand the world by discovering how it is *ordered*. Understanding consists in knowing the order of things, the *general patterns* or *regularities* that can be discovered to describe the world: we understand an event when we can fit it into a pattern, a pattern that holds regularly among things.

If we have knowledge of a *general pattern* or *regularity* then we will be able to predict the way things will develop and change. If we have the pattern that "All A are B", then if we come across an A that we have hitherto not observed, then we will be able to predict that it will be a B.

If the newly observed A turns out to be a B, then we shall have *confirmed* that it is a pattern in things, a regularity, that all A are B.

Of course, if we observe that the A is not B, contrary to our expectations, then we shall have to reject as false the judgments that we made that the pattern in the world is that all A are B. The task of discovering the order in things, the patterns or regularities in the world, will have to begin again.

Scientific understanding is provided by knowledge of patterns or regularities. Such understanding, then, is in terms of knowledge that enables us to explain and predict events in the world. It is also important for practical purposes. Such knowledge of regularities enables us to intervene effectively in the course of nature in order to achieve desirable ends.

It is important to note that the world in which the scientist is interested is the world of ordinary experience, the world that we know by means of our senses. The knowledge is *empirical* or *matter of fact knowledge*, it is obtained in the first instance by means of *observational techniques*.

When a scientist notes some apparent disorder in nature, then, moved by his or her cognitive interest in knowing the order in nature, he or she undertakes to discover the order that is there, behind the apparent disorder. The scientist proceeds in on the basis of *the scientific method* to try to discover that order. The scientific method begins with *observation*, and from there by *inference* arrives at judgments about the patterns and regularities that hold in the natural world.

A scientist will not, of course, merely search randomly in the area where he or she discovers apparent disorder. Rather, he or she will usually approach the area with an *hypothesis* in mind. This hypothesis is a *guess* as to what the order or pattern is that will be discovered. The scientist then searches for observational evidence that will either *confirm* the hypothesis or lead to its *rejection*. The observational evidence will either fit the hypothesized pattern or it will not. If it does fit the pattern the scientist has guessed, then the hypothesis is confirmed. If, contrariwise, the observational evidence goes against the hypothesized pattern, does not fit it, then the guess was wrong, and the hypothesis is to be rejected as false.

Nor is the choice of hypothesis to be tested something merely random. When we note some apparent disorder, there is often a good body of background knowledge concerning the area. Surrounding the area where there is an apparent lack of order is often enough a known order. The hypothesis must fit with this background knowledge. If, for example the background knowledge is itself some form of theory, then the hypothesis must fit this theory. Research in the order of the world is thus often *theory-guided*.

Such a theory is itself a record of *order* that has been discovered in the natural world; it itself is a *general pattern* or *set of patterns* of things and events in the natural world that we know by ordinary sense experience. (Often enough, we will also have *patterns of patterns*, *regularities about regularities*.) This knowledge is, like all scientific knowledge, *empirical*, and is justified by *inferences from observational data*.

Science is thus a **body of knowledge**, a **method**, and a set of **cognitive attitudes**.

Science is in the first place a set of propositions, systematically inter-related, that are capable of truly explaining various events. These propositions are the laws and theories of science, **laws and theories that we use to explain and predict**.

But science is more than this. What is characteristic of science, in contrast to certain accounts of religious truth for example, is that the set of propositions that make up the body of accepted science at a certain time is different from the body of propositions that make up science at a later time. Science, as a body of knowledge, is always changing. To put it another way, what is accepted as scientific truth is always accepted only tentatively; it is always open to revision.

This change is systematic, a consequence of the practice of science in conformity with the rules of the scientific method. These rules are a set of standards such that if we act in conformity with them, then our knowledge will improve. Error will be discovered and eliminated, to be replaced with something which, if not for sure the truth, then something which more reasonably approximates to it. And gaps in our knowledge will systematically be sought out and filled with knowledge that is less gappy. The methods of science are those of experiment and of hypothesis, and the methods of statistics for extending our judgments into areas of uncertainty. Thus, science is a **method of research for the improvement of knowledge and the discovery of truth**.

Finally, science is also a set of cognitive attitudes. We accept the propositions that make up science as a body of knowledge, **accept** them as explanatory. We also **adopt** the method of science as the one that we **believe** will best improve the body of propositions that we accept as science. These attitudes are **cognitive attitudes**, attitudes with regard to the propositions that we **ought** to accept as scientific, and the method that we **ought** to use in order to improve our scientific knowledge.

Besides science, there is also pseudoscience. A pseudoscience too is a systematic body of propositions, practices and attitudes. It is not science but gives the appearance of being science. Indeed, it must appear sufficiently close to science that people can be deceived into wrongly accepting it as science. Pseudoscience misleads people into thinking it is science. In order to give this appearance that can lead people to mistake it for science, pseudoscience must share many characteristics with science. But it also has other characteristics that distinguish it from science. These properties are deeper and perhaps less easily discernible. That is why it is often easy to be misled into thinking pseudoscience is science.

This distinction between science and pseudoscience is important for many reasons. The defenders of pseudoscience often argue, for example, that because it is science it should be taught in the schools alongside what is generally accepted as science. This has been argued in detail by what goes under the name of "creation science," which certain Christian fundamentalist groups have urged be given a place in *science* curricula in high schools alongside Darwin's account of the evolution of species by natural selection. They urge this on grounds that "creation science" is genuine science. And on the face of it, it does look very much like ordinary science. It uses the same sort of technical language as science; the theories appear to be as far, or further-reaching, than those of science, and certainly as equally impressive. It is claimed that these theories are supported as well or better than the theories of Darwin and his successors. Criticisms are met by impressive, and often complex arguments. There are special journals and learned societies and professional institutes devoted to the development and propagation of "creation science." But there are deeper differences that serve to show that "creation science" is not real science after all.

One can come up with other examples that share these surface characteristics with "creation science." There is astrology, for example. Here again there are impressive theories, special methods, learned societies, specialist journals. Another example would be the theories of Velikovsky.

What we are concerned with is the demarcation of science from pseudoscience. It is this sort of understanding of science that is necessary if we are to be able to argue successfully that pseudosciences are not sciences, that they should not be accorded the prestige of the latter nor treated on a par with the latter in regard to explanation. We will be able to show, convincingly, that they have no place in the high school science curriculum. We will be able to show that astrology is not only non-sense but dangerous nonsense, that is it unwise in the extreme, for example, to base U. S. foreign policy on the advice given to the President's wife by her favourite astrologer.

To demarcate science from pseudoscience means that we must uncover the deeper structure of science that serves to distinguish it as the genuine thing from the other theories and practices that constitute pseudoscience. We must, further, be able to show why it is rational to accept, as we do, for purposes of explanation and

prediction the laws and theories and methods of science, while it is irrational to accept the theories and practices of pseudoscience.

The general method that we will use in attempting to discover how to demarcate science from pseudoscience is by looking at examples. Through these examples we will extract certain principles. We will try to refine these principles as we proceed, and, ultimately, to show why they are rational. We shall also, of course, look at examples of pseudoscience.

In the case of science, we shall begin with some rather simple examples. These are drawn from Herodotus. These simple examples will enable us to introduce a variety of basic but important points.

We will then begin our discussion of pseudoscience. As examples of pseudoscience, we will look at claims concerning extra-terrestrial visitors, at "creation science," and at Velikovsky's theories. As well as creation science we shall look at another attack on biological theory, that of Lysenko's theories of heredity. We shall also look at astrology, and various forms of alternative medicine such as homeopathy.

We shall then take a detailed look at the emergence of science — the "new science" — from prescientific theories permeated by Aristotelian metaphysics. We shall look in particular at the work of Galileo and Newton. Again, various concepts will be brought out. One of these will be the cognitive ideals of scientific explanation, another will be the different logical forms that laws can exemplify. After this look at theories in physics, we shall turn to a detailed examination of the Darwinian theory of evolution of species by natural selection. The latter will be the occasion for a further examination of creation science.

Before turning to these discussions, one further point should be made.

Sometimes what is treated as pseudoscience comes to be scientifically acceptable. This happened in the case of the geological theory of continental drift. When this theory was first proposed, it was treated by many as pretty close to pseudoscience; today, it is accepted scientific wisdom. This example shows that, under certain circumstances, pseudoscience can become science. What are those circumstances? Essentially it is a matter of evidence becoming available that persuades the scientific community of the truth of the theory in question. Could astrology become in the same way acceptable as science? Could "creation science" make the same journey from pseudoscience to science? As we shall see, there are strong reasons for thinking that this will never happen. There are features of "creation science" which leave it irremediably flawed, features that it does not share with early versions of the theory of continental drift and which inevitably preclude scientific evidence from accumulating sufficiently to lead us to revise our evaluation of "creation science" as pseudoscience.

Endnotes

1   Thomas Kuhn, *The Structure of Scientific Revolutions*, Second Edition (Chicago: University of Chicago Press, 1970), p. 42.

2   *Ibid.*

CHAPTER 1

SCIENTIFIC INFERENCES: SOME EXAMPLES, SOME LESSONS

Hippocrates of Cos, the great physician who lived somewhere around 400 BCE, founded a school of medicine, and he or his disciples wrote a series of treatises that were used for many generations as textbooks — right down, in fact, to the 18th century — and today medical schools still administer a version of his famous oath to their graduates.

What Hippocrates argued was that medicine should be based on a knowledge of *natural causes*. Thus, in the treatise entitled "The Sacred Disease," concerned with epilepsy, the Hippocratic author tells us that

I do not believe that the "Sacred Disease" is any more sacred than any other disease but, on the contrary, has specific characteristics and a definite cause. Nevertheless because it is completely different from other diseases, it has been regarded as a divine visitation by those who, being only human, view it with ignorance and astonishment. This theory of divine origin, though supported by the difficulty of understanding the malady, is weakened by the simplicity of the cure, consisting merely of ritual purification and incantation.[1]

Disease, according to this author, is not a matter of enchantment; it has natural causes. The resort to unnatural causes, causes that are beyond the reach of ordinary means of investigation, is a result of a combination of ignorance of the true, natural causes, on the one hand, and, on the other hand, the very human need for a way to cure or at least control the disease. These together create the fantasy that there is a divine origin that can be controlled by ritual and incantation. But the futility of the "explanation" of the disease is shown by the fact that this sort of "cure" is in fact ineffective. What is needed is further research into the natural causes rather than resort to fantasy, and the supposition that the world is an enchanted place that can be controlled by magic.

Hippocrates provided here the first clear statement of the case for natural science. He was not alone in this period in his insistence upon the need to rely upon naturalistic causal explanations of things. Another was Herodotus. There are some very interesting examples in the latter and these can provide a very useful starting point for any attempt to discover the important logical and epistemological features of explanations that eschew the beings of an enchanted world and rely instead upon *naturalistic* or *empirical causes*. We therefore begin with Herodotus as our first example of attempted scientific explanations. After examining some of his efforts to understand things scientifically, we shall look at some examples of pseudoscience before turning to more recent examples of science with Newton and Darwin.

Herodotus on the Nile

Herodotus describes his journey to Egypt in the second book of his *History*. Among the many fascinating things that he saw, one of the most interesting to him was the river Nile and the regularity of its annual flooding. He wondered about the cause of this phenomenon, and suggested some possible explanations.

Now the Nile, when it over flows, floods not only the Delta, but also the tracts of country on both sides of the stream which are thought to belong to Libya and Arabia, in some places reaching to the extent of two days' journey from its banks, in some even exceeding that distance, but in others falling short of it.

Concerning the nature of the river, I was not able to gain any information either from the priests or from others. I was particularly anxious to learn from them why the Nile, at the commencement of the summer solstice, begins to rise, and continues to increase for a hundred days — and why, as soon as that number is past, it forthwith retires and contracts its stream, continuing low during the while of the winter until the summer solstice comes around again. On none of these points could I obtain any explanation from the inhabitants, though I made every inquiry, wishing to know what was commonly reported — they could neither tell me what special virtue the Nile has which makes it so opposite in its nature to all other streams, nor why, unlike every other river, it gives forth no breezes from its surface.

Some of the Greeks, however, wishing to get a reputation for cleverness, have offered explanations of the phenomena of the river, for which they have accounted in three different ways. Two of these I do not think it worth while to speak of, further than simply to mention what they are. One pretends that the Etesian winds [the northwest winds blowing from the Mediterranean] cause the rise of the river by preventing the Nile-water from running off into the sea. But in the first place it has often happened, when the Etesian winds did not blow, that the Nile has risen according to its usual wont; and further, if the Etesian winds produced the effect, the other rivers which flow in a direction opposite to those winds ought to present the same phenomena as the Nile, and the more so as they are all smaller streams, and have a weaker current. But these rivers, of which there are many in both Syria and in Libya, are entirely unlike the Nile in this respect.

The second opinion is even more unscientific than the one just mentioned, and also, if I may say so, more marvellous. It is that the Nile acts so strangely because it flows from the Ocean, and that the Ocean flows all round the earth.

The third explanation, which is very much more plausible than either of the others, is positively the furthest from the truth; for there is really nothing in what it says, any more than in the other theories. It is, that the inundation of the Nile is caused by the melting of snows. Now, as the Nile flows out of Libya [Central Africa], through Ethiopia into Egypt, how is it possible that it can be formed of melted snow, running, as it does, from the hottest regions of the world into cooler countries? Many are the proofs whereby anyone capable of reasoning on the subject may be convinced that it is most unlikely that this should be the case. The first and strongest argument is furnished by the winds, which always blow hot from these regions. The second is, that rain and frost are unknown there. Now, whenever snow falls, it must of necessity rain within five says; so that, if there were snow, there must be rain also in those parts. Thirdly, it is certain that the natives of the country are black with the heat, that the kites and swallows remain there the whole year, and that the cranes, when they fly from the rigours of a Scythian winter, flock thither to pass the cold season. If then, in the country whence the Nile has its source, or in that through which it flows, there fell ever so little snow, it is absolutely impossible that nay of these circumstances could take place.

As for the writer who attributes the phenomenon to the Ocean, his account is involved in such obscurity, that is impossible to disprove it by argument. For my part I know of no river called Ocean, and I think that Homer, or one of the earlier poets, invented the name and introduced it into his poetry.[2]

We notice one thing right away. Herodotus is not a hunter-gatherer with regard to the facts. He is not simply picking up pieces of information. Science is not the random collection of facts, nor is explanation an arbitrary putting together of facts.

What Herodotus is seeking is an *order* or *connection* among the facts: Herodotus is seeking a connection between the rising of the Nile and some other facts which will explain it. Facts about individual things are in themselves disconnected and isolated. One fact or set of facts will explain another provided that we can establish that there is a connection between them. The problem, then, is to find not only "the facts" but the connections among the facts.

This comment in turn makes a second point. Herodotus begins with a *problem*. Why, he asks, does the Nile flood each year at the summer solstice? We have a *problem* when there is something we *want* but *do not have*. We solve the problem when we figure out how to get, and then get, what we want. Herodotus wanted a certain piece of knowledge that he did not have. Specifically, what he wanted to know, and what he did not know, were the connections among the facts that would explain the fact of the Nile rising yearly at the summer solstice. He would *value* the knowledge of such a connections because it would satisfy his want.

Since Herodotus *wants to know* something, his desire is a *cognitive* desire or *cognitive interest*. The knowledge that would satisfy the desire is *cognitively valuable*. The problem is posed because Herodotus has certain cognitive desires which are not fulfilled: there are things that he wants to know but does not know. The problem itself is thus a *cognitive problem*: the solution to the problem consists in discovering a certain amount of knowledge. In this case, the cognitive problem would be solved if Herodotus was able to acquire a knowledge of connections among facts that would enable him to *explain* or *understand* certain other events in which he was interested.

We have an explanation of certain facts when we know a connection that links these facts to other facts. If we value knowing an explanatory connection, then we have a cognitive interest in that connection. A cognitive interest that is not satisfied yields a problem. The problem will be solved by research that produces the knowledge that one is cognitively interested in.

*

Herodotus examines three possible explanations of the rising of the Nile. One concerns the Etesian winds. The second concerns the possible connections of the Nile to the river Ocean that some suppose to surround the earth. The third concerns the possible melting of snow around the source of the Nile.[3]

We can learn more about explanation by looking at these.

(a) Attempted Explanation # 1

This explanation goes roughly as follows:

The Nile rises when, and only when, the Etesian winds blow. The Etesian winds blow at the summer solstice. So, as we observe, the Nile rises at the summer solstice.

One presumes that the idea behind this is that the Etesian winds blow against the river and back the waters up sufficiently that they flood the land. But we may leave these sorts of considerations to one side. Our interest is in the logic of the set of propositions that Herodotus offers as a potential explanation.

And here we can already observe something important about explanations: they involve the inference of the event to be explained from certain premises. In this case, the event to be explained is the rising of the Nile at the summer solstice. The explanation that Herodotus is considering *infers* the description of this event from two other statements, one about the connection between the blowing of the Etesian winds, and the other stating when the winds blow. These other two statements function as *premises* in an *argument* which has as its *conclusion* the statement describing the event to be explained. In other words, *the explanation* is an argument in which the fact to be explained is inferred from certain premises; by virtue of this inference from those premises, the fact in question is explained.

Of the two premises of the explanatory argument that Herodotus is considering, one is another statement of individual fact, namely the statement that the Etesian winds blow at the summer solstice. The other premise is one that *provides the explanatory connection* between the fact to be explained and the other individual facts mentioned in the premises. This is the statement of fact that the Nile rises when, and only when, the Etesian winds blow. We can refer to the individual facts as the "initial conditions" of the explanation. The other premise establishes a connection that enables one to say that the fact constituted by the initial conditions causes the occurrence of the fact that one is interested in explaining.

The argument goes like this:

(H1)   Whenever the Etesian winds blow, then and only then the Nile rises.

(IC1) The Etesian winds blow beginning at the summer solstice.

(E)   Hence, the Nile rises at the summer solstice.

More briefly,

(H1), (IC1), ∴ (E)

The argument presents the connection between the facts (IC1) and (E) because, *given* (IC1) and (H1), *then* the event (E) *must* occur. The argument is such that, *given the premises, then the conclusion* MUST *occur*. As one says, the premises **entail** the conclusion. An argument in which the premises purport to entail the conclusion, is a **deductive argument**, and one in which the premises do actually entail the conclusion is said to be **valid**. Indeed, this is the defining characteristic of a valid deductive argument:

An argument $P_1, P_2, P_3, \ldots , P_n$ ∴ C **is a** VALID DEDUCTIVE ARGUMENT **just in case that** *if the premises are true then the conclusion* MUST *be true*.

In a valid deductive argument, the premises are said to ENTAIL **the conclusion**.

We now see why it is a requirement of an argument, if it is to be explanatory, that it be a valid deductive argument. For, an explanatory argument has to establish a connection between apparently separate individual facts: no connection, no explanation. It is precisely because the premises entail the conclusion in such an argument that we can say that a connection really does obtain, a connection linking the initial conditions and the event in whose explanation we have a cognitive interest.

In fact, of course, Herodotus rejects this explanation. He rejects it on the basis of another argument — another valid deductive argument — that he goes on to develop. The supposed connection between the Etesian winds and the rising of the Nile does not really hold, and so the attempted explanation must be rejected. The reason, he argues, for rejecting the supposed explanation is that the connection that it purports to establish really does not obtain. For, the *hypothesis* upon which it is based is *false*. And if the hypothesis is false, there is, contrary to what the argument purports, no connection.

Herodotus reasons in this way:

Whenever the Etesian winds blow, then and only then the Nile rises.
The Nile rises at the summer solstice.

Hence, the Etesian winds blow beginning at the summer solstice.

But,   sometimes at the summer solstice the Etesian winds do not blow.

Hence, it is not true that whenever the Etesian winds blow, then and only then the Nile rises.

The hypothesis (H1) of the attempted explanation is false. It follows that there is after all no connection established between the initial conditions (IC1) and the event (E) to be explained.

Herodotus' reasoning in *refutation* of the attempted explanation goes like this:

(H1), (E), ∴ (IC1). But not-(IC1) ∴ [given (E)], not-(H1)

This inference depends upon the principle that defines valid deductive arguments. This is the principle that, if the premises are true then the conclusion must be true. The argument that Herodotus considers is a valid deductive argument. But its conclusion is false: sometimes as the summer solstice the Etesian winds do not blow. Since in a valid argument, if the premises are true then the conclusion is true, and since the conclusion is false, not true, it follows that the premises cannot all be true: at least one must be false. But one of the premises we know to be true: the Nile does rise at the summer solstice even when the Etesian winds do not blow. Since at least one premise must be false, it follows that it has to be the other premise. This is the hypothesis (H1) that the Etesian winds are the cause of the Nile's rising.

This inference is, of course, in the traditional pattern of *reductio ad absurdum*. Under this argument form, the idea is to prove a proposition

not-H

In order to prove this, one first *assumes*

H

and then *deduces* a false statement. This false statement. This might be an explicit contradiction, say some other statement such as

S and not-S

Since this is an explicit contradiction, it cannot possibly be true: it is necessarily false. Or it might be

not-H

which is a special case of the former, since, if we infer "not-H" from "H", then we have

H and not-H

Or, as in the case of Herodotus' argument, it might be some other statement

R

which just happens as a matter of fact to be false.

Herodotus, then, uses a *reductio ad absurdum* argument to establish the falsity of the premise of the argument that was supposed to provide an explanation for the rising of the Nile. He concludes that the argument does not, after all, provide an explanation: *it established no connection or order linking the fact to be explained to other facts*. This makes clear that an argument that purports to be explanatory must not only have premises that entail the statement describing the fact to be explained, but **the premises must be true**. If the premises of an explanatory argument are not true, then the argument does not explain: no explanatory connection is established.

An argument that successfully explains a certain fact must, we have seen, have premises that *entail* a statement describing that fact. We have now also seen that these premises must be *true*. Let us say that an argument in which the premises both entail the conclusion and are true is a **sound** argument. Thus, an argument which is explanatory must be sound.

We can already see an interesting contrast between Herodotus' reasoning and that of von Däniken. Herodotus is looking for an explanation of the rising of the Nile. His explanatory argument goes like this:

Whenever the cause, then and only then the effect

But,   the cause
_____

Hence, the effect

However, for purposes of testing the hypothesis, he reverses the inference, pointing out that the rising of the Nile requires the Etesian winds to blow, and then noticing that this prediction is sometimes false.

Whenever the cause, then and only then the effect

The effect
_____

Hence, the cause
But,   not the cause
Hence, it is false that whenever the cause, then and only then the effect

The point is that sometimes, when the explanation goes from cause to effect, the prediction or test may involve and inference from effect to cause.

Erich von Däniken argues with regard to certain archaeological events that they can be explained only in terms of extraterrestrial beings from intelligent advanced civilizations.[4] The archaeological events are the effects, the presence of the extraterrestrials the cause. He has the same form of causal generalization as Herodotus, and the same form of explanation:

> Whenever the cause, then and only then the effect.

But,   the cause

Hence, the effect

However, as with Herodotus, the *prediction* goes the other way:

> Whenever the cause, then and only then the effect

> The effect

Hence, the cause

But the prediction does not gain independent confirmation; the extraterrestrials from an advanced intelligent civilization have, so far as we can reliably tell, not been observed. There has therefore been no confirmation of the causal hypothesis. Of course, neither has there been any refutation, so we are not in a position to reject von Däniken's explanation as Herodotus was to reject the explanation based on the Etesian winds. The point is that Herodotus sees that it is the prediction of the cause that *tests the hypothesis*. He recognizes that this is what is required of any hypothesis that is to counted as worthy of acceptance in explanations. Von Däniken, in contrast, never stops to put his hypothesis to the test. Untested, it remains a mere hypothesis. It is not, therefore, to be counted as worthy of acceptance in explanations.

Herodotus goes on to extend his case against the explanation of the Nile flooding in terms of the Etesian winds. What he argues is that there are other rivers in Syria and Africa that flow into the Mediterranean like the Nile, and in the same direction. Hence, if the Etesian winds cause the flooding of the Nile, those winds should cause the other rivers to flood. But they do not. So they could not be the cause of the flooding of the Nile.

In this argument he makes use of a *general premise*:

(H1') For any river in the path of the Etesian winds, whenever the Etesian winds blow, then and only then that river rises.

He also appeals to the initial condition

(IC2) The Nile is a river in the path of the Etesian winds.

These give (H1) as a conclusion by way of the inference

(H1'), (IC2), ∴ (H1)

But he has already inferred from (IC1) and (E) that

not-(H1)

From this he concludes, again by *reductio ad absurdum*, that

not-(H1')

Thus, the behaviour of the Nile refutes the general claim (H1'). But so do other rivers. Herodotus infers this by means of the argument

(H1')	For any river in the path of the Etesian winds, whenever the Etesian winds blow, then and only then that river rises.
(IC3)	There are rivers other than the Nile in the path of the Etesian winds.
Hence, (H2)	Whenever the Etesian winds blow, these other rivers rise.
(IC4)	The Etesian winds blow at the beginning of the summer solstice
Hence, (P)	These other rivers rise.
But, not-(P)	
Hence, not-(H2)	
Hence, not-(H1')	

Herodotus thus rejects the idea that there can be a causal relation between the blowing of the Etesian winds and the rising of the Nile on the grounds that a *general statement* linking the blowing of these winds and the rising of rivers can be shown to be false. Herodotus clearly reasons that if the causal relation holds for the Nile then one infers that the same relation holds for other rivers that are similarly situated relative to the Etesian winds. We see in this inference a very important principle about explanations: *causal explanations imply generalizations*. Thus, to say that

This being A causes that to be B

is to imply that

Whenever an A then a B

This makes clear a further important point about explanations: *the order that an explanation establishes to explain a fact consists of showing that the fact is an instance of a* **general pattern**. In other words, *explanation proceeds by subsumption under a generalization*. An explanation will, in other words, in general have the form

(H)   All A are B
(IC)   This is A
_____
(E)   ∴ This is B

We thus now have **three crucial features** that must be exemplified by any argument if it is to explain successfully: *one*, the premises must entail the conclusion describing the event to be explained; *two*, the premises must be true; and, *three*, the premises must contain a general statement linking certain statements of individual fact (the initial conditions) to the conclusion describing the event to be explained. Only an argument satisfying these conditions can establish an order or connection among individual facts that can be said to be explanatory. Such an argument will establish not only that the fact occurred, but *why* it occurred, that is, why, given certain other facts, this one *must have* occurred: it must have occurred because, given the other facts, that is the *general pattern* of the way things occur.

We have, in addition, a further **fourth crucial feature** of any argument that is to count as worthy of being an explanation: *we must have evidence that the premises are true*. In the case of the general statement, this means that the premise must have survived empirical testing of the sort proposed by Herodotus: *the general premise must yield a prediction which is successful*. This is also, as we have seen, a criterion proposed by both creation scientists and Velikovsky for the acceptability of a proposed explanation as scientific. In general, then, it is agreed that

An argument is worthy of acceptance as a scientific explanation only if the generalization in the premises has been successfully tested, that is, has led to predictions which have turned out to be true.

Von Däniken offers an argument to explain certain things archaeologists have discovered. This contains a general causal statement to the effect that certain effects could have been caused only by civilizations more advanced than ours. The argument is deductive. The premises may even be true. But we do not know if they are true. We have never observed the initial conditions, nor has the general hypotheses survived any empirical test. We therefore have no reason for accepting the premises as true. We are therefore in no position to count the argument as one that is worthy of being an explanation: we simply don't know whether the three crucial features are fulfilled. That we know them to be fulfilled is the **fourth crucial feature**.

Note that the first three conditions are conditions about the argument itself, conditions which hold independently of who happens either to offer or to accept the argument as explanatory. The fourth condition, however, is not a condition on the argument itself, but a condition on the relation between the argument, or, at least, its premises and *person who accepts the argument as explanatory*: the person who accepts the argument as explanatory must *know* the premises to be true, or, at least, have good reason to believe them to be true. The fourth condition is thus an **epistemic condition** on an argument being worthy of being counted as an explanation. The first three conditions, in contrast, are **constitutive conditions**.

In particular, where the argument purports to be explanatory, then the general premise ought to satisfy the *epistemic condition* of having *successfully survived predictive tests*.

As we shall see, there may be other grounds for accepting general premises for use in scientific explanatory arguments. But it is safe to say that the idea of a successful predictive test is by far the most important criterion for science to count a generalization as worthy of use in explanatory arguments.

Another lesson can be learned from Herodotus' practice. When he considers the explanation of the Nile's rising in terms of the Etesian winds, he recognizes that such an explanation commits one to a generalization. He then uses this generalization to *predict* what will happen to other rivers situated similarly to the Nile relative to the blowing of the Etesian winds. The prediction is that these other rivers will rise as the Nile does when the Etesian winds blow.

Herodotus makes this prediction on the basis of a *predictive argument*. We have the explanatory argument

> Whenever the Etesian winds blow, the Nile rises

already analysed as

(H), (IC1), ∴ (E)

The predictive argument looks exactly the same

> Whenever the Etesian winds blow, any river situated similarly to the Nile rises

or

(H2), (IC4), ∴ (P')

What has to be noted is that this argument *has the same form* as the explanatory argument:

(H)   All A are B
(IC)   This is A

∴ This is B

The difference is *not in the form* of the argument. In both cases, the argument entails the conclusion. In both cases, the inference is based on a generalization and on a statement of initial conditions. In both cases, the premises are true. These conditions must obtain for otherwise there would be no connection established that would enable one to explain the fact in question, or, equivalently, predict the fact in question. The difference lies not in the argument itself but rather in the *cognitive relation* of the *person who uses the argument* to the fact described in the conclusion. In the explanatory argument, the person who uses the argument *already knows* that the fact described by the conclusion obtains. In the case of the predictive argument, the person who uses the argument *does not (yet) know* that the fact described by the conclusion obtains.

Thus, whenever we have an *explanatory argument*, there we have an argument that *could have been used* as a *predictive argument*. Conversely, it would appear that whenever we have a predictive argument based on a true generalization, there we have an argument that will become explanatory when the cognitive relation of the person who uses the argument to the fact predicted shall have changed to one where that fact is not merely predicted but actually observed, or otherwise known independently to obtain. We have thus obtained a **fifth crucial feature** of scientific explanations: *in scientific explanations, explanation and prediction are symmetric*.

We shall see later on that this thesis concerning the symmetry of explanation and prediction must be qualified in certain ways, but, as we shall also see, the qualifications do not qualify it away: the symmetry of explanation and prediction is an essential feature of scientific explanations.

The connection of this feature with the fourth crucial feature of scientific explanations should be evident. The fourth crucial feature is the epistemic condition that we require of an argument that is to be *accepted* as explanatory that we have evidence that the premises are true. In the case of the generalization that occurs in the explanatory argument, this epistemic requirement imposes the condition that the generalization be *tested predictively*. But a generalization can be tested predictively only if it is used in a predictive argument. Such an argument is, however, potentially an explanation. Indeed, it will be such provided that the generalization does pass the test. An argument which is an explanation is a prediction: a prediction provides a test: the test, if passed, justifies accepting the argument as an explanation.

We recognize that we are in fact extracting from the examples that Herodotus gives us a coherent picture of the nature and structure of scientific explanation.

We also recognize a *second important feature of deductively valid arguments*. What we have just noticed is that all arguments *of the form*

(H)   All A are B
(IC) This is A

∴ This is B

are equally good, or, as we put it, equally *valid*. No matter what terms we put for "A" and "B," the result will always be a valid argument, one such that, if the premises are true then the conclusion must be true. In other words, we recognize that this is a *logical form* such that, for any argument that has that form, *if the premises are true then the conclusion must be true*. We thus have a second principle for deductive arguments:

> **An argument** P_1, P_2, P_3, … , P_n ∴ C **is a** VALID DEDUCTIVE ARGUMENT **just in case that *any argument of the same logical form is valid*.**

Of course, having said this, we still have to say more about the notion of the "form" or "logical form" of an argument. Examples such as the one just given yield an intuitive idea of the notion of the "logical form," but if one is to develop the idea of a valid argument beyond the intuitive level, then one has to elaborate, and make explicit, the required notion of logical form.

(b) Attempted Explanation # 2

Herodotus rejects the first attempted explanation because the connection it asserts to hold in fact does not: it has a false premise. Herodotus rejects the second attempt also, but for a very different reason. This is an equally important reason for rejecting certain proposed explanations.

This explanation goes something as follows:

> Whenever a river floods periodically, then and only then it is connected directly to the River Ocean.

> The Nile is connected directly to the River Ocean.

∴ The Nile floods periodically.

About this Herodotus asserts that it is "involved in such obscurity, that it is impossible to disprove it by argument."

The important point here is that a proposition that is obscure in its meaning cannot be used to give an explanation. If a statement is obscure in its meaning, then one cannot say what evidence would tell for its being true nor what evidence would tell in favour of its being false. In fact, given its obscurity, one cannot even say exactly what facts will make it true and what facts will make it false. In that sense, *a sentence that is obscure in its meaning cannot be said to be true and cannot be said to be false: it has no truth conditions, and is therefore neither true nor false.* Herodotus' point is that a statement which has no truth conditions, a sentence for which there are no facts that would make it true nor any facts that would make is false, says nothing about the world, nor, therefore, does it establish any connections among those facts. And, in the absence of any capacity to make connections, it cannot possibly explain, since, after all, one explains if and only if one has established a connection!

It is of course true that some accept the story of the river Ocean. After all, one can find it in Homer. Herodotus, however, rejects any such appeal. The fact that a statement occurs in a hallowed myth such as Homer was to the Greeks or the Bible is to Christians is a reason for some for accepting it as true. This is not so for Herodotus. For him, a statement is worthy of acceptance in an explanation only if there are *empirical grounds* for accepting it as true. Like Hippocrates, Herodotus rejects the notion that the world is an enchanted place and that things are to be explained by appeal to things divine, the gods and other sorts of beings to be found in Homer and the religious traditions of his country.

The attitude of Herodotus to myth and legend, to poetic and religious texts, is very different from the defenders of the pseudosciences that we looked at above. Thus, the "creation scientists" make conformity to the Biblical story of creation a test of truth. If something contradicts the Bible it is to be rejected as false. For, the Bible is the inerrant word of God.

Herodotus also differs from von Däniken and Velikovsky.[5] These latter two offer accounts that purport to explain ancient myths and legends, von Däniken in terms of visits by ancient astronauts and Velikovsky in terms of catastrophic changes in the solar system that caused, among other things, the parting of the Red Sea for Moses and the apparent stopping of the sun for Joshua at the Battle of Gibeon. These two, unlike the creation scientists, do not take the ancient legends as literally true. Yet they do take them as a test of truth: if a theory conforms to these legends and tales, then it is true; if the theory cannot explain these tales, then it is inadequate at the least, and probably false. von Däniken and Velikovsky take the myths to *confirm* their theories. As von Däniken puts it with regard to his use of "ancient traditions,"

> I am simply referring to passages in very ancient texts that have no place in the working hypothesis in use up to the present.[6]

Astronomers behave quite differently. They do recognize that the ancient Babylonians, for example, had long records of observation of the motions of the heavens. Yet present day astronomers do not take these as data to which their theories must be made to conform. Nor do they believe that it is incumbent upon them to find explanatory theories that will account for the story, either in detail or outline, of the sun standing still when Joshua so commanded. To be sure, if *currently available evidence* supports an hypothesis about the occurrence of an event, say an exploding star or supernova, and there is an ancient report of that event, then we are entitled to take that report as true. But note that it is current evidence that is taken to support the truth of the ancient report. It is not supposed that the ancient report stands by itself as evidence. If the only evidence that a certain event occurred by a report of the Babylonians, then the event would never come to be counted as part of the body of current scientific knowledge. An astronomer might speculate, perhaps, about the event, but he or she would not treat that event as a fact needing an explanation nor a fact that somehow challenges current theories. Herodotus' attitude towards the legends of the poets is that of the modern scientist.

Often, in fact, it is clear that those who make hallowed myth the test of truth do so because they fear science. We can see why this should be so already quite clearly in Herodotus' attitude. Non-naturalistic explanations sanctified by myth simply do not count as explanatory. The attitude towards myth is dismissive: it is of no concern to science. It is clear that this attitude of dismissiveness can be viewed as a threat. This is evident in the case of the defenders of creation science. Thus,

> If I lose faith with Genesis, I'm afraid I'll lose faith in the rest of the Bible; and if I want to commit larceny, I'll say I don't believe in the part of the Bible that says "Thou shalt not steal." Then I'll go out and steal. The same thing applies to murder.[7]

Or again,

> No Adam, no fall; no fall, no atonement; no atonement, no Saviour. Accepting Evolution, how can we believe in a fall?[8]

The creation scientists invoke science in support of their views, yet it is invoked not in the interest of furthering scientific knowledge, nor in the interest of improving our scientific explanations of things. It is introduced, rather, in order to defend the authoritative basis of their religious and moral views, and prevent from being treated in the dismissive way that they see as undermining it. The problem with which they are concerned is not that of coming to grips with two competing scientific theories, but rather with the problems posed by the dismissive attitude of science towards the authoritative basis of their religious beliefs.

(c) Attempted Explanation # 3

The third attempted explanation that Herodotus considers invokes the hypothesis that the Nile rises periodically due to melting of snow in the interior of Africa near its source. The explanatory argument goes something like this:

> (Ho) Whenever there is snow in the interior of Africa and it melts, then and only then the Nile rises
>
> (C) There is snow which melts in the interior of Africa

Hence, (E) The Nile rises

Here the explanation is of the form

$$(Ho), (C), \therefore (E)$$

moving from cause to effect. Here again Herodotus reverse the order to infer the cause from the effect and predict the former from the latter:

$$(Ho), (E), \therefore (C)$$

This prediction of (C), that there is snow in the heart of Africa that melts each year, then becomes a test of the hypothesis.

Now, Herodotus has never been to the heart of Africa. Nor has he ever met anyone who has ventured that far up the Nile. He is therefore in no position to provide observational data that could confirm the prediction or lead to its rejection. If, therefore, Herodotus is to confirm or reject the prediction then he must find some other way of deciding whether the facts are or are not as (C) predicts them to be. Since Herodotus has no direct evidence, he requires *indirect evidence*, that is, observational data derived from other sources from which he can infer either the truth of (C) of its falsity. This is how he proceeds.

Herodotus, having raised the possibility of snow in the heart of Africa, now goes on to argue that there are reasons to suppose that there is no such snow. This being so, the causal hypothesis in this third attempted explanation also fails to pass the test and the attempted explanation must be rejected.

Herodotus first argues in this way, that the interior of Africa must be hot:

> (H1) For any region, if hot winds blow from it, it is hot.
>
> (IC1) The interior of Africa is a region
>
> (IC2) Hot winds blow from the interior of Africa

Hence, (C1) The interior of Africa is hot

Given that the premises are true, Herodotus reasons that, as this deductive argument entails its conclusion, the conclusion must be true. He then goes on to argue in this way, using the conclusion of this argument as a premise in his next:

> (H2) For any region, whenever snow falls in that region, then that region cannot have a hot climate
>
> (C1) The interior of Africa is hot

Hence, (C2) Snow does not fall in the interior of Africa

And now he notices that

> (C2) entails not-(C)

Hence,

> not-(C)

In other words, the prediction turns out to be false, and the hypothesis (Ho) is refuted by an inference of the now familiar *reductio ad absurdum* form.

Herodotus therefore concludes that this third attempt at explanation also fails.

There is an important message that should be brought out at this point. Duane Gish has, in objecting to the Darwinian theory of evolution, made the argument that "for a theory to qualify as a scientific theory, it must be supported by events, processes, or properties which can be observed..." He then states that the theory of evolution fails to meet this criterion:

> It is obvious ... that no one observed the origin of the universe, the origin of life, the conversion of a fish into an amphibian, or an ape into a

man. No one, as a matter of fact, has ever observed the origin of a species by naturally occurring processes. Evolution has been *postulated*, but it has never been *observed*.[9]

But, as the example from Herodotus makes clear, while one may not be able to test a theory directly by observing the facts that it predicts, there may well be indirect ways to draw inferences about those facts. We cannot observe atoms, but by means of electron microscopes we can draw inferences about them from what we do observe. For that matter, we cannot observe germs. But we can use optical microscopes to draw inferences about them. We can even use that information to help in the search for cures for diseases. We cannot observe the interior of the sun. But we can use well-confirmed theories — theories that are confirmed directly — and instruments — the laws of the operation of which have been confirmed directly — to draw inferences to facts that, while still unobserved, can nonetheless put the theory in question to the test. Herodotus has not observed the central areas of Africa, but he can use a well-confirmed hypothesis to draw inferences about what he has not seen, and then use the conclusions of these inferences to put the theory in which he is interested to the test. Thus, Gish is simply wrong when he states that, since there are many parts of the evolutionary story that have not been, and could not have been observed, by any human — they were not then in existence, nor any of the members of their species! — therefore the theory could not be put to the test and, consequently, could not be scientific. Gish has, unfortunately, simply not thought enough about the procedures that science uses to put its theories to the test.

It is clear that in principle there is nothing wrong with the sort of test to which Herodotus puts the third alternative explanation of the Nile's annual flooding. *In fact*, however, all is not well with the test that Herodotus uses to reject the theory that the Nile floods because of snow melting in central Africa. What we have to note is that there is in fact an *auxiliary hypothesis* that Herodotus must make, namely,

(AA)   There is no small subregion which is highly elevated.

Herodotus in fact must assume that this auxiliary assumption is false, for, if there were an elevated subregion, it would be colder than the rest of the region for part of the year. During that time, snow would fall which, when it melted, could provide the water that causes the Nile to rise. But if Herodotus must make this auxiliary assumption, then this shows that with (H1) we have not accurately rendered the generalization that provides the connection among, on the one hand, the facts that appear as initial conditions — that is, all those facts, including the ones described by the auxiliary assumption — and, on the other hand, the fact described by the conclusion.

The argument

$$(H2), (AA), (C1), \therefore (C2)$$

won't do, since it is clear that (AA) is *irrelevant* to the conclusion: if (C2) already follows from (H2) and (C1) without (AA), then nothing happens when we add — or delete — (AA) as a premise. In order to secure the relevance of (AA) — and it clearly *is* relevant — then we must construe our general hypothesis not as (H2) but as

(H2*)   For any region, whenever snow falls in that region, then that region cannot have a hot climate *unless* snow falls in a highly elevated subregion

We now have the argument

(H2*)
(C1)
Hence, (C2) or (C2')

where

(C2) or (C2')

is the proposition that

Either snow does not fall in the interior of Africa or snow falls in a highly elevated subregion

Since (C2') just is the negation of the auxiliary assumption (AA), this then functions as a disjunctive premise in the argument

(AA) not-(C2')

$$(C2) \text{ or } (C2')$$

———————————————

Hence, (C2)

And we proceed as before: (C2) entails not-(C) and the hypothesis (Ho) fails to pass the test: it is rejected.

But of course, this argument goes through only if the auxiliary assumption (AA) is true. In fact, as we know, it is false. It follows that Herodotus' rejection of the third attempted explanation of the Nile's rising fails: it attempts a *reductio*, but fails because the argument uses a false premise, and is, therefore, unsound.

From what we know today, the third explanation is after all, and contrary to Herodotus, the correct one. The melting of the snow annually in the Ethiopian highlands provides the water that accounts for the annual rising of the Nile River. Herodotus' *implicit* auxiliary assumptions turned out to be false. If there is any moral here, it is surely that we must try to make our auxiliary assumptions explicit, so that we do not overlook them when we are checking the truth values of our premises in order to ensure that our explanatory/predictive arguments are sound.

NOTE ON AUXILIARY ASSUMPTIONS
Some authors[10] suggest that any explanation based on an hypothesis H, e.g.,

$$H, IC \therefore E$$

can be saved from elimination when what it predicts turns out to be false, provided that we simply add some auxiliary assumption

$$H, AA, IC \therefore E$$

This will not do, however. If H and IC entailed E prior to the adding AA, then they continue to do so, and H must for that reason still be rejected. But if AA really is essential to the deduction in the second version, then the hypothesis cannot be the same hypothesis H, but a new one H', perhaps similar to the old hypothesis H but nonetheless different and with different logical properties. Or, if AA really is essential to the deduction of E from IC and H, then the original H could not have entailed E

given IC. We saw above, in our discussion of Herodotus' third attempted explanation, that the addition of auxiliary assumptions required the hypothesis to be modified. Since it is not always easy to find a new hypothesis that will at once entail E and also take account of some auxiliary facts, it is simply not true that any hypothesis can be saved by the addition of auxiliary hypotheses.

The conventionalist assumption that one can always figure out a way to save an hypothesis from rejection in the face of apparently falsifying instances or predictions is simply not reasonable.

Cognitive Norms

In the beginning of Herodotus' cognitive quest for the cause of the annual flooding of the Nile, Herodotus clearly **does not know** the cause of the annual flooding. At this point, he is willing to say this. That, indeed, provides the starting point of his research, for, although he does not know, yet he wants to know. He clearly allows that all attempted explanations may well turn out to be false or somehow inadequate. That would leave him at his starting point where he **does not know**. This willingness to *admit ignorance*, to *suspend judgment* concerning certain matters of fact, embodies a point about science that bears emphasis.

Science adopts the **cognitive norm** that *in the absence of evidence one ought to neither affirm nor deny a proposition*. Herodotus' attitudes conform to this norm: in the absence of evidence he is prepared to make no affirmation as to the cause of the annual flooding: he is prepared to *suspend judgment*.

This makes clear that in the set of cognitive norms of science, there are **three cognitive attitudes** that may be adopted towards a proposition, namely, *affirmation* (or acceptance), *denial* (or rejection), and *suspension of judgment*. One could also speak of *belief*, *disbelief* and *suspension of judgment*. The point is that the cognitive norms of science require *three cognitive attitudes*.

In general, one affirms a proposition (or accepts it, or believes it) when it is true, and denies a proposition (or rejects it, or disbelieves it) when it is false. But adoption of such an attitude depends upon **evidence**. Thus, one affirms a proposition when one *has evidence that it is true*, and denies a proposition when one *has evidence* *that it is false*. And, of course, one suspends judgment with regard to a proposition when one has no evidence that it is true and no evidence that it is false, or if the evidence for the truth of the proposition is exactly balanced by the evidence for the falsity of the proposition.

This last point, about equality of evidence, generates a refinement in the norms concerning the adoption of cognitive attitudes. If there is evidence both for the truth of a proposition and for the falsity of that proposition, but the former is stronger than the latter, then one ought to affirm the proposition *but only in a guarded way*. Where the total evidence available points conclusively to the truth of the proposition, one ought to affirm the proposition *with complete certainty*. Where the total evidence available points in part to the truth of the proposition and in part to its falsity, but more strongly towards the former, then one ought to affirm the proposition *but only with the degree of certainty appropriate to the evidence*. One can make a similar point about the degree of certainty of a denial depending upon the strength of the evidence. In general, *one adopts a cognitive attitude with the degree of certainty that is appropriate to the evidence available*. In due course we shall see that the degree of certainty can be understood in terms of probabilities.

But however the last may be, it is clear that the cognitive norms of science require three cognitive attitudes: affirmation, denial and suspension of judgment.

We may contrast this with the set of attitudes defended by Jesus in the Bible. Jesus proposes that *faith* is a virtue, that it is better to believe than not to believe. First, so far as concerns science, this confuses the issue, since it groups together as one attitude of not believing the two very different attitudes of *denial* and of *suspension of judgment*. And second, again so far as concerns science, it is simply bad as a cognitive norm, since it is *not* always better to believe. Sometimes it is better to deny than to believe, specifically when the evidence is against the proposition in question. Thus, from the standpoint of science, it is better to deny much of creation science than to affirm it. And sometimes it is better to suspend judgment than it is to either affirm or deny. Thus, when it came to the issue of the cause of the Nile's flooding, Herodotus is prepared simply to suspend his judgment.

Many pseudosciences fail to heed this norm of scientific inference. Thus, H. M. Morris has said that "Life has not been created in the test tube,"[11] and infers that, since there has been no scientific explanation worthy of belief, we must move to some non-scientific alternative if we are to achieve an account in which we can believe. The alternative is, of course, "creation science." The Christian norm that belief is better than non-belief may well demand such a move. But it is certainly not demanded by the norms of science. To the contrary, what the cognitive norms of science demand is that, in the absence of empirical evidence one way or the other, one ought simply to suspend judgment, recognizing plainly that one cannot — yet, at least — explain the phenomena in question. As one author has stated it,

> ... to attack an area of science because it has *not yet* reached a given stage, and then to argue a need for resorting to supernatural explanations because specific scientific answers are not reported, consolidated, nor agreed upon, is what is referred to in an America idiom as a "copout."[12]

Other pseudoscientists adopt the same anti-scientific cognitive norm as the "creation scientists," that it is better to believe than not to believe. Thus, von Däniken considers "an array of unexplained mysteries," and then invites us, by means of an apparently rhetorical question, to accept an unconfirmed hypothesis rather than simply suspend judgment:

> Do they [the unexplained mysteries] make sense as the remains of prehistoric space travellers?[13]

Or here is another attempt to force an answer with a fusillade of rhetorical questions:

> Is it all mere coincidence? Are they all merely individual fancies, strange whims on the part of our ancestors? Or is there an ancient promise of corporeal return that is unknown to us? Who could have made it?[14]

Pseudoscientists often use a ploy to try to reverse the onus of proof with regard to the claims that they make. Many, like von Däniken, claim that extraterrestrial visitors have come to Earth. The reasonable response is to ask them to "prove it," to provide evidence for the truth of their claim. They then notice the paucity of their evidence and reply with the charge: "But you can't disprove it!" This is an attempt to shift the onus of proof from themselves to the person who has challenged their right to believe the proposition they have asserted. The sceptical challenger is asked to disprove their claim. But of course the sceptic is usually not in a position to do that. Often, if it is hard to prove a claim, it is even harder to disprove it. The suggestion is that if it cannot be disproven, then there is a right — not just a political right but a *cognitive right* — to believe it. It is put in such a way as to assume that, since there is no reason for disbelieving, therefore belief is satisfactory. If grounds for disbelieving cannot be offered, then there is no reason for them not to believe. The sceptic now has the onus of proof; they have absolved themselves of any need to provide grounds for their belief.

This, however, is to assume — with Jesus — that there are only two attitudes, belief and disbelief. The alternative to disbelieving is not belief. To the contrary, there is not just one alternative to disbelief, but two: either belief or suspension of judgment. Even if the sceptic can find no grounds for disbelief, so that he or she has no right to disbelieve, it does not follow that the person who accepts that there are ETs who have visited us can ignore the onus of providing proof for his or her belief. The absence of grounds for disbelieving does not negate the absence of grounds for believing. And in the absence of grounds for either believing or disbelieving, the correct cognitive attitude, so far as science is concerned, is that of *suspension of judgment.*

The person who is sceptical of the claims that we have been visited by ETs must resist the attempts of the believer to reverse the onus of proof. But it is to be resisted by resisting as a cognitive norm the principle that "if you can't disprove it, I have a (cognitive) right to believe it." This principle of belief formation is not among those that are part of the practice of science. One who accepts it is not in conformity to the standards of science and the use of evidence in science. That is one of marks of pseudoscience.

The cognitive norms and cognitive virtues of the scientist are thus quite different from those defended by Jesus and accepted by "creations scientists" and other pseudoscientists. These anti-scientific norms and virtues can only corrupt if they are introduced into science. That is why "creation science," why such stuff as the "chariots of the gods," why astrology, and so on, which all adopt this anti-scientific cognitive norm, are in the end corruptions of science.

Endnotes

1   "The Scared Disease," in *Hippocratic Writings*, ed. with an introduction by G. E. R. Lloyd (Harmondsworth, Middlesex: Penguin, 1978), p. 237.

Hippocrates was somewhat cynical about the motives of those who referred to epilepsy as the "sacred disease." He continues:

It is my opinion that those who first called this disease "sacred" were the sort of people we now call witch-doctors, faith-healers, quacks and charlatans. These are exactly the people who pretend to be very pious and to be particularly wise. By invoking a divine element they were able to screen their own failure to give suitable treatment and so called this a "sacred" malady to conceal their ignorance of its nature. By picking their phrases carefully, prescribing purifications and incantations along with abstinence from baths and from many foods unsuitable for the sick, they ensured that their therapeutic measures were safe for themselves.... They also employ other pretexts so that, if the patient be cured, their reputation for cleverness is enhanced while, if he dies, they can excuse themselves by explaining that the gods are to blame while they themselves did nothing wrong; that they did not prescribe the taking of any medicine whether liquid or solid, nor any baths which might have been responsible.... (*ibid.*)

Today one might be equally cynical about some of the purveyors of "alternative medicine."

2   Herodotus, *History*, trans. George Rawlinson (London: Murray, 1859), 4 vols., Vol. II, pp. 24-29.

3   Herodotus actually goes on to offer his own explanation, but we will not examine this attempted explanation. (As it turns out, Herodotus is mistaken in this attempt.)

4   E. von Däniken, *Chariots of the Gods?* trans. Michael Heron (New York: Bantam Books, 1971).

5   I. Velikovsky, *Worlds in Collision* (New York: Macmillan, 1950).

6   E. von Däniken, *Chariots of the Gods?* p. 67.

7   Ray Ginger, *Six Days or Forever* (Oxford: Oxford University Press, 1958), p. 111.

8   *Ibid.*, p. 63.

9   D. Gish, *Evolution? The Fossils Say No!*, third edition (San Diego: Creation-Life Publishers, 1979), p. 13.

10 For example, I. Lakatos, "Falsification and the Methodology of Scientific Research Programmes," in I. Lakatos and A. Musgrave, *Criticism and the Growth of Knowledge* (Cambridge: Cambridge University Press, 1970).

11 Henry Morris, ed., *Scientific Creationism* (San Diego: Creation-Life Publishers, 1974), p. 49.

12 S. W. Fox, "Creation 'Copout,'" *Nature*, 292 (1981), p. 490.

13 von Däniken, *Chariots of the Gods?* p. 12.

14 *Ibid.*, pp. 86-7.

CHAPTER 2

PSEUDOSCIENCE: SOME EXAMPLES

There are a wide variety of pseudosciences at which we might look. Only a few must suffice. In the following examples of pseudoscience, besides trying to give an outline of what these positions argue, only an initial critique will be provided. Some discussions are longer than others. Where the discussions are shorter, it is safe to anticipate that there will be more extended discussions later, after a better view of science has been obtained. But in every case, we will take up further relevant issues as we proceed.

Example I: UFOs and Ancient Astronauts

It was shortly after World War II that UFOs were discovered. They first seem to have been sighted in Sweden, where there about a thousand sightings in 1946. The term 'flying saucer' came a little later. On 24 June 1947, Kenneth Arnold, a business man from Boise, Idaho, was flying a private aeroplane near Mt. Ranier, Wash. He subsequently reported seeing a group of objects flying along in a line "like pie plates skipping over water." Subsequent newspaper reports referred to what Arnold sighted as "flying saucers," and the term has stuck, even though not all UFOs are described as being of this shape.[1]

A *UFO*, or *unidentified flying object*, has been defined as

...the stimulus for a report made by one or more individuals of something seen in the sky (or an object thought to be capable of flight but seen when landed on the earth) which the observer could not identify as having an ordinary natural origin, and which seemed to him sufficiently puzzling that he undertook to make a report of it to police, to government officials, to the press, or perhaps to a representative of a private organization devoted to the study of such objects.[2]

This definition indicates merely that whatever a UFO is, it is a stimulus. It could be that the UFO is not in fact a real physical thing; it could be some atmospheric phenomenon. Or it might be some visual impression of a real physical thing distorted by atmospheric conditions. Or distorted, perhaps, by some problem with the observer's vision. It may even be a mental delusion existing only in the mind of the observer without any physical stimulus.

The definition also allows for deliberate deception. In the case of delusion, the reporter is not aware of the lack of a real physical stimulus; in the case of deception, the reporter knows that he or she is making false claims about what was seen.

The words 'could not identity as having an ordinary natural origin' are crucial: the stimulus generates a UFO report precisely because the observer could not identify the thing or entity seen.

If subsequent investigation provides an ordinary interpretation of what was seen, a naturalistic explanation for the appearance of the stimulus, then the UFO becomes an IFO — an *identified* flying object.

Many UFOs have become identified flying objects. They have been identified as atmospheric phenomena, as weather balloons, as high flying aircraft, as brightly shining planets low on the horizon, and so on. There have been a residual number of cases, however, where it has not been possible to provide a plausible interpretation that would account naturalistically for what was seen.

These residual cases must be left simply as "unexplained".

That conclusion does not sit well with many people who go on to suggest that these UFOs that remain Unidentified must be understood as extraterrestrial probes from elsewhere in the universe. The first publication of this suggestion that UFOs are visitors from other planets seems to have been contained in an article in the magazine *True* (1950), in an article by Donald E. Keyhoe entitled "Flying Saucers Are Real."[3]

The hypothesis that the residual UFOs are sent by, or perhaps even piloted by, extra-terrestrial intelligences (ETs) cannot be deemed totally impossible. If it were discovered to be true, it would be one of a most significant discoveries ever made by humankind. The ETs might be visiting us with benevolent intentions, or they might be hostile. But whatever, it is evident that their discovery would bring about great changes not only in the world in which we live but also in how we think of ourselves and of the universe.

The significant issue, however, is whether there is any evidence to support the hypothesis that the residual UFOs are to be accounted for by ETs. It would naturally be very exciting if the hypothesis were true. Many people would, for that reason, like it to be true. But the fact that it would be exciting if it were true does not make it true. It would be exciting if Elvis were alive, but that does not make it so. And wishing something to be true, does not make it true. I often wish I had more money, but, alas! wishing it were so does not make it so. But, though this is indeed the case, it is also the case that, as the old saying goes, the wish is the father of belief. Many people do follow the rule that because they wish something to be true, then they will believe it to be true.

That, however, is not the method of science, which insists upon empirical evidence: no hypothesis is to be accepted for purposes of explanation unless there is good observational evidence to support its truth. The emphasis must be upon both *good* and *observational*. When a scientist is confronted with an hypothesis, he or she insists upon evidence. This must especially be so in the case of hypotheses which, if true, would be very exciting. For it is just in these cases that the world, the scientist included, most wishes to be deceived. A highly critical attitude must therefore be adopted towards supposed evidence in these cases. The evidence to which appeal is made to support the ET hypothesis is tenuous at best. A flying saucer has never landed on a university campus to open and allow to emerge its ET passengers eager to tell the astronomy department from whence they had come, and the physics department the mechanisms that had carried them from there to here. The astonished astronomers and physicists would no doubt have searching questions. Certainly, in such circumstances, the ET hypothesis would be considered to be confirmed, and, indeed, confirmed beyond all reasonable doubt. But in the absence of such clear evidence, caution is the required attitude. Such reasonable caution on the part of good scientists is often mistaken for unreasonable resistance, some sort of *a priori* opposition that insists upon twisting or distorting evidence in order to show that the ET hypothesis could not possibly be true. This mistaken view of the attitude of scientists is sometimes embedded in a more comprehensive theory that there is some sort of conspiracy going on to cover up the truth about ETs.

Erich von Däniken, in his *Chariots of the Gods?*[4] and other books, has argued that there is scientific evidence for the ET hypothesis. One of the arguments that he uses to justify this in the face of cautious scientific scepticism is the argument that "anything is possible."

Yet *nothing* is incredible any longer. The word "impossible" should have become literally impossible for the modern scientist. Anyone who does not accept this today will be crushed by the reality tomorrow.[5]

Or again,

Impossible? Ridiculous? It is mostly those people who feel that they are absolutely bound by the laws of nature who make the most stupid objections.[6]

The argument seems to be this. To be sure, the evidence available is at best tenuous, at worst erroneous; nonetheless, you should grant that there is still something to what I have argued; for, after all, anything is possible.

It is, however, not a good argument.

In what sense can one say that "everything is possible"?

One sense of 'possible' is *physically possible*. An event is physically possible if it is allowed by the known laws of nature. A theory is physically possible if it is compatible our known scientific framework. Thus, it is physically possible for humans to walk on Mars; and the hypothesis that there are ETs is, like the theory of black holes, physically possible. However, the notion that ETs came here from other planets in fairly short periods of time is, contrary to what von Däniken suggests, something that is not physically possible: it is contrary to the known fact of the theory of relativity that no object can travel at speeds in excess of the speed of light.

In this sense of 'possible' it is simply not true that "anything is possible."

There is another sense of 'possible' which is broader than that of physical possibility. This is *logically possible*. A theory is logically possible just in case that it contains no contradiction. It is logically impossible for something to be both red and not red at the same time, or for something to be both made and not made of green cheese. But it is logically possible for the moon to be made of green cheese: the notion contains no contradiction. Yet of course it is false: the moon is not made of green cheese. This hypothesis, while logically possible, is not physically possible.

In this sense of 'possible', almost anything is possible. Yet that gives us no reason for thinking that it is true: *possibility does not imply actuality.*

What von Däniken wants to argue is that *because the ET hypothesis is possible, therefore it might be true.* He means to argue for something stronger than logical possibility. Yet it must be weaker than physical possibility. His point that "The word 'impossible' should have become literally impossible for the modern scientist. Anyone who does not accept this today will be crushed by the reality tomorrow" suggests that what he wants to say is that as theories change what was physically impossible at one time becomes physically possible at a later time. Thus, prior to Einstein, curved space was physically impossible, given the Newtonian theories. After Einstein, and the theory of relativity, curved space was possible. Thus, proposals that do not fit with current scientific theories will turn out to fit with theories developed in the future. Von Däniken's proposals do not fit with current scientific theories, but that is no reason to think that they will not fit future scientific theories. After all, "anything is possible." So, his proposal that ETs can travel faster than light is possible.

But is it really true that anything is possible in this way? Is it really possible, for example, that our current theories of the moon might turn out to be false and that we shall come to allow that it is really made of green cheese, contrary to what was thought in the primitive 20th century? The answer is clearly negative. The evidence is sufficiently strong, that we are not going to accept as reasonable any suggestion that theories could change sufficiently to allow it to be physically possible for the moon to be made of green cheese.

Von Däniken — and other pseudoscientists who also argue on the basis of the principle that "anything is possible" — is correct in thinking that later theories will in some places contradict earlier theories. The evidence will have accumulated to show that there is something wrong with the earlier hypotheses, that they do not correctly record the patterns that are there in nature. The new theory will better record the patterns of nature. In that sense it will contradict the older theory. But theories are complex structures, and to be wrong at one point is not to be wrong at all points. After all, the older theory had been used successfully to explain and to predict. To be sure, it was not everywhere successful, as it turned out, but that does not mean that it was not elsewhere successful. Einstein showed that Newton's theory was wrong. But it was wrong only at certain points, at speeds near the velocity of light and in locations near extremely massive bodies. For the rest, Newton's theory was entirely successful, and in fact the two theories yield essentially the same predictions in these areas.

If a new theory T' replaces an older theory T, then T' will yield successful predictions where T does not, but will yield the same predictions where T successfully predicts.

Since we have solid evidence for theories that determine that the moon is not made of green cheese, it follows that it will never turn out that we will arrive at a theory in which it is physically possible that the moon is made of green cheese. Nor will there ever be a theory for which it is physically possible that objects, e.g., space vehicles, can travel faster than the speed of light.

There is, we see, a solid reasonable basis for scientists to reject many claims of pseudoscientists. These arguments, are, however, not confronted. Rather, it is proclaimed that "anything is possible," and a failure to recognize this is then attributed to irrationality on the part of scientists: "It seems," von Däniken asserts, "as if narrow-mindedness was always a special characteristic when new worlds of ideas were beginning."[7] Insult your opponents, is the order of the day; do not confront their arguments.

The rhetorical ploy is clear also. No one wants to be narrow-minded: narrow-mindedness is hardly a virtue. But to reject von D.'s hypotheses is to be narrow-minded. And so the reader, anxious to be cognitively virtuous, is swept along into accepting those hypotheses, regardless of what the evidence might be.

It is essential to recognize that rhetorical ploys of this sort are often the backbone of pseudoscientific argument. In the absence of good evidence, use rhetoric!

The appeal to the principle that "anything is possible" has another aspect about which care should be taken. It attempts to shift the burden of proof from the pseudoscientist to the other side, to the sceptic. In effect, the argument is that since anything is possible, the claim that something is impossible must be defended. Until impossibility is demonstrated, the proposed hypothesis is possible and must therefore be taken seriously as a contender for the truth about the way the world is.

However, as we have just noted, it is not true that anything is possible, it is not true that just any proposed hypothesis is must be taken as a reasonable contender for recording the truth about the patterns in things. As we just noted, if a proposed theory negates a well established theory in areas where the latter has proved to be predictively successful, then the proposed theory is simply not possible.

To put the point another way, if a proposed hypothesis contradicts a well established theory at some point where that theory has been predictively successful, then *that fact* shows that the proposal is not only physically impossible but further that it is also unreasonable to suppose that it will ever become physically possible: no new theory will ever contradict an older theory at points where the older theory has been predictively successful.

Back of the attempt to shift the burden of proof, is the notion that open-mindedness requires that all possibilities be considered. When the scientist refuses to consider the hypotheses proposed by the pseudoscientist, he or she is said to be closed minded, obstinate and dogmatic. In fact, the suggestion is also made, the scientist is being foolish. For the hypothesis might after all turn out to be true: "anything is possible." But one is closed minded, obstinate and dogmatic, only if one has no good grounds for rejecting the proposed hypothesis. It turns out, most often, that the scientist after all does have good grounds for rejecting the pseudoscientific hypothesis. These grounds are the evidence, the predictive successes, of the current theory. In fact, open-mindedness does *not* require the scientist to consider all hypotheses as reasonable. The scientist should take as reasonable hypotheses only those that are not excluded by the evidence. If there is evidence available to support the current theory, as of course there will be, and that evidence tells against the hypothesis that is proposed, then that evidence has thereby provided grounds for treating the hypothesis as unreasonable. It is excluded not because one is closed-minded but by the fact that one wants to do the reasonable thing and not waste one's time on nonsense.

Von Däniken proposes the hypothesis that the earth has been visited in ancient times by extraterrestrial astronauts. The ancient pyramids of Egypt, the Stonehenge monument in England, and the giant stone heads on Easter Island can be explained only by supposing the human workmen, coming to do the project with only very limited cultural and technological attainments, had engineering advice from a much more advanced race of beings. The large markings on the Nazca plain in Peru can only be explained as an airfield for space craft. After all, these markings consist of straight lines that go on for many kilometres; since they are so large, these lines and the patterns they make could be seen only from a craft looking at the earth from a very great height — and, of course, the ancient Peruvians were earth-bound, and could not have seen, nor, therefore have planned, those patterns.

The pattern of his argument is much the same in each case: *he* cannot conceive how otherwise the events or facts could be explained. Take the last case.

The evidence consists in the lines on the Nazca plain.[8] More specifically, as he states, on this plain there is a set of very long, indeed gigantic, lines. Some of these lines intersect, some are parallel to do each other, some just come to do a sudden end. He then *interprets* this evidence. These lines, he claims, are arranged in the design of an airfield for space craft built on the instructions of the ETs that came in those craft. The justification for the interpretation is the fact that these lines give von Däniken the clear-cut impression of being an airfield. In his words,

> Seen from the air, the clear-cut impression that the 37-mile-long plain of Nazca made on me was that of an airfield.[9]

He adds that "the theory that aircraft could have existed in antiquity is sheer humbug." Since the lines are gigantic it must have been laid out to do be viewed from a very long distance in space, something that is possible only for a vehicle in space piloted by some form of intelligence. That intelligence would, of course, have to do be ET.[10]

There are a number of hypotheses that could be advanced to do account for the lines on the Nazca plain. If we accept the principle that "anything is possible", as von Däniken does, then we should probably include in the list

> The lines were created by the Wicked Witch of the West.

And

> The lines were created by angels falling from heaven.

But we should also include the hypothesis that

> Ancient peoples created the lines for reasons that we to do not know.

As for the argument that the technology was not available to do lay out straight lines over long distances unless there was some guidance from space, alternative besides the hypotheses of space craft are available. Here is one that von D. does not consider:

> How did the Nazcas achieve such exactitude? Along some lines the remains of posts have been found at intervals approaching a mile. Perhaps sighting stations with men standing in line behind them? Perhaps.[11]

Another hypothesis is one provided by Maria Reiche, who has studied the lines for many years. She speculates that the Nazca artists executed the patterns, both the straight lines and the large scale figures whose design could be appreciated only from a high elevation, by first sketching them on small plots of land, and then using a complex system of strings and central piles of rock to do make large-sized "blow-ups." Some of these small designs that might have functioned as dirt "sketch pads" are still visible near many of the larger figures.[12] So here is another hypothesis that von D. does not consider:

> Those who lived on or near the Nazca plain created their patterns using small scale drawings that were then, by simple geometrical means, blown up into larger figures and patterns.

Another hypothesis that is possible — whether it is true or not is another matter — is that the Nazcas *did* have the ability to see their designs from the air. In November 1975 two members of the International Explorers Society flew a crude hot-air balloon over the Nazca plain. It had an envelope constructed of finely woven fabric similar to textiles recovered from desert graves uncovered at Nazca. The gondola was a basket made of tortora reeds from the shores of Lake Titacaca. The lines and fastenings were made from native fibres. Using smoke and hot air from coals in a clay pot, the balloon was taken to a height of some 600 feet.[13] We therefore have the following alternative to the ancient astronaut hypothesis:

> The ancient Nazca inhabitants constructed their designs so they could be appreciated from the

aerial vantage point provided by a hot air balloon.

The experiment by the IES members does not yield any sort of evidence that this hypothesis is true. Nor is there any compelling evidence for Reiche's hypothesis that the patterns were created by blowing up smaller patterns. But they to show that there are alternatives to the astronaut hypothesis. Until von D. provides evidence that these hypotheses, as well as the others that could be proposed, have been *eliminated*, he cannot claim that he has support for his own hypothesis.

It is worth noting that both the Reiche and the IES hypotheses would have greater plausibility than von D.'s. What the proponents of both these hypotheses took the trouble to to do was to find data that would show that they *could* be true, not in the sense that "anything is possible" but in a way that is consistent with other things that we know about the technological capabilities of the ancient Nazca peoples. In contrast, there is absolutely no evidence that ancient astronauts ever landed on the Nazca plain. There has never been any evidence found in that area that could, for example, be reasonably construed as a part of an extraterrestrial vehicle. The only evidence that von D. adduces is the very facts — the lines and designs — that his hypothesis is designed to explain.

What we have to understand is the pattern of reasoning that he follows in order to conclude that his hypothesis is the one that ought to be accepted.

We have a number of alternative hypotheses:

$$H_1 \text{ or } H_2 \text{ or } H_3 \text{ or } H_4 \text{ or } H_5 \text{ or } \ldots$$

If we have something like

$$H_1 \text{ or } H_2 \text{ or } H_3$$

and we have eliminated one of the alternatives, H_3 say, then we can conclude that

$$H_1 \text{ or } H_2$$

But we cannot conclude that H_1 alone is true until we have eliminated *all* the alternatives. to put it more formally, the argument

$$H_1 \text{ or } H_2$$
$$\text{not } H_2$$

$$\therefore H_1$$

is a *valid argument* while

$$H_1 \text{ or } H_2 \text{ or } H_3$$
$$\text{not } H_2$$

$$\therefore H_1$$

is not. The test is that there is no way in which the former can have true premises and a false conclusion, while in the latter case it is easy to find instances where the premises are true and the conclusion false.

Now, it is true that some of the hypotheses alternative to the ancient astronaut hypothesis *have* been eliminated as unreasonable. Thus, archaeologist Paul Kosok proposed that the lines are part of an ancient astronomical observatory. He indicated how certain lines were directed at points on the horizon where the sun rises on the longest and shortest days of the year. Other lines pointed at other astronomical phenomena. It functioned as a sort of almanac for farmers anxious, in the dry climate, to predict the return of water to the valley streams.[14] However, a later study partly financed by the National Geographic Society showed that, while some lines did point to rising and setting points on the horizon and to certain stars, nonetheless the lines did this no more than could be expected by chance.

But the point is that von D. Does not provide evidence that eliminates either Reiche's hypothesis or the IES hypothesis. He therefore cannot reasonably conclude that his own hypothesis is true.

In fact, von D. does not even try to give a complete enumeration of all alternatives. Clearly, if he eliminates some, but there are others that he has not listed but has not eliminated, he cannot safely conclude that his own hypothesis is true.

However, von D. has a shorter way to do things. He does not bother to collect data that might eliminate the Reiche and IES hypotheses. His method is simple. Both these hypotheses — and many more could be generated — are all ruled out once we accept the *interpretation* that von Däniken proposes. This is why he does not feel the need to adduce empirical data to eliminate the alternatives. Indeed, this is why he does not feel the need to try to discover what all the various alternatives are. He knows beforehand, given his *interpretation* of the facts that they will be eliminated. It is thus his *interpretation* which is crucial.

What is the justification that is offered for this interpretation? The justification is that upon looking at the lines, von D. *has the clear-cut impression* that they are the markings of an airfield. Since the interpretation has the effect of excluding all hypotheses save the ET hypothesis, von D. is in effect arguing that upon looking at the lines he *can conceive no other hypothesis to be true*. He appeals to the same *methodological rule* in other contexts. Discussing a being in a wall painting in Tassili, in the Sahara, he says, "Without overstretching my imagination, I got the impression that the great God Mars is depicted in a space or diving suit."[15] What we recognize is that Von Däniken's *rule of method* is this:

if it is *inconceivable* that any alternative to hypothesis h is true, then accept for purposes of explanation hypothesis h.

Any hypothesis accept on the basis of this rule is impervious to actual counter-evidence. There are real problems with von D.'s hypothesis about the Nazca lines. Even in terms of his own hypothesis of space visitors, why would these ancient astronauts need anything akin to a long *runway*? Surely a space vehicle would, like the Mars lander, or the lunar modules, come down vertically, not requiring the long runway that our aircraft need. More importantly, there are data about the Nazca sites that tell against the landing strip hypothesis: the soil of the Nazca plain is soft and sandy, and hardly the kind of solid surface that a landing strip for a heavy object would require. One the persons who has worked on exploring the Nazca lines is Maria Reiche, and she scorns the thought that the lines may have been markings for airfields for visitors from outer space in ancient times. What she points out is that "Once you remove the stones, the ground is quite soft. I'm afraid the spacemen would have gotten stuck."[16] But does evidence of this sort lead von D. to revise his hypothesis, perhaps even reject it as inconsistent with such data? The answer is clearly NO. One the test of inconceivability of alternatives has been

applied, no empirical evidence is deemed to be relevant. In this sense, von D.'s methodological rules determines *prior to any empirical evidence* that an hypothesis should be accepted. The method may therefore be called the **a priori method**.

The question is, of course, why should we accept this rule of method? Why should we allow that the test that alternatives are *inconceivable to me* is a good test for the truth of *empirical* hypotheses? Maybe the fact that some alternative is inconceivable to me is merely a sign that I have limited imagination. For many centuries people claimed the any alternative to Euclidean geometry was inconceivable. No doubt it was. Nonetheless, we now allow not only that such alternatives are conceivable but also that one of them is true!

Take another example, that of Sumerian astronomy. As von Däniken correctly points out, the ancient Sumerians had a very sophisticated astronomy. His *interpretation* of this fact is that the Sumerians were given this knowledge by ancient ET astronauts. The reason he gives justifying this interpretation is that there was no culture prior to our own that was sophisticated enough to obtain this knowledge on its own.[17] Here again we have the same result: the interpretation of the evidence excludes every hypothesis save the ET hypothesis. And why should we accept this interpretation? Because von D. cannot conceive how any civilization less advanced than ours could have developed sophisticated astronomy. As we have suggested, however, this is not a sound principle for guiding one's inferences. Certainly, in this case at least, it leads to a conclusion which seems to be patently false. Various civilizations — Chinese, Egyptian, Mayan — have developed sophisticated astronomy, independently of each other, and prior to ours. In fact, history would seem to show that at its earliest stages, the astronomy of our civilization derived from that of the ancient world, and more remotely from both Sumerian and Egyptian astronomy.

It is worth noting that there is something condescending, if not racist, about von Däniken's suggestion that no civilization prior to ours was capable of developing sophisticated astronomy.

There is another way in which hypotheses accepted on the basis of von Däniken's methodological rules are impervious to empirical data. We can see this by considering another one of his fantastic claims.

Von Däniken relates the Biblical story of the flood and relates it to related stories found among Dead Sea scroll manuscripts. He then raises the question,

> Does not this seriously pose the question whether the human race is not an act of deliberate "breeding" by unknown beings from outer space? Otherwise what can be the sense of the constantly recurring fertilization of human beings by giants and sons of heaven, with the consequent extermination of unsuccessful specimens?[18]

There is an earlier species of apelike creatures on the earth when the astronauts arrive. The "spacemen artificially fertilized some female members of this species…". They repeated their experiment several times "until finally they produced a creature intelligent enough to have the rules of society imparted to it." Hence, the "first communities and the first skills came into being; rock faces and cave walls were painted, pottery was discovered, and the first attempts at architecture were made."[19] Nor is artificial fertilization the only thing that occurred: there was also, apparently, sexual intercourse, for we are told that the "gods" from space "mated with primitive peoples."[20] In a later book, *Gods from Outer Space*,[21] the story is changed somewhat. It is now claimed that the breeding took place by "artificial mutation of primitive man's genetic code by unknown intelligences."

> In that way the new men would have received their faculties suddenly — consciousness, memory, intelligence, a feeling for handicrafts and technology.[22]

This story is odd. It seems not to have been fully thought through. For example, it is claimed that the new beings had consciousness and memory, which, therefore, their predecessors must have lacked. But cats and dogs have consciousness and memory. So our immediate apelike progenitors must have been a singular group of organisms, less alert and adapted to the world than are dogs and cats!

There is a curious ignorance of genuine science — something that it characteristic of pseudoscience: push carefully into the work of the pseudoscientist and one will discover an ignorance of genuine science! In this case,

the problematic claim is that the spacemen manipulated the genetic code of our ancestors. But when scientists undertake genetic engineering, what they manipulate is not a genetic code but DNA, which is the chemical material in which the information necessary for development and reproduction is encoded.

Worse still is the suggestion that the astronauts had sexual intercourse with our apelike ancestors to produce our more immediate progenitors. Unless the genetic material are very similar, no reproduction could occur. Nor could the other suggestion, that of artificial insemination, do any better: it too requires similarity of genetic material if reproduction is to occur. So our apelike ancestors would have already to be similar genetically to both the astronauts and to ourselves.

Modern genetic theory requires that organisms have similar genetic material if they are to successfully reproduce. It is hard to conceive how the astronauts, originating from some place other than the planet Earth, could have a genetic code that resembles our own. In that case, there could be no reproduction.

Von Däniken acknowledges that there is a problem here. He has run into a conflict between his own theory and modern genetic theory. Now, we have seen that in science it is a safe rule of procedure that

> If a new theory T' replaces an older theory T, then T' will yield successful predictions where T does not, but will yield the same predictions where T successfully predicts.

This means that

> Any new hypothesis should be consistent with well established theories in the area.

This is a neat formula but as it stands, it is inaccurate. If taken seriously, it would never allow the introduction of a new theory that might replace the old. Taken literally, it would mean that we could automatically reject Einstein's theory on grounds that it conflicted with Newton's theory. The important point is that, while the two theories did disagree, this disagreement was not in the area where Newton's theory had proved to be predictively successful. Einstein's theory was contrary to Newton's only in the areas where Newton's theory failed to be predictively successful. Thus, our formula is more accurately expressed as

Any new hypothesis should be consistent with those parts of theories in the area which have been well established through predictive success.

The problem for von Däniken is that this rule points immediately to the rejection of his suggestion concerning the origin of human intelligence: it conflicts with well established parts of modern genetic theory.

But does he reject his suggestion? Recognizing that a cross between different species, e.g., man and animals, or, worse, apelike creatures and ETs, is impossible, he introduces one of his favourite rhetorical devices and by way of a rhetorical question invites an answer that rejects the established theory. "But do we," he asks, "know the genetic code according to which the chromosome count of the mixed beings was put together?"[23] Unfortunately, the suggestion is simply nonsense. There is no neat relationship between chromosome count and genetic code; the two are very different things. Genes are located on chromosomes. For reproduction to occur, the chromosomes of the two species would have to be of the same shape and number and the genes on the chromosomes would have to have the same basic arrangement and be located on corresponding chromosomes. More importantly, even if we to not know the genetic codes and do not know the basic structure of the chromosomes, we *do know* enough to be able to say that it is extremely unlikely that species that evolved on different planets, in different environments, from different beginnings would have genetic material sufficiently similar as to allow for reproduction. Von Däniken's notion that human intelligence arose through a process of interbreeding between spacemen and pre-human apelike ancestors is inconsistent with accepted theory and must therefore be rejected as unreasonable. But once again, von D. does not let the facts get in his way.

Nor is his other suggestion, that there was a sort of genetic engineering that produced human beings, any more reasonable. This implies that there is a genetic break between ourselves and our immediate ancestors such that the two species could not interbreed. But there is no evidence in the fossil or archaeological record that there was such a break.

Again, however, rather than discussing the factual data that might be relevant, von Däniken simply ignores it. He ignores the methodological rule of science that we have just stated, a rule which simply says that in evaluating an hypothesis one must take into account the data the support available theory. In ignoring this rules, and in refusing to acknowledge that the facts of genetic theory might be relevant to evaluating his hypothesis about the origin of intelligence in human beings, von Däniken is simply conforming to his own ***a priori principle*** that an hypothesis is to be considered acceptable, and worthy of use for explanation, not on the basis of observational or empirical data but on the basis of the fact that alternatives are *inconceivable to him*.

There is one special set of facts to which von Däniken appeals as evidence that his hypotheses are acceptable. These are the records of the visits by ancient astronauts that our ancestors recorded in their myths and legends. There are various paintings, drawings and carvings that seem to some observers to be, obviously, pictorial representations of astronauts and spaceships. The Bible, like other myth cycles of the ancient near east and elsewhere, is full of stories that can only be construed as accounts of visits by extra-terrestrial astronauts. The destruction of Sodom and Gomorrah should be understood as a nuclear explosion that was set off by "angels," that is, visitors from outer space. Ezekiel's vision of wheels within wheels, and of the "living creatures" that accompanied them should be understood as an eyewitness account of extraterrestrials and their spaceships. These data render the ET hypothesis highly credible, one to which much more serious research efforts should be devoted.

If the urge to discover our past is not sufficient to set modern intensive research work in motion, perhaps the slide rule could be usefully employed. So far, at all events, no scientist has been asked to use the most modern apparatus to investigate radiation at Tiahuanaco, Sacsahuaman, the legendary Sodom, or in the Gobi desert. Cuneiform texts and tablets from Ur, the oldest books of mankind, tell without exception of "gods" who rode in the heavens in ships, of "gods" who came from the stars, possessed terrible weapons and returned to the stars. Why to we not seek them out, the old "gods"?[24]

In invoking ancient texts, von Däniken is not alone. The Bible is a treasure trove which is regularly mined by those who defend the thesis that the residual UFOs are to be explained by the ET hypothesis. Morris K. Jessup, in his book, *UFO[s] and the Bible* , makes the point that seems to UFO enthusiasts the most relevant:

> Much of the past skepticism regarding the validity of the Bible has sprung from the improbability of the events and forecasts recorded therein. The existence of space-intelligence ... and the probability of a super-race using navigable contrivances, fits all conditions which we have been able to attribute to UFO[s], and thus rationalizes scriptural events.[25]

Among the many texts to which appeal is made is *Exodus*, ch. 13, verse 21, states that "...the Lord went before them by day in a pillar of fire, to give them light: to go by day and night." This is regarded as evidence that God, or perhaps ETs , sent a spaceship to guide the Israelites during the 40-year journey to the Holy Land. But the most often cited passage suggesting to enthusiasts a UFO sighting which occurs in the Bible is in the vision of *Ezekiel*, ch. 1, verses 1-20:

> 4: And I looked, and behold a whirlwind came out of the north, a great cloud, and a fire infolding itself, and a brightness was about it, and out of the midst thereof as the colour of amber, out of the midst of the fire.
> 5: Also out of the midst thereof came the likeness of four living creatures ... they had the likeness of a man.
> 6: And every one had four faces, and every one had four wings.
> 13: ... their appearance was like burning coals of fire, and like the appearance of lamps: it went up and down among the living creatures, and the fire was the fire bright and out of the fire went forth lightning.

> 15: Now as I beheld the living creatures, behold one wheel upon the earth by the living creatures, with his four faces.
> 19: And, when the living creatures were, the wheels went by them: and when the living creatures were lifted up from the earth, the wheels lifted up.
> 20: ... for the spirit of the living creatures was in them.

Von Däniken writes of this that

> The description is astonishingly good. Ezekiel says that each wheel was in the middle of another one.

How does he know that it is "good"? Was he there? Von D. uses the mechanism of mere assertion in order to suggest to us, the readers, the we do indeed have a real description. But here is the full quotation:

> The description is astonishingly good. Ezekiel says that each wheel was in the middle of another one. An optical illusion! To our present way of thinking what he saw was one of those special vehicles the Americans use in the desert and swampy terrain. Ezekiel observed that the wheels rose from the ground simultaneously with the winged creatures. He was quite right. Naturally the wheels of a multipurpose vehicle, say an amphibious helicopter, do not stay on the ground when it takes off.[26]

There are various problems with the reasoning here. Some of what Ezekiel says does not fit with what von D. wishes to claim. Well, just dismiss it — "An optical illusion!" — and go on. Do not bother to stop to give evidence why one should accept this reason for dismissing the unwanted data! Moreover, just what does an amphibious helicopter look like? Is it a helicopter with pontoons as well as wheels? What does a swamp vehicle look like? Like a helicopter? To the contrary, where the blades of a helicopter are mounted horizontally on a vertical shaft, swamp vehicles have a rear-mounted fan that sits vertically on a horizontal shaft. It is simply not at all clear what sort of vehicle Ezekiel is supposed to be seeing. But von D. hastens on, confident that Ezekiel is seeing some sort of vehicle or other.

But the more general pattern of the argument is what should be noted. There is a pattern here that we shall see repeated in other sorts of pseudoscience, Velikovsky for example. The pattern is this:

One takes some set of myths or ancient stories. These are then construed (with no argument being given for doing this!) as reports of actual occurrences. One then develops an hypothesis which, were it true, would explain those purported events. The hypothesis attempts to explain those events in terms of something for which we have no direct, present day evidence, e.g., ETs. The myth then taken as providing evidence that confirms the hypothesis.

This pattern of reasoning is absent from science. To be sure, appeal is sometimes made to ancient stories. Thus, there are ancient stories of solar eclipses, and of comets. One of the latter appeared at the time of the Battle of Hastings, and is recorded on the Bayeaux Tapestry that was woven to tell the story of William's victory. Present day astronomers have no doubt that the stories of solar eclipses are true; we to know *on the basis of currently available data* that there are solar eclipses. Astronomers also know that the comet noted on the Bayeaux Tapestry was in fact real. It was Halley's Comet. *Given presently available data*, together with the theory confirmed by those data, astronomers have been able to calculate the orbit of Halley's Comet and its period. Knowing the period, they have been able to deduce that the comet did return at the time of the Battle of Hastings in 1066. That is, scientists are quite willing to interpret the stories and myths of the past, provided that they can to so on the basis of a theory supported by presently available data. But the theory is not developed to fit the ancient stories and myths. For the scientist, the latter are not *data*. In the hands of von Däniken and other UFOers they are data.

However, they become data for the ET hypothesis only if they are given the interpretation that excludes other hypotheses. In order to back up the reading of the ancient myth or story, it has to be assumed that there were spaceships piloted by ET astronauts. But this is precisely what the story or myth is supposed to prove. So it comes down to the rule that we ought to accept the hypothesis because no alternative is conceivable. An unsound rule, as we saw, one which provides no genuine scientific support for an hypothesis.

Appealing to the Bible may add an extra element of *appeal to authority*: since it is in the "good book", it must be true. But even if we accept that the Bible has some sort of authority, it still does not dictate the interpretation of its texts that must be placed upon them if they are to be read as reports of ancient astronauts.

"The unexplained": As we have said there is a residual number of cases of UFOs for which we can provide no naturalistic explanation. These cases are therefore among what purveyors of pseudoscience often refer to as "the unexplained," as something *mysterious*. Thus, what von Däniken presents us with is "an array of unexplained mysteries."[27] Now, the point about a mystery is that it is a thing or event that is beyond reason. Those who accept such mysteries of the Christian religion as the triune nature of God or the incarnation of God as Jesus Christ do so on the basis of faith, not of reason. Nor is every unexplained event a mystery. I find a penny on the street. How it got there is unexplained. It is, however, possible to guess: it is most plausible to assume that someone dropped it. Because there are rationally plausible hypotheses, any of which, if true, could explain the event, that event is not a mystery, beyond reason. But there might be situations in which there is no plausible naturalistic explanation that comes to mind. This often happens in science. There is a part of nature exhibiting apparent disorder. The scientist may have a sense from background knowledge of the general sort of hypothesis that might be relevant without being able to give any specific description of the order that might be there, behind the apparent disorder. The situation of disorder is *unexplained*. It is not assumed that it is *unexplainable*. It is just that it is inexplicable given current knowledge. A mystery is not of this sort: it is *unexplainable*. Even so, not every event that is apparently unexplainable is thereby a *mystery*. Where the cause of something is unknown, we have a secret cause, and the event might thereby be a riddle to one. But a secret or a riddle, while involving something inexplicable, is not a mystery. To make a mystery of something is to keep it a secret in order to make an impression. What distinguishes a mere secret or riddle from a mystery is that the latter has a significance for human being that the former lacks: mysteries are momentous, riddles and secrets relatively trivial and unimportant. The triune nature of God and the incarnation of God as Christ are not riddles, they are mysteries. They are mysteries because they are significant.

The pattern of von D. and other UFOers is clear. They take an unexplained event, an event for which we have not yet discovered the correct explanation and for which we may be in a position to discover the correct explanation. This is then proclaimed to be a mystery, something beyond reason, or at least, beyond the customary reason of everyday science. Yet the event must be explained. Unlike those who accept the Christian mysteries, von D. and other UFOers attempt to use reason to shed light upon the mystery. Yet reason strives to explain it, strives to throw light on that which is mysterious. A mystery demands an explanation, but precisely because it is a mystery and apparently beyond reason, one is permitted to offer fantastic hypotheses in the attempt to provide explanations. Only it is a new reason, a radically different reason than that of ordinary science, one no doubt that they see as revolutionary. Note, however, that what ends up a mystery begins as something that is merely a riddle. A strange light in the sky may be a riddle, it may lack any explanation that we can think of, but is not thereby a mystery, something of momentous import for humankind. It is the proposed explanation that turns the riddle into a mystery. The radically new form of reason yields a conclusion that makes the unexplained event *momentous*. The residual unexplained UFOs are after all to be explained, they are to be explained by ETs, and when they are so explained then they are of course momentous: we have made one of the great steps ever taken by humankind, that of making contact with intelligent species other than our own!

There are at times anomalies in science, events that cannot be explained, or, at least, explained by current theory. Thus, in the mid-nineteenth century, physical theory held that the earth moved through a stationary "ether" which was the medium which carried light waves. Since the ether was stationary, when the earth moved

through it there would be as it were an ether "wind". A light signal sent out in the direction of the "wind" should move faster than one sent out in the opposite direction. The physicists A. Michelson and E. Morley set out to measure the difference and to their surprise found that there was none. This negative result was a real anomaly, an enigma for which theories then available could not account. Naturally enough, this was seen as an area of apparent disorder, and hypotheses were developed which, if true, could explain the anomaly. For example, it was proposed that the earth dragged some of the ether along with it as it moved. If that were so, then no "wind" would be felt and any experiment conducted on the earth would fail to detect a motion through the medium. None of these hypotheses was successful; no one was able to discover an order behind this apparent disorder. It was only in 1905 that Albert Einstein proposed an alternative theory that solved the problem. It did so by proposing a radical change in the physical assumptions that science was making about the propagation of light and the nature of space and time. In particular, this theory postulated that the velocity of light was constant in all directions. With this postulate, what had been an anomaly emerged as a natural result that supported the new theory. Subsequently more and more experimental tests have confirmed the soundness of Einstein's new theory, and in particular the postulate of the constancy of the speed of light in all directions.[28]

Scientists do not go looking for mysteries or anomalies. These appear when nature does not conform to the hypotheses that the scientist has hitherto accepted on the basis of good tests. Pseudoscientists, in contrast, look for mysteries. The scientific anomaly violates reasonable expectations. The mysteries of the pseudoscientist do not violate reasonable expectations since the pseudoscientist has no reasonable expectations. Given the significance attached to the rule that "anything is possible," how could she have any expectations, reasonable or otherwise?

The scientist hesitates before trying to explain the anomaly. Various proposals are made. Initially, the attempt is made to explain the anomaly in terms of the current theory. Scientists, unlike pseudoscientists, do not rush out to abandon at the slightest pretext the old theory. Moreover, this attitude is eminently reasonable. After all, the older theory has been *strongly confirmed* and in the past *problems which it faced have seen solved.* Given past experience, then, it is reasonable to expect the old theory to eventually provide a solution to the problem, to resolve the anomaly. *Past experience makes caution reasonable.* The pseudoscientist exercises no such reasonable caution.

But sometimes the older theory does not yield solutions to the anomaly. The puzzle remains unresolved. So proposals continue to be made. In turn they are put to the test. Eventually one turns out to solve the problem. It may involve, as did Einstein's proposal, a deep change in the structure of one's theory. But in any case, the fact that the proposal solves the anomaly testifies to the acceptability of the hypothesis. In contrast, the pseudoscientist hastens to find the most fantastic hypothesis possible, the more far-fetched the better. No thought is given to how it might fit in other scientific theories for which there is good evidence. No thought is given to how it might square with the data that testified to the truth of older theories, or alternative suggestions that might be proposed by more cautious scientists. To the contrary, the hypothesis is accepted on *a priori* grounds that alternatives are inconceivable (to the pseudoscientist). Moreover, of course, it must be far-fetched, since only in that way can the puzzle or anomaly be turned into something much more grand: a mystery.

The important point here is that Einstein's theory is acceptable not merely because it solves the anomaly but because it has *many and widely varied confirmations.* It is not accepted *solely* on the ground that it resolves the anomaly. To be sure, it does do that: it replaces the apparent disorder with order. But that is *not the sole* evidence that renders the theory acceptable. There are many more experiments that do this. In order for a theory to become acceptable to scientists, it must not only explain the anomaly but also the non-anomalous. It must predict not only the apparently anomalous data but also both square with the data that confirmed the older theory and *also* predict new facts not predicted by the older theory.

In contrast, what evidence are we offered that the ET hypothesis is acceptable? Simply and only the fact that, if true, it would solve the mysteries that it is designed to explain. This is hardly the practice of good science.

Example II: Scientific Creationism

At one time religious fundamentalism attacked the Copernican theory of the movement of the earth about the sun. At that time the fundamentalism was that of the Roman Catholic Church. More recently, it has been Darwin's theory of evolution and of the origin of species by natural selection that has come under attack. Only, this time the opposition has come from Protestant fundamentalists.

Darwin's theory states that present forms of life have evolved from earlier, and simpler forms. Algae were succeeded by algae but also by fish; fish and algae were succeeded by algae, fish and amphibians; the latter by algae, fish, amphibians and reptiles; the latter were succeeded by algae, fish, amphibians, reptiles, birds and mammals. At a still more recent point, humans entered the picture and gradually came to assume their present form. This development from simpler to more complex has left its traces in the fossil record; from this one can re-create the past biological history of our planet. This record establishes the great age of the earth. There are, moreover, the methods of radioactive dating which establish and use criteria independent of the fossil record to date the rock strata as very old. These data, then, establish that there has been ample time for evolution to occur gradually, and establish also the *fact* of evolution. What mechanism explains the developments we find laid out in the fossil record? The mechanism that effected this development, was, Darwin argued, convincingly to most biologists, natural selection. If we add to natural selection an adequate account of heredity, and specifically the modern theory of genetics, then we have a reasonably adequate explanation of the evolutionary development that is recorded in the fossils.

The Bible, in contrast, in the book of Genesis, asserts that God created living things according to their kinds, all sorts of vegetation on the third day; the sun and moon were created on the fourth day; the fifth day, fishes and birds came into being; cattle, creeping things, and beasts of the earth came on the sixth day. It says nothing about fossils. Many who defend the truth of this Biblical account of origins argue that fossils came into being at the time of the Noachic flood. They also suggest that some forms of life that had originally existed, such as trilobites and dinosaurs, became extinct at the time of the flood. But, as people pre-date the Noachic flood, they must have co-existed with the dinosaurs.

There are those who accept this story on religious grounds. So do most "scientific creationists." But the latter also argue that, independently of their religious faith, there are *scientific grounds* for accepting the truth of the story as told in the Bible and for rejecting the Darwinian account.

Here the fossil record is of central importance. What of this evidence? Does this not tell decisively against the Biblical account?

Scientific creationists argue in **three ways**. They argue, **first**, that the whole process of dating by means of the fossil record is circular and therefore vacuous, while other means of dating, for example, radiometric dating, cannot be relied upon to give sound results. They argue, **second** that there are gaps in the fossil record that imply that the evolutionist take of gradual development of certain forms from earlier forms must be wrong. And they argue, **third**, that even where there is the appearance of age, e.g,, in tree rings, it is irrelevant. Let us look at them in turn.

Henry Morris, a leading scientific creationist, has presented the argument for the circular nature of the evolutionist reasoning about the fossil record as follows:

> Creationists have long insisted that the main evidence for evolution — the fossil record — involves a serious case of circular reasoning. That is, the fossil evidence that life has evolved from simple to complex forms over the geological ages depends on the geological ages of the specific rocks in which these fossils are found. The rocks, however, are assigned geologic ages based on the fossil assemblages which they contain. The fossils, in turn, are arranged on the basis of their assumed evolutionary relationships. Thus the main evidence for evolution is based on the assumption of evolution.[29]

As he puts it elsewhere, "The fossils speak of evolution, because they have been *made* to speak for evolution."[30]

The first problem to note with this objection is that basic outlines of the history as laid out in the fossil record as it appears in sedimentary rocks was established by geologists in the early part of the 19th century, well before the appearance of the theory of evolution. How did these geologists establish the relative sequence and thereby the relative ages of the deposits?

Their inferences were based on three laws. The first is the "Law of Initial Horizontality," which states that *sedimentary strata are initially deposited horizontally*. There are two bases for inferring this law. One is empirical observation on how sediment is deposited on the sea bed. The second consists in the theory of hydrodynamics. The latter consists of a body of laws, themselves based on inferences from other laws, including those of mechanics, and inferences from observations. These laws of hydrodynamics entail that deposits will be made in horizontal layers.

This theoretical basis for accepting the Law of Initial Horizontality should be noted:

> **Patterns are accepted as laws not only on the basis of confirming observational evidence but also on the basis of other laws — theories — where these theories themselves consist of laws that have been confirmed observationally.**

This is important, since it means that the evidence that confirms the theory in turn supports the laws derived from the theory. At the same time, the empirical data that confirm the law in turn support the theory from which the law is derived. *If, therefore, one wants to argue that the law is false, one must also reject the theory that implies the law; one must then not only explain away the observations that confirm the law, one must also explain away the further data that confirm the theory; and conversely, the support deriving from the theory provides strong grounds for not rejecting the law.*

Laws thus cannot be rejected on a one by one basis. **Coherence** of a law with a body of confirmed theory means that one cannot reject the law without also rejecting the theory. But if the theory is itself strongly confirmed, then that provides strong grounds for not rejecting the law.

One of the marks of pseudoscience is the failure to note that when it rejects a law of science it comes into conflict with a strongly confirmed theory. Acceptance of this theory provides grounds for rejecting the astounding hypotheses of the pseudoscience. It is not dogma nor closed-mindedness that leads scientists to resist astounding new theories, it is simply the fact that they know a strongly confirmed background theory that provides good grounds for rejecting the pseudoscientific proposal.

There is a further assumption which must be made. This is the assumption that geologists refer to as *actualism*. This is the assumption that

natural laws, including the laws of physics, are unchanging or uniform through time.

One cannot evade the force of a confirmed theory by arguing that the laws of nature have changed:

laws of nature are unchanging patterns, uniformities that hold for all times and all places.

Thus, we take it for granted that sediments were deposited in the geologic past according to exactly the same patterns as we observe them being deposited today.

The second law on which the inference to the order of sedimentary depositions is based is the "Law of Superposition," which states that *younger undisturbed strata invariably overlie older undisturbed strata*. This law is accepted on the basis of observations, background theory and actualism. We have never observed older sediments being deposited on top of younger sediments, and for such to occur there would have to be a violation of strongly confirmed physical laws such as the law of gravity.

When geologists apply these two principles to undisturbed strata, they can immediately establish a preliminary chronology based on the strata alone, with no reference to fossils.

But, once the preliminary ordering is available, it is then possible not only to describe the sequence of fossils in the strata but also to assign to them the temporal sequence of the strata in which they are found. Thus, *the temporal sequence of the fossils is derived from the rocks*. Contrary to what Morris argues, *the assumption of evolution is not used to establish either the order of the fossils or the order of the rocks*.

Once the initial order is established, it can be extended by the third law that is used. This is the "Law of Biotic Succession," which states that *fossils in the strata will always occur in the same sequence regardless of geographic location*.

The inference to this third law is made on the assumption of actualism together with observational evidence. Geologists confirmed this law by tracing laterally, wherever possible, fossil-bearing strata. It was discovered that no matter how far the strata were traced, the same strata had the same fossils and the sequence of fossils in successive strata was invariably unchanging.

Thus, because the sequence of fossil succession is invariable, one can safely conclude that rocks with the same fossils are the same age.

Once the sequence of fossils has been established in undisturbed strata, it can be extended to disturbed strata. Reworking and disturbance of strata by thrusting or overturning of beds cause fossils to appear out of sequence with those in undisturbed beds. But these phenomena are well understood, both theoretically and in practice, and do not present any serious problems.

We can therefore safely conclude that, contrary to Morris, establishing the order of fossils in the stratigraphic record is not a circular inference, but one that is soundly based on observational evidence.[31]

The method of dating by stratigraphic methods gives only an *ordinal* ranking of the ages of rocks and fossils. It puts in them in an order of earlier and later. But it does not establish *how much later*. That is, it does not give a *cardinal* ordering. This has been achieved through *radiometric dating*.

This method depends upon the fact that the isotopes of certain elements spontaneously decay into other elements. Thus, to take one example, the isotope potassium 40 (^{40}K) spontaneously decays into two different products. Most ^{40}K atoms decay into calcium 40 (^{40}Ca). The remainder decay into argon 40 (^{40}Ar). ^{40}Ca is a fairly common isotope, and it is hard to distinguish ordinarily occurring ^{40}Ca and radiogenic ^{40}Ca. But ^{40}Ar is not common.

Samples of crystallized rock that have not been exposed to the air or water will not have gained any potassium nor have lost it save by radioactive decay. Nor, since Ar is both rare and inert, will it have lost or gained any argon save by radioactive decay. Hence, if we know the rate at which ^{40}K decays into ^{40}Ar, it will be possible to calculate the age of the rocks. But the decay of isotopes is well understood theoretically in terms of quantum mechanics. In particular, this well confirmed theory establishes that the rate of decay is constant. Moreover, the rate is something that can be determined experimentally. The age of the rocks can therefore be determined.

The creationist, H. Slusher, has objected that the rate of decay is not known exactly.[32] This is true. However, if we suppose an error of 4% and date a rock as one billion years old then we have set the date within the range 1040 million and 960 million years old. It would be nice to reduce the error by more exactly determining the rate of decay, but the point for the creationists is that such an error will certainly *not* reduce the age of the rock to something less than 10,000 years as the creationist argument requires.

It is also objected by creationists that the systems in question are open systems from which argon may have leaked or into which argon may have diffused.[33] However, it is possible to separate out those samples which have for practical purposes been closed systems and the gains and losses minimal. However, if all the argon that resulted from radioactive decay were lost, then there would be a relatively small amount left in the rock and the age reading would be anomalously *young*! If the proposed defect were really serious, then the results would support the creationist case rather than refute it!

Other points can be made for and against radiometric dating.[34] Three points should be made.

First, there are in fact several methods for such dating. these are established and tested independently. As it turns out, they are remarkably consistent with one another. Thus, to reject one requires one to challenge the independently tested other methods. Once again, coherence among the various scientific theories testifies to their truth, and makes it difficult to reject either one of them or all. One should note that in addition, where there are independent tests, e.g., tree ring dating to test the radiocarbon method, then once again the results confirm each other.

Second, if one rejects the method of potassium-argon dating then one must reject not just the test but also the theory, quantum mechanics, which explains why the test works and justifies its use. Again, one can appeal to

coherence to establish that the creationist rejection of the scientific position is weak.

Third, one should note the varying quality of the evidence offered to counter the scientific claims. Scientific creationists, like other pseudoscientists, believe that they can triumph by the sheer quantity of arguments they offer. Individual pieces may be defective, but since everything they have is shot at the target some of it, they appear to reason, is bound to hit home. This is the technique of the **blunderbuss argument**. One can find the same technique used by von Däniken. Unexplained cave drawing after unexplained cave drawing, strange artifact after strange artifact, are brought forward, often presented in apparent ignorance of what is really known about them. In the end, von Däniken seems to think, if enough examples are tossed out, some at least will be telling. UFOers do the same: sighting upon sighting is reported, most often based in anecdotal reports by untrained observers, often where neither control nor examination of the original data is possible. "But the observers were honest men and trustworthy," we are told. Alas, honest and trustworthy persons can be mistaken. (We discuss observation in Chapter Five, below.) Like all witnesses, they need to be subject to rigorous cross-examination before their testimony can be accepted as reliable. Such cross-examination is simply absent from the discussions of UFOs by enthusiasts. The trouble with the blunderbuss argument is that from the fact that there are many little arguments it does not follow that any of them is any good. Even with many, all might be bad. This is in fact usually the case with pseudoscience.

But is it worthwhile trying to reply to all the little arguments? The general consensus among scientists is that it is not. On the basis of their experience and the **coherence** of their theories and laws, they can judge that most of the claims of pseudoscientists is bound to be *nonsense*. This judgment about the merits of the case of the pseudoscientist is neither dogmatism nor *a priori* rejection: it is a reasonable judgments based on the best evidence available.

The **second argument** that is used by defenders of creation science to attack evolutionary theory is that the fossil record contains many gaps. If later species originated from earlier species, then there will have to have been a series of organisms linking the two. But in many cases such chains cannot be found in the fossil record. This is evidence against the evolutionist claim that later species evolved gradually from earlier species.

Now, in the first place, the existence of gaps does not by itself imply that the evolutionary story is false. There may well be other explanations for the gaps besides the non-existence of the links in the chain. Perhaps, for example, the sedimentary deposits were eroded away, or destroyed by volcanic intrusions. Perhaps they lived in areas such as rain forests which are particularly unsuited for the development of fossils. And so on.

But in the second place, creationist often set the game up in such a way that it will be impossible to find intermediate forms. They consider that a transitional form must be intermediate *in all respects* between the species which it links. Thus, Henry Morris writes that

> At the very least, there must have been a tremendous number of transitional forms between *Archaeopteryx* and its imaginary reptilian ancestor. Why does not one every find a fossil animal with half-scales turning into feathers, or half-fore limbs turning into wings?[35]

Archaeopteryx is not a direct ancestor of modern birds. Nonetheless, it is morphologically similar to both reptiles and modern birds, that is, similar to each *in some respects*. This places it as an intermediate between both. Morris so defines what it is to be an intermediate or transitional form that such a form will never be found. But evolutionary theory does not requires that there be forms that are transitional in Morris' sense, that is, intermediate in *all* respects. What we do find, as we find in *archaeopteryx*, are forms that are transitional in just the way that Darwinian theory requires there to be transitional forms. Thus, for example, although it is often denied by creationists, there is solid evidence for a transitional intermediate forms between cynodont reptiles and mammals.[36]

In short, the fossil record is just the way one would expect on the basis of evolutionary theory, gappy but containing many transitional forms intermediate between earlier and later species.[37] Creationists find problems in that record for evolutionary theory often because they impose requirements stronger than those required by that theory. In other words, they misconstrue what the theory requires and then pronounce the theory false when

it fails to meet those requirements. This technique of **silently re-writing** the theory in order to find a version of it that can easily be shown to be mistaken is clearly illegitimate.

There are cases, however, which even the determined evolutionist cannot explain away. Consider the fact that light travels at a finite speed, and note that stars are many light years away from us. That means that the light they emit takes a very long time to reach us. If, as the creationists maintain, the stars and the rest of the universe were created along with the earth only a short time ago, then only the light from the nearest stars would be reaching the earth by the time the first human appeared. As time went on, light from more distant stars would begin to arrive and they would become visible. Then light from still more distant stars would arrive and they in turn would become visible. Thus, the number of visible stars would gradually be increasing. From a very few, the stars would gradually as it were turn on in the night sky, becoming increasingly numerous as time goes on. It is a fact that this does not happen. Indeed, a new star, what we would now call a supernova, is an exceedingly rare event, but sufficiently spectacular to be recorded. Only a few were ever noted by startled astronomers. This means that the universe cannot be as young as the creationists maintain.

They must therefore explain it away. This is the creationist's **third argument**: *signs of age are irrelevant*.

Here is how it is done. "This problem," Henry Morris writes, "seems formidable at first, but is easily resolved when the implications of God's creative act are understood." We have to recognize that man "was absolutely central in all His plans." In particular,

> The sun, moon, and stars were formed specifically to 'be for signs and for seasons, and for days, and years,' and 'to give light upon the earth' (Genesis 1: 14, 15). In order to accomplish these purposes, they would obviously have to be visible on earth.

But if they are to be visible then they would have to be created in a way that gives *the appearance of age*.

> ...real creation necessarily involves creation of 'apparent age.' Whatever is truly created —

that is, called instantly into existence out of nothing — must certainly look as though it had been there prior to its creation. Thus it has an appearance of age.[38]

The *empirical evidence* thus *testifies to the truth* of the scientific theories that imply a extremely long history for the earth.

Morris in not the first to use this form of argument to try to defeat the claims of geology and of evolutionary theory regarding the age of the earth and the developments were find recorded in the fossils. One such attempt was made by in the 19th century by Philip Henry Gosse, in a book entitled *Omphalos* (navel). Gosse was a naturalist, but also a member of Plymouth Brethren, a fundamentalist and strictly puritanical sect insisting upon a strict and literal interpretation of the Bible. His fundamentalist reply to the evidence of the fossil record was the suggestion that God created the earth with the fossils already in it. Gosse proposed that God created Adam with a navel in order that he look as if he had been born in the usual way with an umbilical cord. In the same way God had created the entire universe with the *appearance* that it had a long previous history. Any creative act of the deity would have brought about a world that had evidences of a past. If Adam were indeed like a human being but created as an adult, he must have been created with adult teeth, but vestiges recalled baby teeth. If trees were created in full bloom, they must have been created with rings. If God created mountains and valleys, he must have created deposits of sand and clay that would suggest their having been laid down by eons ago by slow geological processes rather than having been created by a short while previously. He could equally have created rock strata with fossils already in them. Equally, though Gosse was in no position to know about radioactivity, the deity could have created the world with all the details there suggestive of a longer time period for the history of the earth than the creative act actually required.

This of course raises the problem of what sort of creator would create a world in which vast parts of it were capable of misleading the ordinary observer into thinking it was much older than it actually is.[39] Why would the creator make such a world and include in it beings endowed with reason who would naturally infer a conclusion that was false, namely, that the world is much older than it is? If the creator made a world designed to fool us, that would show that he or she was clever, but equally that he or she was not very nice.

On the face of it, the result is not very flattering to the deity. Certainly, it is hardly the traditional picture to which Gosse was undoubtedly committed. One textbook of scientific creationism puts it this way: "This would be the creation, not of an appearance of age, but of an appearance of evil, and would be contrary to Gods [sic] nature."[40]

To this the response that could be made is that the apparent, but misleading, vestiges of age are meant not to fool us but to test our faith. They therefore are not evil but in fact serve a positive purpose, and consequently are consistent with the traditional notion that the deity is not only powerful and wise but also good.

This response, however, must be based on grounds that one knows *prior to any data about the record* that God is no deceiver, but that he or she — wisely, perhaps — puts in place facts which will test, and therefore strengthen our faith. These facts are designed to cause doubt, but we will be meritorious if we continue in spite of these doubts to maintain our faith.

This defence, however, requires appeal to **non-empirical** and *a priori* arguments. Those who now call themselves "scientific creationists" find these arguments inappropriate. For, these arguments, since they are non-empirical and *a priori*, are *not scientific*. If one is going to offer a defence of creationism as an acceptable *scientific* theory, then one had best use arguments that count as scientific!

It follows that some way other than that of Gosse is required if the basis of the theory of evolution in the fossil record is to be challenged. *Scientific* grounds for the challenge must be given.

It should be recognized, however, that creation scientists do attempt to use arguments based on principles of scientific inference. As we have seen, any proposed hypothesis, to be reasonable, must be consistent with strongly confirmed theories in the area. Creation scientists often appeal to this principle to attack the Darwinian theory of evolution. The argue that this theory is inconsistent with the well established principles of the physical science of thermodynamics.

The science of thermodynamics begins with two basic laws. The First Law of Thermodynamics states that *in* closed systems the total energy is constant. This First Law states the principles of conservation of energy.

The Second Law of Thermodynamics states that *the entropy of a closed system tends to a maximum*. This law states that processes going in a certain direction are possible while others are not. In closed systems the direction of a process is always one in which entropy increases. Entropy is a known quantity that measures the amount of order in a system: the greater the entropy the greater the disorder in the system. Thus, the Second Law states that in closed systems the amount of disorder is always increasing. If, therefore, we have a closed system with an area of localized order, changes in the system will result in the end with that area of order disappearing into disorder.

Henry Morris has stated the argument of the creationists this way:

> There is ... firm evidence that evolution could never take place. *The law of increasing entropy* is an impenetrable barrier which no evolutionary mechanism yet suggested has ever been able to overcome. Evolution and entropy are opposing and mutually exclusive concepts. If the entropy principle really is a universal law, then evolution must be impossible.[41]

He illustrates his point with a photo of a waterfall with a caption stating that

> Evolutionists have fostered the strange belief that everything is involved in a process of progress, from chaotic particles billions of years ago all the way up to complex people today. The fact is, the most certain laws of science state that the processes of nature do not make things go uphill, but downhill. Evolution is impossible![42]

The title of the chapter is "Can Water Run Uphill?"

The argument is that if evolution were true it would involve an increase in order, from incomplexity to complexity. But such an increase in order would imply a decrease in entropy. However, the latter is excluded by the Second Law: in closed systems any change is in the direction of increased entropy. Evolution as described by Darwin's theory is thus inconsistent with the theory of thermodynamics, the Second Law in particular.

The structure of Morris' argument is clear. It is also clear that it is sound only provided that there is in fact an inconsistency between evolutionary theory and thermodynamics. But is there?[43]

The two laws apply to closed systems. *Open systems* are quite different. Heat is a form of energy. It can be transferred from one system to another, provided the systems are open. Thus, a burner on the stove can heat the water in a kettle. In the kettle energy is not conserved; to the contrary, it increases. But this is not a violation of the First Law since the kettle is not a closed system.

Entropy, too, can be transferred from one system to another. This means that there can be a decrease in entropy in an open system, provided that there is a corresponding increase elsewhere in the universe.

Localized entropy reduction are a very common phenomenon. For example, when a snow flake develops there is an increase in order, and, what is the same, a decrease in entropy. The Second Law does not rule out the localized formation of snow flakes in the open systems of the upper atmosphere. Again, when a baby develops into an adult there is an increase in order, that is, a decrease in energy. This is possible because the organism increases its own order only by creating disorder elsewhere. Just as it takes on energy from its surroundings, so it sloughs off entropy to its surroundings. Thus, for food an organism will ingest highly material with highly structured proteins while it will excrete waste material with less ordered molecules. The entropy of the intake is less than the entropy of the material returned to the surroundings. In this way it loses entropy to its surrounding environment.

If Morris' argument were successful, it would rule out not only evolution but also snow flakes and the growth and development that occurs in organisms as they grow from fertilized eggs to reproducing adults. But such process occur, quite consistently with the science of thermodynamics and with its Second Law. Similarly, evolution is quite consistent with the Second Law. The relevant point is the simple one that the biosphere of the Earth is not a closed system.

Moreover, water *can* run uphill. The figure illustrates how this is so. It represents a self-operating ram pump of the sort developed in Britain and put into operation in the late 1700s.[44] The pump involves a very simple arrangement of conduits and self-operating flap valves.

So long as there is water in the reservoir, the energy derived from the downhill flow is sufficient to enable the water to pump itself *uphill* into storage tanks that may be well over a hundred feet in elevation.

Water from a low lying reservoir R flows through the conduit pipe K putting pressure on the spring-hinged "clack" value C. This pressure increases until C snaps shut. The flow of water continues so that there is a pressure surge created by the shutting of the clack valve. This pressure surge forces a small amount of water past the flap valve F, thereby compressing the air A in the coupling tank CT. The pressure in the conduit pipe quickly equalizes and the surge ends. Then the air A expands, keeping F closed and forcing water up the stand-pipe toward the elevated tank E. The clack valve then opens again, allowing the water flow through K to build up until the clack valve starts the process over by snapping shut. Much water is lost as spill-off S in order to elevate a small amount to E. Nonetheless, a ram pump operating at 50% efficiency can elevate 72 gallons per day a net height of 90 feet.[45]

The physical basis of the mechanism is that a backwards or uphill process can occur provided that it is coupled to a more dominant downhill process. The ram pump does not occur in nature, of course, but it shows that it is simply false to say that water cannot flow uphill. The principle on which the ram pump works is simple and it permits there to be spontaneous "uphill" processes in nature, e.g., evolution.

Interestingly enough, even though Morris has a doctorate in hydraulic engineering, he chooses not mention this rather widely known "hydraulic ram principle," and to insist simple mindedly that "water can't run uphill."

In any case, we now see that, although creation scientists have tried to use sound principles of scientific inference, their arguments fail for not understanding the science in question.

Since scientific creationism claims to base its central claims, e.g., the young age of the earth, the separate creation of each species, and so on, upon the methods of empirical science, then, to live up to this proclaimed standard, it must provide empirical evidence for the hypotheses that it proposes to defend. Thus, scientific creationists argue not only that there is very little evidence to support either the theory of evolution or the theory that this occurred by natural selection. They also argue that there is scientific evidence that species came suddenly into existence a relatively short while ago, a few thousands of years at most. They further argue that there has not been any sort of evolution, and, indeed, that the modern theory of genetics shows such evolution is highly

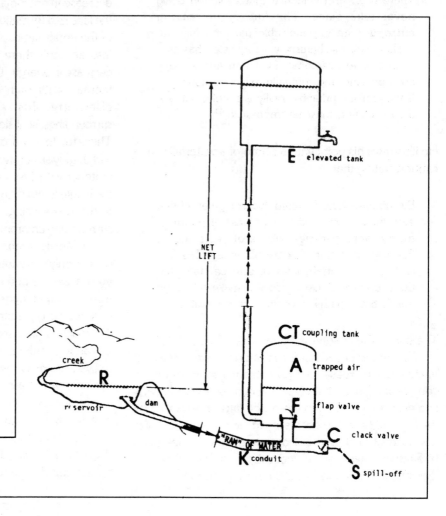

Figure 2.1
THE HYDRAULIC RAM
PRINCIPLE[46]

improbable, so improbable that the theory of evolution by natural selection must be rejected.

Amongst the evidence cited is the fact that many species of animals and plants, e.g., algae and crocodiles, have remained essentially unchanged throughout their histories, even on the evolutionist account. But this identity between living and fossilized representatives is hard to fit into a scheme which emphasizes evolution, change, and development; to the contrary, on the basis of these forms it is reasonable to generalize that all forms, the extant and the extinct together, all began about the same time.

Creation scientists make a variety of other arguments. Thus, they point out that so far as we can tell, most genetic mutations make an organism less fit for survival. Mutation thus cannot do the job that evolutionists require of it, to introduce sufficient favourable variations that natural selection can lead to the development of new species. Moreover, there are many gaps in the fossil record. It is difficult to discover in the record transition species. If eyes developed according to the theory of evolution, for example, they must have developed out of organs or structures that were not eyes and which did not in the same way make the organism fit for survival. But evolutionists have not been able to detect such incipient structures in the fossil record. It is similar with regard to other organs such as wings or legs. "All of which is a specific confirmation of the predictions of the creation model."[47] Here we see creation scientists invoking as a criterion for the worth of their theories a criterion that is in fact often used by scientists: **predictive success**. Nor are the creation scientists the only ones to appeal to this criterion. As we shall note shortly, Immanuel Velikovsky develops a quite different theory of origins. This theory that Velikovsky proposes is incompatible with that of the creation scientists, and, indeed, incompatible with the theories generally accepted by science; its proponents nonetheless appeals to the same criterion of predictive success as do creation scientist and as do scientists themselves. We shall have to explore legitimate appeals to this criterion and illegitimate appeals, and try to discover what it is that makes some such appeals legitimate and others, if not wrong, then at least weak.

The creationist story does not attempt to include in its theory the actual cause of the origin of the species. That cause is supernatural, and therefore could not be the subject of scientific discourse. As Duane Gish, another leading champion of creation science, has made the point:

> We do not know how the Creator created, [or] what processes He used, *for He used processes which are not now operating anywhere in the natural universe.* This is why we refer to creation as special creation. We cannot discover by scientific investigation anything about the creative processes used by the Creator.[48]

But not just anything goes: in spite of appearances, the imagination has no room to run wild. Alternative creation stories, e.g., those of first nations in Canada, are not to be countenanced. For, as Henry Morris has indicated,

> It is *not* necessary to speculate, however, since the Biblical record has provided a clear description of the causes, nature, and results of *true* catastrophism: The Noachic Flood ... We cannot verify it experimentally, of course, any more than any of the various other theories of catastrophism, but we no not need experimental verification; God has recorded it in His Word, and that should be sufficient.[49]

The truth can be known, but it is not to be known by science, not even "creation science."

Creation scientists are greatly concerned about the teaching of evolution and of the theory of evolution by natural selection because, in their view, it is not only false but more importantly socially pernicious and indeed dangerous. It is a source of many of what they view as society's ills.

> Evolution is thus not only anti-Biblical and anti-Christian, but it is utterly unscientific and impossible as well. But it has served effectively as the pseudoscientific basis of atheism, agnosticism, socialism, fascism, and numerous other false and dangerous philosophies over the past century.[50]

Teaching creation science, they argue, will help counter these evil consequences of the teaching of evolution. It will in effect put God back in the classroom. It will do so, not by teaching religion, but by teaching science, the true science of the scientific creationists and not the false or pseudoscience of the Darwinians.

In effect, the argument is that evolution itself is a religion, so-called "secular humanism," and the attacks on evolutionary science are therefore part of a religious crusade. But because of the clause in the American constitution that requires the separation of church and state, clear religion cannot be taught in schools. Creationism must therefore proceed to insert itself into the schools not as religion, but under the same cloak that evolution has insinuated itself into schools, under the guise of science.

The issue is, does it conform to the same standards as science? In the end, the American courts said No. This was the judgment of Judge William Overton.[51] It was not an arbitrary judgment, however. It was based on a clear understanding of the difference between science and religion, or, what is the same in this case, the difference between science and something that is not science disguising itself as science.

One difference between von Däniken's arguments and those of the creation scientists should be noted. Both make extensive use of appeals to the Bible. Von Däniken, who is in this respect parallelled by Velikovsky, takes the Bible to be full of stories that must be interpreted and explained according to the various hypotheses that he proposes. The creation scientists, in contrast, take the Bible to be an *authoritative text*. It is understood to be "inerrant," and therefore a safe guide to the truth. It is not a set of stories that require interpretation in terms of various (pseudo-)scientific theories, it is a settled body of clear truths. The contrast is between a set of data to be explained and a settled body of truth that needs no explanation but to the contrary is the source of the only truly genuine explanations available. Science in the end cannot provide the explanations that we require. For these we must turn to the Bible, the authoritative source of truth.

Creationists thus have a *rule of method* for determining which hypotheses to accept that is distinct from both the method of von Däniken and the method of science. Their *rule of method* is **the method of authority**, which states the norm for acceptance as *accept as true what is asserted by the relevant authority*.

Is this a good norm for the discovery of matter of fact empirical truth? We shall look at this later.

The sorts of plants that are dominant today are those that reproduce by means of seeds, the angiosperms. These came to be widespread in the Cretaceous period. This was very sudden, according to the fossil record, and the apparent gap between these species and ones that preceded them in the fossil record made the whole business an "abominable mystery", as Darwin called it in a letter to Hooker in 1879. The mystery continued up until the early '70's. Duane Gish in his *Evolution? The Fossils Say NO!*, cites a remark by the botanist E. J. H. Corner in 1961 that despairs of the lack of known primitive angiosperms, echoing Darwin's similar despair. Corner's despair was justified in 1961; the gap in the fossil record had still not been closed. But Gish's book was published in 1978. If he had consulted more recent works, for example N. F. Hughes, *Palaeobiology of Angiosperm Origins* (1976),[52] he would have discovered that the mystery had been solved. After an unbroken record of failure to fill in the gaps in the fossil record, the palaeobotanists James A. Doyle and Leo J. Hickey had discovered a sequence of fossil pollens and fossil leaves from the Cretaceous Potomac group of the Atlantic coastal plain.[53] The leaves and pollens in the lowest beds were very primitive, and, as one would expect, as one went to higher beds there was an evolution of reproductive parts towards increasing complexity and a development of leaf patterns that increasingly approached what we observe today. Doyle and Hickey were able to conclude that "the fossil record is now of major significance as evidence for the solution of Darwin's 'abominable mystery' of their origin."[54]

This illustrates a common tendency found among creation science texts. The authors, in their haste to put down the theory of evolution, search for texts by legitimate scientists; they rummage back into the past. Often enough they find texts which fill the bill. There are many unsolved problems in science, and scientists often draw attention, as did Darwin in his letter to Hooker, to these problems. The existence of unsolved problems is then taken by the creationists as evidence that there is a problem that will never be solved, concluding that scientists have themselves provided the refutation of the theory.

What the creation scientists do *not* do is what Gish did not do, namely, look up more recent discussions. As in the present case these more recent developments solve the problem.

I remember how I was once given by a fundamentalist a religious tract attacking the theory of evolution. This tract quoted the eminent geologist James Dawson, Principal of McGill University, to the effect that the theory of evolution had no evidence in its favour. What the tract did not do was point that Dawson was at McGill during the third quarter of the 19th century, when many eminent scientists were still wondering whether Darwin had carried the day. It is not surprising that one could find a 19th century geologist questioning evolution. One will not find such a one today.

It is an easy method to attack the validity of a scientific theory by finding eminent people of good scientific standing *in the past* who questioned the theory as a whole or parts of it. But that is simply not good evidence! What one needs to check out are *the most recent authorities*. Rummaging in the past is good for historians, but it is not the best way to find out if *current scientists* accept the theory.

Here is another example of rummaging in the past.

Doyle and Hickey traced out the evolution of angiosperms from primitive species in the early Cretaeceous period. They concluded that there was "no reason to postulate any extensive prior diversification."[55] This implied a very rapid development and diversification, taking only some tens of millions of years. Earlier palaeobotanists had not thought such a rapid evolutionary change was possible; they had looked for a more gradual development. This meant that they were looking for primitive angiosperms in rocks much older than the Cretaceous, searching instead in the Mesozoic or even late Paleozoic deposits.[56] In effect, earlier palaeobotanists were looking in the wrong places for both the earliest angiosperms forms and for the species intermediate between them and those that are now living.

What occurred in angiosperm evolution was in effect a botanical revolution. There was a long period that preceded the development of angiosperms. Then primitive forms of angiosperms appeared. This started a process of rapid change. The preceding long period of equilibrium was followed by a period of rapid multiplication of the new form of life and a rapid diversification in morphological forms. There are intermediate forms but the transition is not smooth in time, rather the change is extremely abrupt.

This it turns out is in fact typical: there are long periods of relative stasis in evolutionary terms which are interrupted intermittently by brief intervals of rapid evolution. Darwin had himself expected evolution to be everywhere gradual; in this he was wrong. Natural selection is still reckoned to be the mechanism by which species evolve, but the process is now conceived to be one of punctuated equilibria.[57]

Soon after this development in evolutionary theory was accepted, it was seized upon by creationists and used to argue their case against the theory of evolution. The long periods of equilibrium represent the stability of created "kinds". The gaps which appear in the fossil record due to the rapidity of the evolutionary process between periods of equilibrium become "boundaries between kinds." Scientists such as Eldredge and Gould had correctly seen that there is no continuity in the fossil record, that there are no intermediate kinds. But clinging to the dogmatism of evolutionary theory, they deny the obvious. As one creation scientist from the Institute of Creation Research put it, "Rather than forging links in the hypothetical evolutionary chain, the wealth of fossil data has only served to sharpen the boundaries between the created kinds."[58] This misses the point that the theory of punctuated equilibria is not a theory of evolution by jumps but of evolution in spurts associated with speciation. The spurts last many millions of years, and there is no need to postulate any mechanism to explain them other than natural selection.[59]

The creationist argument on this point is tied to their more general claim that evolution is tied to uniformitarianism and creationism to catastrophism, and their argument that the geological evidence supports the latter.

Uniformitarianism was advanced in the late 18th century by James Hutton and in the 19th century by Charles Lyell. They argued that events in the geological past should be explained by appeals to mechanisms that are similar to those we now see at work in changing geological forms, for example erosion, sedimentation, disruption, and uplift. These methods of gradual change were emphasized in order to combat the tendency to explain everything geological by mighty catastrophes such as great floods or violent earthquakes, often in turn explained as the work of God. The uniformitarian view lent support to Darwin's theory of evolution by natural selection since it implied that the Earth was old enough for the evolutionary process to have occurred. As the science of geology developed the uniformitarians gradually came to allow that some changes must have taken place abruptly. Catastrophists also came to admit that, if there were catastrophes, then they were as it were "minor" and that they occurred in a uniform and lawlike manner, rather than executed by special acts of the Deity. So far as the science of geology is concerned, the difference between uniformitarianism and catastrophism is one that now does not exist.[60]

In one sense, creationists are uniformitarians. Geological processes, they argue, all follow the same pattern. But there are occasional interventions by the Deity, e.g., the intervention recorded in the Bible as the Noachic flood. These yield catastrophic changes in the geological structure of the earth. Insofar, then, as creationists recognize that major changes have occurred, they are catastrophists. In fact, they are catastrophists in the pre-Huttonian sense of explaining catastrophes, not as occurring in a uniform and lawlike manner, but only as the result of special acts of God, miraculous and supernatural interventions in the otherwise normal and uniform workings of the world.

The creationist thus takes up a dichotomy that no longer makes sense as a piece of science. They take it up because it fits with the picture of the world that they derive from Genesis. They then try to fit modern developments such as that of punctuated equilibria into this antiquated and outdated framework.

This in fact is typical of many theories in pseudoscience. They are proposed as novel or revolutionary proposals, requiring radical revisions to current science. In fact, however, many of them turn out to be returns to outmoded theories that science has long discarded.

This has been referred to as the tendency for pseudoscience to involve **anachronistic thinking**.[61]

This feature of pseudoscientific thought is related to two others that we have noted. One is the tendency to turn to myths and Biblical stories as evidence for one's theories. This is certainly a pattern found in von Däniken, and, as we shall see, is found in Velikovsky. Officially it is absent from the work of the creation scientists, who argue that they do not rely upon the Bible for the evidence they propose that creation has a scientific basis stronger

than that of evolutionary theory. But unofficially, it is hardly an accident that the story they claim has a scientific basis is in all essential respects identical to the story told in Genesis.

The other feature often found in pseudoscience which is related to anachronistic thinking is the *appeal to authority*. This we of course find in the work of the creation scientists: their sacred authority is the text, written long ago, of the Bible. One finds the tendency to appeal to ancient authorities also present in the works of some astrologers.

Example III: Velikovsky

The Bible records that Joshua made the sun stand still. It also records a parting of the waters of the Red Sea. In the story of Sodom and Gomorrah, it records a great conflagration. At another time, it tells us, manna came from heaven to the earth. Many other cultures have similar stories, of deluge and conflagration, of darkness settling over the earth, and of ambrosia from heaven. Immanuel Velikovsky, a Russian psychoanalyst, attempted to explain these in his books *Worlds in Collision*[62] and *Earth in Upheaval*.[63] His argument was that we can understand these records in the legends and myths of humankind if we suppose that there occurred during historical times a few thousand years ago a series of celestial catastrophes to which these records bear witness.[64]

Upon Velikovsky's theory, a large comet was expelled from the planet Jupiter about 1500 BCE, and passed very close to the earth, causing great catastrophes, and then, returning some 52 years later, causing still further catastrophes.

On the first pass, the comet's tail passed over the earth. This caused a rain of petroleum, as well as dust and cinders from the tail which darkened the earth for several days. The close approach of the comet affected the earth's rotation, perhaps stopping it but certainly slowing it down. This jolt caused tidal waves, hurricanes, volcanic eruptions. The earth became hot, to the point that the sea boiled, and various geological strata were re-arranged. The earth's axis was altered, pointing in a new direction, changing the order of the seasons, and shifting the polar regions. Electrical interchanges between the earth and the comet caused the reversal of the earth's magnetic field. The close proximity of the comet moved the earth away from the sun, changing its orbit and causing the year to become longer.

This, if true, enables us to understand why there are now deserts where land was once fertile. If true, it enables us to understand why there are coalbeds in the Antarctic. If true, it explains why there are rock beds which have reversed magnetic polarity indicating a reversal of the earth's magnetic field. If true, it explains why various rock strata with creatures from different climates, e.g., desert and jungle, all together or in rapid succession.

Above all, if it is true we can explain the early records of humankind.

This first encounter can explain, Velikovsky argues, many of the events recorded in the Bible with respect to the Exodus of the Israelites from Egypt. The various plagues — blood, vermin, hail, etc. — as recorded in the Bible were not miracles but natural events, the effects of the contact of the earth with the comet's tail. The parting of the waters of the Red Sea occurred when the earth's rotation was halted. The manna that fell from heaven to sustain the Israelites in the wilderness is also to be understood not as a miracle but naturalistically, as carbohydrates falling as a result of the contact with the comet's tail.

Fifty two years later the comet returned to make a second pass. It again interrupted the earth's rotation. This happened just when Joshua commanded the sun to stand still at Gibeon. The event recorded in the Bible is thus again not a miracle but something that can be explained naturalistically.

The comet is in fact still with us: it is the planet Venus. In fact, according to Velikovsky, we must suppose that in the 8th century BCE there was another series of cosmic catastrophes. Venus still had a very elliptical orbit, and this brought it close to Mars. The approach knocked Mars out of its orbit, and set it off on a path that brought it close to the earth on at least three occasions. The close encounters once more moved the Earth further from the sun; its orbit was lengthened and the year became its present 365 1/4 days. At this point, Mars and Venus settled into their present orbits. Where Venus had been a hot comet when ejected from Jupiter, by passing close to the sun it cooled into its present solid state. With this the solar system had settled into its present stable form, writing *finis* to the cosmic drama.

Velikovsky, like von Däniken, makes much of being able to give naturalistic explanations of many ancient legends, traditions and records. Of course, none of these unambiguously records the events he describes. This in spite of the fact that some of the ancient peoples, the Babylonians, for example, kept meticulous and accurate records of astronomical events. Velikovsky attempts to explain this by a further hypothesis, that the events he describes were simply so horrific and shattering that humankind at the time collectively repressed them, leaving only the distorted traces that we have in the legends and traditions. Velikovsky refers to the theories of Freud and Jung, and states that "In the light of these theories, we may well wonder to what extent the terrifying experiences of world catastrophes have become part of the human soul and how much, if any, of it can be traced in our beliefs, emotions and behavior as directed from the unconscious or subconscious strata of the mind."[65]

The primary thrust of Velikovsky's argument is that many of the miracles and catastrophes reported in the Old Testament are literally true and that they can be explained by catastrophic near-collisions with wandering comets and planets. In this respect, though the explanations are different, the main thrust of his argument is much of a piece with that of von Däniken. It also bears resemblances to the arguments of the creation scientists, since he attempts to strengthen his case first presented in *Worlds in Collision* by arguing in the later *Earth in Upheaval* that catastrophism rather than uniformitarianism is the correct position to take in geology. Geologists, one of Velikovsky's supporters has argued, "no longer free to appeal to the uniformitarian notion that the record is incomplete, ... will have to pay more serious attention to the fact than alterations between strata are abrupt."[66] In the case of Velikovsky as in that of the creation scientists, there is more ado here than is warranted. Collisions and catastrophism have been part and parcel of modern astronomy as they have been part and parcel of modern geology. There is nothing contrary to recent science in the idea that many catastrophic collisions with comets have occurred. The issue is whether they have occurred as Velikovsky describes.

It is the Biblical evidence that is primary. The rest is a motley lot indeed. Thus, as has been pointed out,[67] Velikovsky relies for his mythology of the Greeks upon the works of Ovid, a Roman writer (43 BCE -17 CE), despite the fact that there are older and less corrupt sources. This is rather as if one relied upon painters of the renaissance for one's information about the ancient world. Ovid, and another of Velikovsky's main sources, Apollodorus (140 BCE), were using already ancient myth for their own purposes, re-working and re-telling it into themes very different from what was being done by those who presented the earlier, and, because earlier, presumably more accurate, versions. Velikovsky also relies heavily on rabbinical elaborations on the Biblical stories, as these can be found in rabbinical sources such as the Midrashim and Haggadah. He also relies on the Fathers of the Church,[68] all of whom, clearly, are late commentators on the Bible rather than reliable authorities. Be that as it may, Velikovsky's use of myths and legends as sources of historical fact seems to gain force from the claimed fact that the same sort of evidence comes from widely separated cultures. It is argued that these reports and myths are records of the same event, and since the cultures are, it is reasonable to suppose, independent, they mutually support one another as independent accounts of the same fact. However, as has been pointed out, if the claim that these stories record the same event, then the usually accepted chronologies must be revised. What evidence is offered that such revision is reasonable? It lies in the fact, Velikovsky argues, the chronology before roughly 687 BCE is unreliable. As one of his supporters has put it, there are two assumptions: "first, that no chronology using retrograde calculations of the positions of heavenly bodies is reliable earlier than -687; second, that the principal clue for synchronizing histories of ancient nations should be the breaks caused in all of them by the catastrophic events."[69]

However, how do we know that the usual datings based on retrograde calculations are unreliable? We know this because they are based on the assumption that year lengths and other standard measures of time have remained unchanged over the centuries. However, Velikovsky claims, this assumption is not justified. We are therefore able to revise the accepted chronologies to make the records report simultaneous events. Why, however, ought we to accept that the usual reckonings of temporal patterns are unreliable? We know this because we know that various catastrophes have occurred, precisely the catastrophes that Velikovsky is

telling us about. These catastrophes led to several changes in the orbit of the planet Earth, leading to changed lengths of the year. And how do we know that these theories proposed by Velikovsky are true? We know this because we have the several independent testimonies from the myths and legends of the past!

The claim is made that in Velikovsky, "the historiography of the ancients is always given first place as evidence by which to reconstruct the sequence of the past."[70] But this historiography must be interpreted; it is not pellucid that tales told by Ovid fit with tales from the Old Testament. They can be made to fit only if one rejects many of the standard procedures that are used, where possible, to interpret the stories. Why should these procedures be rejected? Because they are unreliable, Velikovsky asserts. But what evidence is there that they are unreliable? The evidence lies in the fact that Velikovsky's theory about catastrophic events is true and establishes that other procedures are needed. These other procedures establish clearly that the various independent ancient sources report various catastrophic events. But why ought we to accept Velikovsky's theory? Because, we are told, the ancient sources reliably report the occurrence of these catastrophes.

In short, the worth of the evidence presupposes the soundness of Velikovsky's theories. At the same time, the soundness of the evidence depends upon accepting the truth of the theories. The circularity is patent.

Velikovsky's theory comes into conflict with Newtonian theory. He asserts that an approach by the planet Venus caused the Earth to stop turning, which in turn made it appear to the ancient sources that the sun had stopped. Now, the Earth, spinning on its axis, is a gigantic massive flywheel with vast rotational energy. If any spinning top is stopped, its angular momentum must be transformed into some form of energy. This point can be tested can be tested with a spinning grindstone. If one first cranks the grindstone to high speed and stops it by letting the crank strike one's forearm, the sudden — and painful — jolt felt on the bone of the forearm is evidence that a spinning top cannot be stopped without a rapid transformation of the angular momentum. Moreover, the larger and heavier the grindstone, the greater the momentum. In order to stop the Earth, one would need a direct impact comparable to one's forearm stopping the spinning grindstone. But an impact sufficient to stop the Earth would require an object considerably larger than the object whose impact was sufficiently disruptive to have caused the extinction of the dinosaurs. Its impact would have had the effect of extinguishing all life on Earth.[71] The energy of the impact would, for example, have had to have raised the temperature of the oceans to above the boiling point. As Carl Sagan has pointed out, this fact seems to have been missed by Velikovsky's ancient sources.[72]

It is clear that, given Newtonian physics, scientists can invoke the rules that we have previously noted, and in particular the rule that

> Any new hypothesis should be consistent with those parts of theories in the area which have been well established through predictive success.

Since Velikovsky's theories conflict with well confirmed parts of Newtonian theory, it is safe to reject Velikovsky.

Worse follows for Velikovsky. Once the Earth has stopped, it must be started up again. This would require another force, one equal to the force that stopped the Earth. Venus of course has moved off, so it cannot do the job. Velikovsky is remarkably vague on this issue. He suggests that somehow magnetic forces from the sun could do the job. In fact, however, the magnetic forces coming from the sun are not anywhere strong enough to get the Earth moving again. Moreover, if these forces could do actually do the job of re-starting the rotation of the Earth, then, since the magnetic forces of the sun act continuously, they would continue over time to accelerate the Earth, causing is to rotate at an ever-increasing rate. But of course the speed of the Earth's rotation is not increasing.[73]

Again, the conflict with standard, well-confirmed theories is clear. Since the Newtonian theory of mechanics, including celestial mechanics, is well confirmed, and therefore acceptable, it follows that Velikovsky's theory, which contradicts this well confirmed theory, must be rejected.

It has been argued that Velikovsky can avoid these conclusions if one could develop a single theory that would unify both Newton's gravitational theory and the theory of electromagnetic forces.

> Is it fair that a synthesis which Einstein after decades of work was not able to conclude

satisfactorily be demanded of Velikovsky before his evidence from other disciplines is even considered? The space probes have only shown that a more comprehensive celestial mechanics, based on a physics in which electromagnetism and gravitation are explained by common laws, would have been necessary even if Velikovsky had never raised the issue. It should also be obvious that gravitation can in fact be cogently described in terms of some more fundamental forces, this does not mean that Newtonian physics need be "thrown out"; Velikovsky never said it should be.[74]

Now, it may well be that in the future some theory will be developed that will unify electromagnetic and gravitational theory. But any such unifying theory will yield the same predictions as those that follow from Newtonian mechanics. Recall the principle that we noted earlier:

> If a new theory T' replaces an older theory T, then T' will yield successful predictions where T does not, but will yield the same predictions where T successfully predicts.

It is said that "The specific laws the book [*Worlds in Collision*] was thought to contradict are those of the celestial mechanics which assumes the solar system to be electrically sterile and on that assumption successfully predicts planetary positions."[75] Adding further hypotheses about the electromagnetic forces in the solar system, or about a system of forces that somehow unify gravitational and electromagnetic forces is not going to add anything that contradicts the successful predictions of planetary motions — and also the motions of artificial satellites, the Mars voyagers, comets, the circumjovial moons, and so on — that have been made by Newtonian celestial mechanics. Even if the proposal for a unifying theory turns out to be correct, such a theory will not eliminate the contradictions. For the new theory will agree at the crucial points with current Newtonian theory. So whatever new theory develops will still contradict Velikovsky's theory.

It has to be argued by the Velikovskians that the actual predictions of Newtonian theory are false, and therefore that the theory that yields those predictions is false. This in fact is what Velikovsky himself argued. He simply disputed the claim that the usual Newtonian theory was a set of "known laws": they are in fact false and therefore not laws!

What, then, is the evidence that is offered that Newtonian laws are false? Where do these usual theories break down in their predictions? The answer that Velikovsky gives is that the theory breaks down because it does not predict the various cometary and planetary motions that account for the catastrophes recorded in myths and legends. *The facts that conflict with Newtonian theory are precisely those planetary and interplanetary events that are asserted by Velikovsky's theories.*

But why ought we to accept the claim of Velikovsky that his claims about planetary and cometary motions are true? Why ought we to accept these claims and conclude that standard Newtonian celestial mechanics is false? What Velikovsky says is that

> If, occasionally, historical evidence does not square with formulated laws, it should be remembered that a law is but a deduction from experience and experiment, and therefore laws must conform with historical facts, not facts with laws.[76]

Velikovsky's theories ought to be accepted rather than those of Newton because there are historical data that support the former and contradict the latter. But what are these historical data? This is the historical evidence that can located in the myths and legends of the past. However, as we have seen, *these "data" confirm Velikovsky's theories only if those theories are accepted as the basis on which to interpret those "data".*

Thus, Velikovsky's claim that his theory can escape being rejected by virtue of its conflict with Newtonian mechanics can be made good only if one has already accepted that theory. Once again the circularity is patent.

On the basis of his hypotheses, Velikovsky made a number of predictions that have turned out to be true. Among these were the prediction that Venus is hot, that the atmosphere of Venus contains hydrocarbons, that Jupiter emits radio signals. Since making *successful*

predictions is the mark of a good scientific hypothesis, Velikovsky's hypotheses seem to satisfy the criterion for being science. Since Velikovsky's theories passed the test for being science, he regarded the attempts by various scientists to classify his views as pseudoscience as the result of a pernicious conservatism, jealousy and perversity.

Consider the issue of the temperature of Venus. In the first edition of *Worlds in Collision*, Velikovsky stated that the comet that was to become Venus was in a state of "candescence."[77] At that time, it was a common opinion among astronomers that Venus had a surface temperature not unlike that of the Earth. In the 1965 edition, Velikovsky in the new preface argues that since he had predicted successfully "an incandescent state of Venus"[78] his theory has, in the first place, done everything that is to be expected of a good scientific theory, namely, make successful predictions, and has, in the second place, made a successful prediction which falsified the standard scientific theory.

We should first note that there is a curious double standard. Velikovsky also argues that Mars should, by virtue of its encounters with the more massive Venus and Earth, have a high temperature. As a consequence of this heating, he states, "Mars emits more heat that it receives from the sun."[79] However, this is quite wrong. It was already known in the 1940s, before the 1950 publication of *Worlds in Collision*, that the temperature of Mars is just what one would expect given the calculations about the amount of sunlight absorbed by the surface. These conclusions have since been confirmed by both Soviet and American spacecraft, and by the landings on the planet. Had Mars proven to be unexpectedly hot, then Velikovsky, if the pattern holds, would have proclaimed another successful prediction. But when Mars turns out to have just the temperature that accepted views said it would have, this is not taken as a refutation of Velikovsky.[80] In contrast, the successful prediction of the temperature of Venus is proclaimed as a strong confirmation of Velikovsky and a refutation of the standard scientific position.

As it turns out, however, it is not much of a refutation, nor much of a confirmation.

Take the former first. It is important to realize that, in 1950 when *Worlds in Collision* was first published, estimations of the surface temperature of Venus were very much a matter of guesswork. This was due to the cloud cover of Venus which obscured the surface and could not be pierced by then available instruments. The standard picture of the solar system therefore involved two theories, Newtonian celestial mechanics, call it N, which was (and is) strongly supported by many predictive success, and another theory, call it C (= "cool"), which consists in the theoretical and observational materials that generate a much less secure estimation of the surface temperature of Venus. The standard theory is therefore

N & C

Let E be the estimated surface temperature of Venus. We therefore have

If (N & C) then E

But as it turned out, the estimation E is false; the Venerian space probes established this beyond any reasonable doubt. We therefore have: not E. This implies that the theory that predicts E must itself be false. The inference is simple, of the form traditionally known as "modus tollens":

If (N & C) then E
not E
_____

∴ not (N & C)

Now note that the conclusion is the negation of a conjunction: what is falsified by the unsuccessful prediction is the conjunction (N & C). From the fact, however, that (N & C) is false, it does not follow that N is false. When a conjunction is false, all that follows is that at least one conjunct must be false, though it could turn out that both are false. Technically, the sentence

not (N & C)

is logically equivalent to

either not N or not C [81]

The issue is this: When it turns out that (N & C) is false, do we conclude that N is false or C is false or both?

The answer that we give depends, of course, upon the evidence available. In this case it is clear. We have massive independent evidence that Newtonian celestial mechanics is true. In contrast, we have only weak evidence that C is true. We therefore reasonably conclude that when (N & C) is false we may reasonably continue to accept N while rejecting C as false.

It is hardly the case, then, that Velikovsky's successful prediction of the temperature of Venus has resulted in the refutation of the usual scientific account of the solar system. In particular, the successful prediction of the temperature of Venus does not force the scientific community to reject that theory N which, as we have seen, is in conflict with Velikovsky's theory, and which, because it has been strongly confirmed, requires the rejection of the latter.

So Velikovsky's successful prediction does not in serious way call into question the standard scientific account of the solar system.

Turn now to the other issue, whether the successful prediction confirms Velikovsky's theory.

What must be noted is that in the 1950 edition of *Worlds in Collision*, he nowhere states *precisely* what he took the temperature of the surface of Venus to be. As we noted, what he claimed is that the planet is in a state of "candescence." This is hardly precise. Moreover, Venus is supposed to have cooled fairly rapidly into its present solid state after its close encounter with the sun. So it is hard to say exactly what was predicted by Velikovsky's theory. Yet, in the new 1965 preface he states that he predicted "an incandescent state of Venus," and that this yields strong support for his theory at the expense of the standard theory. We have seen that it hardly touches the standard theory. Does it support Velikovsky's theory?

Scientific theories must make predictions. After all, any scientific theory involves generalizations, statements about patterns that hold universally. If these predictions are successful then that constitutes *further evidence* that the theory is true. Thus, it is true, as Velikovsky claims, that

Successful predictions confirm a scientific theory.

Different predictions have different probative value, however.

Suppose someone, Jones say, claims to have considerable knowledge of race horses, and is able to predict the outcome of races. On the basis of his knowledge, he claims that one of Nashua, Freeport, or Gingerman will win. His prediction is that

(1)   p or q or r

But Charlie, who also frequents the race track, claims that his knowledge of horses is better than that of Jones, and that on the basis of what he knows he predicts categorically that Nashua will win. He predicts

(2)   p

It turns out that Nashua wins. Both predictions were successful. Both confirm that those who made them had knowledge of horses. Jones, however, had to bet on p, q and r where Charlie had to be on p alone. It turns out, then, that Charlie actually wins more than Jones; Charlie has not had to bet on a loser.

The difference between (1) and (2) is that (2) is *more precise* than (1). The fact that Charlie wins more than Jones shows that the more precise the prediction is, the better it is. To put it another way, less precise predictions *provide less information*.

The same can be said of the theories that yield those predictions. Jones' knowledge of horses is less precise than that of Charlie. As a consequence, Jones' prediction is less precise than that of Charlie. Because the more precise prediction yields more information — Charlie wins more — the theory that yields the more precise prediction is, in terms of information, *the better theory*.

This is not to say that Jones' theory is of no use at all. To the contrary, it does yield successful predictions: Jones did win. It is just that it is not as good, in terms of information, as Charlie's theory.

We thus note the following points:

Any theory, to be useful in explanation and prediction, must be true. But a theory that is more precise, and yields more information is *better*, **given our cognitive interests, than one that is less precise, yielding less information**.

We shall have more to say about this standard of better and worse later on, when we come to discuss theories that are informationally "gappy."

What we can say at this point is that Velikovsky's theory, because it yields only very imprecise predictions about the temperature of Venus, is informationally "gappy." It is therefore not among the better scientific theories. In particular, Newtonian celestial mechanics yields very precise predictions. It is therefore far less informationally "gappy" than Velikovsky's theory. As we shall see, there is an important sense in which Newtonian celestial mechanics has *no* informational gaps. In that sense, it represents the ideal of scientific explanation. Velikovsky's theory falls far short of the explanatory ideal that is achieved by Newtonian theory.

But there is another important point. If there are 10 horses running in the race. That means, roughly speaking, that there is an *a priori* probability of 1/10 with regard to each horse that it will win. Since Jones bets on three horse, there is a prior probability of

3/10

that his prediction correct. In contrast, since Charlie bets on only one horse, the prior probability that his prediction is correct is

1/10

This is true in general: *a less precise prediction has a greater prior probability than a more precise prediction*.

Now suppose that Jones and Charlie really do not have any great knowledge of horses. The theories on which they base their predictions are in fact false. The prior probability that Jones' prediction is correct is greater than the prior probability that Charlie's prediction is correct. Even if Jones is wrong, he is still quite likely to win on his bets than is Charlie. Since Jones is more likely to win even if his theories are false, actually winning does not imply as strongly that his theories are correct than does Charlie's actually winning tends to imply that his theories are correct. This, too, is true in general:

a less precise prediction, because it has a greater prior probability, confirms a

theory to a lesser degree than a more precise prediction.

This is why scientists prefer theories that have successfully made ver precise predictions: successful *precise* predictions strongly confirm the theory.

Now, Velikovsky's prediction of the surface temperature of Venus was very imprecise. This successful prediction does tend to confirm the theory, as he correctly states. But because the prediction is imprecise the supports that he gains from the prediction is very weak. Nonetheless, he infers that because his theory is confirmed, however weakly, and that he therefore can reasonably infer that Newton's theory is false. This inference depends upon the inconsistency of his theory, call it V, with N, Newtonian celestial mechanics. The two theories are indeed inconsistent: we have both

> If N then not V

and

> If V then not N

Velikovsky infers

> If V then not N

(a) V

∴ N

But of course, there is another inference that is possible:

> If N then not V

(b) N

∴ not V

Which of these ought one, rationally, to prefer?

Both (a) and (b) are valid argument forms. As for the premises, in both arguments the major premise, the conditionals, are true: as we have seen the two theories are indeed inconsistent one with the other. So the question of which inference to accept turns on which of the minor premises to accept, V in the case of (a), N in the case of (b). Now, both are confirmed, at least if we take for granted that V did make a successful prediction about the surface temperature of Venus. However, the prediction that confirms V was very imprecise. In contrast, the predictions that confirm N are very precise. Thus, N is much more highly confirmed than V. Since the evidence available much more strongly confirms the minor premise of (b) than it does the minor premise of (a), it is the latter argument that ought, rationally, to be accepted.

Velikovsky of course prefers (a). But rationally it is (b) that is to be preferred. Rationally, then, in spite of some successful predictions on the part of Velikovsky's theory, it is reasonable to reject it and continue to accept the standard Newtonian theories and the usual account of the history of the solar system.

Example IV: Astrology

Astrology claims that such things as stellar and planetary positions influence human events on Earth, determining when it is propitious to take various sorts of action, and also determining whether or not persons have various character traits such as generosity or aggression.

Astrology seems to have begun in ancient Mesopotamia, with stellar and planetary objects functioning as omens, rather as other things such as deformed births and comets are often taken to constitute omens. Comets were often held, for example, to foretell the fall of kings; hence the portrayal of Halley's Comet on the Bayeaux Tapestry commemorating the defeat of King Harold of the Saxons by William of Normandy. But the observers in Mesopotamia came to notice and record regularities among the planetary and stellar positions. They already were of course familiar with the changes on Earth regularly consequent upon changing positions of the Sun and Moon, e.g., the seasons and the tides. This led the ancients to speculate that the other heavenly objects had similar, though more subtle, impacts on earthly and human matters. Among other things, they came to the view that there are correspondences between, on the one hand, the constellations that lie along the path of the Sun through the stars (the "signs" of the zodiac) and, on the other hand, the character of persons born when the sun, in its yearly apparent motion along the celestial equator, is apparently located in those signs.

Thus, for example, it was suggested that if a person is born when the Sun appears to be in sign Libra is cooperative and of a harmonious disposition. In time, people began to cast horoscopes based on knowledge of the time and place of the births of person. (The earliest known horoscope dates from 410 BCE.)

Astrology undoubtedly originates in magical thinking. Thus, Libra = the balance = balanced or harmonious personality. To say that, however, is not to criticize it. What starts off as weak science or even non-science can become science over time, provided detailed theories are developed and bodies of evidence built up which establish their scientific credentials. Wegener's theories, starting off as weak science in time became accepted theories. So it is no criticism to say that astrology originates in magical thinking, as a large group of scientists once did in a denunciation of astrology that was sufficiently pompous to ensure its systematically being ignored.[82] It is moreover simply wrong to point out that the heavens have no impact on the Earth. The seasons and the tides are clear, and more recently we have discovered the impact of sunspots and solar flares on terrestrial atmospheric and electromagnetic phenomena. However, it is also true that we cannot conclude from this that astrology has a basis in reason, contrary to what P. Feyerabend has suggested.[83] The fact that astrology originates in magic does not imply that it is now non-scientific, but equally one cannot infer that therefore astrology does have a known basis in fact nor even that astrology is not still rooted in magical thinking. Indeed, if it turns out upon examination that there is no solid scientific basis for astrology, then one might well come to the conclusion that it is one of the forms that magical thinking takes nowadays as it did in the past.

Astrologers conceive the universe to be a sphere with the Earth at the centre. From this geocentric perspective, the Sun moves around the sphere of the stars on a yearly basis, and the planets in various other, more or less regular, periods. The path of the Sun through the fixed stars takes it through 12 constellations thought to be astrologically significant (Aries, Taurus, Gemini, Cancer, Leo, Virgo, Libra, Scorpio, Sagittarius, Capricorn, Aquarius and Pisces). One's zodiacal sign is the constellation in which the sun is located when one is born. Each sign signifies different personality characteristics, e.g., Libra indicates cooperation and harmony.

The signs are grouped in several ways. There are, first, the polarities: the signs are alternately positive and negative. There are, second, the quadruplicities: the signs are divided into cardinal, fixed and mutable. There are, third, the triplicities: the signs are divided into the categories of air, fire, water and earth. These groupings are ways of classifying the sign with regard to the personality traits that are associated with it.

Astrologers also utilize the daily rotation of the Earth about its axis. The heavens are divided into 12 sectors or houses according to the daily rotation. The Sun and the planets all travel through these houses on a daily basis. Each house is associated with a particular aspect of life — the second, for example, with money and possessions.

The planets are also associated with specific influences on character. Thus, for example, Mars is connected with assertiveness, combativeness, and decisiveness. Each planet is held to "rule" one or more signs. Mars, for example, rules Scorpio. A person's sign is that constellation which the sun is in when the person is born; if a planet that rules that sign is also in it when the person is born then it is supposed to have a strong effect on the person's character.

Other significant features of the heavens that have an impact on character and events are the angular relationships, called "aspects," between the sun and the planets. Thus, sextiles (60° angles) and trines (120° angles) denote harmonious personalities, whereas squares (90° angles) and oppositions (180° angles) denote inharmonious personalities. Other aspects are conjunctions (0°), semi-sextile (30°), semi-square (45°), trine (120°), sesquiquadrate (135°), and quincunx (150°).

How many elements are there that must be taken into account when interpreting a horoscope? There is the Sun, the Moon, the planets Mercury, Venus, Mars, Jupiter, and Saturn. There are 12 signs of the zodiac and 12 houses. There are 5 major aspects, and many minor ones. All told there are around 500 different factors. Of these anywhere from 30 to 40 are present in the average chart. No factor can be used in isolation from the rest. There is sufficient complexity that a definitive interpretation any chart is in fact impossible. There are simply too many interacting factors. What happens when a client goes to an astrologer is first, the creation of the horoscope. This involves numerous interacting factors, each with its own meaning, and meanings which can change in the presence of other signs. The astrologer then interprets this chart, and applies it to the client's situation. There is then feedback from the client, in the light of which the astrologer may well modify his or her interpretation.

Taken this way, an interpretation is confirmable but not falsifiable. There are so many factors that the astrologer can always find something that fits the circumstances and personality of the client. An exposition of the chart by a sensitive astrologer of a Jovian personality, of what happens when Mars is in each house, can often involve behind the jargon perfectly sensible and indeed insightful commentary on human behaviour in general and the client's personality and behaviour in particular. But the application to the particular client will not be falsifiable because, through the consultation process, contrary evidence from the client comes to be incorporated into the interpretation of the astrologer. At the same time, the client puts relevant data into the interpretation in such a way that the statements of the chart as interpreted are confirmed for the client's case. What happens, of course, is that the client goes away satisfied. His or her particular situation has been described in a new language, and he or she has been provided with a way of talking about his or her own personality and situation — a systematic way that replaces a situation where one's self-knowledge lacked structure and coherence, a way (in other words) that enables one to become more self-aware and certainly re-assured. The astrological jargon may well help: by connecting truths about the sort of person the client is, and how he or she relates to the surrounding world, to the apparently objective structure of the heavens, it may enable the client to gain a distance and objectivity from his or her own problems that otherwise would be difficult to achieve. Such objectively is often essential if one is to get a fair handle on oneself and one's problems. In any case, the client ends up satisfied and with a certain increased degree of self-knowledge, though it is packaged in the jargon of astrology. And the astrologer comes away convinced once again that "astrology works."

Thus, we find Edith Custer, editor of an astrology journal, has written that "Whether the scientific world accepts or rejects astrology makes it no less a valid tool for me to work with....I know it works and I am satisfied with that."[84] However, does it follow that it is *true*?

The answer to that question depends, as the philosophers say, on what you mean by 'works.' If you mean "helpful," then there is no doubt that it works. It provides people with tools for thinking things through about themselves and their lives in difficult situations. The astrologer, speaking in terms of the birth chart, is objective and non-judgmental. The astrological ideas have a certain charm and even beauty. In our society help often does not come cheap; astrology usually does. It is clearly a good deal.

Moreover, insofar as the astrologer can provide really useful advice, it is likely that there is a lot of sound psychology — sound psychological truths — in which he or she tells the client. The self-knowledge that is embodied in the language of astrology is undoubtedly mostly knowledge, and therefore mostly true. Except that, so understood the knowledge is not knowledge about the supposed causal effects of planetary positions on human character and on the events of our lives. In fact there is a major absence of evidence that astrology works in the sense of its stating truths about the causal impact of planetary motions on things terrestrial.

Thus, in commenting on the fact that there are about 500 relevant factors according to traditional astrology, and on the fact that only 30 to 40 appear in ordinary birth charts, G. Dean has said that

> ... to claim, as many astrologers do, that tradition is the result of millennia of empirical observation, is to claim that the meaning of each of 500 factors can be deduced when any 40 can be present at the same time. This is clearly untenable. Conversely, if the factors were so easy to observe, why is it that today there is not convincing evidence for any of them?[85]

In fact, Dean and his colleague A. Mather, at once astrologers but also striving to be scientific, after reviewing over 700 astrology books and 300 scientific works on astrology, put the essential point this way:

> Astrology today is based on concepts of unknown origin but effectively deified as "tradition." Their

application involves numerous systems, most of them disagreeing on fundamental issues, and all of them supported by anecdotal evidence of the most unreliable kind. In effect, astrology presents a dazzling and technically sound superstructure supported by unproven beliefs; it starts with fantasy and then proceeds entirely logically. Speculation is rife, as are a profusion of new factors (each more dramatically "valid" than the last) to be conveniently considered where they reinforce the case and ignored otherwise.[86]

As they indicate, it is clear that astrology not only has no evidence to support it but that there is evidence that shows that astrology as traditionally understood is false.

To be sure, this has not prevented people from accepting it. Fantasy, after all, does often have an appeal. And there are moreover the illegitimate inferences to its truth from the fact that it "works." Typical are the comments of Linda Goodman,

> Alone among the sciences, astrology has spanned the centuries and made the journey intact. We shouldn't be surprised that it remains with us, unchanged by time — because astrology is truth — and truth is eternal.[87]

This, however, is not the language of science, where conclusions as to what are the truth are always tentative. It is, rather, the language of *authority*, the authority of tradition. Dean and Mather concluded, in effect, that it is the method of authority that is behind the beliefs and practices of astrology, not the method of science: as we saw them express it, "Astrology today is based on concepts of unknown origin but effectively deified as 'tradition'." Moreover, as we saw them go on, "astrology presents a dazzling and technically sound superstructure supported by unproven beliefs; it starts with fantasy and then proceeds entirely logically."[88] So there is not only authority but also the **method of fantasy**: *accept as true that which you want to be true*. But there is still more; another, equally non-scientific method also plays a role. Dean and Mather comment on this state of affairs in this way:

> The current chaos in astrology is largely the result of a chronic infatuation with symbolism at the expense of reason. This is because the majority of astrologers reject a scientific approach in favour of symbolism (based on dubious tradition), intuition, and holistic understanding.[89]

That is, astrologers rely on the **method of resemblance**[90] rather than the *inductive method* characteristic of scientific inquiry. The *method of resemblance* infers a causal connection from the fact that things resemble each other: its methodological rule is: *if A's resemble B's then infer that A's cause B's.* These non-scientific methods of authority, fantasy and resemblance are supplemented by various devices that can be used in *ad hoc* fashion to save various hypotheses from being rejected as false. Dean and Mather put this last point in this way, as we saw: there "…are a profusion of new factors (each more dramatically 'valid' than the last) to be conveniently considered where they reinforce the case and ignored otherwise." All this is put in the context of "entirely logical" developments of the basic non-scientific premises. These developments include good doses of mathematical calculations, so that the scientifically arbitrary assumptions are cloaked in a veneer of rationality. It is non-science disguising itself as science. It is, in other words, pseudoscience.

There are data which significantly call into question the traditional astrology. One of these is the precession of the equinoxes. As the Earth moves about the sun it spins on its axis and, like a top, the direction in which its axis point gradually moves or "precesses" along a huge cone-shaped path. This has the result that the astrological signs of the zodiac, which are usually measured relative to the vernal equinox point, have become displaced with regard to the constellations for which they are named. Thus, the Sun is said to be in Scorpio from Oct. 24 to Nov. 22. But in fact on these dates the sun is to the west in Virgo. When the Sun enters Sagittarius on Nov. 23 it is in fact in the constellation Libra. So, is someone born on Nov. 5 a Scorpio or a Libra? How one answers makes a big difference since the effects of the two regions are very different according to the astrological tradition. The astrological problem is whether one ties the horoscope to a zodiac that moves as the vernal equinox moves with the precession of the axis (a so-called "tropical" zodiac), or one that remains anchored to the traditional set of fixed constellations (the "sidereal" zodiac). There has been no agreement upon what is the correct version. Some astrologers have adopted a position that attempts to reconcile the traditional sidereal scheme with the fact of precession by suggesting that the signs remember the influence of the constellations that corresponded to them 2000 years ago (when the traditional system was first established). This shows clearly that there are no clear data in astrology that would enable one to decide between alternative hypotheses.

Again, the astronomer S. Schmidt pointed out that the Sun passes through not only the usual signs of the zodiac but two other constellations, Cetus and Orphiuchus, and argued that these should be included as part of the system of astrological signs.[91] We therefore have competing with the usual 12 sign zodiacal system a newer 14 sign system. But again, there seems to be no data that can clearly decide between these competing hypotheses.

Again, astrologers claim that the planets influence human character. Traditionally there were just seven heavenly objets, the Sun, the Moon and the 5 planets. But we now know that there are three new planets, unknown to the ancients, Uranus, Neptune and Pluto. Astronomers were able to detect the latter prior to their being observed through the effects that they have on objects nearer to us in the solar system. Neptune was detected because of its effects on the orbit of Uranus, and Pluto by its effects on the orbit of Neptune. Astrologers suppose that planets have influence upon human character, yet these three planets were undetected by astrologers. Perhaps, as some have suggested, planets have no influence until they are detected! Again, this is evidence that the planets in fact have no influence on character.

Not only are there no data that can justify the assertion of various astrological hypotheses, but there is no theory that can guide research. The absence of such a theory is evidenced by the fact that there is no research tradition in astrology — in fact, this is, according to T. Kuhn, precisely what makes astrology a pseudoscience.[92]

More strongly, the traditional astrology is inconsistent with the well-confirmed theory of genetics. This theory argues that insofar as character traits are determined at birth, prior to any learning, then that determination is a result of genetic structure. That is, the character traits are determined prior to birth. In contrast, traditional astrology has them being determined at the time of birth. Modern biology and genetic theory thus provides grounds for rejecting astrology of the traditional sort, much as the well-confirmed theories of Newton provide grounds for rejecting the theories of Velikovsky.

Example V: Homeopathy, Naturopathy and Other Alternative Forms of Medical Practice

Homeopathy is a form of medical practice that was developed some two centuries ago by the German physician, Samuel Hahnemann (1755-1843).[93] It continues to have practitioners today, following the principles laid down by Hahnemann. It has its own colleges, textbooks, professional societies, professional periodicals, sets of medicinal drugs, pharmacies and manufacturers. They treat all forms of disease — epidemic infections, traumas, chronic and degenerative diseases — everything that brings anyone to an ordinary physician. Homeopathic practitioners consider themselves to be diametrically opposed to conventional medicine (which they refer to as "allopathic" medicine).

Hahnemann was dissatisfied with the medical practices of his own day. In searching for alternative practices, he was led to formulate the principle that

similia similibus curantur,

or like cures like, that is, diseases are cured by agents capable of producing symptoms resembling those found in the disease under treatment. He arrived at this principle by examining the effects of Peruvian Cinchona bark (from which quinine is derived) on fever, something which he had recently endured. He took strong doses himself and soon started to exhibit the typical symptoms of malaria, precisely that disease for which quinine provides a cure. He concluded that quinine acts curatively in this disease because of its capacity to elicit malarial symptoms in a healthy person. Generalizing from this he concluded by that selecting a remedy which will match the symptoms of the patient, the disease is driven out of the body.

Hahnemann went on the test other common medicinal substances on himself and others, a process which has come to be known among homeopaths as "proving." There are now over 1500 medicinal substances in the materia medical of the homeopaths.

The "law of similars" is understood both by Hahnemann and by his present day followers to be a law of nature. Since laws are general and unchanging patterns, adherence to this claimed law means that homeopaths have not changed the fundamentals of their practice since they were first formulated by Hahnemann.

Hahnemann also experimented with the size of doses. Substances in the materia medica were dissolved in water or alcohol and then vigorously shaken (succussion). They were then diluted further, and shaken once again. This was repeated a number of times until no physical traces could be detected by chemical means. Homeopaths follow Hahnemann in holding that the process of proving which they use, that is, testing on healthy persons, establishes that the substances retain their curative powers even in very dilute solutions.

Hahnemann was concerned about diseases in the sense of the set of symptoms which create the dis-ease in persons. As he put it, "… in every individual case of disease the totality of the symptoms must be the physician's principal concern, the only object of his attention, the only object to be *removed* by his intervention in order to cure, i.e., to transform the disease into health."[94] The first task for the homeopathic physician, then, is a thorough physical examination followed by a careful assessment of the mental state, since mental symptoms are often most important. One then selects the remedy in accordance with the "law" of similars: "… The only effective therapy … uses in appropriate dosage against *the totality of symptoms* of a natural disease a medicine capable of producing, in the healthy, symptoms as similar as possible."[95] The homeopathic physician uses only a single remedy: : "In no case being treated is it necessary to give a patient more than a *single simple* medicinal substance at one time …".[96] Allopathic physicians give several medicines because they attempt to deal with the symptoms independently of one another: "… the old school has always tried to combat and wherever possible suppress through medicines *only one* of the many symptoms that diseases present…". This is a "short-sighted method" since "a single system is no more the whole disease than a single foot a whole man." Treating symptoms independently of one another not only does no good but can also cause harm.[97] As for the dosage, this should be the minimum. As Hahnemann argues, "the dose of the highly potentized [diluted] homoeopathic remedy beginning the treatment of a significant (chronic) disease can, as a rule, not be made so small that it is not stronger than the natural disease, that it cannot at least partially overcome it, that it cannot at least partially extinguish it …, [and] that it cannot start the process of cure."[98]

It is clear that there is a real problem with the claim of Hahnemann and his homeopathic successors that the "law" of similars really is a law. Is it really plausible to hold that all diseases can be cured by medicines that cause the symptoms of the disease in the healthy? Asprins cure headaches, but when I do not have a headache, taking an asprin does not cause me to come to have a headache. The evidence that homeopaths offer for their "law" is the medicines that they have developed through the tests of "provings." Hahnemann arrived at his law by generalizing from the one case of quinine. But homeopaths have now subjected a variety of other medicines to provings, and these results all confirm their "law." Each of these tests yields a generalization to the effect that

(x)   Whenever A is administered to a person with disease D, then that disease is cured

The medicine A conforms to the condition that

(xx)   Whenever A is administered to a healthy person, symptoms similar to D appear in that person

The suggestion by homeopaths is that a necessary and sufficient condition for a medicine satisfying (x) is that it also satisfies (xx). The "law"of similars is a generalization about all medicines:

($)   Whenever there is a substance f and disease d such that a generalization of the form (x)

holds of f and d, then and only then a generalization of the form (xx) also holds of f and d

While it may well be true that there are laws of the sort (x) and (xx) for the medicines that have successfully been proved by homeopaths, it does not follow that the "law of similars" ($) holds for *all* medicines. To the contrary, it would appear that there are counterexamples to ($), e.g., asprin.[99]

It would seem, then, that the claim that *only* homeopathic remedies can cure diseases is mistaken.

But there is a deeper problem: Can any homeopathic remedy ever really cure a disease? Conventional medicine argues that it cannot. The reason is the fact of that extreme dilutions are used to prepare homeopathic medicines. Dilutions of the order of 10^{-200} or more are common. At dilutions of this order, it is as likely as not that not a single molecule of the curative substance can be found in the sample. But if none of the substance is present in the medicine that is taken, how can it have any effect on the patient?[100] Certainly, current chemical theory states that in such dilute forms substances can have no effects. We therefore have a conflict between homeopathic claims, on the one hand, and a well confirmed scientific theory, on the other. Once again we can invoke the principle to which we appealed in rejecting the pseudoscientific views of von Däniken, Velikovsky, creation scientists, Lysenko and the astrologers. This is the rule that

Any new hypothesis should be consistent with those parts of theories in the area which have been well established through predictive success.

It would seem, then, that we can safely reject homeopathic theories as pseudoscience.

Homeopathic thinkers have, one is not surprised, noticed this argument, and have tried to counter it. The obvious answer, could it be developed, would be a theory that shows how the diluted remedies do really work. Answers of this sort have been proposed by various persons. For example, there is a proposal by P. Callinan,[101] to the effect that the vigorous shaking (succussion) induces electrochemical patterning of the diluent, which then replicates at every further stage of succussion, with a different patterning for each of the different curative substances. The suggestion is that succussion

produces energy storage in the bonds of the diluent in the infrared spectrum which "downloads" in contact with the water in living systems. Perhaps this information then spreads like a "liquid crystal" through the body water, modifying receptor sites or enzyme action.[102]

This is no theory, however. Note the 'perhaps', note the quotation marks around 'liquid crystal': the speculative element is clear, there is no confirmation. But worse: it is hard to see where there could be any confirmation. 'Downloads' is a mere metaphor taken from computer science. In the latter, if one knows the engineering and the programming, one can give a precise characterization of what happens when downloading occurs. In the context of passing information about a substance on to the successive diluents it tells us nothing — except that it is supposed to occur. But in reality it masks the fact that we have no idea what the mechanism is that might bring about the transmission of information. The use of 'liquid crystal' has the same point. What exactly is a 'liquid crystal' supposed to be? A crystal surely, but how could a crystal be liquid? The former implies a rigid pattern, the latter implies flow rather than rigidity. How can the former be the latter? What exactly is the energy that is stored? How can different curative substances have different forms or amounts of energy that get released by succussion? How exactly do the chemical bonds in the molecules store this energy in different ways? And when it is stored, then, once again, what is the mechanism for "downloading"? And further, how is it yet again "downloaded" to receptor sites or enzyme processes? We have a lot of words that have a sense of scientific jargon about them — e.g., energy, crystal, and so on — but they have little empirical content. And in the absence of empirical content, how could the "theory" ever be confirmed? The test of a theory is successful confirmation. A theory that contains terms that are at best vague, and at worse empty but fine sounding metaphors cannot yield predictions. The theory that

All bumbies are gooches

will not do as **scientific theory** precisely because we have no idea in empirical terms what either a "bumbie" or a "gooch" is, and in the absence of such empirical content the theory, that is, the "theory," cannot be made to yield empirical predictions. The same point holds for Callinan's suggestion: it is simply non-empirical because its concepts have not been tied down in such a way as to permit empirical verification.

A theory, to be scientific, must be sufficiently articulated in empirical terms that it can yield predictions that will, if successful, confirm it, or, if unsuccessful, disconfirm it.

Callinan's "theory" fails to pass this test.

The other resort of the defenders of homeopathic medicine is non-empirical theories. Thus, one commentator has suggested, with regard to the highly diluted medicines, that

> If there is no physical substance left in the remedy, it must obviously be acting on the vital field or force of the individual, thus stimulating the vital energy to eliminate the disease.[103]

This "vital force" is of course non-physical, something known by some sort of "intuitive understanding"[104] rather than by ordinary experience, yet at the same time related to the scientific. The "vital force" is a "force" or appears as an "energy field" or is somehow beyond these made of "ether" — a concept which sounds scientific, but which has been outside any serious scientific vocabulary for perhaps eight decades. Thus, we are told by one defender of homeopathy that

> It appears that there is a wealth of information already gathered which illustrate the vitality or electrical factor in health and disease. Our main need is for scientists to make a synthesis and evaluation of the work already done. If this is carried out, the fact of an energy field, or etheric field underlying the physical body, is likely to become a scientific fact.[105]

The concept of energy is, of course, an important concept in physics. It is also well-defined. Energy takes various

forms — heat, mechanical, atomic, and so on. Each of these concepts is carefully related to our ordinary experience. This is done in such a way that each of them can be measured. We even know that energy and matter are in a way interchangeable: the way this goes is governed by Einstein's famous equation

$$E = mc^2$$

But in the homeopathic literature, where references to "vital forces" and "energy fields" can be found in abundance, we are never told how we are to detect by ordinary means, let alone measure, the "vital force" or "energy field" to which reference is trying to be made. The concepts are not hooked semantically to the ordinary world that we know by sense experience. As a consequence, we can never determine by ordinary observation whether these things are present or absent, or whether statements involving them are true or false. In other words, statements involving these concepts cease to be empirical. Explanations in terms of such concepts are hardly better than attempts to explain the flooding of the Nile by reference to the mythological River Ocean. And since statements about "vital forces" or "ethers" and so on involve non-empirical concepts, they cannot be either confirmed empirically or refuted empirically. They are simply not part of science. Nor, of course, do they yield any empirical predictions that can be verified independently of the theory.

The use by homeopaths of terms such as 'energy' and 'force' create a sort of analogy with ordinary physics and chemistry, so that it can be made to seem that what is being proposed is merely an extension of ordinary science. Scientists do of course advance new theories, and when they do they are careful to establish not only similarity to, but consistency with already accepted theories, at least in those areas where accepted theories have established predictive success. The new theory will thus be similar to the old theory. But the new theory, to be consistent with the old theory, will have to make the same predictions in those areas where the old theory has been predictively successful. That means that the new theory will have itself to be an empirical theory, yielding empirical predictions. In turn, that means that the theory will be capable of empirical confirmation or refutation. In contrast, the theories offered by homeopaths are non-empirical, and incapable of yielding empirical predictions or of being put to an empirical test. The homeopathic theories thus are *not* similar to the older theories in the area, those of conventional science. The homeopathic theories *sound* similar, but they are not *similar in logical structure*. They sound similar because they use words that sound the same. But the fact that the theories are homophonic does, to repeat, imply that they are similar in logical structure.

The argument that because theories sound similar therefore they are similar in logical structure is simply invalid. Unfortunately, it seems to be an argument that appeals to homeopathic practitioners. Nor only to them: it is widespread in many forms of pseudoscience. It can be called the **argument from spurious similarity**.[106]

Talk of "vital forces" can be found in the very origins of homeopathy, the writings of Hahnemann, who tells us that disease is "a state of being of the organism dynamically untuned by a disturbed vital force, as an alteration in the state of health."[107] Conversely, "In the state of health the spirit-like vital force (*dynamis*) animating the material human organism reigns in supreme sovereignty." This vital animates the body, maintaining the parts in harmony. This vital force is no ordinary thing, however. It is immaterial and spiritual.

> Without the vital force the material organism is unable to feel, or act, or maintain itself. Only because of the immaterial being (vital principle, vital force) that animates it in health and in disease can it feel and maintain its vital functions.[108]

Thus, "diseases ... *are not* and *cannot be* mechanical or chemical changes in the material substance of the body, ... they do not depend on a material disease substance, but are an exclusively dynamic, spirit-like untunement of life."[109] It is because diseases are an untunement of the vital force that small doses of medicine are sufficient to effect their changes: they need to affect only the vital spirit, and not the material substance of the body.

> Natural substances that have been found to be medicinal are so only by virtue of their power (specific to each one of them) to modify the

human organism through a dynamic, spirit-like effect (transmitted through sensitive living tissue) upon the spirit-like vital principle that governs life.[110]

Homeopathic theory, then, supposes that there are mysterious powers in the body. They are powers that *animate* the body, cause it to move and change in certain ways rather than others. These ways in which the body is moved and made to change by these powers yield a healthy state of being. But a healthy state is one in which there is no dis-ease. The vital powers therefore *aim at the good* of the person.

For our purposes, it is important to note two things about these powers.

The first thing that should be noted is that these powers are *immaterial*, and they therefore cannot be detected by ordinary means, either by our senses or by instruments. The activity which is the exercise of these powers can be inferred from its supposed effects. Of course, such things as atoms are also known only through their effects: they affect our instruments in certain ways, and from these effects we infer their presence. But the instruments are themselves ordinary things, operating according to well-known and empirically confirmed laws. The atoms and other such things as we know only through instruments are themselves material things, and their material presence is inferred by the use of material instruments which in turn materially affect our senses. At no point is there a *leap* from material effects to something *immaterial*. In the case of the homeopathic vital force there is such a leap.

The second thing that should be noted is that the activity which is the exercise of these powers is one that *aims at an end*.

We shall refer to explanations in terms of immaterial powers whose exercise aims at an end as **teleological explanations**.

From the point of view of science, teleological explanations are simply not scientific. This is because such explanations invoke immaterial and mysterious powers, *non-empirical entities that cannot be known through sense experience*. Teleological explanations were once commonly invoked in claims to understand the world. Their greatest exponent was Aristotle. As we shall see, their spell was broken by Galileo, who vigorously defended scientific explanations based simply on matter-of-fact regularities.

In order to know that there are immaterial powers in things, some way of knowing them is needed. Since they cannot be known by sense or by its instrumental extensions, it follows that there must be some non-sensible way of knowing them. Such a non-empirical way of knowing is sometimes called an "intuition," or, since the intuition is one that grasps the powers that are the reasons why things behave as they do, they are also called "rational intuitions." The set of powers that animates a material thing is often referred to as the "Nature" of the thing. Rational intuition therefore grasps the Natures of things. Those Natures determine the motions and changes that occur in things. These motions and changes are towards certain ends, where these ends constitute the well-being of the material body. The Nature therefore also determines the well-being of the good of the body. But appeals to Nature or Natures to explain things are *empirically vacuous*, and in fact, relative to the cognitive interests of science, have no *genuine explanatory powers*.

It is worth noting that many objects exhibit *goal directed behaviour* . When the fox chases the hare, and the hare runs in its apparently erratic pattern in order to escape, we see two cases of goal directed behaviour. There is, first, that of the fox, which aims to capture the hare, and there is, second, that of the hare, which aims to elude the fox. Goal directed behaviour can be observed; we know it by sense experience; its existence is an empirical fact. Such behaviour can be explained empirically. We have come to know many of the mechanisms by which both the hare and the fox monitor the behaviour of the other, and many of the mechanisms that enable the one to chase and the other to flee. Understanding these mechanisms is part of the science of biology.

Goal directed behaviour is something that can be studied by the methods of science. There is no *a priori* reason why goal directed behaviour needs to be given a teleological explanation, that is, one in terms of mysterious Natural powers which aim to achieve the end at the behaviour is directed. Nonetheless, it is clear that there will always be a temptation to search for teleological explanations of goal directed behaviour. This will especially be true when we know there is goal directed

behaviour but do not have any confirmed scientific explanation for such behaviour, or where any scientific knowledge that we do have is extremely gappy with little present prospect for research being able to fill in the gaps. *Where our scientific knowledge is gappy then there will always be a temptation to give ourselves the illusion of knowledge by filling in the gaps by introducing immaterial Natural powers.*

It is precisely this that was done by Hahnemann. The result is a medical theory that is not scientific. Since the scientific knowledge of physiology and of disease that was available at the time of Hahnemann was still very gappy, it is not surprising that he would invoke non-scientific immaterial spirits, and rely upon non-scientific teleological explanations. It is more surprising, now that our knowledge is much less gappy, that there are many who still rely upon such explanations, e.g., all the current practitioners of homeopathic medicine, and all those who are their patients. But perhaps it is not so surprising for the latter. Medical knowledge, though not as gappy as it once was, is still gappy, and often fails to cure. People become desperate for a cure, and some knowledge and understanding of what is happening to them. No one is easier to deceive than someone who wants to be deceived. People who are desperate for a cure will therefore be apt to accept teleological explanations and to rely on cures supported only by teleological theories: because they want the theories to be true, they will accept them as true. This might be called the **method of fantasy**. It is clear that it is a poor way to discover the truth about matters of empirical fact; it is equally clear that there are many who, when they are desperate, will resort to the method. It is also clear that there are many who are prepared, whether sincerely or insincerely, to give them what they want. These are the practitioners.

The supreme exponent of the teleological way of thinking was the greatest physician of the ancient world, Galen.[111] For him, as much later for Hahnemann, disease and health are defined by reference to a Nature. Disease, for Galen, is an Unnatural state of the body, which impairs its Natural functioning; health, in contrast, is a state in which the Natural functioning of the body proceeds unimpaired. Since Nature is purposive and aims at the good of the living thing, she does her best to restore Unnatural states to their healthy condition; this is the famous *vis medicatrix naturae*, the healing power of nature.

Galen accepts much of the traditional physics deriving from Aristotle. There are four basic qualities, the Hot, the Dry, the Damp and the Cold. These combine to form the four elements. Earth is Cold and Dry; Water is Cold and Damp; Air is Hot and Damp; Fire is Hot and Dry. In the superlunary or celestial regions there is a fifth element, the quintessence, called "aether." Living things are composed of the four elements, organized in different ways according the Natures of the different creatures. Some aether occurs in terrestrial regions in living creatures, providing their innate heat and their power of growth and reproduction. According to Galen, the seat of the aether that provides the innate heat is in the heart. It is also located in the semen, which has, like the sun, the power of generating life. The four elements play little role in Galen's thought, but the four qualities are important. They are associated, in the bodies of animals having blood, with the four bodily fluids or *humours*. These are yellow bile (hot and dry), blood (hot and damp), phlegm (cold and damp), and black bile (cold and dry). The first three occur in bodies, but the fourth is purely fictitious, introduced simply to complete the number to four. Both body and mind are controlled by the four humours, though the doctrine of four mental temperaments, the sanguine, the choleric, and phlegmatic and the melancholic, is a development that came after Galen.

Galen did not dissect human bodies; the religion of the age did not permit it. He therefore learned what he knew of anatomy from the dissection of animals, and in particular of Barbary apes. The empirical knowledge of the internal organs from these sources was mixed up with the mythical. There are two kinds of heat, the ordinary variety on the one hand, which consumes things, and innate heat, which makes the body grow instead of consuming it, and which has the power of generation. It is situated in the left ventricle heart and in semen. From semen it is passed on to the embryo. From the heart it is distributed to all parts of the body by the arteries. The function of respiration is to moderate the innate heat. Pneuma or breath derived from inspired air is mixed with arterial blood in the left heart, and is distributed with it. The innate heat is never lost, but is augmented by nourishment. Part of the function of innate heat is to facilitate digestion, the process by which food and drink in the gut is converted into useful blood. This is done in

the liver, from whence the veins conduct the digested food and drink to all parts of the body. The notion that blood circulates is entirely foreign to Galen.

If teleology is the first principle of Galen's system, the second is that of *balance*, which derives it seems originally from Alcmeon of Croton. On this doctrine, health consists in a balance of opposites, while disease consists in a preponderance of one side of a pair of opposites over the other. This yields Galen's system of treatment (against which Hahnemann was in part reacting): *opposites are the cures for opposites*. If the patient is too hot, he or she is treated by cooling; if dry, by dampening; and so on.

Residues accumulate when more nourishment is taken in than the body requires. Inflammation and putrefecation originate from residues. There are natural execratory organs, but these for some reason may be weak and unable to expel the residues. When this occurs, the residues will accumulate. The body should be kept free from residues in the first place; food and drink should be taken in moderation and the diet should be balanced. Constipation or the suppression of a menstrual period was regarded a very serious matter because material that ought to be eliminated was being retained. Bloodletting was frequently used to eliminate residues. It was also called for if there was too much blood in the system (a "plethora"). Fever itself is a disease, an excess of heat, and since food and drink serve to augment heat, the theory calls for starving patients suffering with fever. When used by some ancient physicians, Galen complains, the patient was starved to death.[112] Galen notes that one has to choose between feeding the patient and augmenting the fever. However, he had an alternative available, venesection. If Galen treated fever with venesection, he did not also use starvation.

People are, however, naturally different. Some are, by their Nature, cold, others are hot. It is important in any treatment to judge the Natural temperament of the patient. Nor is this a simple matter, since different parts of the patient may have different temperaments. Every patient must be individually assessed in terms of both Natural temperament and physical type, and in terms of his or her way of life, past and present. There are no easy rules of thumb in treating patients.

Galen was famous as a physician, which means that whatever his theory and whatever his medicines and treatments, he was successful often enough that people kept coming to be treated by him. That means that some of the procedures that he used must have worked. His system is totally fantastic. The doctrine of the four qualities and the four elements has no empirical basis at all. One of the humours which he indicates are present in all human bodies, black bile, simply does not exist. He has no sense of the circulation of the blood, and his account of the anatomy of the human body is confused insofar as it is based on the dissection of Barbary apes. Everything is tied together by the supposition of a teleological Nature. But his success shows that in his system, fantastic though it may be from an empirical point of view, must have had embodied within it several pieces of empirical knowledge.

This combination is far from absent even today. Hahnemann's system of homeopathic medicine is an example of a similar fantastic system of medical theory. So is what has come to be called "naturopathic medicine".

This movement was brought to the United States by Benjamin Lust, who started using the term 'naturopathy' in 1902.[113] This was a doctrine of "natural healing" and aims to include "the best of what is now known as nutritional therapy, natural diet, herbal medicine, homeopathy, spinal manipulation, exercise therapy, hydrotherapy, electrotherapy, stress reduction and nature cure."[114] Naturopathy accepts the Galenic principle of *vis medicatrix naturae*, or, what amounts to the same, a non-scientific teleology.

> Naturopathic medicine is 'vitalistic' in its approach, i.e. life is viewed as more than just the sum of biochemical processes, and the body is believed to have an innate intelligence that is always striving for health. Vitalism maintains that the symptoms accompanying disease are not directly caused by the morbific agent, e.g. bacteria; rather, they are the result of the organism's intrinsic response or reaction to the agent and the organism's attempt to defend and heal itself. Symptoms, then, are part of a constructive phenomenon that is the best 'choice' the organism can make, given the circumstances. In this construct, the role of the physician is to aid the body in its efforts, not to take over the functions of the body.[115]

The vitalism is explicit. Yet the conclusion that one ought to adopt a vitalistic theory does not follow from the fact — *the empirical fact* — that the body often behaves in ways that are goal directed towards the end of its own well being. Of course the body has mechanisms for maintaining its well being. This in fact is one of the things that we would expect if we accept the Darwinian account of the origin of species. If species are developed through the action of natural selection, then we should expect organisms to have within themselves mechanisms which serve to maintain the well being of the organism, to enable it to survive and reproduce. Thus, white blood cells fight invasive bacteria, to pick but one example of such a mechanism. Mechanisms that function to maintain the well being of organisms have been explored by scientists, and we understand many things about how they work. This understanding is straight forwardly empirical: the mechanisms are biochemical in nature. There is no reason to think that there is anything more involved in these processes than the sorts of biochemical mechanisms that science has already discovered to be at work in maintaining the well being of an organism. To be sure, our knowledge is still gappy; there are many processes for which we have as yet no scientific understanding. But from this we need not infer that therefore scientific understanding of these processes is impossible. We need not infer, in other words, that, in the absence of a presently available naturalistic or empirical explanation, we must turn to non-empirical, vitalistic theories, if we are to understand these processes. To the contrary, the fact that we have in the past been very successful in discovering scientific explanations of many of the mechanisms by which organisms maintain their well being testifies the our being likely in the future to be able to discover scientific explanations in those areas where our knowledge is still gappy.

When the first medical schools were established in Europe in the Middle Ages, physician-teachers began to practice dissection of human bodies, despite continuing religious restraints. They gradually came to realize the limitations of Galen's anatomy, recognizing that what he knew of human anatomy was limited by the fact that he had inferred it from the anatomy of dogs and apes. They continued, however, to accept his vitalism. Anatomical knowledge continued to advance, however. Eventually, in the work of William Harvey, the circulation of the blood was discovered through a careful series of experiments. This discovery completely undercut much of the Galenic system, though not the vitalism. The latter, precisely because it is not empirical, is compatible with almost any scientific account of the body. What undercut the vitalism was the recognition that *there are mechanisms that serve to maintain the well being of the body*. There was no need to go beyond the empirical to some "vital spirit" to understand why the body tended through itself to maintain its well being. Moreover, the success of **the method of experiment** convinced many that they had available the tools by which their scientific understanding could be improved.

Vitalism was simply by-passed. More strongly, it was recognized that vitalism was a real hindrance to doing science. This argument was developed in detail by the scientists who came after Harvey, and in particular by the great chemist Robert Boyle. The latter, in his *A Free Enquiry into the Vulgarly Receiv'd Notion of Nature*,[116] attacked specifically the appeal to Nature or Natures as something that at once was no more than an appeal to ignorance and at the same time something that interfered with the progress of experimental science. It involved an appeal to ignorance because statements regarding the supposedly explanatory Natures were vacuous and untestable because they referred to non-empirical entities. The "Nature ... is so dark and odd a thing, that 'tis hard to know what to make of it, it being scarce, if at all, intelligibly propos'd, by them that lay the most weight upon it."[117] At the same, Boyle points out, scientists "observe divers Phaenomena, which do not agree with the Notion or Representation of Nature...".[118] Boyle cites the phenomenon of a vacuum as contrary to many of the things that people have said about Nature as a causal and explanatory force. "Nature abhors a vacuum" it had been said, and on that basis the impossibility of a vacuum had been established. It was left to Boyle so to improve pumps that he could create a vacuum in a laboratory. In other words, when one appeals to Natures to explain, one not only settles for explanations which are bad because they are obscure but also settles for explanations that are in fact wrong, misdescribing the empirical facts. Boyle also cites the alleged explanations of the motions of material bodies near the surface of the Earth in terms of the occult qualities of Gravity and Levity. Bodies of the former sort,

that is, with that sort of "Innate Appetite,"[119] move in straight lines towards the centre of the Earth, while bodies of the other sort move in straight lines away from the centre towards the heavens. Boyle points out how this doctrine makes very little sense with regard to the motion of a pendulum.[120] Boyle also argues in detail[121] that various propositions that are supposed to be established in regard to the notion of "Nature", e.g., "Nature does nothing in vain," either explain too much or too little, and in fact in general can be understood, when taken as scientific and referring to patterns of behaviour of objects in the world of sense experience, as making assertions compatible with empirical science.

Traditional Chinese medicine is very much of a methodological piece with the medicine of Galen. This form of medical theory was, like Galen's, developed in a culture that did not permit dissection. The anatomy is therefore almost wholly speculative. It distinguishes five "solid organs" and six "hollow organs." The "solid organs" store "vital essence" while the "hollow organs" transform the "essence" and discharge waste. The "heart"is the centre of emotional activity and thought processes. Sexual activity is controlled by the "kidneys." The "lungs" are responsible for the condition of skin and hair. The "heart" is intimately related to the tongue, the "spleen" to the lips, and the "kidney" to the ears. Further, anger and frustration can do pathologic damage to the "liver," while an abundance of joy will damage the "heart."[122]

This view is presented as a reasonable alternative to conventional medicine in the West: as one of its practitioners puts it, "The 'heart,' rather than the brain, *as the West believes*, is the center of emotional activity and thought processes."[123] But it is not simply a matter of "belief." The Chinese belief is the same as that of Aristotle, and had already been known to be false by Galen. Western science has over years of scientific research accumulated overwhelming evidence that it is the brain is the physical basis for our thought processes, both our reasonings and our emotions. We have *strong evidence* that the Western belief is *true*. And if we have good reasons for accepting the Western belief, then we therein have reasons that strongly require the rejection of the Chinese belief as *false*. It is not a matter of what scientists in the West *believe*; it is a matter of what they *know*; and what they know entails that what traditional Chinese medicine believes is false.

The traditional doctrine takes the universe to be composed of the pair of opposites, Yin and Yang. Thus, that which is hot is Yang while that which is cold is Yin. Every object, every action can be thought of as constituted by a preponderance of either Yin or Yang. "Nothing exists," we are told, "that is neither Yin nor Yang, and all natural events are influenced by the constantly changing relationships of these two formless aspects of all things."[124] There are five elements — wood, fire, earth, metal and water — which are subdivisions of Yin and Yang. Among the things of the world is the human body; this, too, is composed of Yin and Yang. The solid organs are Yin, the hollow organs are Yang.

The human body is animated by a "vital energy" referred to as "Qi." Blood and other bodily fluids carry Qi through channels to all parts of the body.

> Qi means that which differentiates life from death, animate from inanimate. To live is to have Qi in every part of your body. To die is to be a body without Qi. For health to be maintained, there must be a balance of Qi, neither too much nor too little.[125]

Qi is, in the first place, transmitted from parents to offspring. It is, in the second place, obtained through nourishment. And it is, in the third place, obtained through respiration. Health and illness are understood in terms of balances.

> All of human pathology can be seen in terms of balances and imbalances. A balanced state corresponds to health. Any excess or deficiency corresponds to illness. When the body is in a state of equilibrium, internally and with respect to the external environment, then it possesses a 'positive vitality,' a form of Qi that protects the body and defends it from 'pathogenic factors.'[126]

Two points are clear. Like Galen's medicine, traditional Chinese medicine makes use of the notion of a balance among opposites. And like Galen's medicine, traditional Chinese medicine makes use of teleological explanations: the vital energy Qi (or *chi*) "protects and defends" the body. Any concept of a teleological non-material force or power is non-empirical, as we have

seen. Since Qi is such a concept, it follows that the basic theory of traditional Chinese medicine is untestable and therefore unscientific.

But, we are told, there is a difference between the Chinese theory and that of Galen: "The Chinese system has far outlasted the Greek; twenty-four centuries after its inception it continues to influence the health practices of a billion people."[127] There is the suggestion here that merely because it has survived it must have a core of truth. Perhaps. But there was a core of truth in what Galen taught and practised. That core has survived. What has been lost is the non-empirical and empirically false parts of Galen's theory. The traditional Chinese medicine may have a core of truth, but it has not yet lost the non-empirical theory and the mistaken views about the causal pathways by which things happen within the body. The suggestion is that precisely because it is *traditional* it must be true is little more than an **appeal to authority**. But the **method of authority** is no sure guide to empirical truth: Galen was for many centuries an authority, but it was gradually found out that he had been very wrong about many things, and while he is now honoured as a great physician he is no longer a guide to truth.

Often part of the argument for various kinds of alternative medicine is that everything must be taken into account when treating disease. It is for this reason that the term 'holistic' is often used to describe alternative medical practices.[128] It is sometimes suggested straight off that absolutely *everything* must be taken into account.

> Chinese medicine appreciates the relationship between man and nature. Man does not exist in a vacuum. Human life and death are but a minuscule part of the universe and can be influenced by every other aspect of the universe.[129]

It is not clear what this might mean. The causal texture of the universe is intricate and complex, but it can, by the methods of science, come to be understood. One thing is clear: it is simply *not true* that everything that happens is relevant to our state of health or well being. Indeed, if it were true, then the practice of medicine would simply be impossible: *no one* can take account of *everything*.

More modest claims are sometimes made for the necessity of a "holistic" approach to medical practice. Thus, we are told that "Fundamental to holistic medicine is the recognition that each state of health and disease requires a consideration of all contributing factors: psychological, psychosocial, environmental, and spiritual."[130] It is suggested that one of the major differences between Western and traditional Chinese medicine is that the latter, unlike the former, attends to "the relation between one's life-style or mental state and the disease process." [131]

> According to traditional [Chinese] theory, health and longevity depend not only on environment, genetics, and fate but also on style of living, thoughts, and emotions.[132]

But this is hardly new. Insofar as the factors relevant to disease are known, they are taken into account. And in cases where such things as life style are known to be relevant, they are taken into account. Eisenberg is just wrong when he asserts that Western physicians are not much interested in such factors as life style and emotions.[133] A person with coronary heart problems may be treated with antiplatelet drugs, but they will also be told to reduce stress, watch their diets and quit smoking.

Nor is it true that psychological states are ignored in conventional Western medicine, contrary to the claim that is often made by practitioners of alternative forms of medicine. We are told that "The notion that health depends on behavior and thought is a cornerstone of Chinese medical thought."[134] Another tells us that "all states of health and all disorders are considered to be psychosomatic."[135] That mind and body interact is clear: mental states affect physical states and physical states affect mental states. I hit may thumb with a hammer and that alteration in my physical state effects an alteration in my mental state — I feel pain. I decide to raise my arm, and this alteration in my mental state effects an alteration in my mental state — my arm goes up. But there is nothing new in this and certainly nothing that is foreign to conventional Western medical practice. It is banal that mind and body interact. Unfortunately, this banality is sometimes ignored, with disastrous consequences. But that fact, regrettable as it is, does not imply a need to reject conventional medicine in favour

of some nonsensical alternative. As for the claim that *all* disease is psychosomatic, that is simply false. Smallpox is not in any reasonable sense psychosomatic. Of course, one's psychological state will determine how a disease is endured, but that is a very different matter. In any case, what is called for is research into the causal processes through which mind affects body in disease and disease of the body affects mind. We must try, *using the scientific method*, to come to understand the causal patterns. Resorting to non-empirical theories because they are old or traditional or because they sound good is hardly the way to further this study.[136]

"But it works," we are told. Thus, it is claimed with regard to traditional Chinese medicine that "...sick patients showed marked improvement over time."[137] Similar arguments are advanced in defence of homeopathic practices. As one author puts it,

> ... the homeopath is most convinced by experience. Results of trials, as well as their own observations of what works in particular patients and what does not, form the foundation of homeopaths' belief in the effectiveness of their system.[138]

In both cases, the claims that patients get better may well be true. However, if the patients do get better, then it is not because the theory was any good: it isn't.

The argument that "it works" is, however, fraught with difficulty. Even laboratory tests that seem to support such practices as homeopathy are subject to controversy.

Take one example. In June 1988 the prestigious scientific journal *Nature* published a study from a group headed by a French researcher, Jacques Benveniste.[139] Benveniste and his colleagues studied the effects of very dilute antibody solutions on basophil degranulation for a number of years. In the beginning, Benveniste undertook the series of experiments in an effort to prove wrong homeopathic theories about the effects of very dilute solutions. But as it turned out, he was unable to this: although the solutions were so dilute that the likelihood of there being antibody molecules in the solution was very small, nonetheless effects were observed. The results were submitted to *Nature* for publication. At first the editors were sceptical indeed. For, if Benventiste's results were correct, then there was a conflict with established scientific theory. The Benventiste team admitted that they could not explain their results. They pointed out that "we demonstrated that what supports the activity at high dilutions is not a molecule. Whatever its nature, it is capable of 'reproducing' subtle molecular variations..." they then went on to say that

> The precise nature of this phenomenon remains unexplained. It was critical that we should first establish the reality of biological effects in the physical absence of molecules.[140]

But it is worse than this. It is not only that there is no theory that explains the results. Rather, all available theory testifies that the results reported cannot occur. The results, in other words, conflict with accepted theory. Since the latter is so strongly supported by many data accumulated over a couple of centuries of research, the fact of conflict with accepted theory calls the results into question. As the editors put it in an Editorial in the same number of *Nature*,

> Where ... would elementary principles such as the Law of Mass Action be if Benveniste is proved correct? The principle of restraint which applies is simply that, when an unexpected observation requires that a substantial part of our intellectual heritage should be thrown away, it is prudent to ask more carefully than usual whether the observation may be incorrect.[141]

It has been claimed that "the criticism that homeopathy cannot work is not an empirical one...".[142] But this is just wrong: the argument is eminently empirical, deriving from the weight of evidence that supports established theory, a theory that has the consequence that homeopathy *cannot work*: it cannot work because it is contrary to well confirmed theory.

On the ground that the results were contrary to accepted and well supported theory, the editors of *Nature* had asked for the results to be replicated in other laboratories. This he arranged. In the end, the experiments were replicated seventy times at four different universities over the course of five years. The laboratories where the replications were made were, however, chosen by Benveniste rather than the editors

of *Nature*,[143] which made the supposed replications less convincing than they might have been. The editors then published the study. But they appended an "Editorial Reservation" in which they indicated that they would arrange for a group of independent investigators to observe repetitions of the experiment.[144] These investigators were sent, and they reported their findings in a subsequent number of *Nature*. It was not favourable. The experiments were described as a "delusion," and that there was an "insubstantial basis for the claims" that were made on the basis of the experiments.[145] These investigators concluded that "the hypothesis that water can be imprinted with the memory of past solutes is as unnecessary as it is fanciful."[146] Benveniste replied that the assessment was as seriously flawed as the team claimed his own research had been.[147] But he did not meet the point of the visiting team that Benveniste and his group "seems not to have appreciated that its sensational claims could be sustained only by data of exceptional quality."[148] The visiting team, whatever defects in its approach that Benveniste might have found, certainly did establish that there were real questions about the quality of the research. Others made the same point.[149] Given the weight of evidence to which the editors of *Nature* pointed in favour of the accepted theory which deemed Benveniste's results not possible, the experimental evidence from Benveniste's laboratory would have to have been beyond reproach. It was not.

The editors of *Nature* were in fact criticized for publishing the original study on just these grounds. In their letter to the editors, H. Metzger and S. Dreskin argued that

> We believe that the approach chosen by *Nature* is regrettable. We feel that all ideas no matter how revolutionary, deserve to be heard. However, when new data are proffered that grossly conflict with vast amounts of earlier, well-documented and easily replicated data, a different editorial standard is required. Before the *imprimatur* inherent in publishing them in a leading scientific journal is granted, the new results must be reproducible by disinterested individuals familiar with the field. That is a fundamental principle of scientific objectivity. It's a shame really. It still takes a full teaspoon of sugar to sweeten our tea.[150]

Conflict with established theory was important but it was not, it should be noted, the only grounds for concern. The critics, and Benveniste as well, emphasize the importance of **replication**. Why is this important?

Suppose that we perform an experiment in laboratory X and confirm with our results that all A are B. If it is a genuine matter of fact pattern that all A are B, then wherever an A is produced then a B will thereby also be produced. But if we cannot elsewhere reproduce or *replicate* the results, then that calls into question whether it is really true, universally, that all A are B. Rather, what it suggests is that there is some unknown factor, say C, which is present in the lab X but not elsewhere, and which is such that it is not true that all A that are B but instead true that

> All A & C are B

or even true that

> All C are B

(Of course, if it is true that all C are B then it follows that all A & C are B.) Since the factor C is not identified, the investigators at X have mistakenly inferred that all A are B. And since C is not present in other labs, the latter do not find that when they produce an A it is followed by a B.

We see, then, that

replication is essential if we are to take experimental results as confirming that some general pattern holds among things or events.

Benveniste's results failed this test. Although those results were replicated at laboratories of his choice, they could not be replicated everywhere. From this it could be concluded that

> ... despite the elaborate controls of the previous study, the results do not describe a new scientific principle, but instead must represent, at best, some peculiarity of the assay, and at worst, an intriguing artefact.[151]

There were other defects. Others working in the area suggested that there were features of the experimental situation that were not taken into account, and that if they had been then it would not be necessary to infer that observed results had been caused by the substance that had been diluted away. The observed results could have been due to vortex turbulence caused when the diluted solutions were shaken, or perhaps by substances already in the diluting solution.[152]

Here again, as in the case of replication, *alternative explanatory hypotheses have not been excluded*. Replication serves to exclude accidental features of the laboratory or the experimental setup or even the experimenters (do they cheat?). But if one is to conclude that only one's own hypothesis is the true hypothesis, then one must take steps to exclude all alternatives. We have seen this before. It is a regular objection to a wide variety of pseudosciences that they offer their hypotheses without taking the trouble to exclude possible alternatives. We noted this, for example, in the fantasies of von Däniken. The same, quite reasonable objection, applies to Benveniste's work.

It has been said that "those doing relevant homeopathic research typically do not understand the reason they get the result they do. Most of their energy has been put into verifying the results themselves."[153] In fact, however, not enough care has gone into verifying the results. What we find here, as we shall find in the case of Lysenko, is that investigators rely upon the **method of simplistic induction** to infer their results rather than looking for all the alternatives and attempting to obtain data that exclude alternative hypotheses.

The *method of simplistic induction* conforms to the simple rule:

> from all observed A's are B's infer all A's are B's

Such an inference is not safe, however. The story is told about the ancient Chinese who set off firecrackers whenever an eclipse occurred. They had the theory that the eclipse occurred because a dragon was flying near the sun or moon, whichever was being eclipsed. The firecrackers made a loud noise which frightened the dragon.: upon hearing the noise it flew away and the eclipse ended. Using the rule of simplistic induction, they inferred that setting off firecrackers caused the eclipse to end. What was wrong with their inference is clear. They did not attempt to *eliminate the alternative hypothesis* that the eclipse would have ended anyway, in the absence of setting off the firecrackers.

Here is another example, derived from a story by Bertrand Russell. It concerns a chicken who, on the basis of repeated experience of being fed by the farmer regularly each morning, inferred that on every morning the farmer would feed him. The chicken failed to take account of all alternatives. For, unexpectedly to the chicken, the farmer one morning came not to feed but to wring his neck.

Both this example and the former illustrate the fallacious nature of an inference based on simplistic induction. In general we may conclude that simplisitic induction, induction by simple enumeration as it is also called, is an unsafe rule to follow given our cognitive interest in inferring to *true* matter of fact regularities. *The rule of simplistic induction is an unsafe methodological rule because it fails to take into account, and eliminate, alternative hypotheses.*

Use of this rule, in spite of its unsafe nature, is widespread. It accounts for the amazing success of the many cold "remedies" that are sold over the counter. A person catches a cold, his or her friend suggests a certain remedy which is then taken, the cold goes away. The cold going away is then attributed to the effect of the nostrum that had been taken. The hypothesis that colds go away anyway in a week or so has not been examined or eliminated. The conclusion that the nostrum cured the cold is therefore unwarranted. But it is regularly made: people continue to swear by their favourite "remedies."

It is precisely inferences according to the rule of simplistic induction that practitioners of alternative medicines appeal to when they assert that their remedies work. The "proving" that homeopaths do to justify their materia medica is little more than induction by simplistic enumeration, at least once one gives up the theoretical claim that "similars cure similars."

As for the homeopathic physician who can tell from his or her patients that cures are effected, one must say that he or she rarely if ever takes the care that is required if one is to exclude alternatives. In particular, little effort is made in the ordinary circumstances of treatment to exclude the possibility that the results are nothing more than yet another example of the placebo effect. Thus, it has been pointed out that

In fact, homeopathic treatment almost seems designed to evoke the placebo response. The time spent in close collaboration with a helpful professional, the attention paid to the mental, emotional, and physical aspects of an individual, and the prospect of being given just the right remedy all heighten the expectation that relief is at hand. And because homeopathic theory holds that treatments will often make symptoms worsen before they get better, even someone who stays sick for days or weeks but eventually recovers could maintain the faith that the treatment itself had worked.[154]

There have been many anecdotes about the efficacy of homeopathic medicine. There are been some studies that have tended to confirm its claims. But many of these have been, like Benveniste's, flawed in various ways.[155] The few studies that seem to support claims of homeopaths are, taken by themselves or together, hardly convincing.

Few people would claim that there is sufficient scientific evidence to prove that homeopathy is effective. A single study showing that a remedy seems to help a problem is not reason enough to use the remedy widely, or, for that matter, to accept an entire system like homeopathy.[156]

Homeopaths were quick to latch onto Benveniste's results, and to proclaim that these data provided support for their medical practice and for the theory that supported that practice. In fact, as we now recognize, a single study, especially one that is not perfect, cannot make plausible the extraordinary and often unscientific claims of either homeopathy or other versions of so-called alternative medicines.

Example VI: Lysenkoism

The theory of evolution by natural selection presupposes that there are variations among members of a species and that these variations are hertitable. This can be observed; it is empirical fact. We also know that there are mechanisms in organisms that function to serve the needs of the organism for survival and reproduction.

We of course do not know the details of these mechanisms, though we do know much more than we formerly knew — sufficient to make it reasonable to believe that there are physico-chemical mechanisms in those areas where we do not know the details. That is, it is reasonable to believe that *there are* such mechanisms, even if we have not located the specific details of how these processes proceed. In particular, there are mechanisms for reproduction and these are such as to ensure the inheritance of variations.

What happens in the environment in which natural selection occurs affects the variations that we observe in organisms. Plant an acorn in one environment, stony and windy, and it is a stunted oak that will result; plant an acorn in a more fertile environment and a much larger oak tree will result. Feed a pet from the a litter only the minimum that it requires for survival and it will be smaller than another pet from the same litter that has been well fed and pampered. So environment does affect the observable characteristics of organisms. Of course, heredity is relevant too: an acorn does not grow into a pine tree. Heredity clearly places limits upon the effects of environment. Nonetheless, environment interacts with heredity to bring about the organism that we observe and the set of characteristics that it exemplifies.

There is no reason *a priori* to think that the environment in which natural selection occurs never affects the nature of the characteristics which an organism passes on to its offspring. To the contrary, there is some reason to think otherwise, that is, to think that the environment helps determine what characteristics a parent passes on to its offspring. Why does the giraffe have a long neck? Because the food a giraffe eats is high in trees and, better to reach this food, earlier giraffes with short necks stretched them as far as they would go. This stretching lengthened the neck, and these longer necks were passed on to their offspring until in the end the length of necks corresponded to the height of the food and stretching was no longer needed. Hypotheses of this kind, about the role of use and disuse, were advanced by J. B. P. A. de Monet, Chevalier de Lamarck, in the 18th century, and it is clear that they have a certain amount of plausibility. Charles Darwin also accepted a very speculative theory of heredity which allowed for such environmental effects of use and disuse on the heritable characteristics of organisms.

However, by the end of the 19th century the use of advanced techniques of optical microscopy had identified the chromosomes in the cell as the chemical means through which reproduction occurs. Within these chromosomes was contained the information that led the production of new cells, new organisms that reproduced the kind from which these chromosomes were derived. If the environment was to affect which characteristics were inherited, there would have to be a mechanism by means of which information was transmitted from the environment to the chromosomes. For, only if the latter were somehow changed would there be any effect one the characteristics that the chromosomes caused to develop in the offspring. But no such mechanism has ever been discovered. Biologists therefore concluded that there is no effects transmitted from the environment to the chromosomes that carry the information with regard to the characteristics transmitted to the offspring. A. Weismann was the first to articulate this theoretical perspective clearly. The position of Lamarck and Darwin was judged to be false.

The theory was subsequently developed with the proposal that there were entities at specific sites on the chromosomes that were determinant of each of the characteristics of things. These entities were referred to as "genes." The genes determined the "genotype" of the organisms. Information from the genes was transmitted to the parts of the organisms as they it developed, and information transmitted by this process, whatever the specific details might be, interacted with environmental factors to determine the observational "phenotype" of the organism.

This theoretical perspective was challenged by Trofim Lysenko, a Russian agronomist of peasant background who worked during the Soviet period.

During 1928 and 1929 he performed certain experiments on Winter wheat. To make it sowable in the spring, he suggested that germinating seeds be buried in snow before planting. The experiment was performed and it reportedly led to greatly increased yields. This process was referred at "vernalization." The term was later extended to cover anything done to a crop before planting in order to alter its development to suit local growing conditions. This was explained in terms of the following theory. Each plant, it was argued, goes through distinct developmental stages or phases, each characterized by certain requirements for development.

...the individual development of the plant takes place in successive phases each of which has requirements differing from those of other phases.[157]

The fact of that vernalization works, it was proposed, could be explained if it was assumed, first, that the plant contained within itself the possibility of developing in different directions, and, second, that at the end of each developmental stage the heredity of the plant could be as it were cracked so that a different characteristic would appear.

> Minute environmental differences are able to modify hereditary constitution, therefore even the so-called pure lines are not uniform, but consist of a multitude of physiological variants corresponding to small differences in environment.[158]

Those characteristics that are brought out by the environmental conditions in which the plant is growing are said to be dominant, that is, dominant in that environment, while those that do not show up but which could be made to appear if the environment were different are said to be recessive. The possibility of developing a certain character is carried by some minute, "limiting," factor, one of which is dominant in an environment, the other recessive. These factors "should not be regarded as Mendelian genes."[159] Ordinary genetic theory, that is, Mendelian genetics, supposes that there are dominant and recessive genes, but also holds that whether it is the dominant gene that produces a characteristic or a recessive gene that does so is something which happens independently of the environment: no information from the environment in which natural selection occurs affects the way characteristics are inherited by an organism. Lysenko's theory is thus directly incompatible with ordinary genetic theory.

Whatever the mechanisms are that are at work, they are such that the environment brings out from amongst all the factors present that one which yields the characteristic which ensures the best adaptation of the organism to that environment.

... the worst limiting factor is always neutralized by the better one, encouraging the most appropriate course of development.[160]

There is, however, a tendency to preserve the same kind in reproduction.

> Individual development is a dynamic cyclical process of continual changes. At each state of development the combination of the plant with the assimilated nutrients gives rise to a compound, different from the preceding stage, which again combines with fresh nutrients. Moreover living matter has a property that may be called conservation, which is manifested by a tendency to repeat the developmental cycle of its ancestors. This conservation acts through the capacity of the plant to select the most appropriate nutrients for itself, and thus a developmental cycle similar to that of the preceding generations is maintained.[161]

The breeder is able to select the appropriate nutrients and produce new types of developmental cycles. "In such a case the conservation of the plant is shattered, and the plant is rendered more sensitive to factors inducing a changed cycle." Such factors include adverse environmental conditions and grafting.[162] It is such adverse environmental factors that are at work in the process of vernalization.

It is clear that there is a tendency towards Aristotelian assumptions in this theory. Plants are such that they always develop in the way that is best in the environment in which they are situated: they aim at the *best*. Thus, as the theory says, the "better" factor "always" neutralizes the "worse," "encouraging the most appropriate course of development."

Of course, there may be mechanisms which have developed as the result of natural selection and which account for these effects. Unfortunately, such mechanisms were never discovered; indeed, Lysenko and his colleagues never even ventured any plausible hypotheses as to what such mechanisms might be like. "A serious general criticism," it was pointed out, "which applies to all aspects of the nutrient theory is that it has not been investigated physiologically, in spite of Lysenko's constantly reiterating Timirjazev's dictum that genetics is a branch of physiology."[163] Worse, the existence of such a mechanism seems to be incompatible with what is known about the physiology of organisms. Non-Soviet critics noted that "Lysenko has concentrated on a small number of dubious experiments and erected upon these his own genetical theory, making no attempt to cover the enormous body of well-established data collected by Mendelian geneticists the world over."[164] An Aristotelian teleology of striving for the best was allowed to mask the absence of a secure theoretical basis in plant physiology. "No explanation is given of the way in which the conservatism operates, and it is difficult to avoid the conclusion that the idea was introduced solely to gloss over the deficiency of the theory of nutrients. Conservation is hardly a property that admits of experimental investigation. It is an essentially unverifiable hypothesis devised to explain the likeness of parents and offspring, and is a metaphysical notion...".[165]

The theory simply stated in fact consists primarily of one "law," referred to as "Michurin's Law," after a predecessor of Lysenko. This "law" asserts "the dominance of locally adapted characteristics."[166] This law asserts that the environment in which natural selection occurs helps to determine which characteristics are inherited. It thus conflicts with the standard theory of heredity which has developed since the work of Weismann. The latter theory has considerable empirical support. This means that scientists outside the Soviet Union were in a position to apply the principle we have already noted that

> Any new hypothesis should be consistent with those parts of theories in the area which have been well established through predictive success.

and reject Lysenko's theory.

Of course, if there were strong evidence in favour of Michurin's Law, then that could balance the evidence in favour of the standard theory. What then would be required would be the devising of a set of experiments that could decide between the two theories.

Now, as has been pointed out, to establish Michurin's theory "it would be necessary to plan a large series of experiments with plants exhibiting dominance reversal, and to correlate the expression of dominance in the various instances with the adaptability of the hybrids to a series of controlled environments."[167] These experiments were never in fact conducted. A few controlled experiments did tend to support the Law. However,

> Such isolated instances, even though confirmed, would hardly suffice to establish a general genetical law. Only an extended series of experiments could possibly demonstrate the validity of this generalization and even were this done, the facts already known which appear to conflict with Michurin's statement would require explanation.[168]

There were, to be sure, various uncontrolled experiments of a wide variety, all variations on the theme of vernalization. A wide campaign was instituted to improve grain harvests through the vernalization process (1929-35). A similar process was widely instituted to improve potato production (1935). Again, there was a similar process for improving forestation (1948-52). Then there was another, also similar, one for improving maize (1948-52). The argument advanced in favour of Michurin's Law was simply that "it works in practice."

But as we have seen, this is a weak argument. It fails to take into account the fact that there may be alternative hypotheses that could equally well account for the observed phenomena. A central criticism of all of Lysenko's experiments is that they failed to allow that the original population had included different varieties and that the various treatments merely sorted out a pre-existing variety. The apparent genetical transformation that Lysenko convinced himself he had achieved, in other words, might well have been due simply to selection rather than the inheritance of acquired characteristics. Certainly, Lysenko did not control for this possibility.[169] Since he did not allow that this might be the cause of his results, he did not establish his own theory about the cause, that is, Michurin's Law. Simply relying on "it works," that is, simply relying on the method of simplistic induction, is not a safe rule for arriving at hypotheses acceptable for purposes of explanation and prediction.

In fact, as has been pointed out, all the variants on vernalization were adopted on a large scale because, if true, they promised to provide inexpensive means for fulfilling the goals of Soviet agricultural policy. There was likely an element of the **method of fantasy**, which acts on the rule that the wish is the parent of the belief, i.e., upon the methodological rule

accept as true those hypotheses that you want to be true

Lysenko's proposals for improvements in farming practices promised inexpensive but dramatic enhancements in agricultural production. They were all introduced by government order, without testing under proper controls. All proved unsuccessful. All were quietly phased out when this became apparent.[170] In the event it turned out that they *didn't* work.

Lysenko carefully guarded himself against having his experimental results overturned. It has been pointed out that even with his limited experimental results, Michurin's so-called Law was a gross over-generalization. It was then saved from falsification by contrary results, cases where the environment did *not* affect the inherited characteristic, by appeal unknown environmental factors which interfered with the expected process.[171] It is not true, as some have suggested, that a theory that allows for the action of unknown factors is "worthless," and that Lysenko's theory in particular will not do "unless the type of conditions able to induce mutation can be specified."[172] It is often possible to have a theory in which all the relevant factors are not *specifically* laid down. Such a theory will assert that *there are* certain specific factors that operate without asserting *specifically* what those factors are. Such a theory is *gappy*, since it does not say specifically exactly what the relevant factors are. But a gappy theory is not as good cognitively as we would like — after all, it would be better if we actually say specifically what all the relevant factors are — , nonetheless a gappy theory is still a piece of knowledge, and, in spite of its gaps, often quite useful. Thus, the gappy law that

Water, when heated, boils

has been found to be a reliable piece of knowledge, telling us more or less what we need to know if we are to be successful in making tea.

A gappy law of the sort proposed by Lysenko asserts something to the effect that

(*) For all x there are f such that if x is A but not f then and only then x is B

Because the claim that (*) makes is universal, it is not conclusively verifiable. And because of the existence claim ("there are …") it is not conclusively falsifiable. The latter means that such a law is condemned by Popper's falsifiability criterion for science. Because of the existence claim (*) could not be falsified by a single counterexample. But gappy laws are common in both science and common life — e.g., "water when heated boils" —, and we should therefore reject Popper's criterion for separating science from non-science. There is, however, another problem with (*) as it is stated. *This is the fact that, as stated, it is tautological.*

Suppose that a scientist ventures the hypothesis that

(&) For all x, if x is A then and only then x is B

and then finds that individual *d* is such that

(^) *d* is A and *d* is not B

This falsifies the hypothesis. He then attempts to undo the damage at it were by arguing that the reason that this *d* which is A but not B is not B because there are unknown factors *f* that interfere, that is, which are such that when they are present in A's then those A's are not B. That is, he proposes that instead of accepting (&) — which we can't do because it has been falsified by *d* —, we instead should accept (*). What this scientist is proposing is that we built into our hypothesis an *exceptive clause* that enables us to explain away exceptions to the generalization that all A's are B's. If we find an A which is not B we attribute this exception to the fact that the unknown factors are present.

The difficulty is that it is trivially true that one can always find conditions that will account for the exceptions. That this is so is a simple matter of logic. Thus, with regard to (*), we can suggest, if nothing else comes to mind, that the characteristic of

(%) being *d*

is the required exceptive condition. We could have, instead of (*), the following modification of (*):

(**)   For all x, if x is A but not *d* then and only then x is B

From (**), the exceptive statement (*) trivially follows. What (**) shows is that (*) is *tautological*, that as a matter of simple logic it is always possible to find a predicate which will turn the counterexample into an exception.

The problem with (*) as it stands is that it allows *anything* to explain away the counterexample. With an *unlimited variety* of possibilities available, it is trivially possible always to find *something or other* that will do the job. It is this which makes (*) tautological. And since it is tautological, *it cannot — logically cannot — be falsified*: it cannot be falsified, not because it contains an existence claim, but because it is true as a matter of its logical form alone.

This means that (*) could *not* be the *logical form of a law*, not even a gappy law. A tautology, precisely because it is a necessary truth, something true no matter what the facts are, cannot a law. A law is a *matter of fact truth*, and therefore, while it need not be falsifiable by a single counterexample, *it must place some restriction on what the facts are*. In that sense there must be some conditions or other such that, if they obtained, it would follow that the statement of law was false.

If something like (*) is to be statement of law it must place **generic** *restrictions* on what is allowed to explain away the counter-examples. Rather than (*) we must have a statement to the effect that

(!)   For all x there are f such that f is **F** and such that if x is A but not f then and only then x is B

where 'F' is a *genus* which places *constraints* on what will be allowed as producing A's which are not B. Unlike (*), (!) is not tautological. If it turns out that for *every species* under the genus **F** there are counterexamples which are A but not B, then (!) would be false.

The difference between (*) and (!) is that the former allows an *unlimited variety* of possible exceptive conditions where the latter allows only a **limited variety** of possible exceptive conditions. Where the variety is unlimited it is clearly and simply necessary that one can always find something that will do the job: where there is

no limitation, anything will do, and where *anything* will do one can always find something. What is needed to have a law is something that *limits* what will do. More generally

Any general statement that asserts the existence of some relevant factors without saying specifically what those factors are, if it is to be a law, must place some *generic restrictions* on what those factors are.

If there is no such generic limitation, the general statement will be a tautology and therefore not a law. This principle has been called the **Principle of Limited Variety**.

If a proposed hypothesis does not conform to the Principle of Limited Variety the result is what has been called an "elastic hypothesis." Lysenko's hypotheses were of this sort.

> [Elastic hypotheses] are hypotheses which, in virtue of their form, may be extended to cover all acts either known or yet to be discovered, irrespective of the truth value of the hypothesis. The concordance in this case between fact and hypothesis arises, not from a real correspondence, but because the theory is constructed in such a form that it can always be extended to fit any concrete situation. In most cases, this possibility arises because the theory refers to unknown factors which can always be invoked to cover a discrepancy.[173]

This makes the important point that elastic hypotheses can explain everything. Whenever one finds an exception to one's favourite rule, an elastic hypothesis with respect to the rule, an hypothesis that contravenes the constrainst imposed by the Principle of Limited Variety, permits one to assert that there is *some factor or other* that explains away any violation of that favourite rule. *Except*, because it is trivially true that one can always find some such factor if there are no limitations upon what one might invoke, it follows that elastic hypotheses of this sort *are simply not explanatory*, that is, since they are tautologies they could not possibly provide any explanation in the scientific sense.

In general, then, we may conclude that insofar as Lysenko's theory is empirical, it is false, and insofar as it is true, it is not empirical and therefore not scientific.

It is clear that Lysenko's views are pseudoscience rather than science. They are based on fantasy and very weak simplistic induction, with all this disguised as genuine science. One can therefore raise the issue why Lysenko et al. were able to convince themselves and others that what they had was genuine science.

Lysenko always placed his work in the context of *dialectical materialism*. This system of thought embraces five principles:

(1)   Everything that exists is material.
(2)   Matter is eternal.
(3)   Matter is always changing.
(4)   Matter comprises opposing elements whose interaction is the cause of change.
(5)   Material change is historical.[174]

These principles are taken to be more or less self evident by Marxists such as Lysenko. Exactly what matter is is never clearly specified, nor does the Marxist doctrine that mind "reflects" matter solve the problem of what mind is and how it is connected to matter. These principles have very little to do with Lysenkoism. In contrast, the third principle is important to Lysenko and his followers because it is taken to imply that the proposition of standard genetics, that genes are relatively unchanging entities, must be false. At the same time, however, they are certainly committed to the clear empirical fact that many things endure unchanging, e.g., rocks. The contradiction is apparent — which did not, however, deter Lysenko from using the argument against "Western" genetics. The final principle was not of much importance in biology, but the fourth played an important role in Lysenko's argument. It could be used, for example, to justify a priori the idea that change could be forced on plants by processes such as vernalization.

These principles themselves are not subject to empirical investigation. But they were used to justify the acceptance and rejection of certain theoretical propositions which themselves are empirical and capable of experimental investigation. Lysenko and his colleagues thus justified their position on the basis of the **a priori method**, the methodological rule that

a scientific proposition is accepted or rejected in case that it conforms or conflicts with a basic set of propositions that are taken to be self evident, that is, which are such that one cannot conceive their contraries to be true.

They were to be followed in this way of thinking by later pseudoscientists such as von Däniken. Thus, the biologist "encountering the claim that his data must conform to the philosophy of dialectical materialism as interpreted by Lysenko ..., is not likely to accept this philosophy without considering its validity. He will discover that certain questions, which are for him a matter for experimental investigation, are forejudged and regarded as established *a priori*."[175]

Besides taking the principles of dialectical materialism to be true *a priori*, Lysenko also uses the **method of authority**. Truth is determined not by the data and experimental elucidation but by appeal to the opinions and preconceptions expressed by accepted authorities. We can see this at work in statements such as the following:

> Only the materialist theory, raised to unprecedented hieght in the works of Lenin and Stalin, has enabled biologists to develop Michurinist materialistic biology, which is free from all idealism, and thereby to perceive the development of living nature as a specific form of the movement of matter, to know living nature and its laws such as they really are.[176]

In this sort of appeal to authority, Lysenko and his followers are akin to the creation scientists. An *authority* in the former Soviet Union — as in theology elsewhere — is taken as stating the truth; whatever the authority asserts is accepted as true without further question. Unfortunately — again as with theology elsewhere — the authority is seldom thoroughly pellucid and seldom simple. So the original starting point, what the authority says or is alleged to have said, must be interpreted. So, besides the authority we have to cite the glosses of later interpreters, later authorities. These glosses might alter the original meaning, but that, again, is the way of authority. Lawyers and students of mediaeval philosophy will recognize the pattern.

As with theology, so with Marxism as practiced in this way: besides the authorities there are ranged against them the heretics and their heresies. In fact, to label a viewpoint as "heretical" is more damaging than the experimental proof that a vague and over-generalized hypothesis cannot account for the known facts.[177] In the former Soviet Union, as with an earlier Christian church, it could be deadly.

Lysenko rose gradually to be a leading Soviet era biologist and himself someone who would be cited as an authority. Prior to his ascendency, there had been a significant group of geneticists who worked with, and helped in the further development of standard, Mendelian genetics. The most significant figure in this group was Nikolai Vavilov. In 1939 Lysenko was elected a full Academician in the U. S. S. R. Academy of Sciences and appointed a member of its governing presidium. In August 1940, Vavilov was arrested; this was followed by the arrests or disappearances of other Vavilovites. All died in prison or in the camps of the Gulag in the early 1940's. It is now known that Lysenko and his followers were directly or indirectly involved in these arrests. In late 1940, immediately following Vavilov's arrest, Lysenko moved from Odessa to Moscow to take Vavilov's place as director of the Academiy of Sciences Institute of Genetics. But he was still not in full control of Soviet biology. After World War II there was a revival of standard genetics and widespread criticism of Lysenko's views. But in 1948, under orders from Stalin, all this was eliminated.

> Orthodox geneticists accept the fact that nature cannot be altered by nurture except unpredictably and slowly over many generations. The laws of Mendel stand in the way. Lysenko replies that the needs of Soviet agriculture require that nature shall be altered by nurture, and quickly; and if the laws of Mendel stand in the way, the laws of Mendel must go. In brief, to criticise the theory of Michurin-genetics was to sabotage Soviet agriculture. Therefore it was treason.[178]

Genetics in its standard form was depicted as a capitalist, idealist, bourgeois enterprise linked to fascism. As a consequence, most Soviet geneticists were fired from their jobs, institutes and laboratroies were closed or re-organized, and Lysenko took full control of Soviet biology. This dominance continued under Khrushchev, with whom Lysenko established his position by playing on their common peasant background, by embracing Khrushchev's agricultural policies, and by convincing him that their failure was due to obstruction by powerful bureaucrats. At the same time, however, scientific opposition began once again to be heard. It was only with the fall of Khrushchev in 1965, however, that Lysenko was removed as director of the Institute of Genetics. Nonetheless, that was not the end of his influence, since his "Michurinist" biology had been firmly implanted in Soviet bureaucracies and ideology by Stalin.[179] It was only with Gorbachev and the beginning of *glasnost* in 1987 that there was any serious evaluation of Lysenko's theories, and the methods used to secure their acceptance.

We can now recognize that Lysenkoism secured its triumph not by sober argument and appeal to careful experiments but rather by an appeal to authority and an exercise of brutal power.[180] We should not be complacent, however. The ideal of free rational discussion is still under threat. The best that we can say is that things in both the West and in the states that succeeded the former Soviet Union are now less worse than they were under Stalin. As it has wisely been put,

> For any scientist to speak nonscience, and to use the prestige of his scientific position to expound nonscientific views, is to be guilty of the ultimate treason in the long battle to free the human mind.

> The fight against revealed authority, against enthroned opinion, and against the use of power to force acceptance of *ad hoc* assumption as "revealed truth" is by no means ended. It goes on here [in the West], as it must [then, 1949] go "underground" in the Soviet Union if the minds of that fine courageous people are ever to be free. It goes on when we try to resolve the paradox of preserving academic freedom even for those pledged to destroy it…. [S]cientists who pontificate without adequate knowledge, our trustees and executives who engage in mass witch hunts, all these give aid and comfort to that ultimate enemy of science and intellectual freedom.[181]

Endnotes

1   E. U. Condon, "Summary of the Study," in E. U. Condon, *Final Report of the Scientific Study of Unidentified Flying Objects* (New York: E. P. Dutton, 1969), pp. 11-12.

2   *Ibid.*, p. 9.

3   *Ibid.*, p. 25.

4   New York: Bantam, 1970.

5   *Chariots of the Gods?* p. 30.

6   *Ibid.*, p. 84.

7   *Ibid.*, p. 30.

8   For a discussion of these, see Loren McIntyre, "Mystery of the Ancient Nazca Lines," *National Geographic*, vol. 147, May 1975, pp. 716-728. See also "Mystery on the Mesa," *Time*, March 25, 1974, p. 94.

9   *Chariots of the Gods?* p. 32.

10 *Chariots of the Gods?* p. 32.

11 McIntyre, "Mystery of the Ancient Nazca Lines," p. 720.

12 "Mystery on the Mesa," p. 94.

13 "Nazca Balloonists?" *Time*, December 15, 1975, p. 50.

14 "Mystery on the Mesa," p. 94.

15 *Chariots of the Gods?* p. 31.

16 Loren McIntyre, "Mystery of the Ancient Nazca Lines," p. 718.

17 *Chariots of the Gods?* pp. 41-42.

18 *Ibid.*, p. 43.

19 *Ibid.*, p. 53.

20 *Ibid.*, p. 51.

21 New York: Putnam, 1968.

22 *Gods from Outer Space*, p. 26.

23 *Ibid.*, p. 146.

24 *Ibid.*, p. 40.

25 M. K. Jessup, *UFO[s] and the Bible* (New York: Citadel Press, 1956), p. 98.

26 *Chariots of the Gods?* p. 39.

27 *Ibid.*, p. 12.

28 Both Einstein's theory and the scientific revolution that it generated will be discussed in greater detail in Chapter Three.

29 Henry Morris, *Circular Reasoning in Evolutionary Geology*, ICR Impact Series, no. 48 (San Diego: Institute for Creation Research, 1977).

30 Henry Morris, *Scientific Creationism* (El Cajon, Calif.: Mater Books, 1974), p. 96.

31 For further discussion of relevant issues, see, Steven Schafersman, "Fossils, Stratigraphy, and Evolution: Consideration of a Creationist Argument," in Laurie R. Godfrey, ed., *Scientists Confront Creationism* (New York: Norton, 1983), pp. 219-244.

32 H. Slusher, *Critique of Radiometric Dating* (San Diego: Institute for Creation Research, 1973), pp. 29-31.

33 Henry Morris, *Scientific Creationism*, p. 145f.

34 See Davis Young, *Christianity and the Age of the Earth* (Grand Rapids, Mich.: Zondervan, 1982), Ch. 7; and Stephen G. Brush, "Ghosts from the Nineteenth Century: Creationist Arguments for a Young Earth," in Laurie R. Godfrey, *Scientists Confront Creationism*, pp. 49-84.

35 Henry Morris, *Scientific Creationism*, p. 85.

36 See A. W. Compton and F. A. Jenkins, "Origins of Mammals," in J. A. Lillegraven, Z. Keilan-Jaworowska and W. A. Clements, eds., *Mesozoic Mammals: The First Two-Thirds of Mammalian History* (Berkeley: University of California Press, 1979), pp. 59-73.

37 See also J. Cracraft, "Systematics, Comparative Biology, and the Case against Creationism", in Laurie R. Godfrey, *Scientists Confront Creationism*, pp.163-192; and Laurie R. Godfrey, "Creationism and Gaps in the Fossil Record,", *ibid.*, pp. 193-218.

38 Henry Morris, *The Remarkable Birth of Planet Earth* (Minneapolis, Minn.: Dimension Books, 1972), pp. 61-62.

39 Bertrand Russell was once asked what he would say to the deity if it turned out that, upon dying, he was ushered into the presence of the Lord. Russell replied that he would ask of God why He left so little evidence of His existence.

40 Henry M. Morris, ed., *Scientific Creationism*, p. 210.

41 Henry Morris, *The Troubled Waters of Evolution* (San Diego: Creation-Life Publications,), p. 11. See also John C. Whitcomb and Henry Morris, *The Genesis Flood* (Philadelphia: Presbyterian and Reformed Pub. Co., 1961).

42 Henry Morris, *The Troubled Waters of Evolution*, p. 110.

43 For greater detail on this argument, see John W. Patterson, "Thermodynamics and Evolution," in Laurie R. Godfrey, *Scientists Confront Creationism*, pp. 99-116.

44 See A. M. Green, *Pumping Machinery* (New York: Wiley, 1911).

45 Cf. Frank D. Graham, *Audel's Pumps, Hydraulics, Air Compressors* (New York: Theo. Audel & Co., 1943).

46 From John W. Patterson, "Thermodynamics and Evolution," in Laurie R. Godfrey, *Scientists Confront Creationism*, p. 107.

47 Henry M. Morris, *Scientific Creationism*, p. 54.

48 Duane Gish, *Evolution? The Fossils Say No!*, 3rd ed. (San Diego: Creation-Life Publishers, 1979), p. 42 (his italics).

It worth noting that the Public School Edition uses the word 'Creator', whereas the original edition uses the word 'God.'

49 Henry Morris, *Biblical Cosmology and Modern Science* (Nutley, N. J.: Craig Press, 1970), p. 30.

50 Henry Morris and Martin Clark, *The Bible Has the Answer*, cited in Judge William J. Overton, "The Decision of the Court," in A. Montagu, ed., *Science and Creationism* (Oxford: Oxford University Press, 1984), p. 368.

51 Judge William J. Overton, "The Decision of the Court," in A. Montagu, *Science and Creationism*.

52 N. F. Hughes, *Palaeobiology of Angiosperm Origins: Problems of Mesozoic Seed-Plant Evolution* (Cambridge: Cambridge University Press, 1976).

53 J. A. Doyle and L. J. Hickey, "Pollen and Leaves from the mid-Cretaceous Potomac Group and Their Bearing on Early Angiosperm Evolution," in C. B. Beck, ed., *Origin and Early Evolution of Angiosperms* (New York: Columbia University Press, 1976), pp. 139-206.

54 *Ibid.*, p. 198.

55 *Ibid.*, p. 198.

56 See C. B. Beck, "Origin and Early Evolution of Angiosperms : A Perspective," in C. B. Beck, *Origin and Early Evolution of Angiosperms*, pp. 1-10.

57 The classic statement of this view is Niles Eldredge and Stephen J. Gould, "Punctuated Equilibria: An Alternative to Phyletic Gradualism," in T. J. M. Schopf, ed., *Models in Paleobiology* (San Francisco: Freeman, Cooper & Co., 1972), pp. 82-115.

58 Gary Parker, *Creation, Selection and Variation*, ICR Impact Series, no. 88 (San Diego: Institute for Creation Research, 1980), p. iii.

59 See Laurie R. Godfrey, "Creationism and Gaps in the Fossil Record."

60 See Davis A. Young, *Christianity and the Age of the Earth*, Ch. 10.

61 Daisie Radner and Michael Radner, *Science and Unreason* (Belmont, Calif.: Wadsworth Publishing Co., 1982), pp. 29-32.

62 New York: Macmillan, 1950.

63 New York: Doubleday, 1955.

64 A brief statement of Velikovsky's theories can be found in the article "Collisions and Upheavals," *Pensée*, 2 (1972), pp. 8-10. See also William Mullen, "The Centre Holds," *Pensée*, 2 (1972), pp. 32-35.

65 Velikovsky, *Worlds in Collision*, p. 383.

66 Mullen, "The Centre Holds," p. 33.

67 Cf. Cecilia Payne-Gaposchkin, "Worlds in Collision," *Popular Astronomy*, 58 (1950), pp. 278-86.

68 Cf. Cecilia Payne-Gaposchkin, "Worlds in Collision," p. 279.

69 Mullen, "The Centre Holds," p. 34.

70 *Ibid.*

71 Payne-Gaposchkin, p. 285.

72 Carl Sagan, "An Analysis of 'Worlds in Collision'", *The Humanist*, 37, Nov.-Dec. 1977, pp. 11-21, at p.17.

73 Payne-Gaposchkin, p. 285.

74 Mullen, p. 32.

75 *Ibid.*

76 Velikovsky, *Worlds in Collision*, p. vii.

77 Velikovsky, *Worlds in Collision* (New York: Doubleday, 1950), p. 77.

78 Velikovsky, *Worlds in Collision*, p. xi.

79 *Ibid.*, p. 368.

80 Cf. Carl Sagan, "An Analysis of 'Worlds in Collision'," p. 19.

81 Traditionally this logically equivalence is known as De Morgan's rule.

82 "Objections to Astrology: A Statement by 186 Leading Scientists," *The Humanist*, vol. 35, Sept./Oct. 1975.

83 P. Feyerabend, "The Strange Case of Astrology," in his *Science in a Free Society* (NLB: London, 1978).

84 E. Custer, Editorial Comment, *Mercury Hour*, no. 21, April 1979, p. 15.

85 G. Dean, "Response to Professor Abell," *Zetetic Scholar*, 3 & 4 (1979), p. 93.

86 G. Dean and A. Mather, *Recent Advances in Natal Astrology: A Critical Review 1900- 1976,* (Para Research: Rockport, Mass., 1977), p. 1.

87 Linda Goodman, *Linda Goodman's Sun Signs* (Bantam Books: New York, 1971), p. 475.

88 Geoffrey Dean and Arthur Mather, *Recent Advances in Natal Astrology: A Critical Review 1890-1976* , p. 1.

89 Dean and Mather, p. 2.

90 Cf. Paul Thagard, "Resemblance, Correlation, and Pseudoscience," in Marsha Hanen, Margaret Osler, and Robert Weyant, eds., *Science, Pseudoscience, and Society* (Waterloo, Ont.: Wilfrid Laurier University Press, 1980), pp. 17-27; and "Why Astrology is a Pseudoscience," in P. D. Asquith and I. Hacking, eds., *Proceedings of the Philosophy of Science Association (1978)*, pp. 66-75.

91 S. Schmidt, *Astrology 14* (Pyramid Books: New York, 1970).

92 T. S. Kuhn, "Logic of Discovery or Psychology of Research?" in I. Lakatos and A. Musgrave, eds., *Criticism and the Growth of Knowledge*, p. 9.

93 Cf. Harris L. Coulter, "Homeopathy," in J. W. Salmon, ed., *Alternative Medicines* (New York: Tavistock Publications, 1984), pp.. 57-79.

94 Samuel Hahnemann, *Organon of Medicine*, trans. J. Künzli, A. Naudé and P. Pendleton (Los Angeles: J. P. Tarcher, 1982), pp. 12-13.

95 *Ibid.*, p. 70.

96 *Ibid.*, p. 197.

97 *Ibid.*, p. 13.

98 *Ibid.*, pp. 202-203.

99 Cf. Oliver Wendell Holmes, "Homeopathy," in D. Stalker and C. Glymour, eds., *Examining Holistic Medicine* (Buffalo, NY: Prometheus Books, 1989), pp. 221-244.

100 Cf. *ibid.*

101 P. Callinan, "The Mechanism of Action of Homeopathic Remedies — towards a Definitive Model," *Journal of Complementary Medicine*, 1 (1985), pp. 35-56.

102 D. T. Reilly *et al.*, "Is Homeopathy a Placebo Response: Controlled Trial of Homeopathic Potency, with Pollen in Hayfever as Model," *Lancet*, 2 (1986), pp. 881-996, at p. 885.

103 Judy Jacka, *A Philosophy of Healing*, (Melbourne: Inkata Press, 1979), p. 49.

104 *Ibid.*, dedication inside front cover.

105 *Ibid.*, p. 14.

106 Cf. D. Radner and M. Radner, "Holistic Methodology and Pseudoscience," in D. Stalker and C. Glymour, eds., *Examining Holistic Medicine*, pp. 149-160, at pp. 153-55.

107 Hahnemann, *Organon of Medicine*, p. 14.

108 *Ibid.*, p. 15.

109 *Ibid.*, p. 31.

110 *Ibid.*, p. 17.

111 For a good discussion of Galen, see Peter Brain, *Galen on Bloodletting* (Cambridge: Cambridge University Press, 1986), Ch. 1.

112 Brain, p. 128.

113 M. T. Murray and J. E. Pizzorno, *An Encyclopaedia of Natural Medicine* (London: Macdonald & Co., 1990), p. 5.

114 *Ibid.*

115 *Ibid.*, p. 6.

116 London, H. Clarke, 1686.

117 *A Free Enquiry into the Vulgarly Receiv'd Notion of Nature.*, p. 129.

118 *Ibid.*, p. 136.

119 *Ibid.*, p. 147.

120 *Ibid.*, pp. 148-9.

121 *Ibid.*, sect. viii.

122 David Eisenberg, *Encounters with Qi: Exploring Chinese Medicine* (New York: Norton, 1985), pp. 42-43.

123 *Ibid.*, p. 42; emphasis added.

124 *Ibid.*, p. 37.

125 *Ibid.*, p. 44.

126 *Ibid.*, p. 44.

127 *Ibid.*, p. 46.

128 Cf. C. Glymour and D. Stalker, "Engineers, Cranks, Physicians, Magicians," in D. Stalker and C. Glymour, *Examining Holistic Medicine*, pp. 21-28.

129 *Ibid.*, p. 40.

130 K. R. Pelletier, *Holistic Medicine: From Stress to Optimum Health* (New York: Dell, 1979), p. 13.

131 Eisenberg, p. 153.

132 *Ibid.*, p. 152.

133 *Ibid.*, p. 143.

134 *Ibid.*, p. 47.

135 K. R. Pelletier, *Mind as Healer, Mind as Slayer: A Holistic Approach to Preventing Stress Disorders* (New York: Dell, 1977), p. 318.

136 Cf. Austen Clarke, "Psychological Causation and the Concept of Psychosomatic Disease," in D. Stalker and C. Glymour, *Examining Holistic Medicine*, pp. 67-106.

137 Eisenberg, p. 133.

138 Carol Bayley, "Homeopathy," *Journal of Medicine and Philosophy*, 18 (1993), pp. 129-145.

139 E. Davenas *et al.*, "Human Basophil Degranulation Triggered by very Dilute Antiserum against IgE," *Nature*, 333 (1988), pp. 816-818.

140 *Ibid.*, p. 818.

141 The Editors of *Nature*, "When to Believe the Unbelievable," *Nature*, 333 (1988), p. 787.

142 Bayley, "Homeopathy," p. 139.

143 H. Metzger and S. Dreskin, Correspondence: "Only the Smile is Left," *Nature*, 334 (1988), p. 375.

144 "Editorial Reservation," *Nature*, 333 (1988), p. 818.

145 J. Maddox *et al.*, "'High-dilution' Experiments a Delusion," *Nature*, 334 (1988), pp. 287-290.

146 *Ibid.*, p. 287.

147 J. Benveniste, "Dr. Jacques Benveniste Replies," *Nature*, 334 (1988), p. 291.

148 J. Maddox *et al.*, "'High-dilution' Experiments a Delusion," p. 290.

149 P. M. Gaylarde, Correspondence: "Only the Smile is Left," *Nature*, 334 (1988), p. 375.

150 H. Metzger and S. Dreskin, Correspondence: "Only the Smile is Left," p. 375.

151 See Correspondence, "Evidence of Non-Reproducibility," *Nature*, 344 (1988), p. 589.

152 See K. S. Suslick J. and L. Glick, Correspondence: "Only the Smile is Left," *Nature*, 344 (1988), p. 376.

153 Bayley, "Homeopathy," p. 139.

154 "Homeopathy: Much Ado about Nothing," *Consumer Reports*, 59 (1994), pp. 201-206, at p. 205.

155 "Homeopathy: Much Ado about Nothing."

156 *Ibid.*, p. 206.

157 C. Zirkle, "The Theoretical Basis of Michurinian Genetics," *Journal of Heredity* 40 (1949), pp. 277-78. This paper translates an exposition of the theory from the Czech of Karel Hruby from which this quotation is taken.

158 *Ibid.*

159 *Ibid.*

160 *Ibid.*

161 *Ibid.*

162 *Ibid.*

163 Cf. P. S. Hudson and R. H. Richens, *The New Genetics of the Soviet Union* (Cambridge, England: Imperial Bureau of Plant Breeding and Genetics, Cambridge School of Agriculture, 1946), p. 69.

164 *Ibid.*, p. 69.

165 *Ibid.*, p. 68.

166 *Ibid.*, p. 35.

167 *Ibid.*

168 *Ibid.*

169 *Ibid.*, pp. 40-41.

170 Cf. Mark B. Adams, Article "Lysenko," in *Dictionary of Scientific Biography*, ed. F. L. Holmes, vol. 18 (New York: Scribner's).

171 Hudson and Richens, p. 34, p. 41.

172 *Ibid.*, p. 41.

173 *Ibid.*, p. 67.

174 *Ibid.*, p. 52ff; also pp. 4-5.

175 *Ibid.*, p. 55.

176 T. D. Lysenko, "Stalin and Michurinist Agrobiology," in C. Zirkle, "L'Affaire Lysenko: Spring 1956," *Journal of Heredity*, 47 (1956), pp. 47-56, at p. 56.

177 Cf. R. C. Cook, "Lysenko's Marxist Genetics: Science or Religion?" *Journal of Heredity*, 40 (1949), pp. 169-202.

178 E. Ashby, "Science without Freedom?" *The Listener*, vol. 40, Nov. 4, 1948, p. 678.

179 On the role of Soviet ideology, see D. Joravsky, *The Lysenko Affair* (Cambridge, Mass.: Harvard University Press, 1970), Ch. 1.

180 See Conway Zirkle, "The Involuntary Destruction of Science in the USSR," *The Scientific Monthly*, 77 (May, 1953), pp. 277-283; "L'Affaire Lysenko: Spring 1956"; and *Evolution, Marxian Biology, and the Social Sciences* (Philadelphia: University of Pennsylvania Press, 1959).

   See also the following exchange: E. Ashby, "Science without Freedom?" *The Listener*, vol. 40, Nov. 4, 1948, pp. 677-8; J. B. S. Haldane, "Letter to the Editor," *ibid.*, Nov. 18, 1948, p. 707; S. C. Harland, C. D. Darlington, R. A. Fisher, and J. B. S. Haldane, "The Lysenko Controversy: Four Scientists Give Their Points of View," *ibid.*, Dec. 9, 1948, pp. 873-5.

181 R. C. Cook, p. 202.

CHAPTER 3

SCIENCE

Example I: The Birth of Science

Herodotus' reasoning concerning the flooding of the Nile conforms nicely to the norms of science. It is indeed a paradigm case. But such patterns of reasoning did not become widespread until the 17th century, when science became firmly established as an acceptable basis for forming our beliefs and explanations about the world. The crucial figure here was Galileo, though one has also to mention the names of Kepler and, above all, Newton. And, of course, there were many figures, only barely less important, who contributed to this development.

However, if we are to understand science, it will help if we also take a look at pre-science, the sorts of explanations that thinkers offered for natural phenomena prior to when they began to offer scientific explanations. There are two cases at which we will take a look: the explanations offered for heavenly phenomena, and the explanations offered for terrestrial phenomena, projectile motion in particular. After that, we will turn to Galileo's scientific explanation of projectile motion and Kepler and Newton on celestial phenomena.[1]

(a) Pre-Science: Explaining Planetary Motion

The ancient world bequeathed to the modern a clear scheme for explaining the motions of the planets and the sun. This account reached its fullest development in the work of the Greek astronomer Ptolemy. He published his results in the *Almagest*.

The phenomena to be explained were the positions of the (then known) planets, Mercury, Venus, Mars, Jupiter, and Saturn, but including as well the moon and the sun. There was, first, the daily motion of all the heavens around the earth, the regular motions of the sun and fixed stars as well as the planets, creating day and night. There was, second, the regular yearly motion of the sun relative to the fixed stars. Third, there were the regular motions of the planets against the background of the fixed stars. Among the latter, there was, fourth, the fact of stations and retrogressions — some of the planets appear to slow down, stop (at a station), move in the opposite direction (retrograde motion), stop again, and then move off in its original direction (see Figure 3.1).

Finally, there was, fifth, the fact that Venus in particular appears brighter at certain stages in its motions.

The explanation scheme for these phenomena was based on two principles.

Principle of Circularity: The planets, including the sun and the moon, move through the heavens on paths that are either circles or compounded of circles and at uniform speeds.

Principle of Geocentricity: The earth is the stationary centre of the celestial motions.

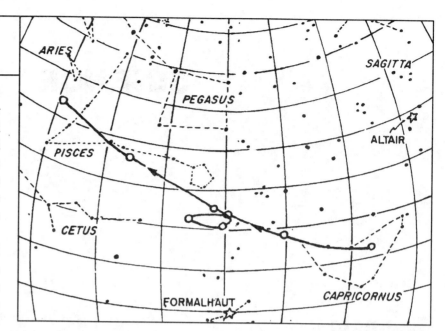

There were two criteria for adequacy: **one**, the paths attributed to the planets, including the sun and the moon, had to conform to these two principles; and, **two**, the paths had to permit the prediction of the future positions of these objects relative to the fixed stars.

The daily motion of the celestial objects was explained by the hypothesis that the whole vault of the heavens turned about the earth once every 24 hours. Consistent with the Geocentricity Principle, the earth was the stationary centre of this motion of the vault of the heavens.

The stars revolved completely every 24 hours, but the other objects went through the motion at a slightly slower rate. This accounted in a general way for their motions against the background of the fixed stars. In particular, it accounted for the yearly motion of the sun relative to the fixed stars.

The hypothesis, consistent with the Circularity Principle, that the path on which the sun moved was circular. This motion had to be uniform, that is, at constant speed. As it turned out, the second condition, accurate prediction, could not be met if it was assumed that the circle describing the motion of the sun had as its centre the centre of the earth. In order to square with the prediction criterion, the centre of the motion of the sun

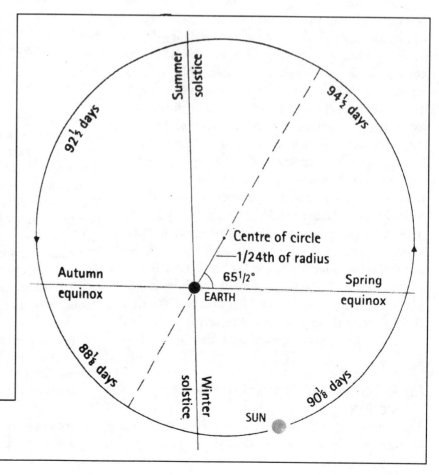

Figure 3.2
THE ECCENTRIC DEVICE[3]

Figure 3.3
THE EPICYCLE DEVICE[4]

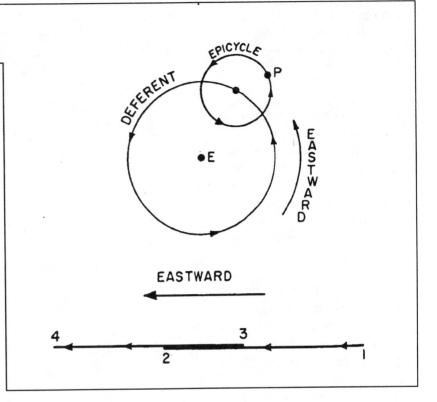

had to be placed slightly away from the centre of the earth. The centre of the sun's circle had to be placed slightly *eccentric* to the centre of the universe, i.e., the centre of the earth. This part of the hypothesis was known as the *eccentric device* (see Figure 3.2). It was in fact used not only for the descriptions of the sun's path but also for the paths of the other planets.

Another device was used to account for stations and retrogressions. In this case the path was hypothesized to be compounded of *two* circular motions. The planet was taken to move uniformly about a small circle (called an *epicycle*) the centre of which uniformly moved on a larger circle (the *deferent*) the centre of which was near the centre of the earth, that is, the centre of the deferent was eccentric to the centre of the universe (see Figure 3.3). This was the *epicycle device*.

The path attributed to the object upon this device was a looping path (see Figure 3.4), but the device did enable the ancient astronomers to account for the phenomena of stations and retrogressions (see Figure 3.5).

Figure 3.4
PLANETARY PATHS ON THE EPICYCLE DEVICE

Viewed form the Earth E, the planet moves forward from 1 to 2, moves in retrograde motion from 2 to 3, and then moves forward again from 3 to 4.

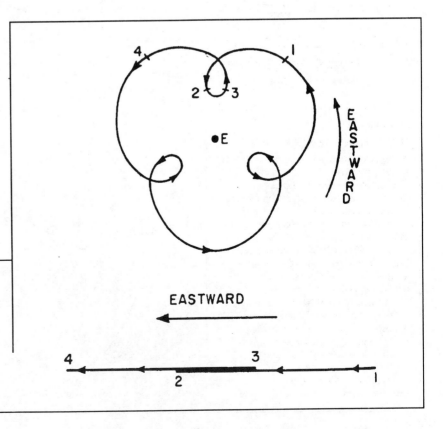

Figure 3.5
ACCOUNTING FOR STATIONS AND RETROGRESSIONS[5]

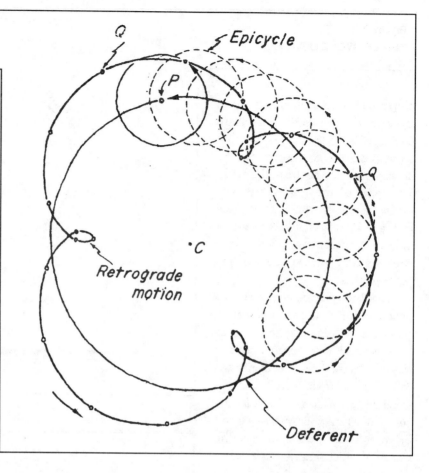

Even when epicycles were placed on top of epicycles, however, it was not possible to square the hypotheses with the prediction criterion. Yet another device was used to bring the hypotheses into line with predictions. This was the *equant device*. The various devices so far indicated were able to get the planets, sun and moon into roughly the right positions. But if it was assumed that "uniform motion" meant that the radius from the centre to the planet swept out equal angles in equal times, then it was not possible to get the planet into the correct position at the correct time. This assumption was given up by Ptolemy. Instead, he set his hypotheses so that the radius from another point, called the *equant point*, to the planet swept out equal angles in equal times. The eccentric

Figure 3.6
THE EQUANT DEVICE[6]

The equant point was a device invented by Ptolemy to account for apparent changes in the speed of the planet. Relative to the equant point, the planet moves through equal angles in equal times. The angles a, b, and g are equal. The planet therefore moves through the arcs AA', BB', and CC' in equal times. But since these arcs are of different lengths, the planet moves at different speeds.

The equant point is on the line that joins the centre of the circle and the eccentric point, but on the side of the centre opposite the eccentric point. The distance of the equant point from the centre is the same as the distance of the eccentric point from the centre.

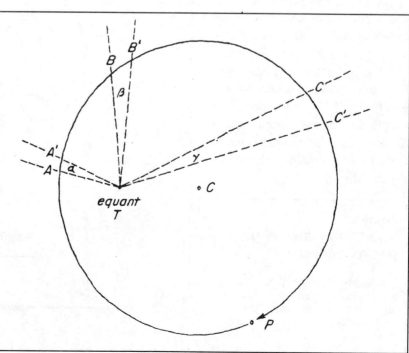

point was situated relative to the centre of the earth, and the equant point was on the same line, but on the opposite side of the earth, the same distance from the centre of the earth as was the eccentric point (see Figure 3.6).

With this device, one could secure fairly good agreement between the positions predicted by the scheme of hypotheses and the actual observations of the positions of the celestial objects.

As for the changes in apparent brightness of Venus, that was easily accounted for in terms of its being closer at certain times to the earth than at other times.

This system of hypotheses came down from the ancient world to the renaissance. At that time, however, it was seriously challenged for the first time by Copernicus.

Copernicus did not attack it on grounds that it failed to yield accurate predictions. In fact, he used the table of observations that Ptolemy had used and had included in the *Almagest*. Rather, he argued that the Ptolemaic system of hypotheses did not conform to the requirements of the Circularity Principle. Copernicus proposes that "The motion of the heavenly bodies is uniform, eternal, and circular or compounded of circular motions" (*On the Revolutions of the Heavenly Spheres*).[7] The equant point device violates this principle: it requires celestial bodies to move through unequal distances in equal times (see Figure 3.6). Thus, with reference to the motion of Mercury, Copernicus states that

...there were three centres: namely, that belonging to the eccentric which carried the epicycle; secondly, to the circlet; and thirdly, to that circle which more recent astronomers call the "equant". Passing over the first two centers, the ancients allowed the epicycle to move uniformly only around the equant's center. This procedure was in gross conflict with the true center [of the epicycle's motion], its [relative] distances, and the prior centers of both [other circles] (p. 278).

This is because, Copernicus argues,

... a simple heavenly body cannot be moved by a single sphere nonuniformly. For this

nonuniformity would have to be caused either by an inconstancy, whether imposed from without or generated from within, in the moving force or by an alteration in the revolving body. From either alternative, however, the intellect shrinks. It is improper to conceive any such defect in objects constituted in the best order (p. 11).

This is an important argument. We shall examine it in greater detail below with regard to how it stands from the perspective of the methodology of empirical science.

In order the eliminate the need for the equant point device, in order to eliminate the need to compromise the Circularity Principle, Copernicus proposed instead to abandon the Geocentric Principle. To the contrary, he argued, the heavenly system is *heliocentric*, and the earth moves about the sun on a circular orbit, as well as turning daily about its axis. Of the objects in the system, the moon alone continues to have a geocentric orbit.

Copernicus offered another argument, besides the need to preserve the Circularity Principle, to justify the heliocentric hypothesis. He points out that

At rest ... in the middle of everything is the sun. For in this most beautiful temple, who would place this lamp in another or better position? For, the sun is not inappropriately called by some people the lantern of the universe, its mind by others, and its ruler by still others. [Hermes] the Thrice Greatest labels it a visible God, and Sophocles' Electra, the all-seeing. Thus indeed, as though seated on a royal throne, the sun governs the family of planets revolving around it (p. 22).

The argument is clear: the sun *ought to be* at the centre of the universe and therefore it *is* at the centre.

By allowing the earth to move, Copernicus could account for several sets of facts all at once. By having the earth turn upon its axis, Copernicus was able to account for the observed daily changes in the heavens. These daily motions of the heavens were, on Copernicus' account, not real, as they were on Ptolemy's, but only *apparent*, the result of a different real motion, that of the earth about its axis.

Similarly with regard to the stations and retrogressions. These are now understood to be merely apparent motions that result from the real motion of the earth about the sun (see Figure 3.7). As the earth moves about the sun it moves faster than a planet such as Jupiter. As it catches up and goes beyond Jupiter, the latter *appears* to slow down, stop, and then move backwards, later to resume its forward motion.

The resulting heliocentric system has at least two advantages over the geocentric system of Ptolemy. Both involve the increased *unity* that Copernicus achieves. In the first place, where Ptolemy had no way to determine the distances of the planets from the sun, Copernicus does have such a method: the relative distances fall out automatically from the heliocentric hypothesis. In the second place, the planets all move in concert with the sun. Mercury, in particular, is always close to the sun. In the Ptolemaic system, all this is purely accidental. In the Copernican system it falls out naturally from the fact that they all move about the sun and therefore follow it in its apparent motions throughout the year.

There are other predictions that are entailed by the Copernican hypotheses. One of these is that Venus, since

it now goes about the sun, should, like the moon, display phases. On the Ptolemaic scheme, in contrast, with Venus going about the earth, there should be no phases observed. The planet is too far away for any phases, if they exist, to be observed by the naked eye. But when Galileo perfected the telescope, he was able to observe these phases, confirming empirically at least this aspect of the Copernican system, and refuting the alternative Ptolemaic account of the motion of Venus. This does not establish heliocentricity, since it is compatible with Venus going about the sun and the sun going about the earth. But it does refute the Ptolemaic claim that the centre of the earth is the centre of the motions in the planetary system.

But what about the Geocentric Principle? How was that defended by the ancients? How was Copernicus able to defend its rejection? As we shall see, the issues that are involved here are connected with the defence of the Circularity Principle given by Copernicus. It will help in understanding both these principles, however, if we glance at the work of Kepler.

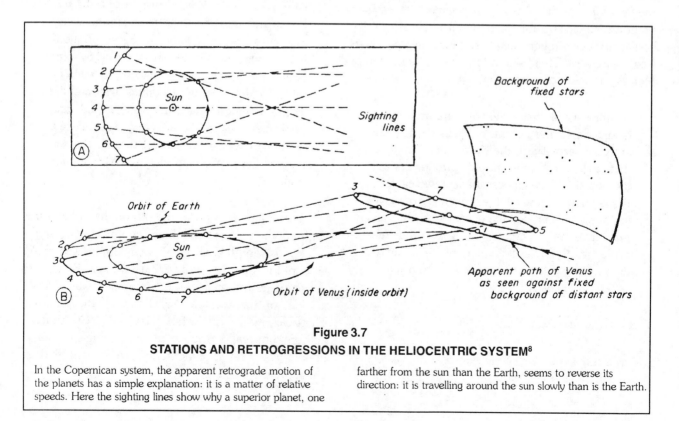

Figure 3.7
STATIONS AND RETROGRESSIONS IN THE HELIOCENTRIC SYSTEM[8]

In the Copernican system, the apparent retrograde motion of the planets has a simple explanation: it is a matter of relative speeds. Here the sighting lines show why a superior planet, one farther from the sun than the Earth, seems to reverse its direction: it is travelling around the sun slowly than is the Earth.

(b) Kepler's Astronomy

Copernicus with his new system was able to obtain what he took to be better agreement with the Circularity Principle than did Ptolemy, and was able to obtain as good an agreement with the observational data then available. Or rather, to the data that Ptolemy had used and has recorded in the *Almagest*. Copernicus was as much a bookish person as any renaissance humanist, and if something was recorded in a text deriving from the ancient world of the Greeks and Romans, then it was taken for granted that it was acceptable.

In fact, however, the data were not all that good. They were relatively sparse, only a few sightings used for each planet, and even these observations were not as accurate as could be made with naked eye observations, unassisted by a telescope. This was clearly established shortly after Copernicus with the work of the great astronomer Tycho Brahe. Using carefully designed instruments, Tycho took astronomical observations to the limits that could be achieved prior to the introduction of telescopes. Thus, he carefully measured the positions, or rather apparent positions of comets and was able to establish conclusively that comets were phenomena that occurred in the super-lunary regions of space. This was important for two reasons. First, Aristotle had held that these entities were atmospheric phenomena, occurring in sublunary regions of space. To show that they were superlunary was a challenge to the authority of the ancients, Aristotle in particular. Second, many had held that the spaces between the planets was filled by solid crystalline spheres. Given the looping paths required by the epicycle device, it was not easy to see how the doctrine of solid spheres could be squared with the Ptolemaic hypotheses; but many nonetheless accepted the idea. Since it was clear from Tycho's data that the paths that comets followed cuts through the regions supposedly filled by the solid spheres, that doctrine could no longer be accepted. This was again another blow to the notion that ideas that had come down from the ancients had to be accepted.

But Tycho also amassed a great number of observations. His pursuit of data was not only careful but systematic. In particular, he amassed a vast number of observations on the orbit of Mars. These data, those on Mars in particular, made clear that the empirical condition

on astronomical hypotheses, that they yield accurate predictions, was simply not fulfilled by the Copernican system.

Tycho's student, Johannes Kepler, to whom Tycho bequeathed his data, set out to remedy this defect. At first, under the impact of tradition, Kepler accepted the Circularity Principle, but eventually, under the need to make his hypotheses conform to the empirical standard of yielding correct predictions, that is, conforming to the data that Tycho had gathered, he came to abandon this Principle. What he came to accept was that his hypothesis must conform to the data *and to nothing else*: there was no further principle, such as the Circularity Principle, which could function as a standard of truth for rational acceptability of astronomical hypotheses.

Once Kepler had given up the Circularity Principle, he was free to try a wide variety of curves — which he did. He concentrated on Mars, because that was the planet for which Tycho's data, everywhere superb, were at their best and most numerous. He reckoned that if he could solve the problem for Mars, then he solve it for the other planets. After much labourious work, he was finally able to conclude that the form of Mars' orbit was that of an ellipse, with a certain specific eccentricity e_1, and with the sun at one focus (see Figure 3.8).

More specifically, he was able to establish that

Any place p such that Mars is located at p is also such that it is situated on a path the form of which is an ellipse with eccentricity e_1, and with the sun at one focus

This is a *generalization* about the places at which Mars is located, a *law* about the positions of that particular planet.

Kepler immediately generalized. If the orbit of Mars were an ellipse, then, since Mars is a typical planet, other planets, too, must have elliptical orbits:

For any planet there is a form f which is an ellipse with the sun at one focus such that, for any place p such that that planet is located at p is also such that it is situated on a path the form of which is f.

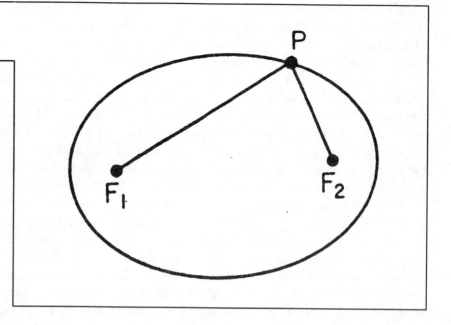

Figure 3.8
ELIPTICAL ORBIT

This is Kepler's First Law: the planets, the earth included, move in elliptical orbits about the sun which is located at one of the foci of each ellipse.

With this law to guide him, it would not be difficult to discover the specific elliptical form that described the orbit of Jupiter, say. Note that Kepler's First Law is a *law about laws*: it describes the *sort or kind* of law that one should look for, and asserts that there is such a law, there, to be found. It is then the task of the researcher to find the specific law of the sort that the laws about laws asserts is there, to be discovered.

Kepler needed another law besides this, however. Merely to get the planets in certain positions was not enough. They had to got to their positions at the correct time. For this a speed law was need. This would, of course, replace the part of the Circularity Principle that asserted that celestial objects always moved with uniform speed. What Kepler discovered about Mars was that

The line joining Mars to the sun sweeps out equal areas in equal times.

This was of course immediately generalized to the law about laws that

For any planet, the line joining the planet to the sun sweeps out equal areas in equal times.

This is Kepler's Second Law (see Figure 3.9).

The radius from the sun to the orbit sweeps out equal areas in equal times; the areas PP'S are equal. The planet therefore moves through the arcs PP' in equal times.

Finally, after considerable further empirical trials and errors, Kepler located his Third Law, again a law about laws. This relates the period of the planet, that is, the time that it takes to go about its orbit, and its distance from the sun. This law states that

There is a constant K such that for any planet, if T is its period and a is its distance, then

$$K = T^2/a^3$$

"K" is called Kepler's constant.

As with Copernicus, so with Kepler: they both in their own way saw the sun as somehow the mystical centre of the system of planets. However, there is also an important difference. For Copernicus, the mystical value of the sun became an argument for imposing a certain form upon the paths upon which the celestial objects moved. Copernicus of course had the empirical criterion: prediction had to be correct. But he also had this further criterion, the mystical value of the sun — and also, of course, the Circularity Principle. All these were for Copernicus criteria invoked to establish that his system of hypotheses was true. For Kepler, in contrast, it performed no such function. To be sure, there is little doubt that these sorts of mystical arguments helped Kepler to *discover* the laws that carry his name. But consistently Kepler appeals to nothing more than observational data to *justify accepting the hypotheses as true.*

Figure 3.9
KEPLER'S SECOND LAW

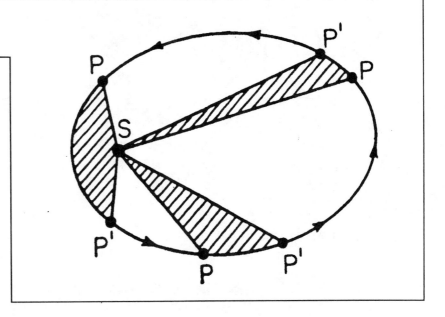

This points to an important distinction between **the context of discovery** and **the context of justification**. Acts of the imagination, mystical visions, dreams, whatever, may help us to formulate hypotheses that are worthy of test. These may in fact turn out to be true hypotheses about the world we observe. This happened in Kepler's case. But what makes any such hypothesis *cognitively* or *rationally acceptable*, what justifies their acceptance as true hypotheses about the matter of fact world that we experience by means of our senses, is precisely the empirical, observational data that can be marshalled to support the hypotheses.

This is not to say that only imagination or mystical visions or whatever leads to scientific discovery. To the contrary, research is often guided by laws and theories. In particular, as we have noted in Kepler's case, a *law about laws* is often a useful guide as to which hypotheses are worthy of further exploration. The law about laws itself has observational justification, making it worthy of acceptance as true. Kepler's First Law is such a law about laws. It itself is empirically justified. From it we can infer that the orbit of a new planet or one whose orbit we do not know is elliptical in form. That means we do not have to bother with trying out all the myriad of other forms that are possible; we can restrict our search to elliptiform shapes. In this case, the same sort of rational empirical consideration that justify accepting hypotheses as true also guide our research, help us in the discovery of new laws; but in other cases what helps us in the discovery of new laws is something non-rational, e.g., mystical regard for the sun. There is, therefore, not a rigid distinction between the context of discovery and the context of justification. But, *there is a distinction*, and that is what is important.

(c) Pre-Science:
Explaining Terrestrial Motions

Ptolemy argues for the Geocentric Principle. Those who propose that the earth moves

> ...would have to admit that the revolving motion of the earth must be the most violent of all ... seeing that it makes one revolution in such a short time [that is, one day];. the result would be that all objects not actually standing on the earth would appear to have the same motion, opposite to that of the earth: neither clouds not other flying or thrown objects would seem to move in the direction of the west and the rear.... Yet we quite plainly see that they do undergo all these kinds of motion, in such a way that they are not even slowed down or speeded up at all by any motion of the earth.[9]

If I were to drop a stone from the top of a tower towards the earth, the motion of the stone would continue in a straight line. But the earth would have moved. The stone would therefore strike the earth at a point *behind* the base of the tower (see Figure 3.10). Experience shows, however, that such a stone does not strike the earth behind the point directly below that from which it was dropped. The earth therefore cannot be in motion.

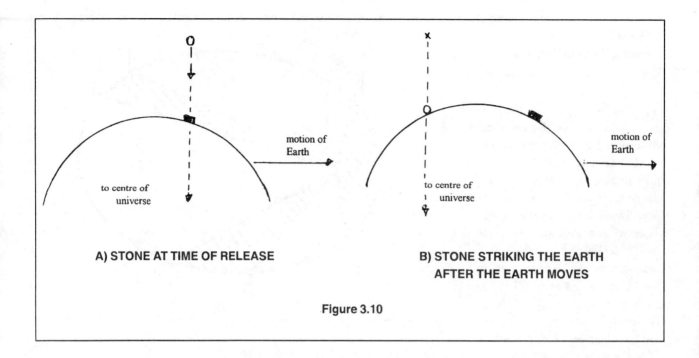

A) STONE AT TIME OF RELEASE   **B) STONE STRIKING THE EARTH AFTER THE EARTH MOVES**

Figure 3.10

This is Ptolemy's argument. It depends, clearly, upon the observed facts of the motion of freely falling heavy objects. Ptolemy characterizes the motion this way:

> ...the direction and path of the motion (I mean the proper, [natural] motion) of all bodies possessing weight is always and everywhere at right angles to the rigid place drawn tangent to the point of impact (*Almagest*, p. 43).

Note that he here speaks of the "proper" motion of freely falling objects. This way of talking is of a piece with that of Copernicus, when he argues that there cannot be any irregularity in the motions of celestial objects because it "... is *improper* to conceive any such defect in objects constituted in the *best* order" (*On the Revolutions*, p. 11) or when he argues for heliocentricity with the rhetorical question "For in this most beautiful temple, who would place this lamp in another or better position?" (p. 22) The argument is that because the motion or location is *proper* or *right*, because, in other words, *it ought to be*, therefore that is how it *is*. This is in fact a crucial feature of pre-scientific theorizing.

Of that, more directly. But another point about Ptolemy's argument must be noted.

This is the point that the argument clearly presupposes that the stone in free fall moves in a straight line towards a certain fixed point that does not move. This point is not simply the centre of the earth. For, in that case, if the earth moved the centre towards which the stone gravitated would also move. Thus, the falling stone would strike the earth at a point directly under the point of release even if the earth were moving. Hence, for Ptolemy's argument to be sound, he must presuppose that the stone is falling towards a centre that is fixed independently of any possible motion of the earth. This point is the centre of the spherical universe. This *unique place* and not simply a position relative to other objects is causally relevant. It is *proper* for heavy objects to move towards this place and therefore they do move towards this place. This is the relevant law to which Ptolemy appeals in his argument against a moving earth.

Of course, as it turns out the centre of the universe is also the centre of the earth. Because of the law for freely falling objects that have weight, all the parts of the earth gravitate towards the centre of the universe. So objects do in fact move towards the centre of the earth. The point is that they do not so move because it is the centre of the earth but because it the designated place towards which all heavy objects gravitate.

Note that not all motions are proper. There are also *improper* motions. Thus, not all motion of heavy objects is that which obtains under conditions of free fall. There are, for example, projectiles, stones or javelins, which are thrown by some force such as that provided by a human arm. These are heavy objects but have a horizontal motion.

Copernicus indicates that irregular, that is, non-uniform or non-circular motions, would be improper motions in the heavens. Again, as we saw him express it, "It is improper to conceive any such defect in objects constituted in the best order." However, if improper motions do not occur in the heavens, they do occur in the terrestrial regions, where, clearly, not everything is organized "for the best."

For our purposes, in trying to become clear on the nature of pre-scientific explanations, there are three points worthy of attention.

First, the laws describing various motions do not merely describe but determine also what is proper or what ought to be. The laws are not merely descriptive but also normative. We can already see how this distinguishes pre-science from science: given Kepler's commitment to the empirical, a law for him is merely descriptive, not normative. For Kepler, in other words, laws are descriptive matter of fact general statements, describing certain observable patterns of fact. They describe what *is*, not what *is proper* or *for the best*.

Second, precisely because in pre-science what are called laws do not merely describe, they cannot be *merely empirical*. The normative feature is not something we discover by observation, in our sense experience of the world. Thus, pre-science inevitably involves a *non-empirical element*.

Third, because laws in the pre-scientific sense describe what is proper or for the best, *there can be exceptions to these laws*. A law, in the pre-scientific sense, need not therefore describe a genuine regularity, one that holds without exceptions. It follows that success in predictions is not always a mark of lawfulness in the pre-scientific sense. A law may fail in its predictions, but that is not taken as refuting the purported law; rather, all that happens is that it is taken to imply is that this is an *exception to the rule or norm of what is proper*. In contrast, Kepler maintained that success in prediction

was the only criterion that he would accept. If he proposed an hypothesis, and it failed predictively, if what it predicted turned out to be false, then *that alone* was sufficient to lead him to reject the hypothesis. For Kepler, and for science, in contrast to pre-science, a statement of law is a statement of genuine regularity, a pattern that holds universally, without exceptions.

This pattern of explanation for free fall derives from Aristotle, who also systematized the whole account of what is an explanation, that is, an explanation in the pre-scientific sense.[10]

Upon Aristotle's scheme every object has a *nature*. This nature is *metaphysically necessary* to the being of the object; it defines what it is in its *essence*. This nature is a *power*, an *active disposition*, that moves the object in certain defined ways.

Thus, for example, it is the nature of a stone to *gravitate*. To be *grave* is an *active power*. In exercising this power, the object *moves itself*. (Note the contrast to our, more recent and scientific notion of gravity; in the latter there is no notion of *self movement*.) This power is such that

if the object is unsupported then it moves towards the centre of the universe

More generally, let "N" be the nature, "F" the occasion of its exercise, and "G" the end of its exercise. Then we have

(D)   For all objects x, x is N if and only if, if x is F then x is G

We explain the behaviour of an object by appeal to its nature. This nature is *active*: the model is that of human volition. Thus, for Aristotle, all objects are active in the sense in which human beings are active, though some, e.g., human being or dogs, are more active than others, e.g., stones. To say that they are more active is to say that they have more powers, more complex natures. Since the powers are active, modelled on human activities, they are *powers* the exercise of which is *towards an end*. The pre-scientific explanations of Aristotle and his successors such as Ptolemy are thus *purposive*; every explanation is a *teleological explanation*. In the case of stones, the purpose or end at which the stone's activity is aimed at achieving is being at the centre of the universe.

The activity is as it were constant. But it is not always exercised. The stone is constantly striving to be at the centre of the universe. But sometimes it is *prevented* from moving towards that end. Thus, if I hold the stone up at the top of the tower, I am preventing it from moving towards the centre of the universe. That tendency I feel as the weight of the stone. If the impediment is removed, if the stone becomes unsupported, then the tendency will manifest itself in the properties of the stone, it will in fact change places as it moves itself systematically towards the centre of the universe.

The nature, that is, the "N" of (D), is not given to us in sense experience. It is rather, Aristotle argued, given to us in a *rational intuition*. For Aristotle, reason is what grasps the reasons for things, and the reasons for things behaving as they do are their natures. Reason, then, for Aristotle, provides us with special insight into the metaphysical structure of the world. This notion of "reason" is very different from that of the scientist such as Kepler, whose practice it was that reason aimed to discover genuine matter of fact regularities, universal and exceptionless patterns of behaviour. Reason did not aim at insight into metaphysical structures but was a human instrument that restricted itself to the world of sense experience, endeavouring to discover exceptionless patterns of behaviour of objects.

In (D), the "F" and "G" are features of the object known in sense experience. Since (D) relates the nature N to these features of sense experience, where N is *not* given in sense experience, it follows that (D) is not itself an empirical truth, something the truth of which can be discovered in sense experience. We discover its truth not by observation but by reason, that is, the reason that grasps the natures or reasons of things. A statement such as (D) which relates a nature to the empirically observable occasion and end of its exercise is *metaphysically necessary*.

As for understanding the natures of things, this is done, according to Aristotle, by giving a *real definition* of the nature. The nature is a *species*, and the species is defined by giving its *genus* and *specific difference*. Thus, in the case of human beings, the nature is "humanity"

and the real definition is given by "rational animal", where "animal" is the genus and "rational" is the specific difference. The real definition is exhibited in a *syllogism*:

$$\frac{\begin{array}{l} \text{All M are P} \\ \text{All S are M} \end{array}}{\text{All S are P}}$$

"S" and "P" are the subject and predicate of the conclusion, and "M" is the middle term that joins them in the premises. When the syllogism exhibits a real definition, "S is the species, "P" is the genus, and "M" is the specific difference. Thus, the real definition

human is rational animal

is exhibited in the syllogism

$$\frac{\begin{array}{l} \text{All rational are animal} \\ \text{All human are rational} \end{array}}{\text{All human are animal}}$$

In the case of stones we would have

$$\frac{\begin{array}{l} \text{All centre loving objects are material} \\ \text{All stones are centre loving objects} \end{array}}{\text{All stones are material}}$$

Syllogism is thus not only a form of argument but also a logical structure that exhibits the metaphysical structure of the world. It reveals the complex structure of the active dispositions or nature of an object. It reveals, in the genus, those dispositions which the nature shares with other objects, and, in the specific difference, it reveals those dispositions which distinguish it from other sorts or species of object. Thus, for Aristotle and his successors, *understanding the natures of things consists in grasping the ways in which they are similar to and differ from other sorts of things*. Explanation consists in grasping similarities and differences among things.

To summarize: In order to explain individual events, we locate the events in an object and point to the nature of the object as providing a tie that links the events. To understand the nature of a thing, we locate the ways in which it is similar to, and the ways in which it differs from, other natures.

Notice that the explanation of individual events is given by an *entity that unifies the events*. This entity is not among the events to be explained; it lies as it were outside the order of sensible events, outside the order of change and flux. Since the world we know by sense, the world of change and flux, is the realm of the temporal, it follows that the nature of the thing is outside the temporal realm; it is a timeless entity. Thus, upon the Aristotelian account of explanation, particular events are explained by a **timeless entity** that lies **outside the realm of sense experience**.

In contrast, a scientific explanation of individual events is given by a *general pattern that unifies the events*. We unify by pattern, that is, subsumption under a general rule, rather than unify by entity. The scientific explanation, unlike the Aristotelian, requires no entity that lies outside the world of sense experience. To be sure, the pattern that explains is, by virtue of its generality, a *timeless pattern*, one that holds everywhere and everywhen — that, after all, is what is implied by the fact that the pattern is *general*. Note, however, that, while it is a timeless pattern of things or events, all of those things or events are within, rather than without, the realm of sense experience Thus, upon the account of explanation that we have in empirical science, particular events are explained by a **timeless pattern** that is exemplified by entities all of which are **within the realm of sense experience**.

There is one final point about the Aristotelian patterns that should be noted.

An object has the nature it has as a matter of metaphysical necessity. If it loses its nature, it simply ceases to be. Since an object *must* move in conformity with its nature, it makes no sense to say that it *ought* to move in some other way. The nature thus defines how an object *ought* to move, and therefore what, with regard to that object, is *the best*. The nature of a thing thus not only explains why the thing moves as it does but also establishes that it *ought* so to move. As we put it before, the patterns of pre-scientific explanation, the Aristotelian patterns, are not only descriptive and explanatory but also *normative*, defining what is *proper* and *best* for the object.

As we have seen, Copernicus appealed to this form of explanation when he defended the Circularity Principle and when he defended the thesis that the natural place of the sun is at the centre of the universe. As we have also seen, Ptolemy appealed to this form of explanation, when he argued for the Geocentric Principle. Since the natural end of stones is to be at the centre of the universe, the earth cannot move. Copernicus, in order to reply to this latter argument, was forced to deny that being at the centre of the universe was the true or natural end of stones. For Copernicus, the natural motion of all things is circular, and rectilinear motion occurs only when something is as it were out of place.

> …the statement that the motion of a simple body is simple holds true in particular for circular motion, as long as the simple body abides in its natural place and with its whole. For when it is in place, it has none but circular motion, which remains wholly within itself like a body at rest. Rectilinear motion, however, affects things which leave their natural place or are thrust out of it or quit it in any manner whatsoever. Yet nothing is so incompatible with the orderly arrangement of the universe and the design of the totality as something out of place. Therefore rectilinear motion occurs only to things that are in proper condition and are not in complete accord with their nature, when they are separated from their whole and forsake its unity (*On the Revolutions*, p. 17).

As we have seen, the traditional pre-scientific explanation patterns allow that besides the natural motions of things, explained in terms of the natures of those changing things, there are unnatural motions that are contrary to the nature of the changing thing. According to Aristotle, and, following him, Ptolemy, projectile motion is unnatural — or, as Aristotle also called it, "violent." Upon Aristotle's account, there are these tow kinds of change, the natural and the violent or unnatural. Natural change is explained by appeal to the nature of the thing itself; that nature moves the thing to change as it does. Unnatural change is explained by appeal to external forces that restrain the object from moving in its natural way, and impose upon it changes or motions that are contrary to its nature. A cart is a heavy object whose nature it is to gravitate, like the stone, to the centre of the universe. Its horizontal motion along the road is unnatural. It is produced by the power of the horse that is pulling the cart, and causing it to move in this unnatural way. The horse is the external power which, through its strength, can impose upon the cart a form of motion that is contrary to its nature.

Ptolemy, following Aristotle, held that vertical free fall was the natural motion of things like stones. Copernicus argues to the contrary that the natural motion of stones is, like the motion of the earth and the other objects in the system of the world, circular. So the stone that is dropped from the tower participates in the circular motions of the earth, and, moving with the earth, falls at the base of the tower from which it was dropped. As for rectilinear motion, this occurs only when an object like a stone is removed from its natural place which is to be part of the great earthy whole that is our planet. The rectilinear motion of free fall occurs only when there has been a violent change that has removed an object from its natural place of being a part of the earth itself.

This suffices to answer Ptolemy. Or at least, it answers Ptolemy *provided that Copernicus can defend the claim that his rational intuition of the nature of earthy objects such as stones is correct and the rational intuition of Aristotle and Ptolemy as to their nature is erroneous.* In fact, on this point, Copernicus does not argue, he merely asserts. But for that matter, neither do Aristotle and Copernicus argue; they too merely assert.

So long as people accept on the basis of tradition or authority the view of Aristotle and Ptolemy that it is the nature of stones to move towards the centre of the universe, there was no disputing the Geocentric Principle. But Copernicus rejected that tradition, and proposed an alternative natural motion for stones. That enabled him to reject the Geocentric Principle. But it raised deeper methodological question of how we can *know* when our rational intuitions are correct? How do we decide between Aristotle and Copernicus on the natures of stones?

In fact, there was never any answer forthcoming. This was part of the reason that researchers in the early modern period rejected the very idea that one could have an explanation of the sort proposed by Aristotle. They came to reject the very idea of natures as something that was cognitively coherent. The challenge was mounted in detail and very successfully by the first of great practitioners of empirical science, Galileo Galilei.

(d) Galileo's New Science

Galileo was concerned to defend the heliocentric world view that Copernicus had developed. To this end, as well as others, he investigated the behaviour of falling objects and also of projectiles. In these investigations of terrestrial phenomena he sought, as Kepler sought in his investigations of celestial phenomena, to discover mater of fact patterns of behaviour. His standard of judgement was conformity to observational data. At the same time, he rejected any other standard. In particular, he rejected the notion of the Aristotelians that knowledge of the nature of things could provide a standard of truth in determining which hypotheses about the patterns of the behaviour of things are rationally acceptable as laws. After Galileo, there would in effect be no appeal to the natures or essences of things. Such appeals would be rejected as pre-scientific, appeals more to ignorance than to knowledge. Galileo offered philosophical argument against such appeals. At the same time, he so successfully practised the scientific method, was so successful in discovering matter of fact regularities or patterns of behaviour, that the pre-scientific patterns used by the world from Aristotle through Ptolemy and up to Copernicus never recovered their intellectual respectability. In effect, Galileo relegated them to the dustbin of history.

Galileo was concerned to discover scientific explanations of the motions of things like stones near the surface of the earth. He was to use the knowledge that he gained in order to defend the Copernican vision of the universe, and in particular to reply in more adequate fashion than Copernicus had to the argument of Aristotle and Ptolemy that the earth could not move since, if it did, objects would not fall to the ground at a point directly underneath their point of release. Thus, Galileo was concerned about the motion that results when a heavy object is released from a height and allowed to fall freely, without interference, towards the surface of the earth (see Figure 3.11).[11] He was also, as part of his rejection of the sorts of explanations that Aristotle gave, concerned with projectile motion, where an object such

Figure 3.11
FREE FALL

as a spear or a cannon ball is launched with a horizontal velocity and allowed to move horizontally and vertically until it returns to the surface of the earth (see Figure 3.12).

Galileo's interest is in *scientific* explanation. This means that he has a cognitive interest in discovering a generalization that connects, on the one hand, being an object of a certain sort in a certain position, and, on the other hand, the sort of motion that such an object in such a position undergoes. He has a cognitive interest, in other words, in a generalization of the following form:

> Whenever such and such a sort of object is in such and such a sort of position, then it undergoes such and such a sort of motion.

But prior to Galileo this was not the central concern of those who took an interest in the motions of projectiles and objects in free fall.

Prior to Galileo most thinkers held that one explained these two sorts of motions in different ways. Free fall, it was held, was "natural motion" while projectile motion was "unnatural" or "violent." This distinction was based upon the metaphysical "natures" of things.

Upon this view, many motions are explained on the basis of a internal metaphysical principle or "nature." This metaphysical principle moves the object in certain ways. Or, rather, since this metaphysical principle is the very nature or essence of the object, such motion is a consequence of the object moving itself. Motion which flows from the nature of the object moving is "natural" and the pattern describing that motion is a "natural law."

Thus, it is the nature of earthy or massy objects to move vertically towards the centre of the earth (which was, for most, also the centre of the universe). The nature of such objects therefore moves them, when they are unsupported above the surface of the earth, towards the centre of the earth, efficiently in a straight line downwards. The free fall of objects is thus explained as their natural motion.

But not all motion is natural. There is also unnatural motion. This occurs when one object moves a second in a way that is contrary to the natural tendency of the latter. This can happen when one object is naturally more

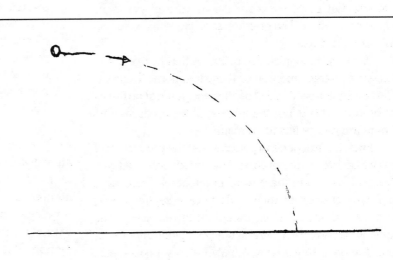

Figure 3.12
PROJECTILE MOTION

powerful than a second. Every object tends to move according to its nature, but a naturally more powerful object can move a naturally less powerful object in a way that is contrary to its natural strivings or tendencies. Thus, for example, a cart is a heavy object. As a heavy or earthy object, its natural motion is the vertical motion of free fall. Horizontal motion is contrary to its nature. But if a horse is hitched to the cart, the latter can be moved by the horse. The horse is, by its nature, more powerful than the cart. It can therefore move the cart in a way that is, for the cart, unnatural. The cart of course naturally resists such motion, to the extent of its natural power. That is why the cart feels heavy when the attempt is made to move it. Indeed, such resistance to unnatural motion may mean that there are some objects that are not, in themselves, powerful enough to move the cart. Perhaps I am not strong enough to move the cart, at least when it is loaded. Otherwise I probably wouldn't use the horse! But in any case, the point is that, upon this view of things, the metaphysical nature of the cart is such that it is more powerful than I, and I therefore cannot force it in to some sort of unnatural motion.

Among the kinds of motion that are unnatural is projectile motion: since it involves a horizontal component, it is not natural motion. But unnatural motion occurs only if there is a second, naturally more powerful object present to impose that motion upon the unnaturally moving object. In the case of projectiles the problem is that once the object is in motion, e.g., the spear has left the hand of the thrower or the ball has left the barrel of the cannon, then there appears to be no external object present that could maintain the moving object in its unnatural motion. The research problem, then, is to find this external force.

We here recognize the central difference between Galileo's interest and that of these traditional theorists. Galileo is interested in discovering the *general patterns* of motion. This is not the interest of his predecessors. There are two points to be made.

First, it is not general patterns that the predecessors were interested in discovering. They were concerned with "natural laws" and these were not general. They were not general because they could have exceptions. Any unnatural motion was an exception to the "law." The latter was therefore a pattern, but not a *general pattern*, one that holds *universally*. What Galileo's predecessors

had to find was that natures of things, and, in the case of unnatural motion, the external force that was causing it.

What Galileo did was reject the idea that the main task of science was to search for these forms that determine natural motion and violent motion, and in particular that the main task for the physicist with regard to projectiles was to discover the external force accounting for the violent motion. As Stillman Drake has argued, Galileo gives up this old task of science (*scientia*) and replaces it with a new task.[12] The old cognitive interest was a metaphysical interest, non-empirical; it is replaced by a new cognitive interest, an interest that can, unlike the old, be fulfilled by empirical research.

In Galileo's dialogue on *Two New Sciences*,[13] Salviati represents Galileo. Sagredo provides wise and intelligent commentary. It is the latter who introduces "the question agitated among philosophers as to the possible cause of acceleration of the natural motion of heavy bodies." He then proposes an answer in terms of an external force impressed upon the object by the substance that initiated the non-natural or violent motion that is contrary to its natural motion as a heavy (or gravitating) object straight downwards.

> ...let us consider that in the heavy body hurled upwards, the force impressed upon it by the thrower is continually diminishing, and that this is the force that drives it upward as long as this remains greater than the contrary force of its heaviness; then when these two [forces] reach equilibrium, the moveable stops rising and passes through a state of rest. Here the impressed impetus is [still] not annihilated, but merely that excess has been consumed that it previously had over the heaviness of the moveable, by which [excess] it prevailed over this [heaviness] and drove [the body] upward (p. 158).

And so on. The traditional philosopher Simplicio immediately raises some objections to this account in terms of active forces. As Stillman Drake points out in his note to this text (p. 158n), this reasoning of Sagredo is precisely that which Galileo himself had used in this earlier thinking concerning free fall and projectile motion. But Galileo's spokesman Salviati, instead of replying to the points raised by Simplicio, now suggests a very different move. He suggests that this old problem simply be abandoned, as a search after fantasies, and that they search instead for a simple description of the patterns in which these objects move.

> The present does not seem to me to be an opportune time to enter into the investigation of the cause of the acceleration of natural motion, concerning which various philosophers have produced various opinions .... Such fantasies ... would have to be examined and resolved, with little gain. For the present, it suffices ... to investigate and demonstrate some attributes of a motion so accelerated (whatever be the cause of its acceleration) that the momenta of its speed go increasing, after its departure from the rest, in that simple ratio with which the continuation of time increases, which is that same as to say that in equal times, equal additions of speed are made. And if it shall be found that the events that then shall have been demonstrated are verified in the motion of naturally falling and accelerated bodies, we may deem that the definition assumed includes that motion of heavy things, and that it is true that their acceleration goes increasing as the time and the duration of the motion increases (pp. 158-9).

What we here see is Galileo changing the question that the physicist was to ask: instead of forces grounded in the metaphysical natures of things, he proposed instead to discover the *exceptionless patterns* or *regularities* of motion. Move, Galileo is saying, from a metaphysical problem that admits of no solution to an empirical problem that can be solved.[14] In fact, as we know, when Galileo gave up the quest for natures and natural forces and searched instead for empirical regularities, he was remarkably successful in discovering regularities in the motions of things. Not only did he discover the precise form of the motion of objects in, as one says anachronistically, "free fall", but he also discovered the regularities that describe the motions of all projectiles. The latter, as he discovered, always moved along curves having the form of parabolas.

In this context, mathematics did not attempt to describe a deeper underlying metaphysical essence as the syllogistic described the deeper ontological structure of the essences of things; rather, mathematics was for Galileo merely a tool to describe the observable motions of things, a tool to record the regularities in a convenient language.[15]

As Stillman Drake has convincingly argued in, for example, his *Galileo: Pioneer Scientist*,[16] Galileo was able to make his great breakthrough with regard to the motions of objects precisely because he gave up the search after metaphysical forces of the sort that Aristotle thought were there. The very notion of what science was up to had changed. But, it is equally clear, not the notion of cause. To insist that science search after regularities *rather than causes*, is not yet to replace the notion of cause as it had come down in the philosophical tradition with the notion of cause as regularity. It is not yet to replace the Aristotelian notion of cause that we find in Aquinas by the notion of cause as regularity that found its first clear statement in Hume. In order for this to happen, the whole notion of Aristotelian causes as active powers, forms, natures, and essences, had to be subjected to criticism and eliminated. Only upon that elimination could the discovery of causes in the sense of regularities be proposed as the goal of science.

The second way to locate the difference between Galileo and his predecessors is by noticing, once again, that the natures of things are *not* empirical entities, known by means of ordinary sense experience. They are, rather, metaphysical entities that cannot be known by sense. They are, instead, known by a rational intuition. There is disagreement as to the metaphysical basis of this intuition. Aristotle held that we arrive at this intuition through an act of abstracting the nature of the thing from its sensible appearances. Others, such as Descartes, followed the line of thought initiated by Plato, and argued that we have a innate knowledge of all the natures of things. In any case, however, all were agreed that it was not by means of sense experience that we grasped the natures of things.

Nor, therefore, could it be by means of sense experience that we could determine which motions are natural and which are unnatural. Nor, further, could sense experience determine when it was necessary to search for an external cause. For such external causes were to be invoked only when unnatural motion was involved.

Galileo, in contrast, is interested in the *general* or *universal patterns* of motion *so far as these can be discovered* in sense experience. Like Herodotus, and like Hippocrates, Galileo insists that *science is naturalistic*. He rejects any attempt to go beyond the world of sense experience to discover metaphysical essences and the natural forces that move things. In searching for general or universal patterns rather than essential natures Galileo changed the cognitive problems that were being asked. In fact, he rejects the language of "natures" as empirically vacuous.

In Galileo's *Dialogue concerning the Two Chief World Systems*[17], Galileo's spokesman, Salviati, challenges his opponent, the traditional philosopher Simplicio to "teach me what it is that moves earthly things downward." To this Simplicio responds,

The cause of this *effect* is well known; everybody is aware that it is gravity.

Salviati immediately retorts:

You are wrong, Simplicio; what you ought to say is that everyone knows that it is called "gravity". What I am asking you for is not the name of the thing, but its essence, of which essence you know not a bit more than you know about the essence of whatever moves the stars around...[W]e do not really understand what principle or what force it is that moves stones downward, any more than we understand what moves them upward after they leave the thrower's hand, or what moves the moon around. We have merely, as I said, assigned to the first the more specific and definite name "gravity," whereas to the second we assign the more general term "impressed force," and to the last-named we give "spirits," either "assisting" or "abiding"; and as the cause of infinite other motions we give "Nature." (pp. 234-5).

Because we in fact have no knowledge of the natures of things it can safely be concluded that we cannot appeal to them in offering explanations of the motions of things.

Disputes involving them are useless. Thus, when Copernicus and his Aristotelian/Ptolemaic opponents disputed over the "true" nature of heavy objects, they were in fact disputing in vain. Copernicus held, in order to defend heliocentrism, that "gravity is nothing but a certain natural desire, which the divine providence of the Creator of all things has implanted in parts, to gather as a unity and a whole by combining in the form of a globe" (*On the Revolutions*, p. 18). His opponents held that gravity was a natural desire to be located at the centre of the universe. There is no way, Galileo is arguing, to resolve the dispute. Best then to leave such disputes aside.

Galileo's claim is that the attribution of a nature to a thing is merely to *re-describe* in a certain way. Recall the Aristotelian formula (D)

For all objects x, x is N if and only if, if x is F then x is G

This is supposed to provide the connection between the observed changes or motions of the object and the nature which purports to explain those changes. Galileo is arguing that "N" just re-names those patterns of change, re-describes them with a new term. Suppose someone asked

Why is Jones an unmarried male?

and the response of Smith was that

Because he is a bachelor.

This would not be an explanation, but merely a *re-description* of the fact that Jones is an unmarried male. For, the statement

For all objects x, x is a bachelor if and only if x is unmarried and x is male

is *true by definition*. In other words, Galileo is arguing that explanations in terms of natures are not explanatory and that (D), while ostensibly a metaphysical necessity, is in fact nothing more than *true by definition*. On Galileo's view, (D) is indeed a necessary truth, because it is true by definition, *ex vi terminorum*, but for that reason is also *vacuous*, and without explanatory power. Molière was later to make the same point when he satirized the Aristotelianism of the scholastics, mocking the idea that one can explain why opium puts one to sleep when it is ingested by appeal to its "dormative power."

Galileo, then, dismissed as word play the notion that one can explain changes and motions by appeal to dispositions and powers. **Dispositions and powers simply do not explain**. He replaced the illusory cognitive goal of the Aristotelians by a more reasonable cognitive goal, that of discovering matter of fact regularities.

This had the consequence that Galileo was able to answer his question, solve his problem, where his predecessors had been unable to solve their's. There had been no agreement upon where the external object was in the case of projectile motion. For that matter, there had been no agreement on exactly what was the essential nature of heavy objects. So much for "rational intuition." Galileo in effect argued that we should put an end to the interminable discussions with regard to the natures of things and, in particular, with regard to exactly what the external force is in the case of projectiles. We should do this by limiting our science to the empirical world. We should restrict our search to explanatory patterns that are wholly naturalistic. As it turned out, this was a re-orientation of problems that was immensely fruitful: it turned out that Galileo could solve his problem, that of giving the general patterns for free fall and projectile motion, where his predecessors had never been able to solve their problems, either about the nature of things or about the nature of the external forces in the case of projectile motion.

Once again we meet a central point about science. As for Herodotus, as for Hippocrates, so for Galileo: **science limits itself to the empirical, aiming to come to know general or universal patterns which will yield naturalistic explanations of the behaviour of things**.

With regard to free fall, Galileo ventured a number of hypotheses before he succeeded in discovering an hypothesis that could pass a predictive test. In particular, he considered three possible hypotheses: that the speed is proportional to the weight of the object, that it is proportional to the distance fallen, and that it is proportional to the time fallen. These would be

$$v = g\,w$$
$$v = g\,s$$
$$v = g\,t$$

where v is the velocity or speed, w is the weight, s is the distance, t is the time, and g is a constant of proportionality. In the first case, g is the velocity acquired by an object of unit weight; in the second hypothesis, g would be the velocity acquired by an object moving through a unit distance; in the third, g would be the velocity acquired by an object in a unit of time, say 1 second. Each of these hypotheses had to be tested. The false would then be rejected as not describing the order or general pattern of the motion of objects in free fall. Conversely, if an hypothesis succeeded in passing a predictive test it would be accepted as worthy of use in scientific explanations.

As for the first, the story — perhaps apocryphal — is that Galileo tested this hypothesis by dropping objects of different weights from the leaning Tower of Pisa. Many of his predecessors had argued that, as the weight we feel when we hold an object in our hand is due to its natural striving to return towards the centre of the earth (or the centre of the universe), a heavier object is naturally striving harder and will, therefore, through its exercise of its greater power, move towards the earth at a greater speed than will an object that weighs less. The test consisted of dropping objects of different weights from the top of the leaning Tower. If, as the hypothesis says, the velocity of the fall increases with the weight, then the heavier object will arrive first. In fact, it turns out that objects of different weights arrive at the same time. What is predicted by the first hypothesis turns out to be false. The hypothesis itself has thus failed predictive testing and is itself false. It is therefore not worthy of acceptance for purposes of explanation and prediction.

As for the second hypothesis, Galileo argued — though in this he was in fact mistaken, as we now know — that it had the consequence that the falling object should travel *instantaneously* through a portion of the path traversed. He also argued that this was impossible, and therefore rejected the hypothesis by a *reductio* argument: since the hypothesis had a consequence that is impossible and therefore false, it follows that the hypothesis itself must be rejected as false, and that is therefore not worthy of acceptance for purposes of explanation and prediction.

Finally, the third hypothesis: this asserts that the increase of velocity is the same for each unit of time. Since the increase of velocity per unit time is the acceleration of the object, it follows that this third hypothesis can equivalently stated as the hypothesis that the acceleration of an object in free fall is constant.

The difficulty is that this hypothesis cannot be subjected directly to a test: there is no instrument or technique available by means of which one can directly measure either acceleration or velocity. At this point, a knowledge of logic and mathematics came to his assistance. He was able to show that the hypothesis had a *logically equivalent* form:

$$s = \tfrac{1}{2}\left(g\, t^2\right)$$

which states that the distance fallen is proportional to the square of the time fallen. To say that the two forms are "logically equivalent" is to say that *each entails the other*. Now, if P entails Q, then if P is true, Q must also be true. Hence, if P and Q are logically equivalent, the one is true if and only if the other is true. Hence, since the two forms of Galileo's hypothesis are logically equivalent, if we confirm one to be true we have *ipso facto* confirmed the other as true. As it turns out, it is possible to confirm experimentally the second form of the hypothesis concerning distances. Thus, an object that falls for 2 seconds travels four times as far as an object that travels one second, a body that travels for 3 seconds nine times as far, and so on. These times and distances can be measured. When they are, we discover that they are related in just the way that the hypothesis predicts that they are related. The hypothesis about distances and times has thus passed the predictive test for acceptance as worthy for use in explanation. But any evidence which tells in its favour has also to tell in favour of its logically equivalent form about velocities. Hence, that hypothesis, too, is also acceptable as worthy for us in explanation.

Galileo has thus found a regularity of the sort of general pattern in which he was interested. The general form, as we noted, was

Whenever such and such a sort of object is in such and such a sort of position, then it undergoes such and such a sort of motion.

What he has discovered is that

Whenever a heavy object is released in free fall, then it moves in a straight line towards the surface of the earth at constant acceleration.

In discovering this pattern Galileo discovered something that had eluded his predecessors. But, then, they were looking in the wrong place. They were searching to understand things in terms of their metaphysical natures. Galileo shifted the question: he sought to understand the motion of things in terms of general patterns. These one finds by means of empirical tests, not by looking for the essential natures and forces of metaphysics.

Galileo's method illustrates a point that one can also see from the attempts of Herodotus to explain the rising of the Nile. Like Herodotus, Galileo had a cognitive problem. Like Herodotus, Galileo's problem concerned the matter-of-empirical-fact connections among things: Galileo wanted to know, what he did not know, to wit, the general pattern to which the motion of objects in free fall conform. In order to discover this, he had to *select an hypothesis*. This hypothesis had to satisfy the condition of being an *empirical* hypothesis; it could not be either a metaphysical proposition, or a proposition based on religion or poetry, purporting to make assertions about what lies outside or beyond the world that we know by means of our sense experience. It must be a proposition such that either it, or some other proposition that can be deduced from it, can be *put to the test*. Moreover, of course, the hypothesis that is selected for testing must provide a possible answer to the question that is moving the inquirer. An hypothesis can satisfy this condition even though it turns out in the end to be false, that is, even though it turns out in the end that either it, or other propositions that can be logically deduced from it, turn out to fail predictive tests. In short, *scientific research must be guided by an hypothesis about the connection one has a cognitive interest in discovering. It is this hypothesis, or consequences that can be deduced from it, that the researcher puts to the test.*

Nonetheless, the hypothesis *must be selected*. For any event in which one is interested, there are in fact innumerable *possible hypotheses*. Thus, for example, in the case of Herodotus, there are many facts about the Nile or about Africa or about Egypt that are *possibly* connected to the rising of the Nile. There is, for example, the fact that the Nile flows through a desert. There is the fact that it flows northwards. There is the fact that the water is coloured brown. There is the fact that the Egyptians offered prayers as the summer solstice approached for the Nile to rise as it had in the past. And so on. Similarly, for falling bodies there are again many possible hypotheses. Perhaps they fall with a speed related to their colour, or to their temperature. Perhaps their rate of fall varies with the latitude. And so on. In each case, and in fact with regard to any cognitive problem about the matter of fact connections among things, there is always an indefinitely large group of possible hypotheses. To undertake any reasonable programme of research, one must select, from among the hypotheses in this indefinitely large range, only a small range for actual testing. Otherwise one's task would simply be impossible: one cannot put an indefinitely large number of propositions to the test.

In order to select a range of hypotheses for testing, one must begin with *background knowledge* about what are *possibly relevant* factors. Herodotus could dismiss the number of prayers or the colour of the Nile or the direction of its flow as very likely irrelevant. In contrast, he knew on the basis of his background knowledge that the melting of snow can often be related to the volume of water in rivers. Similarly, Galileo knew on the basis of his background knowledge that the colour and temperature of an object is irrelevant to the speed of its fall. In contrast, time, distance and weight could not be dismissed as irrelevant. To say that a factor is irrelevant is, of course, to say that one's background knowledge makes it unlikely in the extreme that there is any connection between this factor and the factor in which one is interested. It is unlikely, in other words, that there is any general pattern that connects them. For any suggested pattern linking these irrelevant factors with the one in which is interested, there are other facts that imply the falsity of the supposed pattern. For this reason, our background knowledge allows us eliminate these factors as irrelevant, as factors for which we know, or at least reckon as very unlikely, that there is no general

pattern proving the sort of connection which we are interested in discovering. At the same time, background knowledge will locate another, smaller, set of factors that are not eliminated as irrelevant. It is these that will feature in the hypotheses that the researcher will actually put to the test. Thus, *in research background knowledge picks out a range of hypotheses as hypotheses that are* LIKELY *to be true and therefore worthy of being put to the test*. It is, of course, the one hypothesis (if any) that survives testing that one accepts as true and as worthy of use in explanations.

We have here distinguished between those hypotheses which are worthy of being put to the test and that hypothesis which is worthy of use in explanations. The former hypotheses have been picked out by our background knowledge as relevant. That makes it reasonable to put these hypotheses and not any others to the test. But until they have actually been tested and, moreover, until they have survived such testing, they cannot yet be counted as worthy of use in explanations. For until we have an hypothesis that has survived testing, the *epistemic* condition that any purported explanation must fulfil will not in fact be fulfilled. This is the *epistemic condition*, discussed above, of having *successfully survived predictive tests*.

Galileo's law for free fall, that

$$v = g\,t$$

can be re-formulated in yet another logically equivalent way, to wit, as

$$a = g$$

where "*a*" is the acceleration of the object. That is, Galileo's law for free fall states that for objects near the surface of the earth, the acceleration is a constant, *g*. Galileo generalized this notion. He made the inductive inference that, if unsupported freely falling objects at this height from the earth had the acceleration *g*, then *any object in the same circumstances* would have the same acceleration. That is, he generalized to the hypothesis that

(FB)   Any object in the circumstance of being unsupported near the surface of the earth has acceleration *g* vertically towards the earth

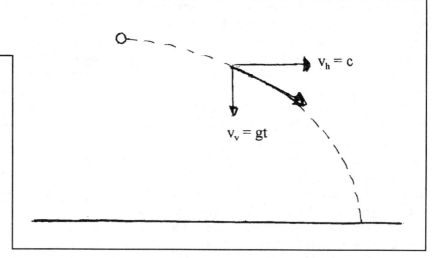

Figure 3.13
PROJECTILE MOTION

In particular, it did not matter if the object had velocity zero, i.e., was starting from rest, or was moving upward, or downward, or even was moving with a motion having a horizontal component, i.e., was a projectile. At a given position not too far from the surface of the earth, the velocity which an object has may vary. What may not vary is the acceleration. Acceleration, not velocity, is regularly connected to the circumstance.

With this hypothesis ready to hand, Galileo reasoned that when a projectile is released, it has two components of motion, a horizontal component and a vertical component (see Figure 3.13).

Galileo reasoned that the acceleration of the projectile moving horizontally was the same as the acceleration of an object released from rest at the same point: it is acceleration that is connected to the circumstance the object is in, and not the velocity. Since the acceleration of an object undergoing free fall is g in the vertically downward direction, the projectile will have the same acceleration. To be sure, its velocity will be different, since its motion has a horizontal component. But its acceleration will be the same. In particular, all the acceleration will be vertically downward and of amount g. There will be no horizontal component to the acceleration. Therefore the horizontal component of the velocity will be constant:

$$v_h = c$$

The vertical component conforms to the law of free fall:

$$v_v = g\, t$$

As is well known, these two equations jointly describe a *parabola*, and, if c and g are known then it describes a very specific parabola. Thus, the path which any projectile traverses is a parabola, and a particular projectile

released under specific circumstances will follow a specific parabola determined by those circumstances. Galileo has again found the sort of general pattern in which he was interested. The general form, as we noted, was

> Whenever such and such a sort of object is in such and such a sort of position, then it undergoes such and such a sort of motion.

What he has now discovered is that

> Whenever a heavy object is launched as a projectile, then it moves in a parabola.

And in fact, as we just remarked, he discovered a much more specific pattern than this, since he can describe the particular parabola the projectile follows in terms of the initial conditions of the projectile, that is, the specific direction in which it was originally launched.

This enabled Galileo to provide a reply to the arguments of Aristotle and Ptolemy for the immobility of the earth. The stone that is released from the top of a tower will move like all objects in conformity with the law of free fall. But it will also partake of the forward motion of the earth; this horizontal component remains unchanged, in conformity with the law (FB). The earth has moved, but, by this law, the stone will have moved the same distance. The dropped stone will therefore land at the base of the tower, directly below the point at which it was released.

In discovering this pattern for projectile motion Galileo again discovered something that had eluded his predecessors. But, then, they were looking in the wrong place. They were searching to understand things in terms of their metaphysical natures. They sought to understand the motion by searching for an external force that moved the object contrary to its nature. Galileo shifted the question: he sought to understand the motion of things in terms of general patterns. These one finds by means of empirical tests, not by looking for the essential natures and forces of metaphysics. There is no distinction in the observed behaviour of things between natural and unnatural motion. This distinction makes sense only in a context in which one allows that there are non-empirical natures of things that can be known by some means, e.g., rational intuition, that goes beyond our sense experience of the world and penetrates to its essential metaphysical core. It is this metaphysical distinction that dictated the search for an external cause for the continuance of motion of the projectile in a non-vertical direction. Give up the metaphysical natures and the corresponding metaphysical distinction between natural and unnatural motion, there will be no necessity to search for the external cause. One will be left free, as Galileo was, to search for the relevant pattern, and in particular not demand that the explanation refer to the continuous effects of an external object. If the pattern does not require an external object, it is for all that a general pattern, and therefore a connection that can provide a scientific explanation.

(e) Newton's Perfection of the New Science

Galileo's law (FB) for objects moving near the surface of the earth,

Any object in the circumstance of being unsupported near the surface of the earth has acceleration *g* vertically towards the earth

applies to *one kind* of mechanical system. It states a regularity about such systems. For such mechanical systems, it states that *acceleration is uniformly connected* to certain *circumstances*.

This law that Galileo discovered and confirmed applies to one *species* of the *genus* mechanical system. Later philosopher-scientists, beginning with Gassendi, were to generalize this law. Roughly, this generalization states that

(IN)   For any sort of mechanical system, for any object in the system, and for any acceleration which that object has, there are circumstances in the system which are necessary and sufficient for that object to have that acceleration.

This law, it should be noted, first, is *generic*: in contrast to (FB), which applies to a single species of mechanical system, (IN) applies not just to one species of mechanical system but to all such species. In addition, it should be noted, second, where (FB) states specifically what the circumstances are to which the acceleration is regularly connected, (IN) merely states that **there are** such circumstances. That is, (IN) states that *there is a specific law* for any species of mechanical system, but, unlike (FB), does not *state specifically* what law it is that is there. Since there are different laws for different species of mechanical system, (IN) could hardly apply to all mechanical systems, if it ascribed to each the same specific law that (FB) ascribes to the specific sort of mechanical system to which it applies. (IN) can generalized from (FB) only if it *abstracts from* some of the specific form of (FB). The law (IN) generalizes to all species of a certain genus by becoming *generic* and *abstracting from* the specific features of those laws, like (FB), which are its instances.

(IN) is, of course, the Law of Inertia. The specific sorts of circumstances which it mentions are, basically, the masses, (relative) positions and (relative) velocities of the objects in the system. In some circumstances — those that we might in certain contexts be called "normal" — the acceleration is zero and the object continues in its state of rest or of uniform motion in a straight line. In other circumstances the object will have non-zero acceleration, and will be moved, as one says, in a non-inertial way. In any case, however, what the Law of Inertia asserts is that for any acceleration, zero or non-zero, *there are* specific circumstances to which that acceleration is uniformly related.

The Law of Inertia does *not* assert *specifically* what the circumstances to which a *specific* acceleration is regularly connected. But it does assert that for any specific acceleration, *there are* circumstances of the required sort, that is, that *there is a general pattern*. It is a generalization, making an assertion about all mechanical systems. But this is a *generic description*. It says something about *every specific sort* of mechanical system. In particular, it asserts something about the *form of the pattern* that will be found in any specific sort of mechanical system. It asserts that the pattern will be one in which the acceleration of the object is uniformly related to specific circumstances falling within the generic description of masses, positions and velocities. *The Law of Inertia is thus a* **pattern about patterns**, or, if you wish, a **law about laws**.

Laws of this sort, laws that are laws about laws, form a very important part of the background knowledge that is used to guide research in science.

Suppose, let us say, that we see an object accelerating, for example, a hockey puck leaving a hockey stick, and that we form a cognitive interest in coming to know the pattern that relates that acceleration to the circumstances. The Law of Inertia asserts that *there is* such a connection, *there to be discovered*. In giving a *generic description* of the circumstances, the Law of Inertia locates the *sorts* of things that we must examine as we endeavour to find the connection. At the same time, of course, it declares other sorts of factor to be irrelevant. The Law of Inertia provides a guide for forming hypotheses that are worthy of putting to the test.

The Law of Inertia was used by Newton to guide his research. It was his First Law of Motion. In his other two Laws of Motion he placed further restrictions on the *form of the patterns* that one finds relating accelerations to circumstances. The Second Law states that the patterns connecting accelerations to circumstances can be described in terms of forces, where a force upon an object is measured by the product of its mass and acceleration. The Third Law states that for *every* force there is an equal and opposite force. We shall more to say about this structure of laws in mechanics directly, but for now this suffices. The central point is that these further Laws of Motion are additional patterns about patterns, laws about laws, that provide a guide for research.

Newton made the Law of Inertia the first of his *axioms* for classical mechanics, the first of the basic principles of this science. But Newton proposed further axioms. These further axioms, like the Law of Inertia, apply to *all species* within the *genus* mechanical system. They too are laws about laws, placing generic restrictions upon the forms exemplified by specific laws for specific sorts of mechanical system.

Newton noted the further constraint that it is not just acceleration but more specifically the product of mass (m) times acceleration (a) that is uniformly related to circumstances. This law, Newton's Second Law of Motion, states that

(SL)   For any sort of mechanical system, for any object in the system, and for any value of $m \times a$ which that object has, there are circumstances in the system which are necessary and sufficient for that object to have that value of $m \times a$.

Here, mass m is a constant which is specified for any object in a mechanical system. Since the product $m \times a$ is related to circumstances, the same circumstances will give a greater acceleration a if the mass m is less, and, conversely, a smaller acceleration a the greater the mass m. Thus, the mass m measures as it were the resistance of an object to the acceleration that is induced on it by the circumstances in which it is located.

Because "$m \times a$" is the state of the object that is regularly connected to circumstances, it may be described as the state that is *required by* the circumstances in which the object is located; it is the state that those circumstances as it were induce upon the object. Newton therefore thought of this characteristic of things, the product

$$m \times a$$

as a measure of the "force" which the circumstances exert on the object. That is, in effect Newton *defines* "force" to be this product:

(SL*) $F = m \times a$

In fact, this is the usual simple formulation of the Second Law of motion, the law that we have called (SL). Although this formulation (SL*) is said to be a "law", it is in effect a *definition* of the notion of "force" within the classical mechanics that Newton developed. Insofar as something like (SL*) is used to state a *law* of motion, then it states

(SL**) For any sort of mechanical system, for any object in the system, $m \times a$ for that object varies directly with the circumstances in which that object is located.

Note about this formulation, as about (SN) itself, the law is generic, and abstracts from specific forms: it asserts that *there is* a specific law for any sort of mechanical system and that this law must fulfill certain generic constraints, but it does not say *specifically* what that law is that it asserts to be there.

It is important to note that, unlike the Aristotelian Natures, there is nothing extra-empirical about Newtonian forces. To the contrary, the notion of force is given a full empirical explication in the Second Law.[18] The Aristotelian notion that forces are somehow active, on the model of human volition, becomes a mere metaphor in the new science developed by Newton. But we can understand how that metaphor lingers on. Mass is a measure of the extent to which objects "*resist*" being accelerated by the circumstances in which they find themselves. This resistance is "*overcome*", however, by the circumstances, which "*induce*" upon it, or even "*force*" upon it, a certain acceleration. To say that the circumstances induce a certain acceleration given the mass is to say nothing more than that those circumstances are necessary and sufficient for that acceleration given the mass. But the volitional metaphor comes naturally to our thought, and with it, alas! the Aristotelian patterns. On the whole in physics, those patterns are nothing more than dead metaphors. But in talk about traditional mechanics, those metaphors sometimes do a job that is more than metaphorical.

Certainly, in the use to which Newton put his axioms, the Aristotelian or volitional notion of force play no role whatsoever.

Before discussing the use to which Newton put his theory of mechanics, we should just add, at this point, for the sake of completeness, his Third Axiom. This states that

In any two body mechanical system, the forces are such that $(m_1 \times a_1) = - (m_2 \times a_2)$

where '1' and '2' are the objects in the system.

We should also note that there is an additional axiom, the Law for the Vector Addition of Forces, which enables one to deduce the laws for many body systems given that we have laws of the conceptually distinguished two body systems that they contain. We shall return this latter law in due course.

At this point, however, what we need to note is that the logical forms of the laws that formed the set of axioms with which Newton worked, that is, their *generic abstract forms*, or, what is the same, the fact that they are laws about laws, enabled Newton to complete his great achievement, that, as he put it in his *Principia Mathematica*, of "demonstrating the frame of the system of the world."

In this task, Newton was able to use the laws that Kepler had discovered concerning the motions of the planets.[19] Kepler had discovered that planets move in ellipses about the sun, with the sun at one focus. This is his first law. He discovered further that, as the planet moves along its orbit, the line joining it to the sun sweeps out equal areas in equal times. This is his second law. The third law states a regular relation between the distance of the planet from the sun and the length of time that it takes to complete its orbit. Newton was able to use his three laws of motion and Kepler's first two laws to *deduce* that the force that is acting upon a planet is inversely proportional to the square to the distance; while using his own three laws and Kepler's third law he was able to deduce that the force that is acting upon a planet is directly proportional to the product of the masses of the sun and the planet. In other words, we have

$$F = G \times [m_s \times m_p] / r_{s,p}^2$$

where m_s and m_p are the masses of the sun and the planet respectively, $r_{s,p}$ is the distance between them, and G is a constant of proportionality. Thus, given, **one**, the background theory constituted by the Three Laws of Motion, and, **two**, the *observationally confirmed Kepler's laws for planetary motion*, Newton can *deduce* that the force that describes the connection between acceleration and circumstances in the planetary system is the gravitational force.

Using this law Newton could go on to provide, in principle at least, a complete description of the patterns governing the motions of the planets. This law enabled him both to explain those motions and to predict the future and past orbits that the planets followed. This by itself was a magnificent achievement, about which we shall have rather more to say later.

There is a moral to be drawn — we have drawn it before, but it is worthy of emphasis:

A scientific theory that contains laws about laws constitutes a powerful guide for research aimed at discovering specific explanatory laws.

Another of the great triumphs of Newtonian theory is worth noting here.

Prior to the new science, comets had been thought to be atmospheric phenomena. This was defended on the authority of Aristotle. More particularly, they were thought to portend great events such as the coming of plagues and the downfall of princes. And so consider the fact that a comet appeared at the time William the Conqueror won the English crown by defeating and killing the Saxon King Harold at the Battle of Hastings in 1066. When some Norman citizens came to make the great Bayeaux Tapestry, portraying William's triumph, they included in their design an image of the comet. Kepler's teacher Tycho Brahe, the great astronomical observer, had been skilful enough in his observations to establish that comets were objects that moved among the planets, that they were not atmospheric phenomena. This dealt a blow to the Aristotelian notion that the spaces between the planets were filled with solid crystalline spheres. It challenging this, and the notion that comets were sub-lunar phenomena, Tycho's observations dealt a blow to the authority of Aristotle. Nonetheless, comets continued to be held to be divine messengers, presaging great events.

It occurred to the Newtonians, however, that these celestial phenomena had naturalistic explanations, and that all that had to be done was to find the patterns that governed their orbits. But the theory that Newton had developed gave them the pattern! The path about the sun had to be either an ellipse, or, if it were very elongated, then either parabolic or hyperbolic in form.

When a comet appeared, Edmund Halley made a number of observations of its orbit, and used Newton's theory to calculate the unobserved parts. The orbit was elliptical, and Halley predicted that the comet would return in 76 years. When the comet did return at the predicted time, within very reasonable limits of error, and in the predicted place, it was considered an event that *very strongly confirmed Newtonian theory.*

We have said that laws and theories must be predictively successful if they are to be used in science for purposes of explanation and prediction. A regularity, to be considered worthy of acceptance for use in explanations, must have *successfully passed predictive tests.* We can now strengthen this point. For those who did not accept Newton's theory, the return of Halley's comet was completely unexpected. Rather, they were irregular events, sent by the deity to foretell important events. The return was to be expected only given Newton's theory. But return it did. The confirmation was therefore very important. The point seems to be that

if a theory passes a predictive test where the event predicted would be unexpected in case the theory is false, then the confirmation is much stronger than if the predicted event were expected even if the theory is false.

The rationale for this norm is fairly simple. It is in terms of the falsifiability of hypotheses. An hypothesis makes a prediction. That prediction puts it to the test: if things turn out as predicted then the hypothesis is confirmed; if things do not turn out that way then the hypothesis is falsified and thereupon rejected. If we expect the event predicted in any case, then not much is at risk when we test the hypothesis. But if the event predicted is otherwise unexpected, then the hypothesis is much more at risk. An hypothesis, if true, places constraints as it were on the world. These constraints are weaker when what it predicts is in any case expected. In that sense, the hypothesis is more inexact. And as we have seen, the more inexact what is predicted, the weaker is the confirmation. The point is, there is more at risk when what is predicted is otherwise unexpected. We may therefore say that an unexpected confirmer confirms more strongly than one that is otherwise expected. The norm is clear:

one ought to accept a theory as worthy of acceptance for purposes of explanation just in case that theory has yielded a successful prediction of an event that would be unexpected if the theory were false.

Another point merits our attention. Having discovered the orbit of what came to be known as Halley's Comet, astronomers could trace the motions of the comet back into the past. As it turns out, the comet that appeared at the time of the conquest of England by William the Norman, and was portrayed on the Bayeaux Tapestry commemorating William's victory, was Halley's Comet. We have thus explained the event recorded on the Bayeaux tapestry in naturalistic terms.

In one sense, this is the same sort of thing that Velikovsky and von Däniken do when they insist that they want to explain the events recorded in the legends, traditions, and myths of humankind. **Except!** Except that they take these stories as data *from which* to infer the theory. In contrast, the procedure of the astronomers is to take *observational data*, none of which is legendary, mythical, or traditional, and construct their theory on the basis of these data. They then use *this* theory, constructed on the basis of observational data available to them, to *interpret* the things recorded in — some — legends, traditions, and myths. Which things as so interpreted? Just those that the theory can be extended to cover. The remainder are left to be the stuff of legends, traditions, and myths — stories that have no place as data used to support the theory one is developing.

Science does not mind attempting to explain the stuff of legends; it must resist the attempt to construe that stuff as data.

Andreas Osiander was a colleague of Copernicus who wrote added to the latter's *De revolutionibus* (*On the Revolutions*) a "Preface" to the reader. A difficulty in the way of accepting Copernicus' work lay in the fact that, if it did assign a motion to the earth, then it was in conflict with the literal truth of the statement in the *Bible* that when Joshua was assaulting the city of Jericho his victory was assured when God miraculously lengthened the day by making the sun stand still. Clearly, the Lord could not make the sun stand still if it was the earth and not the sun that was moving. Copernicus dismissed the

charge that his work was in conflict with the Bible; what he was doing was physics, not religion. He points out that "...it is not unknown that Lactantius, otherwise an illustrious writer but hardly an astronomer, speaks quite childishly about the earth's shape, when he mocks those who declare that the earth has the form of a globe" ("Preface" to the *De revolutionibus*, p. 5). Nonetheless, there was real danger that Copernicus' work would be rejected on a theological grounds, by an appeal to Scripture. So, to protect Copernicus for the charge of heresy, Osiander argued that Copernicus' "...hypotheses need not be true nor even probable. On the contrary, if they provide a calculus consistent with the observations, that alone is enough" (*De revolutionibus*, p. xvi). This charge of heresy was later raised against Galileo when he took up the defence of heliocentrism. In fact, Galileo himself raised the issue, when he has the spokesman of the opposition, Simplicio, state that heliocentrism is absurd because it implies "That when Joshua commanded the sun to stand still, the earth stood still..." (*Dialogue concerning the Two Chief World Systems*, p. 357). Galileo's spokesman, Salviati, responds by separating the Holy Book from physics and astronomy: "...let us, for our part, revere it [the Holy Writ], and pass on to physical and human arguments" (p. 358). The test of truth in the area of physics and astronomy is empirical, based on the human means of sense experience, and not theological, based on faith. This distinction was not enough to save Galileo from being condemned by the Inquisition, and being forced to abjure what was declared to be the heresy of heliocentrism. The Church was, of course, later forced to abjure its condemnation of Galileo. Galileo himself remarked, after his condemnation, in a Note added to the preliminary leaves of his own copy of the *Dialogue*:

> Take note, theologians, that in your desire to make matters of faith out of propositions relating to the fixity of sun and earth you run the risk of eventually having to condemn as heretics those who would declare the earth to stand still and the sun to change position — eventually, I say, at such a time as it might be physically or logically proved that the earth moves and the sun stands still.[20]

In any case, it is clear that Osiander is wrong in his suggestion that Copernicus regarded his hypotheses as mere calculating devices. To the contrary, it is perfectly clear that Copernicus regard his claims about the orbits of the planets to be a correct description of their motions. Why else would he have included arguments designed to refute the defence that Ptolemy offered for the earth being stationary?

Copernicus emphasizes hypotheses, to be accepted as true, must yield truth. In his "Preface" to the *De revolutionibus*, he remarks that

> ...in the process of demonstration, or "method" as it is called, ... [those who defend the geocentric hypothesis] are found either to have omitted something essential or to have admitted something extraneous and wholly irrelevant. This would not have happened to them, had they followed sound principles. For if the hypotheses assumed by them were not false, everything which follows from their hypotheses would be confirmed beyond any doubt (p. 4).

If the hypothesis is true, then it will have only true consequences; in particular, its predictions will all be correct. And conversely, if the predictions turn out to be false, then the hypothesis is false.

Pierre Duhem took much the same line as Osiander: Copernicus did not have sufficient grounds for accepting his hypotheses as true, and therefore those hypotheses had to be understood as mere calculating devices.[21] According to Duhem, hypotheses about the universe can be accepted as true only if, first, they yield correct predictions ("save the phenomena"), and, second, it can be shown that no contrary or competing hypothesis could also do the job.

> To prove that an astronomical hypothesis conforms to the nature of things, it is necessary to prove not only that the hypothesis is sufficient to save the phenomena, but also that these phenomena could not be saved if the hypothesis were abandoned or modified.[22]

It is certainly true that Copernicus was unable to demonstrate that his was *the only* system of hypotheses that could account for the observed motions of the sun, moon and planets. To that extent it is true that he lacked *conclusive* proof that his system was true. Nonetheless, the conclusion does not follow that the system must therefore be nothing more than a mere calculating device. There is an alternative, namely, that the hypothesis be accepted as *probably true*, that is, accepted as the best hypothesis currently available, but accepted with the recognition that new data might be located that require its rejection or modification or that a better hypothesis might yet be discovered.

Duhem's conclusion that Copernicus could treat his hypotheses as nothing more than calculating devices follows only because of the cognitive norm that he proposes. This norm states that nothing can be accepted as true unless it is absolutely certain, that is, unless it is absolutely certain that no contrary data or better formulation will ever be discovered. But surely this norm is *unreasonable*. If it were to be adopted, no hypothesis would ever satisfy our cognitive interests. For all hypotheses are matter of fact generalizations; they make a claim about all members of a population. But all we ever observe is a sample. The data are therefore never conclusive. So Duhem's norm would imply that no scientific hypothesis be accepted as true, not even tentatively. This undoubtedly *is unreasonable*; certainly, Duhem gives us no reason for accepting his norm as one to guide us as we attempt to satisfy our cognitive interests. Not only is Duhem's proposal unreasonable, it is quite contrary to the practice of science.

Our cognitive interests lead us to desire the discovery of a true hypothesis. But the data may only tell in favour our hypothesis being true without confirming it conclusively as true. It does not follow, contrary to Duhem, that we must simply reject such an hypothesis. To the contrary, we can quite reasonably accept it, provided only that in so accepting it we recognize that we may have to replace it with another hypothesis if fresh data, or improved theorizing require it. The appropriate *cognitive norm*, one more reasonable than that proposed by Duhem, is this:

An hypothesis may be accepted as true even if the data do not make it certain provided that the acceptance is tentative and open to revision.

It is true that Copernicus did not prove that only his hypothesis could "save the appearances." Sometimes, however, one can do just that. In fact, Newton in effect was able to do this. In this respect he was able to make his induction much more secure than that of Copernicus.

What Newton was able to demonstrate that *if the planets move in ellipses then the force that moves them MUST be as the inverse square of the distance from the sun*.[23] In this demonstration, Newton was able to *eliminate all competing hypotheses*. Duhem is correct in his suggestion that, if this can be done, then the acceptance of the hypothesis as true is made more secure. In other words, the rule that

An induction in which all competing hypotheses but one are eliminated testifies strongly to the truth of the uneliminated hypothesis.

Still, it does not eliminate all inductive uncertainty. The absolute certainty after which Duhem aspires is not to be found. For, Newton's deduction goes through only given the axioms of mechanics, in effect, Newton's Three Laws of Motion, as premises. And these are empirical, matter of fact generalizations. Since these needed premises are about all mechanical systems, about a population, where all we have observed is a sample, it follows that their acceptance can only be tentative. And if they are tentative, it follows that anything that can be deduced from them must also be tentative.

Still, those axioms, the theory that Newton used, is itself confirmed. When the Law of Gravity is confirmed, then the data that confirm that Law in turn confirm the theory which supports that Law. Thus, the confirmation of the predicted orbit of Halley's Comet was a great triumph for Newton's theory because not only was the orbit confirmed but so was the theory that led to its discovery. The confirmation of specific laws such as that describing the orbit of Halley's Comet also serves to confirm the theory, or law about laws, that led to the discovery of that law. Thus, the confirmation of the path ascribed by Newton's theory to Halley's Comet in turn confirms the theory of gravitation that predicted that law. But the theory of gravitation it itself an instance of the Law of Inertia and the other two Newtonian axioms of mechanics. So the confirmation of the path of Halley's

Comet serves to strengthen the confirmation of Newton's three Laws of Motion. The point is general and is worth emphasizing:

the confirmation of a specific law also confirms the generic theory or law about laws that led to its discovery.

The generic theory is antecedently confirmed. By virtue of the fact that the theory has been confirmed, it makes probable, worthy of test, an hypothesis at the specific level. That is, prior to testing the specific hypothesis is made probable, worthy of testing, by the background generic theory. Then, when the specific law is confirmed in testing, those confirming data also confirm the generic theory that provided it with its prior probability. We have seen these interlocking patterns of confirmation at work in the case of Newton's theory and Halley's Comet. We shall see later that they are also at work in Darwin's theory of evolution by natural selection.

If a theory is confirmed by certain data, that is, if those data testify to its truth, then those same data testify to the falsity of any contrary theory. Thus, for example, the data that testify to the truth of Newton's theory testify to the falsity of a number of Velikovsky's hypotheses. One of the immediate corollaries of Newton's Three Laws of Motion is the law that momentum is conserved. This includes angular momentum. A spinning grindstone has much angular momentum. Bringing it to a sudden stop requires considerable force. Anyone who has let the spinning handle and crank of a grindstone smack into one's forearm, bringing the spinning to a halt knows that it gives one a considerable crack. That crack represents the force that is need to bring the spinning grindstone to a halt. Well, the earth is like the grindstone a flywheel, only much larger, and has considerable angular momentum indeed. Velikovsky has among his various hypotheses, the hypothesis that the earth came to a sudden halt in its spinning one time in its history. This occurred at the time when Joshua was fighting the battle of Jericho, and is supposed to account for the Biblical claim that during the battle the sun stood still. This stoppage was due to a close encounter with Venus. Velikovsky needs a tremendous force to bring the earth to a halt. He vaguely refers to electrical or magnetic forces from Venus could do the trick, but any such forces would simply be too weak to do the job. This was then followed, according to Velikovsky, by the earth starting up again. So another force was needed to put it back in motion, equal to the stopping force but in the opposite direction. He suggests that the sun's magnetic field could do this. But it is not strong enough. Other things could be said. For example, much of the angular momentum that was lost in the stoppage would be transformed into thermal energy, which would in turn cause the earth to heat up, causing many fires, bringing about considerable soot in the air, so that the heating up would have been followed by a massive cooling when the sunlight could not get through. Or again, if the sun's magnetic field could cause the earth to start rotating once it had stopped, this force would continue to act, and would therefore bring about a regular acceleration of the earth's speed of rotation. We would be spinning faster and faster. But that is clearly not happening. The point to be emphasized, however, is that Velikovsky's hypotheses are clearly inconsistent with the Law of Conservation of Angular Momentum. The latter is entailed by Newton's theory. Thus, any evidence which tells in favour of Newton's theory tells against Velikovsky's. Reasons for accepting Newton's theory for purposes of explanation and prediction are reasons for rejecting Velikovsky's.[24]

Conversely, of course, reasons for accepting Velikovsky's theory for purposes of explanation and prediction are reasons for rejecting Newton's. The issue then is this: which theory is more strongly supported?

In fact, Newton's theory has been strongly and repeatedly confirmed. In contrast, there is very little direct confirmation of Velikovsky's theory. What little plausibility it has derives from the reading of ancient and not ancient texts such as the Bible and Greek mythology — that latter very often from the very late Ovid, who is cited much more often than is the earlier Homer who would have lived closer to the events that Velikovsky claims to have happened.[25] In other words, the support for Velikovsky's hypotheses is much weaker than the support that there for Newton's. It follows that we ought to accept the latter and reject the former.

(f) Gapless and Gappy Knowledge

We can use the Law of Inertia to explain why an object moves. What the Law of Inertia asserts is that for

any acceleration, zero or non-zero, *there are* specific circumstances to which that acceleration is uniformly related. The Law of Inertia does *not* assert *specifically* what the circumstances to which a *specific* acceleration is regularly connected. But it does assert that for any specific acceleration, *there are* circumstances of the required sort. We have the following inference:

> Law of Inertia
> (e)   This object has acceleration *a*
> _____
> Hence, There are circumstances in which this object
>   is located and connected to *a*

From the effect, that is, the acceleration, we can infer the presence of the cause, that is, the relevant circumstances. We do not know *specifically* what those circumstances are; the Law of Inertia does not give a specific characterization of those circumstances. Rather, the Law of Inertia gives a generic characterization, and asserts that *there is* some specific sort of circumstance of this generic sort. Since we do not know specifically what the sort of circumstance is, we cannot predict what the acceleration will be. Nonetheless, given that the acceleration occurs, the effect, then we can use the Law of Inertia to deduce the presence of the cause, those circumstances.

It does not follow, however, that we cannot *explain* the effect. For, given that we have used (e) to derive the presence of the cause, we can turn the inference about and use that cause to explain the effect:

> Law of Inertia
> (ee)   There are circumstances in which this object
>   is located and connected to *a*
> _____
> Hence, This object has acceleration *a*

In (ee) we have a deductive argument in which the effect is inferred from the cause by means of the Law of Inertia.

This is good, but in fact in many contexts we can do better than this. To what this better is, we have merely to turn to the theory that Newton developed. Mars moves in an elliptical orbit, and, in not moving in a straight line, it is accelerating. The Law of Inertia asserts that there are circumstances regularly connected to its acceleration, and, as the acceleration changes, so do these circumstances. But it does not say specifically what those circumstances are. Newton, however, provided us with knowledge of what these specific circumstances are. What he was able to establish is that the motion of Mars depends on its own mass and on the masses, positions, and velocities of the sun and of the other objects in the solar system. Given the Law of Gravity, Newton is able to state *specifically* what are the circumstances that necessary and sufficient for Mars having this acceleration *a*. Let us call these specific circumstances "C". This means that with the knowledge that Newton has given us, we can provide the following explanatory argument:

> Law of Gravity
> (eee)   This object is in circumstances C
> _____
> Hence, This object has acceleration *a*

But this argument could also have been used to predict the acceleration of Mars. For, with the knowledge that Newton has provided, we *do* know specifically the circumstances C. Knowing this, we can use (eee) to predict the acceleration of Mars.

In this respect the knowledge that Newton gave us is superior to the knowledge we have in the Law of Inertia. The Law of Inertia gives us a generic knowledge of the case; Newton's explanation provides us with specific knowledge of the cause. In the case of the Law of Inertia we are able to explain where we cannot predict; the explanation is *ex post facto*. In the case of Newton's explanation, we not only can explain we can also predict; thus, explanation in this case is not merely *ex post facto*. An *ex post facto* explanation does not cease to be an explanation, but it is clearly not as good, that is, *not as good from the standpoint of our cognitive interests*, as one that also allows predictions.

The point about the explanation (ee) is that it provides only a generic description of the cause where (eee) provides a specific description. In that sense, (ee) is **gappy** where (eee) fills in the gap.[26] Or, equivalently, the Law of Inertia is **gappy** where Newton's Law of Gravity is not. It is this **gappiness** which has the consequence that (ee) can be given only *ex post facto*. Conversely, it is because (eee) is not gappy at this point

that it can predict as well as explain. We can conclude in general that

from the standpoint of our cognitive interests in explanation, laws that are less gappy are preferable to those that are more gappy.

This is not to say — the point bears repeating — that gappy laws and gappy explanations are somehow not knowledge and not explanatory. To the contrary, that gappy laws can explain must be emphasized against those who suggest otherwise.[27]

However, it is still true that a less gappy explanation and a less gappy law are cognitively preferable to the more gappy. When we have a gappy law, and later discover, as Newton did, one that is less gappy, we replace the older, gappy, explanations by the less gappy. In fact, the more gappy law will be deducible from the less gappy. In that sense, the less gappy law will explain the more gappy.

Notwithstanding this point, there is something that the gappy law can do that the less gappy cannot. This is the fact that the more gappy law can be used to describe and explain things in systems other than the specific systems to which the less gappy law applies. The Law of Inertia applies to motions in the solar system, but the very gappy explanations that it offers of those motions were replaced by the much stronger explanations that could be given once Newton had discovered the Law of Gravity. But there are other sorts of system to which the Law of Inertia applies but where the motions are not explained by the force of gravity. For example, the motion produced when a spring is first compressed and then released is described by Hooke's Law rather than the Law of Gravity. But the Law of Inertia also applies to such systems. It can do this because it describes things in generic rather than specific terms.

It is precisely this generic feature of the Law of Inertia that enables it to guide research. When it is applied to some new sort of system it states, as we have seen, that *there are* specific conditions there to be discovered. That is, it states in effect that *there is a specific, non-gappy law that applies to systems of this specific sort*, and it is the task of the researcher to discover that specific, non-gappy law. Thus, while a gappy law is less desirable for purposes of explanation, the converse is true in the case of research. In the latter case, we need gappy laws. In fact, it is precisely the gappiness that enables the laws of a good theory to guide scientists in their research. What is less desirable from the standpoint of explanation is more desirable from the standpoint of our need for a theory to guide research. Thus,

from the standpoint of our cognitive interest in explanation of individual facts we prefer the less gappy to the more gappy but from the standpoint of our cognitive interest in discovering more laws we need a theory involving gappy laws to guide research.

Newton provided us with very strong knowledge about the motions of the solar system. He provided us with a rule that tells us how, as these circumstances change, as the other objects move, the acceleration of Mars changes. Then, knowing the acceleration from moment to moment, and knowing some initial conditions about the present position and velocity of Mars, we can predict the orbit that Mars will follow. Moreover, we can also predict, or, if you prefer, retrodict, the orbit that Mars followed in the past. Further, he provide us with this knowledge for each object in the system.

In greater detail, Newton provided us with three important pieces of knowledge about the solar system.

(1) *A complete set of relevant variables*: In the case of the solar system these are the masses, positions, and velocities of all the objects in the solar system. This set is complete because we do not need to know anything else about the objects in order to predict their future motions and to retrodict their past motions. We do not need to know, for example, their colour, or their chemical composition. This is not to say that the latter cannot be subject to scientific investigation. The point is simply that in order to deal with the motions we need to know only the masses, positions and velocities.

(2) *Conditions of closure*, or more generally, *boundary conditions*: In the case of the solar system, we know that nothing outside the system has effects on what goes on in the system. It is closed to outside influence. It is closed because ti is sufficiently far away that nothing outside has any influence on the motions of

the planets about the sun. There are other examples of closed systems, e.g., a martini in a thermos bottle. But in general systems are not closed. In that case we need to know how things outside the boundary of the system affect things that happen inside the system. That is, we need to know the (casually relevant) boundary conditions.

(3) *A process law*: Newton provided us with a rule which states *necessary and sufficient conditions for any variable taking on any value within the system*. If we call the values of all the variables are any one time the "state" of the system at that time, then this rule — the **process law** — enables us, given the present state of the system, to predict any future state of the system and to retrodict nay past state of the system. Moreover, the process law enables to deduce what the system must have had to have been like if things were to be different in certain ways from what they actually are. For example, we can deduce what the past must have been like if the orbit of Mars were to have been closer to the earth, or if Venus had not been in its present position at some time in the past. [As it turns out, the motions attributed to the objects in the solar system are quite inconsistent with the motions as predicted and retrodicted by Newton's process law.] We can, in addition, deduce what changes must be made if we are to bring about in the future a different state of the system. Knowledge of the latter sort is important from the viewpoint of intervention: what changes do we have to effect if we are bring about some state that we find for some reason desirable.

These three things — a complete set of relevant variable, conditions of closure or boundary conditions, and a process law — provide us with knowledge that is **gapless**. Conversely, any knowledge in which we do not know all the relevant variables or in which our knowledge is only generic; or any knowledge in which we do not know the boundary conditions or conditions of closure or in which our knowledge is only generic; or any knowledge in which we do not know how these variable and conditions interact or do not know specifically the form in which this interaction occurs; — if we have any knowledge of these latter sorts, then our knowledge is **gappy**.

Now, with **gapless**, or, as we shall say, **process knowledge**, we can predict what will happen in the system, what has happened in the system, what would happen if things were different, what must happen if things are to become different from what current conditions determine them to be, and what must have been different if things were now to be different. *What more could one want for purposes of explanation?* The question is, of course, rhetorical. In makes clear the point that

from the standpoint of our cognitive interests in explanation, the ideal of explanation is gapless or process knowledge.[28]

In stating that process knowledge is what, ideally, we would like to have in order to salsify fully our cognitive interests in explanation of individual facts and processes, we are *not* stating certain things. For example, when we have process knowledge, we have knowledge of a *law*, a *general* pattern or regularity. Since this pattern is general, it is about a population where the data we have that confirm it constitute only a sample from that population. So there is a logical gap, here as elsewhere in science, between data and the law that we accept for purposes of explanation. To state that process knowledge is the ideal of explanation is not to say that it somehow overcomes the inevitable infirmity of inductive inference.

We of course have seen this happen in the case of Newton's process law for the solar system. Scientists tried hard to bring the theory in conformity to the observed positions of the planets. These efforts led to the discovery of Neptune and Pluto. There were perturbations in the orbit of Uranus that deviated from those predicted by the law. In order to account for these, the hypothesis was formed that there was another planet, beyond Uranus, which caused them. The theory that Newton gave to science was powerful enough to enable researchers to predict with very good exactitude the mass, position and velocity of this planet. When telescopes were turned to the predicted location of the planet, it was observed to be there. This was the discovery of the planet Neptune. The process was later repeated with the discovery of the planet Pluto. There were also deviations in the orbit of Mercury from what the process law predicted. To account for this, it was inferred that there was another planet between Mercury and the sun. This was tentatively dubbed "Vulcan." But in spite of careful observation, Vulcan was never observed. Other possibilities were suggested, e.g., that there is special

bulge around the equator of the sun that would exert enough gravitational pull to account for the deviations. This possibility was rejected as a careful survey of the shape of the disc of the sun revealed no such bulge. In spite of the best efforts, Newton's law could not be brought into conformity with the observed orbit of Mercury. In the end, Einstein developed an alternative theory — general relativity — that could account for everything that Newton's could account for, and in addition could account for what Newton's theory could not, namely, the observed motions of Mercury. Newton's theory, in spite of its power, in spite of its being gapless and therefore the ideal of explanation, was shown to be false. That is the nature of induction. However, to say that Newton's theory is false is misleading. It is a *very good* approximation to the theory that we now accept — though still only tentatively accept — as true, namely, Einstein's theory. Newton's theory is not so much to be rejected as false but to be accepted as approximately true. In fact, since Newton's theory was really very successful at predicting, we could hardly accept Einstein's theory unless it yielded all the same predictions as did Newton's theory where the latter did successfully predict. Einstein's theory agrees with Newton's at *every* point where the latter was successful in its predictions, and disagrees with Newton's only at those point where the latter failed to successfully predict.

There is another point to be made about process knowledge. Just as we cannot reasonably expect process knowledge to overcome the infirmities of induction, so we cannot expect it to overcome the limitations that exist on all measuring processes. There are limits to the accuracy of all measurements, and process knowledge shares in these infirmities, too.

Finally, although process knowledge is the ideal of scientific explanation, in fact it is often unattainable. For example, it is humanly unattainable if there are too many relevant variables. This is the case with most biological systems. In such cases, we simply have to the best we can to cope with the inevitable gappiness of our knowledge.

Moreover, from the fact that process knowledge is our cognitive ideal, it does not follow that process laws actually obtain. *Whether or not there is a process law for a certain sort of system is a matter of fact.* If that fact does not obtain, if there is no process law for a certain sort of system, then we must make do with less. We still have the cognitive ideal of process knowledge, but the world has turned out to be a place where our cognitive ideal is not satisfied. Such is life: there is no guaranteed happy ending, not even in science.

When we began our discussion of ordinary causal explanations with Herodotus, we noted that these explanations involve generalities, or, what is the same, matter of fact regularities. These regularities are of the form

(P1)   Whenever C then E.

where 'C' and 'E' are observable characteristics of things or can at least be explained in terms of such characteristics. (P1) is a regularity about individuals. This feature can be made explicit if we re-write (P1) as

(P1*)   For any individual x, if x is C then x is E

Regularities of the sort (P1) = (P1*) enable us to explained particular events by means of arguments of the form

Whenever C then E
Here is a (an instance of) C

So,   Here is an E

Thus, we explain why this match lights by pointing to the fact that it was struck and to the causal regularity, the general pattern, that whenever a match is struck it lights.

If we have a pattern like (P1), then we say that C is a **sufficient condition** for E.

Not all causally relevant factors are sufficient conditions. Thus, although the presence of oxygen is not sufficient to cause the match to light, it is something which, in cases where it is absent, the match will not light. Its absence is sufficient to ensure the absence of the match lighting. So its presence, while not sufficient for the match lighting, is *necessary* for it. Here the relevant pattern is

(P2)   Whenever E then K

or, what is the same,

(P2*)   For any individual x, if x is E then x is K

If we have a pattern like (P2) = (P2*), then we say that K is a **necessary condition** for E.

It is evident that when we compare (P1) and (P2), that with regard to C and E, where C is a sufficient condition for E, E is a necessary condition for C.

We not only have cases where we infer the effect from the cause, but also cases where we can infer the cause from the effect. Thus, the presence of the flu virus is sufficient for a person to catch the flu, but at the same time if we note that a person has the flu then we can safely infer the presence of the flu virus. The presence of the flu bug is sufficient for the presence of the disease, so that the disease is necessary for the presence of the flu virus; but at the same time, the presence of the disease is sufficient for the presence of the flu virus. In other words, the disease is both *necessary and sufficient* for the presence of the flu virus. And equally, of course, the presence of the flu virus is also necessary and sufficient for the presence of the disease.

In this case we have a regularity of the form

(P3)   For any individual x, x is C if and only if x is E

or, what is the same,

For any individual x, if x is C then, and only then x is E

If we have a pattern like (P3), then we say that C is a **necessary and sufficient condition** for E.

We should note strictly speaking, the striking of a match is *not* sufficient for the match to light; after all, we also need the presence of oxygen. This is true in general: if the presence of any condition is to be sufficient for a given condition, then all the necessary conditions for the presence of the given condition must also be present. Given (P2), that K is necessary for E, then it is not so much (P1) that states a sufficient condition, but rather

(P4)   For any individual x, if x is C and x is K then x is E.

We can, of course, simply take for granted that all necessary conditions for a given effect are present. This we normally do with regard to the striking of the match: we simply take for granted that oxygen is present, that the match is dry, and so on. But, strictly speaking, if K is a conjunction of *all* necessary conditions for a condition E, then any statement of sufficient conditions has the form

(P5)   Whenever C and K then E

or, what is the same,

(P5*)   For any individual x, if x is C and x is K then x is E

We should also note that in general a given kind of event will have several sufficient conditions. Thus, we can get the match to light by striking it, or we can get it to light by heating with sunlight focussed on it by a magnifying glass, or we can get it to light by holding it in the flame of another match, and so on. We not only have C as a sufficient condition for E but also C', C", etc. Which means that besides (P1), we also have

(P6)   Whenever C' then E
(P7)   Whenever C" then E

And, of course, if C, C' and C" are separately sufficient for E, then so is their sum or disjunction: C or C' or C". That is, if we have (P1), (P6) and (P7), then we have, equivalently,

(P8)   Whenever either C or C' or C" then E

or, what is the same,

(P8*)   For any individual x, if either x is C or x is C' or x is C" then x is E

We must also recall the point that we just made about necessary conditions: no condition by itself is a sufficient condition apart from the necessary conditions. If K is the conjunction of necessary conditions for E, then we have not so much (P8) as

(P9)   Whenever either (C and K) or (C' and K) or (C" and K) then E

or,

(P9*)   For any individual x, if either x is (C and K) or x is (C' and K) or x is (C" and K) then x is E

Now note that we generally assume that the events in which are interested, certainly those in which we are interested in finding an explanation, have a cause. That is, we assume for any sort of event that there is a sufficient condition for its occurrence. This means that the occurrence of the event implies that at least one of its sufficient condtions is present. But if it implies the latter then the latter is a necessary condition for that sort of event. Schematically, if W is the sort of event with which we are concerned, then the assumption is that W will not occur unless at least one of its sufficient condtions has occurred. If X, Y and Z are sufficient conditions for W, then to say that at least one of them has occurred is to say that "either X or Y or Z" has occurred. Thus, "either X or Y or Z" is a necessary condtion for W. But it is also a sufficient condition. Thus, *the sum of the sufficient conditions for a given conditon is also a necesary condtion for the given condtion.* In other words, *the sum of the sufficient conditions for a given condtion is a necessary and sufficient condtion for the given condition.* Or rather, this is so *provided that we assume that the given event is such that it will not occur unless at least one of its sufficient conditions occurs.*

Thus, if we have (P9*), and we know further that the list of sufficient conditions which it contains is compelte, and if we assume that the event E is such that it will not occur unless at least one of its sufficient conditions occurs, then we also have

(P10)   For any individual x, if either x is (C and K) or x is (C' and K) or x is (C" and K) then, and only then x is E

With reference to (P10), we can say, first, that while each of the (conjunctive) sufficient conditions (C and K), (C' and K) or (C" and K) is a sufficient condition for E, none of them is a necessary condition. Thus, each of these conditions is an *unnecessary sufficient condition* for E. We can say, second, that in the condition (C & K), neither C nor K is by itself a sufficient condition for E, but also that each is required, that is, necessary, if this condition is to bring about E. Thus, each of these conditions is an *insufficient necessary part of this sufficient condition* for E. But that sufficient condition is in fact an unnecessary sufficient condition. Thus, we can say that C is an *Insufficient Necessary part of an Unnecessary Sufficient condition* for E, or, more briefly, C is an **INUS** condition for E.[29]

When we strike a match, we call the striking of the match "the cause" of its lighting. We now see that, with regard to most of those conditions that we normally designate as causes, what we are in fact picking out is not strictly speaking a sufficient condition but rather an INUS condition.

Clearly, as we can see from (P10), there are many INUS conditions. Why, in our ordinary ways of speaking on particular occasions, do we single out one of these INUS conditions as "*the* cause"?

To answer this question, we have to notice that most of the conditions K are what are often called *standing conditions*. Oxygen is normally present; matches are normally dry. Taking these as givens, one can then ask what it was that caused the match to light. The answer to that question is that it was the striking of the match. That is, it was precisely this factor, and not any other, which, when *added to the standing conditions*, made it such that there was then a condition present that was sufficient for the match to light. "The cause" is the INUS condition which, when added to the standing conditions, transformed the latter into a sufficient condition. "The cause" is precisely that condition which *made a difference* between the presence of a set of standing conditions which were not sufficient for the effect and the presence of a set of conditions that were, jointly, sufficient for the effect. Briefly, "the cause" is that INUS condition that *makes a difference*.

Focussing on that which makes a difference in a set of standing conditions is very important from the point of view of being able to interfere intelligently in a process to achieve some end that we find desirable. We want to light a candle. So we obtain a match, and strike on an appropriate surface. The striking of the match causes it to flame and we can then use it to light the candle that we wanted lit. Striking the match is at once that which makes the difference and also something that we *can do*, something that we can bring about by *acting* in certain ways, doing certain things by *manipulating* objects. In this sense, the cause is not only an INUS condition, but also something that we can bring about by manipulating things. We are, in other words, often interested in INUS conditions as causes out of our pragmatic interests in achieving certain ends.

Where knowledge is motivated by pragmatic interests, that is, where we desire knowledge for the sake of application, then a knowledge of an important INUS condition may suffice. But in general that will not do. For, we often want knowledge not merely for the sake of application — although the word 'merely' in this context is not entirely appropriate, as the case of medical cures makes evident; but however the latter may be, we also often desire knowledge for its own sake; we often are motived simply by the sentiment of curiosity or love of truth. In that case, we would want to know all the details of the relevant regularities. We would want to know not just one of the INUS conditions, but the other INUS conditions that together make for a sufficient condition. And we would also want to know all the other sufficient conditions, all those conditions whose sum is not only sufficient but also necessary.

It is precisely this that is available if we have process knowledge of a system. Knowing the relevant boundary conditions, knowing a complete set of relevant variables, and knowing a process law, one knows for any state of the system or any possible state of the system precisely what preceding states would be sufficient. And since one knows this for all possible states of the system, one also knows the sufficient conditions for any part state: it would be any condition sufficient for bringing about a state which has it as a part.

Our ordinary language of causes can reasonably be expressed in terms of sufficient conditions, necessary conditions and necessary and sufficient conditions. This sort of discourse seems distant indeed from Newton's law for the solar system and from the ideal of process knowledge. But it is distant only insofar as the partial is distant from the ideal. For, as we have now seen, process knowledge does in fact give us all the information we could want, ideally, to know, out of either some pragmatic interest or out of idle curiosity, about the sufficient, the necessary and the necessary and sufficient conditions for the system in which we are interested.

The ordinary concept of cause is not fully pellucid. In the first place, to say that C is the cause of E is to say that C is an INUS condition for E. In the second place, to say that C is the cause of E is to say that we can manipulate things to bring about C and thereby bring about E. And third, there is undoubtedly a tinge of Aristotelian anthropomorphism: to say that C, which is distinct from E, is the cause of E is to say that C is a thing which in the exercise of its active powers brings it about that another thing is E. On this last view, the cause is an external force that imposes itself on another thing and constrains it to act against its natural tendencies.

Now, the early defenders of empirical science strongly emphasized the point that natural science gave one control over nature. Thus, Bacon for example argued that knowledge and power are one, and that the promise of science is that of "leading to you Nature with all her children to bind her to your service and make her your slave."[30] This emphasis on control over nature was part of an argument that natural science was useful and could benefit humankind, in contrast to Aristotelian "science" which, being non-empirical, was useless and of no benefit to people. It was important so to argue. For, earlier philosophers such as Aquinas argued that it was a mortal sin to be interested in things of this world, matters of empirical fact, since to be interested in such things meant that one was not interested in knowledge of God and of other things that are central to salvation but transcend the world of ordinary experience. Since, however, it was not sinful to be one's brother's keeper, the argument from the utility of science could put that study in a moral context in which it was no longer characterizable as a sin but, to the contrary, as a virtue.

However, it is also true that knowledge of matter of fact regularities, while knowledge of regularities that could, *if one wished*, often be used to control and manipulate, is also knowledge that could be sought for its own sake, out of idle curiosity, with no desire to control or manipulate.

Some, however, focus on the discourse used by Bacon and others to justify to their fellows that natural science was a good thing to do, and argue that it was in fact conceived in sin, the sin of aiming to control nature. The contrast that is made is to what is argued to be a more morally appropriate attitude towards nature, that of living in harmony with it. The sin in which science was conceived has not been washed away, it is still present, and as with all sin the temptation to commit it must be curbed: the impulse to seek scientific understanding of nature must be leashed and restrained.

This charge of being conceived in sin is made stronger if the ordinary, somewhat confused sense of 'case' is kept in mind when discussing natural science. Natural science seeks causes, and these causes are not only causes that permit interference and control but — recall the residual Aristotelianism — constitute external forces that impose themselves on other things and constrain the latter to act against their natural tendencies. The knowledge science seeks, in other words, is intrinsically coercive.

This latter picture of science, that it is intrinsically coercive, has been linked to the notion of domination and thereby to imperialism, on the one hand, and to sexism, on the other. Science is in its very nature an instrument for domination, and is therefore an agent of western imperialism. It should not, therefore, be used to criticize such things as non-western forms of medicine. Moreover, since aiming to dominate and control are male characteristics rather than female, science is intrinsically sexist and anti-feminist.[31]

A clearer vision of empirical science would unmask this view of science. Once the confused notion of cause that underlies these arguments is deconstructed, it is evident that there is nothing about science as such that makes it either imperialist or sexist. The notion of cause that appears in science is that of *regularity*. It has nothing to do with a notion of cause as somehow something coercive. The latter is simply the residue of Aristotelianism. Nor does the fact that the knowledge of science can be applied to achieve certain ends imply that its knowledge must be used in that way exclusively or, indeed, that it must be used in any way, that is, used in any way that aims to satisfy our pragmatic interests. For, after all, knowledge of matter of fact regularities can be sought simply for tis own sake, out of idle curiosity.

Attacks on science as intrinsically an agent of western imperialism are particularly pernicious. It implies that science, the only method that stands any reasonable chance of uncovering matter of fact truth, will not be used to evaluate the worth of such things as ancient Chinese medical practices. To be sure, science is not infallible, and mistakes have, perforce, been made. Often, the arrogance of scientists has proposed solutions to problems that were not only wrong but coercive. Nonetheless, to repeat, science is the only method for discovering matter of fact truth that so far as we know has any reasonable chance of success. And so, even though bear galls and rhinoceros horns have no scientifically validated medical use, the argument we are considering allows them to escape such criticism: to deny their utility on grounds that there is no scientific evidence either that they work or even could work is simply western cultural imperialism. In the meantime, these animals are hunted for their supposed medicinal parts to the point where there are threats of extinction.

To think of causation in terms of INUS conditions alone, however, tends to encourage the confused ordinary sense of cause. Specifically, it tends to encourage people to think that causation works in terms of simple-minded chains — C is that which makes a difference and brings about E, E is that which makes a difference and brings about D, etc.: for want of a nail a shoe was lost, for want of a shoe a horse was lost, for want of a horse a battle was lost, for want of a battle a kingdom was lost, and so on. This thinking of causation in simple linear terms fits in with the Aristotelian framework: C coerces E out of the background, E coerces D, D coerces ... etc.

To think of chains of INUS conditions is not wrong in itself. Certainly, they can be understood in strict scientific terms, that is, in terms of matter of fact regularities. But to take this as the central point about causation is seriously misleading, certainly a partial view of the cognitive aims of science.

This partial view of the aim of science can be corrected by turning to the cognitive ideal of scientific explanation of individual facts, that is, to the ideal of process knowledge. In order to predict the value of any one variable at a time, it is generally necessary to know the values of all the other variables at some previous time. Thus, in order to predict the position of Mars at some future date, one needs to know its mass, its present position and velocity, the mass, position and velocity of the sun, the mass position and velocity of Jupiter, and in addition the values of these variables for any other planets of sufficient mass to affect the orbit of Mars. Since in this sense the value of any one variable depends on the values of the all the other variables, we can say that *the value of every variable depends on the values of all the variables*. In this sense, in any system of the sort for which process knowledge is the ideal, we have **total interaction** among the set of relevant variables.

With total interaction of this sort, where the value of every variable depends on the values of all the variables, the metaphor of the chain makes little sense at all. This means that that metaphor, encouraged by thinking of causes simply in terms of INUS conditions, that causation is linear and chain-like, is totally out of place in the broader picture of the world as conceived by empirical science.

Evelyn Fox Keller[32] has directed our attention to how thinking of causation in linear terms can be misleading in the context of scientific practice. She notes how many scientists tend to think of genes as crucial entities in human development, and that the task is to trace out how each gene makes a difference to the way the organism develops. This is hardly a complete picture, however. If different genes determined in a linear way the phenotypic characteristics of things, then every organism with the same gene would have the same phenotype. But this is not so. The phenotype that we observe is also partially determined by the environment. Different diets, different climates can affect organisms differently. That is why we have, for example, dwarf pines in inhospitable environments, and why some children are born with alcoholic fetal syndrome. In the development of the phenotype, there is a complicated pattern of interaction between the gene and the environment in which it is located. There is, in other words, a pattern of **total interaction**.

Keller refers to work on DNA. The normal conception, she argues, is one in which "the DNA encodes and transmits all instructions for the unfolding of a living cell..."; on this view "a master control ... [is] found in a single component of the cell..." But, as the work of Barbara McClintock established, control, rather, "resides in the complex interactions of the entire system." Thus, in contrast to the linear model, McClintock's research "yielded a view of the DNA in delicate interaction with the cellular environment."[33]

This pattern of total interaction is often missed when the gene and its effects are conceptualized in terms of the linear model of causation. This means that the researcher who uses the latter sort of conceptualization will be inclined to ignore the broader patterns of interaction. The corrective to this restricting way of thinking of the effects of genes, and the way to liberate science from its constraints is to recall the *explanatory ideal of process knowledge*.

The point is, of course, that the difference between the two approaches to DNA and its effects are not simply a matter of theoretical formulations. It is rather a matter of having a partial picture, on the one hand, and a fuller, more adequate picture, on the other, of the cognitive ideals of science. But with the restricted or partial picture of the cognitive aims of science, research was restricted. A broader, more complex theoretical understanding arose when McClintock approached her subject matter with a more adequate concept of the cognitive ideals of science, one which recognized that one should expect not linear chains of causation but total interaction among the variables.

Evelyn Fox Keller suggests that the difference between thinking in terms of linear chains and total interaction is a matter of masculine in contrast to feminine approaches to subject matter.[34] This is certainly subject to debate. After all, Newton knew that there was total interaction among the variables that characterized the entities in the solar system and did not restrict himself to linear chains of causation. But in any case, the correction is not so much the introduction of feminist ways into science, but to get everyone to recognize that the cognitive ideal is process knowledge, and that in such systems what we encounter is total interaction, not neat linear chains. The correction to a limited vision is not so much feminism as a more adequate philosophy of science, one which recognizes process knowledge as the ideal of scientific explanation.

In doing science we must pass from the ordinary concept of causation to something more adequate. Even if we eschew the Aristotelian embellishments of the ordinary concept, and restrict ourselves to the notion of an INUS condition, we have not gone far enough. We have restricted ourselves to matter of fact regularities, as required by the cognitive aims of science, but we must go beyond even this to recognize that the ideal is process knowledge. This sort of knowledge does yield a knowledge of necessary and sufficient conditions, which we do require for our ordinary purposes. But it also makes clear that the tendency to fall into linear thinking, which is there if we limit ourselves to INUS conditions, is indeed a limitation and that in the real world total interaction is the rule.

There are several formal developments of the sort of law that Newton discovered for the solar system that are worth noting.

The Newtonian process law for the solar system is usually expressed mathematically by a set of ordinary, second order differential equations. When these equations are solved one has the orbit of each object in the system.

The equations can be solved exactly when there are only two bodies in a system. But when there are three or more objects, then exact solutions are not possible. The solutions can, however, be expressed as an infinite series, and, by taking the terms of this series to any number we want, it is possible to get an approximation to the solution that is as accurate as we would want.

Ordinary differential equations are suitable for objects like planets that behave like points in space. Other objects, however, are spread throughout space, objects such as electrical and magnetic fields. A process law for entities such as these is usually expressed by a series of partial differential equations. Finally, there are some phenomena which are such that, in order to predict a future state it is necessary to know not only the present state but also the past history. The magnetic phenomenon of hysteresis is of this sort; to know how a magnetized bar will behave it is necessary to know precisely the history of the process by which it was magnetized. Another example of such laws — they might well be called "historical" laws — is the case of learning in both human beings and in other animals, e.g., white rats. In order to know how a subject will respond R to a stimulus S, it is usually necessary to know what the subject has previously learned. A rat that has learned to run a maze will respond differently than one which has not yet learned the maze. A person responds to a poem differently on the second reading than on the first. Historical laws of these sorts also have a mathematical model: they can be represented by integro-differential equations of the sort first studied in the early 20th century by Volterra.

We saw earlier that Osiander defended Copernicus by proposing that the heliocentric calculations be treated not as providing a true description of celestial motions but as being a set of mere "calculating devices." What precisely is such a device, and why should it be characterized as "mere"? After all, is not the point of science to discover regularities that permit prediction? If we have a set of rules that do that job, then why should we disparage them as mere calculating devices? Or, if they are to be disparaged, what sets them off from other regularities that also permit prediction?

Let us first take another example. Consider an organism O. Presented with some stimulus R, this organism exhibits response R. We might very well be able to find a function f that would determine the response R in terms of the stimulus S:

(r) $R = f(S)$

But to talk about this relation and nothing else is to ignore the organism O. There is a process within O that leads from S to R. We have not

(r') $S - R$

but

(r") $S - O - R$

Knowledge of a law that connects S with R but does not mention the processes within O would leave out many details of the full process that leads from S to R. As knowledge, it would be gappy. We would prefer, from the viewpoint of our cognitive interests, less gappy knowledge of the full process (r").

If we have a law (r) that describes the connection (r'), we will be able to predict the response R given the stimulus S. If the function f does not in any way purport to describe the O part of the process (r"), then it is nothing more than a way of helping us infer R from S. It is, if you wish, a "mere calculating device." This is not to say that it has no explanatory power; to the contrary, insofar as it systematically relates S and R it does indeed have explanatory power. It is just any explanation based on it alone is bound to be gappy: we do know that there are intervening factors represented by O in (r"), and a more adequate, less gappy explanation will have to take them into account.

C. G. Hempel once raised the issue about why the scientific theorizer should worry about the intervening factors.[35] After all, he argued, if science searches for laws that predict, and a law that relates S and R without referring to the intervening variables is a law that predicts, then why not settle for the latter? Why do what scientists everywhere do, and search for the intervening variables? Hempel saw this as a puzzle, for which he could offer no clear solution. In fact, however, there is no real puzzle.

For, prediction is *not* the sole end of science; there are cognitive ends to science other than prediction. As we have just seen, science searches after laws that will fill in our gappy explanations, to replace them with less gappy explanations. The researcher will therefore seek to discover the intervening variables, until he or she discovers a theory that includes them. If Hempel had kept these other cognitive ends in mind, he would have recognized that scientific theorizing about intervening variables was not really a puzzle at all.[36]

To be sure, the *test* for accepting an hypothesis as a law is its success at prediction This is true of any law, gappy or less gappy. But from the fact that successful prediction is the test of accepting a generalization as a law it does not follow that there are no other cognitive aims in science. To the contrary, there are other aims, as we have just argued, namely, gapless knowledge.

The scientist, then, has good reason for not settling for the law (r) and setting out to discover the intervening variables in the process (r"). When the details of the process (r") emerge after further research, when the gaps in the knowledge are eliminated, then of course the law describing this process will *explain* the gappy law (r); the latter will follow from the former.

Osiander was in effect arguing that Copernicus was offering a formula parallel to "f" in (r), which would enable Copernicus to make predictions about the positions of the various celestial objects. Osiander was in effect adopting the Hempelian idea that prediction was the sole aim of science, and that being so, it was unnecessary to hold that the mathematical constructions Copernicus developed represented intervening states of the system. Copernicus, in contrast, held that those constructions did in fact represent the orbits of the planets, that is, did represent intervening states. In aiming to describe these intervening states as well as make successful predictions, Copernicus took a step towards less gappy knowledge. To be sure, he was still far away from the Newtonian achievement of (a reasonable approximation to) process knowledge. Moreover, Copernicus was still full of pre-scientific notions of explanation, e.g., the Circularity Principle and the Aristotelian notion of natures as unanalysable dispositions. Nonetheless, Copernicus did take a step towards the cognitive ideal of less gappy knowledge — a step that Osiander and Hempel have trouble understanding when they allow that the only goal of science is prediction.

(g) Non-causal Theories

Process knowledge is knowledge of causes. To get a better idea of this sort of law, it will pay to contrast the sort of laws that it provides with the sorts of laws to be found in certain non-causal theories. The example that we will use is that of geometrical optics.

Some objects make themselves visible through their own self-illumination. Other objects — called dark objects — become visible only when illuminated by the self-luminous objects. The sun is an example of a self-luminous object; so is a burning candle. These objects make other objects such as cows or buildings visible by illuminating them. A dark object that is illuminated by a self-luminous object can also in turn illuminate other objects by, as one says, reflected light. Let us suppose that we have a flagpole. This is not a self-luminous object. Neither is the ground upon which it stands. Both these dark objects can be illuminated by the sun. However, only part of the earth will be illuminated. Another part, what we call the shadow of the flagpole, will be unilluminated. The area of the earth that is illuminated, and therefore also the area which is not, is determined by the geometrical relations of the source of illumination, the earth, and the intervening dark object, the flagpole. The area of illumination is established by the *Law of Rectilinear Propagation of Light*. Specifically, in the case of the flagpole (see Figure 3.14), the sun considered as a point source of illumination must lie on a straight line extending from the end of the shadow, or, what is the same, the edge of the illuminated area, through the topmost point of the intervening dark object, the flagpole.

If the flagpole is known to be perpendicular to the surface of the earth, then, if the length of the shadow is known, then we may use the propositions of geometry to calculate the height of the flagpole. Or, if we knew the height of the pole, then we could calculated what length shadow it would cast. The Law of the Rectilinear Propagation of Light is thus useful in predicting.

We can also use the law to calculate various other things. For example, we could suppose that the shadow is longer than it in fact is, and then calculate what the height of the pole would have to be to cast a shadow of that length. Or, we could suppose the height of the pole to be different and calculate what the length of the shadow would then be.

This theory is non-causal in two ways. First, the laws of geometrical optics do not mention time. Or rather, they state that for any time the relations between sources of illumination and shadows will be thus and so. Thus, they state relationships that hold in every temporal cross-section of a process. A causal law, in contrast, and a process law in particular, states how a process changes from cross-section to cross-section.

Second, the laws of geometrical optics, in contrast to Newton's laws of mechanics, make no mention of forces. Since we very often cause things to happen by exerting a force, the laws of geometrical optics do not tell us what forces we might exert in order bring about changes. For example, they do not tell us how to exert a force on the flagpole in order to change the length of the shadow, nor do they tell us how to exert a force on the shadow in order to change the height of the pole. Of course, we know that in order to change the length of

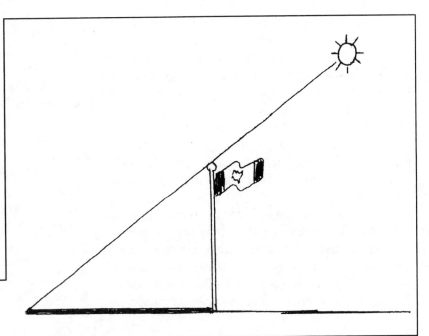

Figure 3.14

the shadow we must change the height of the pole, either shortening or lengthening it. And we also know that we cannot similarly manipulate the shadow in order to bring about a change in the height of the pole. But this knowledge of how to make changes in the system comes from other laws, not those of geometrical optics.

It has been objected that while the height of the pole explains the length of the shadow, it is not true that the length of the shadow explains the height of the pole.[37] The suggestion is that, because we must manipulate the length of the pole in order to change the shadow length while we cannot change the height of the pole by manipulating the shadow length, therefore the height of the pole explains the length of the shadow but not conversely. We therefore seem to have a case of a law, here one that enables us to infer the height of the pole from the length of the shadow, where we do not have an explanation. We seem to have prediction without explanation.

However, upon reflection it seems we do have an explanation. To explain why something is so and so is to give a *reason for its being* so and so. The length of the shadow does give a *reason* for the pole *being* the height that it is. To that extent, the law of geometrical optics that enables us to infer the height from the length of the shadow is indeed explanatory. In other words, the law is explanatory as well as enabling us to predict.

Of course, the law is not a process law. It does not tell us how the system of the flagpole, self-illuminating object (that is, the sun) and the shadow are going to change over time. Since we would like to know this sort of thing, geometrical optics does not fulfill our cognitive interests to the extent we would like. It does not follow that the laws are therefore somehow not explanatory. All that follows is that the explanations they provide fall short of process knowledge. To suggest that somehow they are not explanatory is to confuse the general concept of explanation with one of its species, namely causal explanation.

The objection seems in fact to be based on just this confusion. But since the laws of geometrical optics do not mention forces, this is no objection to their having explanatory power. To be sure, the laws of geometrical optics are not causal laws. Nor are they process laws. But, again, that does not imply that they have no explanatory power.[38]

(h) Theoretical Unification and Composition Laws

Process knowledge is not the only cognitive ideal of science. There is a second. This ideal is that of **unifying theory**.

Theories do unify. That is their job. Thus, the Newtonian theory of mechanics unified celestial and terrestrial motions by explaining both Galileo's law of falling bodies and Kepler's laws of planetary motion.

Newtonian theory could do this because it abstracted from the specific laws a common generic form that they shared. This makes the point quickly:

> **theories unify though abstracting a common generic form for more specific laws in several areas**.

We can see immediately what is the cognitive ideal with regard to theories:

> **the cognitive ideal in the case of theory construction is a theory that unifies all science**.

Science is in fact unified by a common method, and by a common goal. The method is the inductive method; the cognitive goal is knowledge of matter of fact regularities. But beyond that, there is the cognitive goal of a unified theory. The former features of science do not imply that there is a unified theory that unites all more specific laws. Furthermore, whether this latter goal of a unified theory can be attained depends upon what the facts are: are there, or are there not, abstractive generic laws that can unify all laws in science? This question is difficult to answer, but the success so far in developing unified theories gives one some grounds, though far removed from certainty, that there is in fact a unified theory. Certainly, however, we are far from knowing with regard to any theory, e.g., the fundamental theory of physics, that it can actually unify everything else.

Theoretical unification is not only a cognitive ideal. There is another reason why scientists value unifying theories. Think of Kepler. Once he discovered his laws about planetary orbits, it was relatively easy to move on

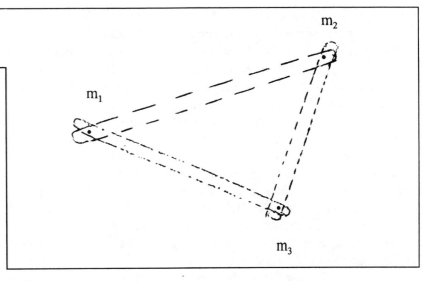

Figure 3.15

to discover the orbit of Jupiter. These unifying laws guided his research into the planetary orbits. This points to the utility of unifying theories:

unifying theories guide research.

Thus, theories are valued not only as satisfying of cognitive ends, but also as means; *theories are tools to guide the researcher in the discovery of new laws*.

One axiom of Newton's theory of mechanics that we have not yet mentioned is very important in this context. This is the **composition law** that is among the basic assumptions of the theory.

A composition law is a law that enables one to deduce the law for a complex system if one knows that laws for simpler systems. In Newton's theory, the simplest systems are *two-body systems*. More complex are three-, four- and *n-body systems*. The composition law enables one to deduce the law for an n-body system, given that one has laws for two-body systems.

The law states that any n-body system is to be *conceptually* decomposed into all possible two-body subsystems. Thus, in a three-body system, one can conceptually distinguish three two-body systems (see Figure 3.15).

One then makes an appropriate assumption about the law that *would* describe the behaviour of the objects in these two-body systems *were* they to be (what they are in fact not) isolated. In Newton's case, this is to assume a force function for the conceptually distinguished two-body

systems. This force function determines the accelerations that the two bodies would have were they in fact isolated (which they are not). Newton's rule then states that the real acceleration is the *vector sum* of the accelerations that the objects would have in the conceptually distinguished two-body systems (see Figure 3.16).

The rule is the well-known parallelogram rule: the real acceleration in the three-body system is given by the vector sum of the accelerations the objects would have were the two-body systems actually isolated. The composition law extends this three-body case to the n-body case.

Several points should be noted. First, in order to deduce the three-body or n-body law one needs to know

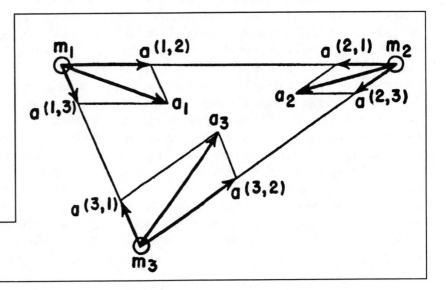

Figure 3.16[39]

not only the relations between the objects in the conceptually distinguished two-body systems, but also all the other relations that define the complex configuration of the three or n bodies. The deduction is possible only if the structure of the configuration is given.

Second, the law for the n-body system is a matter of fact that is not contained in the laws for the conceptually distinguished two-body systems. In order to deduce the n-body law, we need the composition law as an additional premise. This shows that the composition law itself contains factual information not contained in the laws for the conceptually distinguished two-body systems. What this shows, of course, is that the composition law is indeed a factual claim, a matter of fact generalization or *law*, and in particular, a *law about laws.*

Third, that the composition law works for all n is a matter of fact. It could turn out that it works for say n up to 99 but not for n = 100. We have no reason to suppose that the Newtonian composition law breaks down at that point, but that it does not is a matter of fact, and not something to be decided *a priori*.

It is easy to illustrate the power of a composition law in a theory by reference to the case of planetary motions. This is the case of the discovery of the planet Neptune, to which we have already alluded.

Taking into account the planets from Mercury to Uranus, scientists were able predict an orbit for Uranus. The observed orbit did not, however, fulfill these predictions. That falsified the theory, or rather the theory together with the auxiliary hypotheses. Something had to be done to correct the premises which led to the incorrect predictions. Since Newtonian theory itself was so highly confirmed, it was more reasonable to modify the auxiliary hypotheses. The auxiliary assumption that was modified was the assumption that all planets had been taken into account. Rejecting the assumption that there were seven planets all told, it was suggested that there could be eight planets, the seven from Mercury to Uranus plus one as yet unobserved beyond Uranus. The gravitational effects of this unobserved planet would account for the perturbations in the orbit of Uranus, its deviations from the previously predicted path. Scientists started with these effects and inferred the presence of a cause. The scientists were able to calculate what position, velocity and mass the unobserved planet would have to have in order to account for the observed effects. Two

assumptions were needed. One was that the relevant force was gravitational. This force was assumed to act between the members of every pair of objects in the solar system. The other assumption was the composition law that enabled the scientists to deduce what the law would have to be for the nine-body system (the sun, the seven known planets, and the unobserved planet) if they were to be able to account for the observed perturbations in Uranus' orbit.

When telescopes were turned to the place where the previously unobserved planet was calculated to be, it was observed. This planet they named Neptune. Since probability of the existence of such a planet would be very low unless Newton's theory were true, its discovery was a powerful further confirmation for Newton's theory.

The theory told the investigators precisely what new auxiliary hypothesis was needed if it was to be brought into conformity with the observed planetary positions. The premise that enabled the theory to locate precisely where the unobserved planet was located was the composition law.

Clearly, any theory that contains a composition law contains a powerful research tool. At the same time, it becomes through that law a tightly unified theory. Like any abstractive generic theory, it is a set of *laws about laws.* But an abstractive theory with generic laws does not by itself provide deductive relationships among the various specific laws which it covers. However, if there is a composition law, then deductive relationships are established among the specific laws: some specific laws can be deduced from other specific laws. The composition law relates deductively the laws for complex systems to the laws for simpler systems in such a way that the former can be deduced from the latter. This provides a tighter unity than is had with an abstractive generic theory without a composition law.

The composition law for Newton's theory is one of the simplest. There are other theories with composition laws. The law of partial pressures in gas theory is another example. The laws for adding waves together in both optics and acoustics are other examples.

When people speculate about the unification of biology and chemistry, thinking of somehow reducing the laws of the former to the laws of the latter, there are two things that they must have in mind. One is that the concepts of biology can somehow be translated into those of chemistry. That would mean that biological systems were in effect very complex chemical systems. Call this "reduction$_1$." But reduction$_1$ alone is not enough. It could still be that the laws of the complex biological systems cannot be deduced from the laws for the simpler chemical systems. In order to achieve such a deduction, a composition law is necessary. Whether there is such a composition law is, as we have seen, a matter of fact. If there is in fact such a law, then it will be possible to deduce the laws of biology from those of chemistry. Call such a deduction based on a composition law "reduction$_2$."

Clearly, reduction$_1$ does not imply reduction$_2$. If there are properties of biological systems that cannot be defined in terms of the properties of the chemical subsystems and the structural relations among these subsystems, then reduction$_1$ will not be possible. Where reduction$_1$ fails, one has what have been called *emergent properties*. Even if we have reduction$_1$, we might not have reduction$_2$. In that case we have patterns of behaviour, laws, for the complex systems that cannot be deduced from the laws governing the behaviour of simpler systems. In this case one can speak of *emergent behaviour* or *emergent laws*.

The failure of reduction$_1$ does not automatically imply the failure of reduction$_2$. If the emergent property does not interact with the chemical variables then reduction$_2$ is still in principle possible.

It could turn out, for example, that there is a non-causal cross-section law relating the emergent property to some set of non-emergent properties. There would be a parallelism between the emergent property and the non-emergent properties to which this law related it.

The common wisdom in science is that biology can in principle be both reduced$_1$ and reduced$_2$ to chemistry, just as chemistry has been in effect reduced$_1$ and reduced$_2$ to physics. There are no emergent properties in either chemistry or biology.

However, since the properties of minds, such as hoping or loving or believing or supposing or willing, do not seem to be reducible$_1$ to the biological or chemical properties of our brains, it seems that they have to be counted as emergent. Psychologists have argued that an objective science of human being is possible, however, in the sense that the laws of psychology need mention only environmental, behavioural, and biological variables. In this sense, they have been prepared to argue that

psychology is reducible$_2$ to biology. They can maintain this provided that they hold that there are mind-brain cross-section laws that establish a parallelism between the mental variables and the biological.[40]

Consider the following account of a human being which occurs in a discussion of homeopathic medicine. "The soul," we are told,

is the source of the order of life, of harmony, of rhythm and self-organization. The soul "in-forms" the body, it directs and guides it, changes the physical world and gives it life.[41]

But, we are also told,

…spirit itself, outside of space-time relationships, cannot be subject to material conditions and, consequently, cannot be scientifically investigated.[42]

The soul is thus an entity that causally interacts with the body, yet cannot be explicated in terms of the physical parts of the body and their relations; it cannot be reduced$_1$ to the ordinary parts of things that we observe in ordinary experience. The soul is thus an emergent entity. Because this emergent entity interacts with matter, the laws that describe the development and behaviour of human beings cannot be reduced$_2$ to the laws of biology, chemistry and physics.

The soul is a teleological entity: it "directs" and "guides" the body, we are told, that is, directs it to some end or other. The soul provides the form of the body, the organization, which cannot otherwise be accounted for save through its teleological operations. This makes the soul, as described by these writers, as a sort of Aristotelian form or nature, an active power that organizes things into wholes and determines their form. What is true of the soul, according to these writers, is true of Aristotelian forms in general. We have to conceive them to be emergent entities that causally interact with the parts of the entity which they ensoul.

These entities are not in fact acceptable to science. This is because they lie outside the realm of ordinary experience: we cannot observe them. Since we cannot observe them, either directly or by means of instruments or other tests, they cannot form part of any scientific explanation. In the past it was often though necessary to introduce such entities if we were to be able to explain the complex order of living organisms. It was not easy to see how within a small cell, the fertilized egg, the information would be available for the development of something as complex as even an earthworm let alone a human being. It was not easy to see how simple chemical complexes could initiate a process that would generate out of itself and other bits of chemicals complex biological organisms. Hans Dreisch so argued, to give but one example.[43]

It is a strange argument, however. It begins with the fact of ignorance. We do not know how to explain the phenomena scientifically, in terms of naturalistic causes. Therefore, no such explanation is possible and it is necessary to introduce non-empirical entities to do the job. Given the past successes of science in uncovering hidden mechanisms that explain complex patterns of behaviour and action, it in fact seems unreasonable to suppose that there is no hope of success in the cases before us. In fact, of course, since Dreisch wrote, the information carrying chemicals have been located, and many of the mechanisms by which these chemicals produce complex entities which replicate the parents are now reasonably well understood. The relevant chemical, is of course, DNA, and molecular biology has located and studied many of the processes through which in the multiplication of cells, complex organisms develop.

Because the soul, or Aristotelian nature, or "entelechy" as it was called by Dreisch, is a property of the whole organism and because this property of the whole determines the behaviour of the parts, it is sometimes said that on this view "*the whole is greater than the sum of the parts*." Or, rather, it is taken as evident that this formula expresses a great truth and that it is a virtue of any theory that introduces souls or forms or entelechies that they are able to account for this fact, where, it is also claimed, empirical science, which is said to be "reductive," cannot.[44]

In reply to these arguments, it needs to be stressed that, in any reasonable sense, the formula that the whole is greater than the sum of the parts applies to any complex system even on the empiricist view. In fact, several reasonable senses can be given to that formula.

In the first place, complex systems have parts, but the parts stand in certain *relations*. These relations are not exemplified by any of the parts taken in isolation. One cannot describe the complex systems without taking into account those relations. In that sense, the whole is not simply a heap of parts; it is, rather, a *structured* entity. But as we have seen, empirical science recognizes this point. Certainly, when Newton was considering the solar system and its behaviour, he recognized that he had to include in his description of the objects he was considered the *spatial relations* among them. These in fact, the relative positions and the relative velocities, are among the relevant variables that had to be included when formulating the law for predicting the behaviour of the objects in the system.

Secondly, objects in complex systems often behave in ways in which they do not behave when they are in isolation. Thus, the behaviour of Mars travelling about the sun is very different due to the effect of Jupiter from what it would be were that large planet not to exist. That is, the laws for the behaviour of objects in two-body systems are different from the laws for the behaviour of objects in three- or n-body systems. However, from the fact that the behaviour of objects *in* complex wholes is different from the behaviour of those objects in simpler systems, it does not follow that the laws for the behaviour in complex systems cannot be deduced from the laws for the behaviour in simple systems. To the contrary, the such a deduction might well be possible. All that is required is that there be a *composition law*.

There are those who accuse empirical science of ignoring the "fact" that the whole is greater than the sum of the parts, inferring therefrom that we have to introduce souls or forms or entelechies if we are to understand complex biological systems. What has just been said provides an adequate reply to such persons, and shows the invalidity of the inference.

Nonetheless, there are those who do conclude, for whatever reasons, that such souls or forms or entelechies exist. And with regard to those accounts of the world, it is clear that in those theories of the organism the whole, or a property of the whole, does play a role that it does not play in empirical scientific theories. It does not follow that they are true, nor even that they make any real sense, that is, real empirical sense. Certainly, if we insist, as Judge Overton insisted when he found creation science to be non- or rather pseudoscience, and as we also should insist, that scientific theories be

... testable against the empirical world.[45]

then precisely because these theories do introduce non-empirical entities they are clearly not scientific.

Theories that introduce these entities into explanations in biology are often called "organismic" since they claim to be based on treating the organism as a whole rather than, simplistically, we are told, a set of parts. We should not, however, confuse this view, which is anti-scientific, with the view that in studying an organism we must attend not only to the DNA but to the whole organism in the sense of looking at the environment with which the DNA interacts. We earlier noted that this view was defended by Evelyn Fox Keller.[46] This view, too, can be characterized as one which insists that one must look at the whole and not simply at the parts. But what this means is simply that processes except in special cases will involve a *total interaction* among all the relevant variables and that one cannot find out how some variables behave if one ignores others. This sort of "dynamic interactionism" is just good science. To relate it, as Keller does,[47] to organismic views of biology is therefore misleading.

In contrast, there are those who today insist that we must take a "holistic" approach to various things, in nursing, in medicine, in astrology. In some cases, the call makes sense. In looking after a patient, one must think of the whole person. Similarly in medicine. There is not just one entity, the disease, that can be treated in isolation. Rather, it interacts with other parts of the organism, and conversely, things in the environment of the person, e.g., his or her relatives, the kind and sympathetic feelings of the nursing staff, the grumpiness of the physician, and so on, also interact with the disease. In these case, all that is being called for, and rightly so, is that one take account of as many of the relevant variables as one can. But in most cases where a "holistic" approach is called for, the demand is made on grounds that ordinary empirical science is simply inadequate for the job. It is "reductionistic", looks only at "parts" and not "wholes", and so on. Often the call is made on grounds no better than those of Dreisch: we cannot now give an explanation, or at least not an explanation that completely explains certain things, and we must therefore introduce a non-scientific explanation, something that recognizes what empirical science supposedly misses, the significance of the whole.

What would have happened if this attitude had been there when the predictions based on Newtonian theory of the orbit of Uranus turned out not to be those that were observed? It would have been said that "We can't explain the motion of Uranus; the system is simply too complex to admit of a mechanical explanation; there obviously have to be non-mechanical variables interfering with the motion of Uranus; obviously the system has a spiritual soul that interacts and move the planet Uranus in ways that contravene the ordinary laws of nature." But of course, if this attitude towards "reducing" the motion of Uranus to the laws of mechanics for simple systems had in fact prevailed, then we would not have found the correct explanation. Instead of searching for the planet Neptune, scientists would instead have — vainly — searched for the (non-empirical) "spiritual" or "holistic" entity that caused the deviation.

Scientists, however, — and fortunately, given the cognitive interests that it serves — , did not take seriously the notion that the composition law broke down and that there were non-empirical entities interacting. They assumed, to the contrary, that the composition law continues to hold as systems become more complex. It was this assumption, *rather than the "holistic" assumption*, that led to the discovery of the planet Neptune. If the "holistic" assumption had been accepted, Neptune would never have been discovered.

The point is clear: once again it is necessary to insist that theories to be scientific must be

... testable against the empirical world.

and to insist also that simply because we have not yet succeeded in giving a scientific explanation of something, therefore we must retreat to non-science. We should, rather, admit our ignorance, and then use that as grounds for undertaking further research!

In effect, the calls for a new "holism" are calls to introduce non-empirical entities into our theories. It is to call for non-scientific ways of knowing, and non-inductive methods of inference. What are these entities? We are never told for sure, because we have no way of actually testing them against experience. But they are almost always somehow "spiritual" rather than "material" — and we all know how science is vulgarly materialistic.

But it really is nothing more that an attempt to re-enchant the world, to fill it up with also sorts of gods and demons and souls and spirits, all of which Hippocrates so long ago enjoined us to reject, when he rejected the then common view that epilepsy was the "sacred disease."[48]

Example II: Darwin's Science

Darwin succeeded in providing a theory for biology that now constitutes the basic framework for all research and theorizing in the area: it is accepted as true, worthy of use in explanation and prediction, especially in research, by all biologists. It does conflict, however, with the Biblical story of creation. Just as the Copernican theory, and its champion Galileo, were attacked by appeal to the Bible, so Darwin and his theory have been attacked by more recent defenders of the literal truth of the Biblical scriptures. In the past, these attacks attempted to challenge Darwin on the simple ground that it conflicted with scripture. Various fundamentalist groups defended the creation story of Genesis as being true. Since this creationist story, if taken literally, conflicts with Darwin's theory of evolution by natural selection, the latter was rejected as, in effect, heresy. On that basis, it was attempted earlier in this century to exclude the theory of evolution by natural selection from being taught in schools in various states in the United States. Such efforts were eventually declared unconstitutional because they violated the constitutional requirement of the separation of church and state. More recently another line has been taken. This line is that there is scientific evidence that Darwin's theory is incorrect and also scientific evidence in favour of several parts of the creationist account of the origin of living things. This more recent argument has come to be known as "scientific creationism." Since this, it is claimed, is science and not religion, it can be taught in schools in the United States without violating the constitutional requirement for the separation of church and state. Eventually this too went to court when the state of Arkansas made it obligatory that creation science be taught in high schools alongside Darwin's theory.[49] In a landmark decision, Judge W. Overton ruled in a well argued opinion that creation science is not science at all.[50] It is after all pseudoscience, and more specifically religion masquerading as science. It therefore had no place in the schools, he concluded, since, being in fact religion, teaching it in schools violates the constitutional requirement for the separation of church and state.

One must add that it is not only wrong to propose that creation science is *science*, but to give it a place in the schools would make it impossible to do any reasonable sort of biology. For, biology today has the theory of evolution by natural selection as the one theory that provides the organization for its entire theoretical structure.[51] Moreover, we shall here that, as in the case if Newton, this capacity to organize laws in diverse areas into a systematic body of knowledge strongly confirms the theory.

We shall turn to Darwin's theory directly, but in looking at the structure of this theory, we will at the same time look in somewhat greater detail than hitherto the arguments of the creation scientists.

(a) Biological Adaptation: The Givens

There are a simple set of given facts about biological phenomena, that is, about organisms and the environments in which they come into existence, live, survive through feeding, reproduce, and die. These, in the case of biology, as the "appearances" that any good theory must "save."

Of these facts, the most important is this:

(G)   . Organisms, on the whole, behave in goal-directed ways that are plastic and persistent towards the overriding goals of survival and reproduction.

This fact can be elaborated upon in four ways:

(a) Organisms, on the whole, in their normal environments do as a matter of achieve their goals.

(b) For each species of organism there are various conditions ("needs") the fulfilment of which are necessary conditions for organisms of that species to achieve their goals.

Note that from (a) and (b) it follows that on the whole organisms manage to satisfy their needs.

    (c) For each species of organism there are various anatomical and physiological mechanisms the working of which in organisms of that kind in normal environments is necessary and sufficient for the satisfaction of the needs of those organisms.

    (d) For each species of organism there are various anatomical and physiological mechanisms which (i) monitor the environment and the organism itself and (ii) which have the effect of causing and maintaining behaviour that is sufficient in organisms of that kind in normal environments for attaining the goals of survival and reproduction.

To say that there are mechanisms that can explain the satisfaction of needs and guide the organisms towards satisfaction of their goals is to say that

(S)   The members of any species S are well-adapted to achieving their goals of survival and reproduction in their normal environments.

We know furthermore that

(R)   Reproduction on the whole preserves the specific kind, save that every offspring varies in minor ways from the parents and from the norm of the species.

It follows that

(P)   For any species S, that species reproduces itself.

Note that (d) and (R) imply that there are physiological and anatomical mechanisms that are sufficient to ensure reproduction according to kind allowing for minor variations.

(b) Geological Givens

The framework in which Darwin's theory of evolution developed derives from geology. Specifically, geologists have clearly established the following theses:

a.   The Earth has existed for something like 4.6 billion years.

b.   The geologic record is a record of changes over time which are to be explained naturalistically by appeal to forces that can be identified as working now.

Many of these geological processes are very slow, for example, the laying down of sediments on ocean floors. There may well be many sudden catastrophic events that also occur, e.g., volcanic eruptions or meteorite impacts. The point is that these are all to be understood naturalistically, without appeal to supernatural forces. What is excluded are catastrophes the supposed causes of which are supernatural, e.g., a great world-wide Flood as recorded in the Book of Genesis. Here, of course, is one of the great divides between those who call themselves creation scientists, on the one hand, and, on the other, the defenders of Darwin, and, more generally, those who accept the account of the long history of the earth as given by almost all geologists.

c.   The geological record is a record of changes in which simpler organisms preceded more complex organisms.

d.   The environments of most organisms have been in constant change throughout geological time.

e.   The geological record is very incomplete.

The incompleteness of the fossil record does not mean that it is not very full. To the contrary, it provides a sufficiently detailed record that it makes the conclusion that evolution has occurred all but inevitable. It is this record more than anything else that convinces people that evolution has occurred. The only problem is to find a mechanism that explains the fact.

At the same time, this record becomes a central point for creation scientists to challenge.

(c) The Problem Facing a Theory of Evolution

To say that a theory of evolution becomes almost inevitable once one examines the fossil record is not to say that such a theory is not without difficulties. Right from the beginning, there are problems which any such theory must face. The biological facts themselves, those we listed above, raise the central problem with which it must cope.

A theory of evolution of species states that

(TE)   For any organism at t_0, there are organisms at any t prior to t_0 such that the latter are the reproductive ancestors of the former.

From the geological record we can infer the following three things:

(A)   Some species of organism which existed in the past do not now exist.

(B)   Some species now existing did not exist prior to certain times.

By and large, organisms early in the geological, and therefore temporal, sequence are less complex than those toward the present.

The theory (TE) and the facts (B) and (C) from the geological record jointly entail

(D)   Presently existing species of organism evolved from earlier, less complex, species.

This conclusion seems to be in *prima facie* conflict with (R), that organisms reproduce according to kind.

Moreover, the fact (A) from the geological record seems to be in *prima facie* conflict with thesis (S), that species are generally well-adapted to survival and reproduction.

Here, then, are the problems that face any attempt to develop a theory of evolution for biological species:

Any theory of evolution must show [i] how to reconcile (A) and (S), and [ii] how to reconcile (TE), (B), (C), and (D) taken jointly with (R).

To many, this has seemed an almost impossible task. Certainly, so the creationists argue. As it turns out, Darwin's theory can in fact cope with this problem.

But before we turn to that theory, it will pay to examine in detail the scientific creationist account of the geological and biological facts.

(d) The Scientific Creationist Account of the Origin of Species

The creation scientists reject any theory of evolution. They do not challenge many of the empirical data, for example, the geological record of fossils, but they do challenge how these are to be understood. Thus, they argue that the geological record does not record a series of events that happened over a very long period of time. To the contrary, they argue, there is scientific evidence that the period of time is very short indeed, compared with what would be needed if Darwinian evolution were to occur.

The view developed by the creation scientists is seldom fully laid out. In fact, on perusing the literature one finds a certain amount of disagreement as to exactly what the position is. Nonetheless, there is a structure of claims and hypotheses that is fairly clear. It goes something like this.

The first thing to note is that it proposes that the history of the earth and, indeed, of the universe, is to be understood in terms of four stages. There is, *first*, the creation of the universe and its aftermath. This is followed by the *second* period which is ushered in by the fall of humankind and the original sin, since inherited not only by all humankind but by the universe as a whole. This period was followed, *third*, by a universal or world-wide flood, which destroyed almost all life save that which escaped through being on the Ark with Noah. We are now in the *fourth*, post-flood period. In greater detail, the proposed explanatory system goes like this:

I. The Creation
   a. Accomplished by a supernatural being
   b. Everything created from nothing
   c. The creation of the universe, and therefore of the earth, was relatively recent
   d. The earth was perfectly designed for life
      i. Originally protected by a vapour layer

α. Originally a uniform warm climate

β. Cosmic radiation could not penetrate the vapour layer

ii. No wind or rain

iii. The land irrigated by water from underground

e. All kinds created separately

   i. Each kind is unique and fixed

   ii. Each kind is genetically highly variable

f. Humans were uniquely created

g. Originally no decay occurred

II. The Fall

  a. The second law of thermodynamics begins to apply. This law requires that

    i. The originally perfect order begins to deteriorate

    ii. Death, decay and disorder began

  b. People began to populate the Earth, all humans descended from the original couple

  c. The vapour barrier was still in place and allowed greater longevity than is possible now that it has gradually deteriorated in conformity with the second law of thermodynamics

III. The Flood

  a. There was a simultaneous, worldwide cataclysm of a great flood

  b. All land was covered within forty days of the beginning of the great Flood

  c. Flood water came from two sources

    i. The vapour barrier

    ii. Underground reservoirs

  d. The Flood began 1656 years after the creation

  e. The Flood formed and deposited the various layers in the geological record

    i. Some species, e.g., dinosaurs, became extinct

    ii. Weaker species of animals drowned first, stronger species later, leaving behind the layered structure of the geological record

  f. The Flood split the land mass into the present continents

  g. The only survivors of the Flood were aboard one boat

    i. Eight humans

    ii. One pair of most kinds of animals

    iii. All were aboard the boat for 371 days

IV. The Post-Flood Period

  a. Leftover flood energy caused the ice ages

  b. Flood survivors repopulated the Earth

  c. All living species are descendants of the survivors

    i. All animals have great original genetic variability

    ii. They were modified by differential development to fill the Earth with various subspecies

  d. The vapour barrier was destroyed when it provided the Flood waters — decreased longevity resulted

  e. All species degenerate since disorder *must* increase

  f. Present geological processes are different from those of the Flood

It is clear that these propositions mix considerable science with non-science, and we shall have to separate out the various strands. But it is absolutely clear that this creation science set of explanatory proposals conform to the requirements of Christian faith as these requirements are understood by Protestant fundamentalists.

Here is a statement of these **Fundamentals of Christian Faith** taken from the *General Catalogue of Christian Heritage College*, 1977-78, of the Christian Heritage College. This college was co-founded by Dr. Henry Morris. Dr. Morris is also Director of the Institute for Creation Research (ICR), which publishes extensively in the area of creation science, pursuing its research in an attempt to provide a scientific creationist alternative to Darwin and any theory of evolution. The ICR is the research division of Christian Heritage College.

Here are the **Fundamentals**:[52]

- *The absolute integrity of the Holy Scripture* and its plenary verbal inspiration, by the Holy Spirit, as originally written through men prepared of God for this purpose.
- *The tri-une God — Father, Son and Holy Spirit.* There is only one God, who is the

source of all being and meaning, who is structured in three persons, each of whom is eternal, omnipotent, personal and perfect in holiness.

- *The pervasive influence of sin and the Curse.* When man first sinned, he brought himself and all his descendants, as well as his entire dominion (the universe), under God's curse.

- *The redemptive work of Jesus Christ ...* the great Curse finally reached its climax when the Creator Himself accepted and endured its ultimate and greatest intensity.

- *The immanent return of Christ.* Although the price has been paid, and the victory is ensured, the final consummation is to be realized only when Jesus Christ, now in Heaven at the right hand of the Father, returns personally to the earth to destroy all rebellion and to establish the perfect and eternal reign.

Those amongst these propositions that refer to things specifically Christian would of course be rejected by other religions, e.g., Judaism and Islam. But taking the Bible literally, and therefore the creation story in Genesis, is common to fundamentalism in all three religions. So those parts of the scientific creationism that refer to this and are not specifically Christian, as well as the attack on Darwin and the theory of evolution, have been accepted by fundamentalist groups in these other religions. But other aspects, e.g., those, such as the specific role of the second law of thermodynamics, that turn upon the doctrine of original sin, have to be rejected by these other fundamentalist interpreters of the Bible.

It is clear that in the over-all creation science set of hypotheses, there are those that are not only clearly religious but are clearly Christian. But it is also the case that there are aspects which are not specifically religious but are at the same time non-scientific.

Let us first look at the **Non-Scientific Non-Religious Aspects of Scientific Creationism**.

These have to do with the doctrine of creation. The creationist hypotheses accept that the universe as a whole was created, that this was accomplished by a supernatural being, and that everything created from nothing. It is true that Christians, Jews and Muslims all accept this doctrine of creation of the world. They all, in the end, accept these propositions as a matter of faith. At the same time, there are metaphysical arguments that can be offered to defend the doctrine of the creation by a supernatural being of the universe out of nothing.

This thesis is, to be sure, not scientific. The very fact that it invokes a supernatural being excludes it from being scientific, since the scientific search for causes, as we have seen with Herodotus and Galileo, limits itself to the world that we observe in ordinary sense experience, that is, the natural world. But the fact that the proposition is one that is not scientific does not make it religious. Thus, for example, one could hold that there has been a creation of this sort without holding that the creator has any moral qualities. This would not be what the religious believers accept, since they hold that the creator is good and just, as well as omniscient and omnipotent. It would, however, be a possible position.

The argument is often made that, since every event has a cause, therefore the universe as a whole must have a cause. This cause of the universe, since it is outside the universe, is the creator.

Now, this argument is not itself scientific. For, in the *first* place, its conclusion refers to something outside the natural world. The causal principle upon which it turns must therefore make reference to things supernatural. The causes to which it refers cannot be the merely empirical causes to which science refers. The proposition that

(CP) *Every event has a cause, with like effects having like causes and conversely.*

insofar as it is scientific is a generalization about events in the natural order, events that can be known through sense experience. If (CP) is worthy of acceptance, it is because it has been confirmed by our actually having discovered the causes of many events. Such discovery of causes is the discovery of events in the natural order, by means of sense experience. So the causes to which (CP) refers are always natural causes. (CP), taken as a proposition of science and as confirmed in ordinary experience, therefore cannot be used to infer the existence of a supernatural being. Science cannot lead us to the extra scientific.

In the *second* place, insofar as the argument for a first cause appeals to (CP) in order to establish a cause for the natural order as whole, it commits the *fallacy of composition*, as the philosopher Immanuel Kant pointed out. The events to which (CP) refers are events *within* that natural order, that is, things which are *parts* of a whole. But the inference to a first cause of the natural order applies (CP) to the natural order *as a whole*. A proposition that applies quite correctly to each part cannot without further justification be applied to the whole. For wholes often have properties that its parts fail to have, and, conversely, often lack properties which its parts have. It is true to say of each tree in the forest that it is covered in bark. But it is wrong to infer that therefore the forest as such is covered in bark. It is wrong because that conclusion is in fact false, and that falsity shows that it is an invalid inference — a fallacy — to infer that what holds for a part holds for the whole.

We cannot, therefore, appeal to the scientific notion of cause or to the causal principle (CP) understood as a scientific proposition confirmed in ordinary experience, to infer the existence of a supernatural being.

Nonetheless, if one held, for example, that the Aristotelian account of causality, while not scientific, was in fact true, then one might well be able to argue that there is sound inference to the existence of a supernatural being. On the Aristotelian account, a thing has being, that is, *is* in some way other, by virtue of some active power within it; such events are explained by appeal to the exercise of those powers by that thing. It is often argued by those who accept this position that *every* contingent being requires explanation. Every being in the universe is contingent, and receives an explanation by the exercise of some active power. But the universe as a whole is also contingent. So it, too, requires explanation by appeal to the exercise of some active power. The explanation will never be complete unless one appeals to the exercise of an active power by some necessary being, that is, some being whose existence is necessary in the sense of not being dependent upon the existence of any other being. This supreme active power, this active power that explains all other active powers as well as itself, is reckoned by some to be what has traditionally been called God.[53]

This is not the point to analyse this argument. Suffice it to say that it has had a long history; many have found it rationally acceptable, though that is of course open to debate. The point to be made here is rather different.

There may indeed be some causal argument, based on a non-scientific notion of cause, that will establish the existence of a necessary being as the first cause of the universe. The point is that the argument is *not scientific*.

Thus, insofar as the hypotheses of the creation scientists contain the thesis that there is a creator and that She created the natural order our of nothing, then those hypotheses are non-scientific. But, to repeat, this does not by itself make these aspects of the set of hypotheses specifically religious. There are, however, other aspects which are specifically religious.

Let us now turn to the **Religious Aspects of Scientific Creationism**.

There are two main religious features of scientific creationism. The first is the doctrine that prior to a certain point the world was perfect, or at least as perfect as a world of finite beings could be, and that imperfection entered with the Fall, the original sin of Adam and Eve. That sin, it is held, corrupted not only human being but also everything over which human being had been given dominion, which is in fact the universe as a whole. The imperfection that entered the universe consists in the Second Law of Thermodynamics coming to describe the workings of the universe. This law asserts that over time there is a tendency for complex things to degenerate into less complex, a tendency for a non-uniform distribution of energy and matter to transform itself into a more uniform distribution of energy and matter.

The Second Law of Thermodynamics is certainly a law of science; that is one of the scientific aspects of creation science. But the claim, that it began to describe the workings of the universe only after a period of time when is did not correctly describe how things go, is certainly not part of science. Moreover, the claim, that the world changed so that it does correctly describe the world only when Adam and Eve first sinned, is clearly non-scientific. Not only is it clearly non-scientific, it is clearly religious. Indeed, it is a peculiarly Christian doctrine. Other faiths may be sympathetic to the anti-evolution claims of scientific creationism, but a faith such a Judaism does not accept the reality of original sin. The latter doctrine is peculiarly Christian.

The doctrine of the flood is for the most part religious also. It is possible to argue that there is evidence of an empirical sort that a world-wide flood occurred in the past. But that it occurred 1656 years after creation clearly has no basis in science; to the contrary, that belief is rooted in the scriptures, the story in the Book of Genesis.

Moreover, that is equally clearly the basis for the claim that the only survivors of the Flood were aboard one boat, the claim that these consisted of eight humans and one pair of most kinds of animals, and the claim that all were aboard the boat for 371 days.

In the overall body of claims made by scientific creationists, then, there is much that can be accepted only on the basis of religious faith, and, specifically, Christian religious faith. "Scientific creationism", understood as a series of hypotheses to explain the natural order that we now observe, thus has a quite considerable religious component. This did not escape the notice of Judge Overton, when he argued in his judgment that "scientific" creationism was more religion than science.

Now let us turn to the **Science in Scientific Creationism**.

This consists in seven major claims.

1. The Second Law of Thermodynamics prevents large scale evolution of organisms of greater structural complexity from organisms of simpler structural complexity.
2. (a) Many physical observations of the earth and the universe suggest ages that are many times shorter than the commonly accepted 4.6 billion-year age for the Earth and the 10 to 20 billion-year age for the universe.

    Moreover, (b) radiocarbon dating is unreliable.
3. The Earth never had a methane based atmosphere, and even if it had, the fundamental components of life (e.g., amino acids) would not have formed.
4. The geological record shows that the earth's history has been punctuated by many catastrophes, and perhaps even a world-wide flood.
5. The fossil record shows sudden shifts in life-form morphology, not gradual change; therefore, many evolutionists are abandoning Darwin's ideas as totally unworkable. The logical alternative would be creationism, but

evolutionists have a "religious" commitment to evolution and thus cannot accept the obvious.

6. Most of life, especially biochemistry, is too intricate and interdependent to have evolved. The intermediate stages would not have been viable; thus, creative design is seen at all levels of organization.
7. The probability of getting even the simplest self-replicating systems purely by chance is essentially zero.

We can look at these in turn. Some of the points we have made earlier, but it will pay to re-state these briefly.

Re. (1): The Second Law of Thermodynamics prevents large scale evolution of organisms of greater structural complexity from organisms of simpler structural complexity.

This claim is made to refute the idea, inferred from the geological record, that simple biological organisms have been succeeded by organisms of increasing complexity; it is argued by creation scientists that, given this Law, it is *scientifically impossible* that current complex organisms have evolved from simpler organisms. But the claim that the Second Law precludes large scale evolution simply not true. The Second Law of Thermodynamics applies to a system as a whole, and declares with respect to such a system that *overall* there is a tendency for complexity to degrade into simplicity. But a tendency towards overall degradation is quite compatible with *local* growth and *local* increases in complexity. The Second Law of Thermodynamics is thus, as we previously noted, compatible with there being a process — a biological evolutionary process — in which complexity emerges out of simplicity.

Re. (2): (a) Many physical observations of the earth and the universe suggest ages that are many times shorter than the commonly accepted 4.6 billion-year age for the Earth and the 10 to 20 billion-year age for the universe. Moreover, (b) radiocarbon dating is unreliable.

These claims are the scientific basis for the creation science challenge to the evolutionist claim, based on the geological record, that the earth has had a very long history. The claim that the earth's history is very long is based on the facts of geology. Allow the claim is geological, it is a claim that is essential to any theory of evolution insofar as that theory proposes — as Darwin's theory does — that evolution proceeds through a long series of incremental steps. In (a), the creation scientists are claiming that the geological record is not as clear as evolutionists state, and in (b) the creation scientists are claiming that the standard techniques for dating the age of geological formations is faulty and cannot be relied upon.

The great age of the Earth was proposed by geologists well before Darwin developed his theory of evolution by natural selection. Extensive work, especially in Britain, in building roads and canals led to the exposure of many levels of geological deposits. Geologists argued that the uppermost deposits were, generally speaking, the more recent, although one could also find places where the order were reversed. The reversal could be seen because fossils that were lower down in one deposit were in upper layers in other deposits. Careful accumulation of data enabled geologists to sort out these problems of order. Geologists could also see the difference between deposits that are laid down quickly, as in floods, and those that are laid down gradually, which consist of particles of much smaller size. The geologists of the early 19th century concluded that the deposits they were examining had been laid down very slowly, over a span many 1000's of times longer than is allowed by the Biblical story in Genesis. They concluded, moreover, that the earlier deposits on the whole contained simpler organisms. This much was available to Darwin before he began his work.

Scientific creationists must attack this view. They attempt to give the appearance of open-mindedness:

> As a matter of fact, the creation model does not, in its basic form, *require* a short time scale. It merely assumes a period of special creation sometime in the past, without necessarily stating when that was. On the other hand, the evolution model does *require* a *long* time scale. The creation model is thus free to consider the evidence on its own merits, whereas the evolution model is forced to reject all evidence that favours a short time scale.[54]

But this mistakes the logic of science. If all the evidence tells very strongly on one side and there is only questionable evidence on the other, then the only reasonable thing to do is to accept the theory which has the far stronger support and indicate, on the basis of the evidence, how unlikely it is that evidence will be found that will tell strongly against the theory. This does involve the notion that apparently contrary evidence is likely to be found unacceptable. This not *mere* rejection, however, nor is it irrational. To the contrary, *since* the theory is on the basis of good evidence strongly supported, *then* it is reasonable to expect that there will be no contrary evidence or, if there is, that it will be weak.

With the discovery of radioactivity and radioactive decay, scientists were able to develop tests that can date deposits with considerable accuracy.[55] These tests depend upon knowledge of rates of radioactive decay of various elements. There are several independent such tests, each of which has been carefully verified as to accuracy. Creationists attack these on several grounds. Three are worth noticing.

In the *first* place, they claim that there are other methods which imply a much younger earth than that implied by the techniques of measuring radioactive decay. The claim is that these data are as strong as those proposed by defenders of evolutionary theory, and that this contradiction in data requires us to suspend judgment about the theory of evolution and to allow that the creationist thesis of a short time span for the history of the earth is equally worthy of consideration. At the very least, both of them should be taught in the schools.

Thus, the creation scientists have cited the fact that over the past century and a half scientists have recorded a decay in the earth's magnetic field. If this decay is extrapolated into the past, then we are required to attribute an ever increasing strength to the magnetic field, and if we go far enough into the past we require a magnetic field of a strength that no known process in the earth could support. This then places limits on the age which can reasonably be attributed to the earth.[56]

The difficulty with this argument is that the earth's magnetic field is known to leave in the rocks of the earth's crust when they are formed a record of its force and direction. The history of the magnetic field can therefore be traced. It turns out that the earth's magnetic field has gone through many variations and reversals of polarity in the history of the earth. This evidence, available in many respected studies, is simply ignored.

This is in fact a common strategy in scientific creationist arguments: Present an analysis that seems to be every bit as scientific as one that is offered by evolutionists, and then ignore any data that call that analysis into question. The former gives the appearance of science. The latter turns that appearance into falsity: when good science is done, *all* the evidence is taken into account. It is precisely this sort of thing that gives creation science the veneer of science when in fact all real science is absent. This is why creation science can reasonably be classified as pseudoscience.

The *second* line of attack used by creation scientists is to argue that scientists disagree about rates of decay, implying that this throws into question the accuracy of the methods.[57] However, this disagreement about the actual rates of decay is not an absolute disagreement. In fact, while scientists disagree on the *exact* rate, they do agree that there is nothing in the variation that would throw estimations out by the 50,000,000 % that would be required if the earth were to be as young as the creationists claim.

This illustrates a common technique in the rhetorical arsenal of the scientific creationists. They very often can point to places where scientists disagree about the details of their theories. This disagreement is then taken to show that there is no consensus about the theory, and that in the presence of this conflict of evidence then at best we ought to suspend judgement about the theory or, they suggest more often, we ought to reject the theory. The scientific creationists fail to note that scientists can disagree about details without disagreeing in a way that calls their theory into question. The evidence available may tell strongly in favour of the theory overall, while leaving open, awaiting further investigation and debate, details about the theory. Thus, Newtonian scientists could agree about the general structure of mechanics, and agree that the evidence strongly supported it, while also disagreeing about the exact value of the gravitational constant "G" as it appears in Newton's law of gravitation:

$$F = G \times [m_s \times m_p] / r_{s,p}{}^2$$

But there is another, *third*, gambit that is employed, that of simply accepting the results and then dismissing them. Thus, "We reply," it is said to the argument from radioactive dating,

...that the Biblical outline of earth history, with the geologic framework provided thereby, would lead us to postulate exactly this state of the radioactivity evidence. We would expect radiogenic minerals to indicate very large ages and we would expect different elements in the same mineral, or different minerals in the same formation, to agree with each other.[58]

But precisely *why* should this be expected? Why should we expect the creator to place in the rocks data which, when discovered, would lead us to expect that the world is very old, 50 million times older than what the creationist model claims? We are back at the Gosse hypothesis, that God has created a universe which exhibits, but falsely, the appearance of being very old. It is there to test the faith of the true Christians.

Re. (3): The Earth never had a methane based atmosphere, and even if it had, the fundamental components of life (e.g., amino acids) would not have formed.

This is not an argument against the Darwinian theory of evolution. The latter is a theory about the origin of species of living things, not about the origins of life itself. One could in fact hold that life itself requires a supernatural creation while consistently holding that once created it evolved in the way Darwin said. Of course, to say that life requires a supernatural creator is to go beyond science; such a claim would be non-science. And most scientific biologists are convinced that life originated from non-life by natural means. Specifically, it is believed that the basic building blocks of life, the amino acids, originated from the earth's former methane based atmosphere. So claim (3) does challenge the currently accepted scientific framework for the origin of life.

Once again, no strong arguments, *taking account of all available evidence*, are offered for the claim that the earth never had a methane based atmosphere. As for the claim that amino acids and other organic molecules could not have emerged out of non-organic mixtures, this is based on the argument that complexity cannot emerge out of simplicity. In part this is related to the argument about the Second Law of Thermodynamics, which we have already discussed, and in part it is a more philosophical point about complex systems. This latter argument appears elsewhere in the creationist case, and we shall return to it below when we discuss creationist points (6) and (7).

Re. (4): The geological record shows that the earth's history has been punctuated by many catastrophes, and perhaps even a world-wide flood.

According to the Darwinians, evolution is a slow process in which many small changes gradually accumulate until a new species has emerged. But, according to the creation scientists, the geological record shows that there have been many catastrophes in the earth's history. In other words, the history is not one that is "slow and steady," as the Darwinians require, but one that is marked by great and sudden changes.

Naturally, what the creation scientists do not point out is that of course the Darwinians accept the geological point that there have been many catastrophes in the history of the earth — even great floods at times, though no world-wide flood. The Darwinians argue that the existence of catastrophes is quite compatible with the claim that species originate through a slow and steady accumulation of small changes which in the end add up to a big change. They even argue that the *rate of accumulation* may vary from time to time. Thus, there may be a great catastrophe, such as the meteorite collision that is now believed to have led to the extinction of the dinosaurs, which is followed by a sudden spurt of evolutionary development of new species which is then followed by a period of relative equilibrium. The issue is whether evolution through such a "punctuated equilibrium" is compatible with the Darwinian theory. Those biologists who accept the view that the history of biological species is characterized by punctuated equilibria — and this seems now to be most biologists — agree that this is quite compatible with evolution proceeding by natural selection.[59]

Moreover, some of the things that are often referred to as "explosive" changes are that only by way of contrast. It is true that many species of modern phyla of organisms made their appearance during the Cambrian geological period. It was then that the first examples of vertebrates, crustaceans, molluscs, and echinoderms made their appearance. However, this geological period covers 100 million years of earth history. When events are spread over a time span of this order, it is misleading to call their occurrence "sudden". Moreover, many

modern species did not make their appearance until much later. Thus, our own species, *homo sapiens*, made its first appearance only 2 to 4 million years ago.

Re. (5): The fossil record shows sudden shifts in life-form morphology, not gradual change; therefore, many evolutionists are abandoning Darwin's ideas as totally unworkable.

There are two aspects to this criticism. The first has to do with the doctrine of punctuated equilibria in the fossil record, that is, the doctrine that there have been bursts of evolutionary development which have been followed by long periods of relative equilibrium. We have already dealt with this point in discussion of the "creation scientist"'s point (4). The other part of the present criticism is the implication that there are no intermediate forms. This is a further issue. It does not have to do with the *rate* of evolution; it has to do, rather, with the claim that evolution has occurred. Variation in the rate of evolution is compatible with evolution. What is not compatible with evolution is the claim that there are for most forms of life no intermediate species to be found in the fossil record.

The Darwinians claim that all living things have as their ancestors the simpler organisms that the geological record shows to have existed in the past. If this is so, then we should expect to find transitional forms. But the geological record is in fact gappy. The absence of these forms testify to the falsity of evolutionary theory.

> If millions of species have gradually evolved through hundreds of millions of years, the fossil record must contain an immense number of transitional forms — museums should be overflowing with them. The fossil record, however, is an explosive appearance of a great variety of highly complex creatures for which no ancestors can be found and systematic gaps between all higher categories of plants and animals. The fossil record is thus highly contradictory to evolution but remarkably in accord with creation.[60]

In the absence of evidence for intermediate forms, one can only conclude that species are in fact fixed. After all, everyone does accept as a biological given that

(R)   Reproduction on the whole preserves the specific kind, save that every offspring varies in minor ways from the parents and from the norm of the species.

from which, as we have recognized, it follows that

(P)   For any species S, that species reproduces itself.

Since the species are fixed, the logical alternative to evolutionary theories would be creationism, but evolutionists have a "religious" commitment to evolution and thus cannot accept the obvious. The defenders of scientific creationism thus attempt to unmask the defenders of evolution as really basing their beliefs on a kind of blind faith, which, though not Christian faith, is nonetheless faith and not science. It is the faith of secular humanism that would empty the universe of God and value.

It will not stand, however. For, in fact there are many intermediate forms in the fossil record; it provides a wealth of detail that makes the conclusion that evolution occurred almost inevitable. Thus, for example, the transition from reptile to mammal is represented by a dozen intermediaries.[61] Again, there have been many recent discoveries in east Africa of intermediate forms that preceded *homo sapiens*.

For the scientific creationists, in contrast, the history of the earth is a history of catastrophes, of which the last is the great Flood.

...it seems reasonable to attribute the formations of the crystalline basement rocks, and perhaps some of the Pre-Cambrian non-fossiliferous sedimentaries, to the Creation period, though later substantially modified by the tectonic upheavals of the Deluge period. The fossil-bearing strata were apparently laid down in large measure during the Flood, with apparent sequences attributed not to evolution by rather to hydrodynamic selectivity, ecologic habitats, and differential mobility and strength of the various creatures.[62]

That is, the fossil record as understood by geologists and evolutionists is simply an illusion. What appears to be a record of evolutionary change was in fact set down in the 300 days of the Flood. The weaker survived for the shortest period of time, and drowned first, going to the bottom. Those who were stronger survived longer, and their forms can be found higher up in the strata. Some parts of the land were higher than others and flooded last. The animals that lived on these places also survived longer, and this can account for other features of the strata. Some species that were not carried on the Ark became extinct.

This will not do either. There are all sorts of mammals that are very weak and cannot swim, and yet we find none of these below the strata that contain the dinosaurs. The Flood geology of the creation scientists simply ignores most of the data that make the fossil record such strong evidence for evolution. Once again we see how there is a parade of ostensible data covering up the basic fact that most of the data is simply ignored. The result is not science but pseudoscience.

In fact, creation scientists simply fail to provide any plausible account of the biological givens that we noted above could be inferred from the geological record:

(A)   Some species of organism which existed in the past do not now exist.
(B)   Some species now existing did not exist prior to certain times.
(C)   By and large, organisms early in the geological, and therefore temporal, sequence are less complex than those toward the present.

Their alternative proclaims that these facts can all be accounted for in terms of "creation." But that is no explanation: it simply leaves the details of the natural history of our planet veiled in mystery. That may be fine for religion; it is not science.

Two points about the way creation scientists argue are worth mentioning.

First, they argue against evolutionary theory by pointing to the fact that biological scientists disagree about the rate at which evolution has occurred. Since there is disagreement about the rate, some defending the thesis of punctuated equilibria while others attack it, the inference is that the whole theory is therefore to be rejected. After all, if scientists can't agree on it, then it must be wrong!

But this, surely, is to misunderstand the role of **criticism** in the growth of knowledge. The process of scientific research attempts to replace gappy knowledge by less gappy knowledge. In order to do this, one needs to formulate hypotheses about factors that can fill in the gaps in one's knowledge of laws. But these hypotheses compete; they are contraries, and cannot all be true. In order to find out which among the alternatives is true, some have to be shown to be false. It is the point of criticism to discover which among the several hypotheses are false, and which, therefore, are true. The existence of conflicting opinions does not show the weakness of a theory. To the contrary, it shows that the method of science is being implemented as it should be! Conflict and criticism is not a sign of weakness as the creation scientists would suggest, but rather a sign of strength in the search after truth.

Of course, for the creation scientists there should be no disagreement. They have the Bible to guide them inerrantly. Conflict and debate implies that some one at least is in error, and error is a sin: it is better to believe than not to believe, better to have faith than to reject the Bible as the inerrant standard of truth. From their standpoint of Biblical *authority*, debate is a sign of failure. No wonder, then, that they should look at debate as a sign of weakness.

But to the contrary, the very fact that they take debate as a sign of weakness shows that they are not operating within the framework of the scientific method: theirs is pseudoscience, not science.

The point is, naturally enough, that science rejects any appeal to authority as providing a criterion of truth. The only evidence that is acknowledged as relevant in the search after truth about matter of fact regularity and cause in the world is the testimony of observation. Since these data are always incomplete, science is always tentative, and debate is always possible, especially at the moving frontier of research.

This is not to say that everything is debated all the time. To the contrary, in large areas of science there is consensus about the basic structure of the theory. This is true even in biology. For all that there are debates between those who defend the thesis of punctuated equilibria and those who attack, there is in fact agreement that evolution is a fact and that it does proceed much as Darwin said it did, by a process of natural selection.

Because of the complex structure of the theory, there can be disagreement. One can accept the abstract generic theory, while disagreeing about hypotheses at the more specific level. By focussing on the debate and ignoring the consensus, creation scientists show that they misunderstand the logic of theory structure.

Second, the defenders of creation science argue that there are only two "models" for the history of the world, the creation science model and the evolution model.[63] They attempt to criticize the evolution model aiming to show that it is inadequate and then conclude that their own position is correct.

This, however, assumes that there are only two possible hypotheses. But this is not so. As we have seen, there may be several competing alternatives under what they call the "evolution model." To show that one of these is false is not to show that all of them are false.

Moreover, it is not just a matter of competing hypotheses. Even if creationists were able to show that there are significant gaps in the fossil record, it does not follow that we are obliged — that is, cognitively obliged — to accept the creationist claim that evolution has not occurred. All that follows is that **we are ignorant** of the intermediate steps. Failure to find does not entail absence. Indeed, given the exceptional conditions under which fossils are preserved, it would not be at all surprising that there are gaps in the fossil record.

Of course, in the context of the religious background of the creation scientists, it is not surprising that they are unprepared for the alternative of simply admitting ignorance, or, what is the same, *suspending judgment*. For Christians, and Christian fundamentalists in particular, it is better to believe than not to believe. To suspend judgment is, like disbelieving, a matter of *not believing*. For them, in other words, it is a sin to admit ignorance of the causes of things. To avoid admitting ignorance, they insist, one must turn to faith, and the creation science story as rooted in the inerrant word of the Bible. But whatever the rules are for cognitive attitudes in matters of faith, in science there are three attitudes: believing (accepting), disbelieving (rejecting) and suspending judgment. That they do not respect that there are these *three* alternative cognitive attitudes show that the creation sciences are not working within the framework of the scientific method.

This, of course, is not the only case where pseudoscientists argue that it is either ignorance or "my view", and since you do not want to admit ignorance then you had better accept my view. One can find the same sort of pattern in those who argue for the influence of ancient astronauts on human civilization. Von Dänikin and others argue that we cannot understand how or why our ancestors built certain ancient monuments unless we accept that they are aided in this task by extraterrestrial visitors. Since we do very much want to explain these events, and not remain in ignorance, we had better to accept the theory of the ancient extra-terrestrial astronauts. To which the appropriate response is that, to the contrary we must simply admit our ignorance since there is no evidence, beyond the facts of the monuments themselves, to suggest that ancient astronauts visited our planet. Again, the point is, that the appeal to such things as a way of avoiding simply admitting our ignorance is a sign that these "thinkers" are not operating within the framework of the scientific method.

In both these cases of arguments for ancient astronauts and the arguments of the creation scientists, we find that they clearly argue in ways that are not within the framework of the scientific method. To repeat, theirs is pseudoscience, not science.

Re. (6): Most of life, especially biochemistry, is too intricate and interdependent to have evolved. The intermediate stages would not have been viable; thus, creative design is seen at all levels of organization.

The claim is that on the evolutionist's view every intermediate stage leading up to, say, the eye would have to have enhanced the probability of the survival of the organisms that possessed those stages. The suggestion is that any intermediate stage would be an imperfect eyeball. Since the imperfect eyeball would not enhance the survival value of its possessor, the evolutionist's picture cannot be correct. Yet adaptation is a fact. The only recourse is to the idea of design.

We shall return to the latter notion directly. Here, however, the argument with which we must be concerned is that which purports to establish the inadequacy of the evolutionary point of view. Here it must be said that the claim is simply false that intermediate stages, because they would be imperfect, would not be viable. This is clear: short sightedness is an imperfection, but even so,

to see things in a fuzzy way, that is, imperfectly, is better than not to see them at all.

The further creationist claim that it would be impossible for something so intricate as the eye to evolve is also mistaken. Darwin himself raised the issue: "To suppose," he wrote, "that the eye with all its inimitable contrivances for adjusting the focus to different distances, for admitting different amounts of light, and for the correction of spherical and chromatic aberration, could have been formed by natural selection, seems, I freely confess, absurd to the highest degree."[64] But however much it seems unlikely, reflection can show that it is in fact entirely plausible. Appearances or seemings in this case are deceiving. For, it can be shown that it would in fact take only a relatively small number of steps for a flat piece of photo-sensitive skin to be changed into an eyeball, that is, a cup-shaped piece of skin complete with a lens capable of forming well-focussed images.[65]

If the process were truly evolutionary, the changes would have to be small, say no larger than 1% of the eye-to-be, and each would have to improve the organ, which, in this case, means that each would have to ensure that a better image was produced. A model of such a process can be developed in which each stage of the sequence is an eye which resembles a sort of eye that can be found among animals existing today; it is known, for example, that molluscs and annelids display a complete series of eye designs, from simple epidermal aggregations of photoreceptors to large and well-developed camera eyes.

It turns out that a process of the indicated sort would take slightly less than 2000 steps. If we think, as would be reasonable, of each step occurring after 200 generations, we are talking of an eyeball emerging from as it were nothing in 400,000 generations. If we assume each generation of small animal lives one year, then we are talking of less than half a million years. But this is hardly any time at all, if once we take account of the time available in the geological record. So the argument that the eyeball is too complicated to emerge from nothing is no good at all.

One could raise the issue that, if advanced camera lens eyes can so easily be developed, then why are there so many representatives of the other, earlier stages of the model sequence to be found in living creatures? The answer is that the eye does not exist in isolation. A lens eye can be useful only with more highly developed types of neural processing. For a sea worm to develop lens eyes out of its photo-sensitive cells it would in effect have to develop into a fish. Furthermore, the lens eye and the other required advanced features become useful to their possessors only when the ecological environment where visually guided locomotion is beneficial. Even so, there will be ecological niches in which such advanced features are not beneficial, and therefore do not develop.

Re. (7): The probability of getting even the simplest self-replicating systems purely by chance is essentially zero.

The idea is this: although it is possible that a tornado sweeping through a junkyard would re-arrange the litter into a luxury car, the probability of this is in fact vanishingly small. Chance processes are unlikely to produce complex entities. This is no doubt true, but the evolutionary process is not one of pure chance. To the contrary, there are many causal laws which govern the processes through which natural selection works.

There is, to be sure, a chance element in the evolutionary process. For natural selection to work, there must be small variations among individuals. The occurrence of these variations is random so far as concerns the environments in which the forces of natural selection work. Knowing those forces, we cannot predict what variations will occur, nor, therefore, the directions in which evolution might proceed. However, given that a change has in fact occurred, then whether or not it will survive to be passed on to future generations is a straight forward causal process rather than one of chance.

Then there is the origin of life itself. Again, this, strictly speaking, has nothing to do with Darwin's theory of evolution. But be that as it may, the creationist claim is that this, too, could not have happened by chance. However, it turns out that chemical mixtures of methane, ammonia, water and carbon dioxide such as existed early in the history of the earth can yield several complex molecules when exposed to electrical discharges (e.g., lightening) or ultraviolet light. The actual production in the laboratory of self-replicating systems has not yet been accomplished. What is known, however, makes it clear that there is no reason to suppose that anything other than natural forces and interactions brought self-replicating systems into existence. Again, the point that must be emphasized is that the process is in fact not one

of chance. Again to the contrary, the laws that govern the processes that took place to produce living forms are causal processes, not events related only by chance.

The problem is that there are two meanings of chance at work. One meaning of chance is "random process." Neither the process of evolution by natural selection nor the processes that produced living organisms are in this sense chance processes. The other meaning of chance is "unintended event," as when I meet someone by chance in the marketplace: neither of us intended to meet the other there, so we met by chance. Creationists slide quickly — but illegitimately — from a process not being one of chance in the first sense to its not being one of chance in the second sense: if it does not happen by chance then there must be design and intention. Reason demands that we resist this fallacy.

We may safely conclude that science in scientific creationism does not exist. What we have instead is the appearance of science, mere trappings, which, when explored, turn out to mask the real heart of the matter, fundamentalist Christian religion. Creation science is pseudoscience.

(e) Darwin's Theory of Evolution

Let us now turn to the theory of evolution. We shall distinguish two parts. First, there is the theory — or better, the fact — that evolution has occurred. And second, there is the theory about the cause or mechanism by which this has occurred. The latter is provided by Darwin's theory of evolution by natural selection. As we shall see, this theoretical structure is logically tight and has empirical data that strongly support it. In this it contrasts strongly with creation science with its ramshackle structure, a mixture of some poor science, dogmatic assertion, outdated metaphysics, and (mostly) fundamentalist religious faith.

In fact, the theory that Darwin's replaced, while it appealed to the workings of a powerful, human-like agent as a designer, was nowhere near the creationist model. We should look at this earlier theory first.

This older theory took seriously the problem of the eyeball, or, more generally, the fact that organisms are well adapted to survive and reproduce in their normal environments. This is the thesis (S) noted above that

The members of any species S are well-adapted to achieving their goals of survival and reproduction in their normal environments.

This states a fact, a fact about the natural world. Such facts require explanation. In the 17th and 18th centuries, the community of scientists rejected Aristotelian explanations of such facts, and argued instead that one had to make all their inferences in conformity with the causal principle that we noted above:

(CP) *Every event has a cause, with like effects having like causes and conversely.*

This principle must, they insisted, by understood naturalistically. The chemist Robert Boyle was especially forceful and clear in his argument that Aristotelian explanations were vacuous and had no place in science, whether it be physics or biology.[66] In making this argument he was following the lead of Galileo. At the same time, however, he argued that a purely scientific argument could be given to infer the cause of the observed adaptation of species to their environments, an argument based on the principle (CP).

The notion is simple enough.

Consider a watch. By observing certain effects, the motions of the watch, we can infer its function: its periodic motions are a regular measure of time. Inside the watch there exists a mechanism that keeps the motions not only periodic but regular in the periods. The mechanism is in fact very efficient in producing this result. In a similar way, the eyeball has a certain function, namely, enabling its possessor to see things. Inside the eyeball there exists a mechanism — the lens especially, but also the retina, the rods and cones, the iris, and so on . This mechanism is in fact very efficient in producing the effect of sight, that is, of ensuring that the eyeball fulfills its function. In the two cases, the watch and of the eyeball, we have similar effects. (CP) requires that we attribute like causes to these like effects. Now, the cause of the watch is a designer who has solved the problem of how to mark temporal intervals by creating a mechanism that ensures that that function is efficiently fulfilled. By (CP), therefore, we need to attribute a like cause to natural mechanisms such as those in the eyeball that ensure that organisms are well adapted to survive and reproduce in their normal environments.

This argument is straight-forwardly scientific: the inference to the non-observed cause is based on the principle (CP) which has been confirmed in experience. The inference here is parallel to the inference to a cause of the rising of the Nile that Herodotus quite legitimately attempted to make. Or the inferences that were made, based on Newtonian theory, to an (as yet) unobserved planet as the cause of the perturbations in the orbit of Uranus, the deviations of that orbit from the path predicted by the theory.

This scientific argument for a designer as the cause of adaptations in the biological realm was developed in detail by scientists such as Robert Boyle,[67] and later endorsed by theologian-philosophers such as William Paley.[68] The argument was subjected to severe criticism by David Hume,[69] but even he was careful to allow that this causal argument proved its conclusion. Thus, he argued that

> So little ... do I esteem this suspense of judgment in the present case to be possible, that I am apt to suspect there enters somewhat of a dispute of words into this controversy, more than is usually imagined. That the works of Nature bear a great analogy to the productions of art, is evident; and according to all the rules of good reasoning, we ought to infer, if we argue at all concerning them, that their causes have a proportional analogy (p. 216).

Hume's concern was that the theologians seldom paid sufficient attention to the disanalogies.

> But as there are also considerable differences, we have reason to suppose a proportional difference in the causes; and in particular, ought to attribute a much higher degree of power and energy to the supreme cause, than any we have ever observed in mankind (p. 216).

If we do attend to these disanalogies, then we will find that we have little reason to attribute to the cause of order, of adaptation, many of the attributes traditionally attributed to God, for example, moral qualities, and even omnipotence. Hume is willing to allow, however, that it is possible to call this cause "mind".

No man can deny the analogies between the effects: to restrain ourselves from enquiring concerning the causes is scarcely possible. From this enquiry, the legitimate conclusion is, that the causes have also an analogy: and if we are not contented with calling the first and supreme cause a GOD or DEITY, but desire to vary the expression; what can we call him but MIND or THOUGHT, to which he is justly supposed to bear a considerable resemblance? (p. 216)

Hume concludes that the conclusion of the argument is in fact reasonable, provided that the needed qualifications are made:

If the whole of Natural Theology, as some people seem to maintain, resolves itself into one simple, though somewhat ambiguous, at least undefined proposition, That the cause or causes of order in the universe probably bear some remote analogy to human intelligence: if this proposition be not capable of extension, variation, or more particular explication: if it affords no inference that affects human life, or can be the source of any action or forbearance: and if the analogy, imperfect as it is, can be carried no further than to the human intelligence, and cannot be transferred, with any appearance of probability, to the qualities of the mind; if this really be the case, what can the most inquisitive, contemplative, and religious man do more than give a plain, philosophical assent to the proposition, as often as it occurs, and believe that the arguments on which it is established exceed the objections which lie against it? Some astonishment, indeed, will naturally arise from the greatness of the object; some melancholy from its obscurity; some contempt of human reason, that it can give no solution more satisfactory with regard to so extraordinary and magnificent a question (p. 227).

To repeat, with Boyle, Paley and Hume, this is a scientific argument to the cause of the adaptations that we observe in nature. It is this argument that is taken up in part by the scientific creationists, though they of course make the further inference that the agent that cunningly created the observed adaptations also created the world out of nothing. The two inferences are very different, as we have seen. The former is scientific, the latter is non-scientific. And of course, the creation scientists also add a great deal of fundamentalist Christian commitment to a literal reading of the Book of Genesis. But however it may be that the creation scientists add to the argument, it remains, as Hume the sceptic emphasizes, a sound scientific argument. It is this argument that confronted Darwin, who was in fact impressed by the accumulation of evidence by Paley for the adaptedness of organisms to survive and reproduce in their normal environments.

There are two aspects to Darwin's theory that should be distinguished. One is the *theory of descent with modification*. This simply states that evolution has occurred. The second aspect is the *theory of evolution by natural selection*, which describes the casual mechanisms which explain the fact of descent with modification and thereby the origin of species through the accumulation of many small modifications.[70]

Consider first the **Theory of Descent with Modification**. This has a series of component propositions.

a. Life began with a few simple kinds.
b. All things presently living have simpler things as their reproductive ancestors.
c. All present forms of life have evolved from simpler kinds; each species, living or fossil, arose from another simpler species that preceded it in time; the members of the latter species are the reproductive ancestors of the members of the former species.
d. Evolutionary changes were gradual and of long duration.

Note that this is compatible with the thesis of punctuated equilibrium, that evolution occurs in spurts followed by periods containing relatively evolution of species. That is, (d) is compatible with

d*. Evolutionary changes occur during periods of relatively short duration followed by long periods of equilibrium.
e. Over long periods of time new genera, new families, new orders, new classes, and new

phyla arose by a continuation of the kind of evolution that produced new species.

f. Each species originated in a single geographic location.

g. The greater the similarity between two groups of organisms, the closer is their relationship and the closer in geological time is their common ancestral group.

h. Extinction of forms (e.g., species, etc.) is a consequence of the production of new forms or of environmental change.

i. Once a species or other group becomes extinct it never re-appears.

This theory of descent by modification is a re-statement, given the details of the geological record, of the theory of evolution (TE) discussed above.

Turn now to second part of Darwin's theory, **the Theory of Evolution by Natural Selection**. Again, this consists of a series of propositions. In these claims, it is important to notice the evidential support that they have, and the evidential support that they provide to one another. As will become clear, this support is tremendously strong.

Darwin's theory begins with three patterns or regularities about organisms that have been well confirmed by observations.

(T1) For any species of organism in its normal environment, offspring are (on the whole) more numerous than their parents.

(T2) For any species of organism in its normal environment the food supply and the other conditions of life required to satisfy the needs of the organisms are limited.

(T3) The greater [the more restricted] the food supply and the other conditions of life, the greater [the more restricted] the number of individuals that can be supported.

Note that these laws are about *species* and not just about individuals within those species. That is, these laws are already abstract and generic.

From (T2) and (T3), it follows deductively that

(T4) For any species of organism in its normal environment, the number of individuals is (on the whole) constant.

From (T1) and (T4), it follows deductively that

(T5) For any species of organism in its normal environment, fewer individuals survive than result from procreation.

Now recall the well confirmed law that

(G) Organisms, on the whole, behave in goal-directed ways that are plastic and persistent towards the overriding goals of survival and reproduction.

(G) and (T5) deductively imply that

(T6) For any species of organism in its normal environment, there are individuals born that, in spite of their plastic and persistent efforts, do not attain their goals of survival and reproduction.

Given the plastic and persistent nature of the efforts of organisms to survive and reproduce, (T6) implies that life might reasonably be characterized as a "struggle for existence" — taking 'struggle' in a suitably broad and metaphorical sense.

It is important to notice the patterns of confirmation. (T1), (T2) and (T3) all have empirical support in our observations of populations of organisms. (T4) is deduced from (T2) and (T3); the support that the latter derive from observations therefore also supports this proposition. But (T4) is also supported on independent empirical grounds concerning what is known about and animal and plant populations. These two lines of support for (T4) strengthen the evidential structure. For, the observational support for (T4) also serves to support the theoretical structure (T1)-(T3) which deductively implies (T4).

There are parallels to this sort of confirmational support in Newtonian mechanics. There is, in the first place, observational evidence that supports Newton's Laws of Motion, e.g., the evidence that Galileo produced for the Law of Inertia. At the same time, there is observational evidence in support of Kepler's Laws of planetary motion. These laws, together with Newton's Laws of Motion deductively imply the Law of Gravitation.

This law therefore receives observational support from the evidence that confirms Kepler's Laws and confirms Newton's Laws of Motion. The Law of Gravitation also deductively implies Galileo's Law for free fall. The latter is therefore supported by the theory. So the evidence that supports the theory, including such disparate facts as those that confirm Kepler's Laws, also confirms Galileo's Law. But of course, Galileo's Law also has independent evidence that confirms it; this was the evidence that Galileo himself provided prior to the work of Newton. This evidence that confirms Galileo's Law also confirms that theory that entails it. Just as Kepler's evidence supports the theory, so does Galileo's evidence; and just as Kepler's evidence about celestial motion supports, via the theoretical connections, Galileo's Law, so Galileo's evidence about terrestrial motions supports, via the theoretical connections, Kepler's Laws.

In the two cases, that of Newton's theory and that of Darwin's theory, we have the following patterns. A general theory T_1 deductively implies two different laws T_2 and T_3. There are observational data that support T_1. Because deductively implies T_2 and T_3, these data support the latter two laws also. *This support is carried by the entailment or consequence relation*: if T deductively implies T', then data that support T also support T'. At the same time, there are observational data that support T_2 and T_3. These data also support the theory T_1 that deductively implies them. *This support is carried by the converse entailment or converse consequence relation*: if T deductively implies T', then data that support T' also support T. This, to be sure, is not the whole story, but it is good part of it.

These patterns of confirmation impose constraints on how one might modify one's theory if further evidence comes in that requires change. Again, suppose that we have a theory T_1 that deductively implies two different laws T_2 and T_3, and suppose that there is evidence that supports T_1, independent evidence that supports T_2, and still further evidence that supports T_3. Finally, let us suppose that we acquire some data that falsifies T_3. We have to give up T_3. But since the latter is deductively implied by T_1, we have to give up this theory also. However, this theory is supported by independent data. This means we have not so much to give up T_1 as modify it. How do we modify it? In general there will be no neat rule or algorithm that will tell us how to proceed, though in some cases there might be a background theory that

can guide us, as we have seen in the case of Newton's theory with the discovery of Neptune as a result of trying to deal with the unexplained perturbations in the orbit of Uranus. But if there is no rule to guide us, there are still constraints. First, there are the data that confirm T_1. We can say that any modification T_1' of T_1 must be such that these data continue to confirm T_1'. Second, T_2 has not been falsified; indeed, its independent evidential support remains. So T_1 cannot be modified in a way that requires the rejection of T_2. That is, the modified theory T_1' must continue to deductively imply T_2. And the data that support T_2 will also provide support for the modified theory T_1'. The data that support T_1 and the data that support T_2 both provide constraints on the modifications that can be incorporated into the new theory T_1' that we introduce when the consequence T_3 of T_1 is falsified.

Now return to Darwin's theory. There are further evidential connections that should be noted. There is independent observational evidence confirming (T5). But, also, (T1)-(T4) deductively imply and therefore also support (T5) — or at least, the observational evidence that supports (T1)-(T4) also, via the deductive relationships, supports (T5).

There is independent observational support for (T6): we all know from simple observation about the fact that many creatures are "checked" and do not survive into maturity.

This independent support for (T6) also becomes evidential support for (T5), since it is clear that (T6) deductively implies, but is not deductively implied by (T5). In fact, the relation of (T6) to (T5) is that of the less gappy to the more gappy. In this sense, (T6) explains (T5).

Finally, let us note that we have in these laws a set of facts of great interest. Yet they are totally ignored by the creation scientists. Though these facts beg to be fit into a broader theory, no such theory is forthcoming from these thinkers — other, of course, than the appeal to the works of the deity, that is, to a non-explanation.

We now come to another set of important facts. We all know that there are various "checks" on organisms, that is, forces that prevent organisms from surviving and reproducing. We all know a vast number of facts of this sort:

- Foxes eat rabbits that cannot run fast enough
- If dogs catch rabies they die

- Insectivorous birds eat insects that they can discriminate against the background
- If two individuals are competing for the same food, and one is significantly stronger, that one will get the food
- There is a minium amount for food that any individual member of a species requires for survival and reproduction
- *and so on*

Call these facts (F).

We now generalize from (F):

(T7) For any species S of organism in its normal environment, there are forces f in that environment such that, for any member s of S, that the forces f act on s is sufficient for s not to achieve its goals of survival and reproduction, and, moreover, for any member s of S, if s does not achieve its goals, then one or more of the mentioned forces has acted on s.

(T7) is an abstractive and generic law, generalizing from the specific and determinate laws (F).

(T7) states the fact that *life is a struggle for existence*. Contrast this claim with the traditional claim that nature as created is perfect. The existence in the normal environments of organisms of the "checks" to survival and reproduction establishes that there are limits to the extent to which organisms are well-adapted to achieving their goals in their normal environments.

Now consider a particular species S in environment C. We can specify (T7) for this species and environment as

(L) There are forces f in C such that, for any member s of S, that f acts on s is necessary and sufficient to prevent s from achieving its goals of survival and reproduction.

Since we do not know specifically what these forces are, we cannot use their presence to predict which individuals will survive and reproduce. But if an organism fails to survive we can use (L) to deduce that the forces, whatever specifically they are, must have bene there. We can then offer a (gappy) explanation of why the individual did not survive by appeal to those forces. We can explain *ex post facto* even though we cannot predict.

We have discussed this feature of gappy laws above, when we noted that this sort of explanation occurred when one attempted to offer various sorts of explanation with the Law of Inertia. There is nothing puzzling about the case. The only difficulty is that we would prefer to have less gappy knowledge, knowledge that would permit not only *ex post facto* explanations but also predictions. In the case of Newton's theory, scientists were very often able to find specific laws that would permit both explanation and prediction. Unfortunately, the subject matter of biology is far too complex to give us anything approaching a process law. It turns out, however, that there are other ways in which this defect of gappy laws can at least partially overcome.

Even though we cannot predict using laws such as (L), we can partially overcome the limitations of this gappy knowledge by relying upon statistics, sampling survival rates however roughly and estimating probabilities for statistical laws such as the following:

(L') For any population of species S, if an individual member s of that population is in C, then, with probability p, s will achieve its goals of survival and reproduction.

The probability p is a measure of the efficiency of the forces mentioned in the laws in checking individuals. Equally, it is measure of the degree to which S's are adapted to surviving and reproducing in C. It is, in other words, a measure of the *fitness* of S' to survive and reproduce.

Recall the law

(R) Reproduction on the whole preserves the specific kind, save that every offspring varies in minor ways from the parents and from the norm of the species.

(R) mentions minor variations that appear in offspring relative to parents and relative to the specific norm. Darwin's theory incorporates this law. In fact, Darwin's theory makes a stronger claim:

(T8) For any species, there are small variations that appear and these are heritable from a set of parents to individuals of the nth generation, for all n.

Recall now the fact (G)(c):

For each species of organism there are various anatomical and physiological mechanisms the working of which in organisms of that kind in normal environments is necessary and sufficient for the satisfaction of the needs of those organisms.

Given (G)(c), it follows from (T8) that

(T8') For any species, there are small variations and there is a physiological-embryological-chemical mechanism such that these variations are heritable from a set of parents to individuals of the nth generation, for all n.

Some variations are fit-making:

(T9) For any species S, there are individual s in S, and characteristics f which are heritable variations of s with respect to S, and S which are f are more fit than S which are not f

(T10) It will on the whole be reasonable to expect that a population where some individuals lack a fit-making characteristic will evolve into a populations in which all individuals have the fit-making characteristic.

(T9) does not explain the emergence of new varieties or (in the longer run) new species; it only makes it something that is reasonable to expect. This follows from its statistical nature. In order to obtain an explanation, we must turn to the forces that explain the survival and non-survival of organisms. These are of course the forces (F) which are generically described in (T7). Consider any species S. (T7) asserts that there are forces in the environment of S's that can prevent S's from achieving their goals of survival and reproduction.. These forces are *the forces of natural selection* with regard to S. What the theory assumes is that where there are two varieties of a species

distinguished by the one having characteristics that are more fit-making that is their absence, then there are different forces of natural section for the two or at least differentially acting forces. *This assumption is the* **Principle of Natural Selection**:

(T11) For any species S, if there are individual s in S and characteristics f which are heritable variations of s with respect to S, and S which are f are more fit in their normal environment than S which are not f, then there are g', g" such that g' are the forces of natural section with respect to S which are f, such that g' are the forces of natural section with respect to S which are not f, and such that g' and g" are different.

(T11) is an abstract generic law that asserts the existence of a variety of specific laws. It asserts the existence of the laws that, by definition, describe the workings of "natural selection", or, what is the same, according to Darwin, "the survival of the fittest." What the forces of natural selection do is account for the spread in a population of adaptive or fit-making characteristics.

Note, by the way, that what evolves is the population, in the sense of an actually or potentially interbreeding and reproductively isolated group of organisms.

Natural section causes the accumulation of new variations and the loss of non-fit-making variations to the extent that new species may arise.

Darwin argued that the accumulated evidence testified to the claim that all species can have their origin explained by appeal to forces of natural selection. This is the claim

(T12) For any species S', there is an earlier species S" such that all members of S' are reproductive descendants of S", and such that there is a characteristic f which is had by all members of S' and not by all members of S"; and there is a time at which f appeared as a variation in S"; and such that organisms which are S" and f are more fit than organisms which are S" and not f, and such that there are forces g', g" such that g' acted with respect to organisms which are S" and f and g" acted with respect to organisms which were S" and not f and g' and g" are different.

This thesis is the Principle of the Origin of Species by Natural Section. It asserts that natural selection operating on variations is not only *a* sufficient condition for the origin of a species but also *the only sufficient condition* for the origin of any species. But if we assume that everything must have a cause to account for its origin, then the satisfaction of some sufficient condition is a necessary condition for the origin of that thing. Hence, if the satisfaction of a sufficient condition is a necessary condition for the origin of any thing, then if there is only one sufficient condition, then the satisfaction of that condition is also a necessary condition for the origin of the thing. Hence if we assume that the origin of a species must have a cause, then natural selection which is the only sufficient condition must also be a necessary condition. In effect, then, what (T12) asserts is that natural selection is a necessary and sufficient condition for the origin of any species.

If we assume (T12), then the theory of descent by modification (TE) follows as a matter of course. By offering a plausible and highly confirmed (though nonetheless still gappy) explanation from which the theory of descent by modification follows as a matter of course, it becomes possible for the first time to accept the fact that evolution has occurred.

For the first time a detailed scientific hypothesis became available which could demonstrate the compatibility of the fact of evolution with our knowledge of the fossil record and of the adaptedness of species in their normal environments.

It is sometimes said that all this is hypothetical because we have never seen natural selection at work. Now, this objection is mistaken from the start. Just because we have never seen natural selection at work it does not follow that we have no grounds for accepting the hypothesis. We have seen that this sort of objection is illegitimate. Herodotus was able to form hypotheses about parts of Africa which he had never seen, and could do so quite reasonably. Such hypotheses will be reasonable provided that they are built up on the basis of other hypotheses that are supported by observational data. The same holds in the present case. So long as we have built up our theoretical structure on the basis of empirical evidence, then the fact that we have never observed natural selection at work — if it be a fact — is simply irrelevant. And, as we have seen, they evidential structure supporting Darwin's theory is strong indeed.

But in fact it is simply wrong to say that we have never seen natural selection at work. It is true that we have never seen natural selection producing a new species; that process would take too long for us to have observed in the short span of time since we started, in the light to Darwin's discoveries, to observe biological processes. However, it is still true that we have observed natural selection at work.

The first case in which natural selection was at work was cited, quite correctly, by Darwin himself as evidence in support of his theory. This is the example provided by the many cases of *artificial selection*. British cattle breeders and English pigeon fanciers had managed through carefully controlled matings to produce useful, or, in the case of some of the pigeons, fanciful varieties of stock. These varieties were the result of human art. The breeder was part of the environment in which these varieties developed. His or her activities were centrally among the forces of natural selection which worked to produce these new varieties. In the activities of the breeders we can see the forces of natural selection at work producing new varieties. It is reasonable to expect that if the process were to go on long enough then the new varieties would gradually develop further into new species.

It is true that as soon as the fanciful varieties of pigeon were released back into the ordinary environment of pigeons, where the breeders and their art was not at work, the varieties reverted to kind. This is to be expected if Darwin is correct. For the normal environment contains forces that make the ordinary variety of pigeon well-adapted to survival and reproduction. As soon as these forces alone are at work and not the forces of the breeder's art, then they will guarantee that the fanciful varieties, less well adapted to the ordinary environment, will gradually disappear in favour of the variety that is better adapted to survival and reproduction in that environment.

The second case in which natural selection has been observed to work is the case of so-called industrial melanism. There is a species of moth in England the normal colour of which is white. But in central England around Birmingham the landscape came to be dominated by industry that used large amounts of coal for energy. The coal produced smoke, and the countryside became darkened by large amounts of soot that settled out of the smoke. Against that background, the white moths stood out, and could be picked off by predators. In a fairly short time a new variety of the moth developed in which it was covered by dark spots. This new variety was much less visible against the sooty background around Birmingham. The predators that could easily distinguish its unspotted cousins could not nearly so easily distinguish this new variety. It was better able to survive and reproduce in the environment created by industrial development. This phenomenon of moths developing into a melanistic variety is a clear example of natural selection at work.

We could not, of course, in the case of the moths around Birmingham, have predicted the appearance of a new variety with black spots. The coming to be of these minor variations is something that is random with respect to the environment in which natural selection occurs. But, given that such a fit-making variation does occur, then natural selection will lead to an increase of its numbers in the population.

Because variations upon which natural selection works occur randomly relative to those forces, it follows that we *cannot predict* which new species will come to be. We can often identify the forces that favour the survival of a certain variation, but we cannot predict the occurrence of that variation. It follows that with (T12) we can at best achieve *ex post facto* explanations of the origins of species. Why are there dogs? *Ex post facto*, we can use (T12) to deduce both that the required variations occurred in the canine ancestors and that there were forces of natural selection that favoured those variations. We can often identify those forces. We can, furthermore, often identify the forms which were ancestral to the dog. But knowing those ancestors and those forces of natural selection, we still could not have predicted the coming to be of dogs, for we could not have predicted the occurrence of the required variations. Indeed, since those variations are a matter of chance relative to the environment wherein the forces of natural selection work, it follows that the occurrence of those variations is a matter

of accident, that is, accident relative to that environment. Thus, given that we know that there are always variations occurring, we can use (T12) to deduce that new species will emerge. What we cannot predict is precisely which new species will emerge. Thus, in biology the origin of species will always be explainable only *ex post facto*.

Recall, now, that argument of Boyle and Paley that the fact that organisms are well-adapted to survive and reproduce in their normal environments implies, on ordinary scientific grounds, that there is a designer which caused and explains these facts of adaptation. This conclusion was expressed by Hume as the proposition that "the cause or causes of order in the universe probably bear some remote analogy to human intelligence"[71]

What is it that is characteristic of human intelligence? It is human intelligence that designed the watch. What is characteristic, then, of human intelligence is that it can solve problems, and, in particular, problems of design, how to shape a mechanism that can fulfill a certain function. Now, organisms are in fact well designed; they have within themselves mechanisms that enable them to survive and reproduce in their normal environments. This is the point that we previously expressed as the thesis

(S) The members of any species S are well-adapted to achieving their goals of survival and reproduction in their normal environments.

We now see in Darwin's theory of the origin of species by natural selection that the explanation of this fact is *natural selection*. It is therefore natural selection that is the cause of adaptation. But to say that it is the cause of adaptation is to say that it has solved the problem of adapting mechanisms that ensure that in organisms various functions are efficiently performed. But if natural selection has the capacity to solve problems of design then it does share something with human intelligence, namely, precisely this capacity. Natural selection is, in Hume's terms, "the cause or causes of order in the universe". We now see that this cause *does* "bear some remote analogy to human intelligence." Boyle, Paley and Hume are thus quite correct in their argument. The argument from design, as based on the scientific causal principle (CP), is in fact a sound argument, its conclusion one that can reasonably be affirmed: the cause of order in the universe, namely, natural selection, does bear a remote analogy to human intelligence.

This is not the traditional God whose existence Boyle and Paley hoped to prove by the argument. But we know from Hume that this conclusion was unreasonable anyway. As Hume says, the analogy has to be remote. It is indeed.

One important difference between natural selection and human intelligence is worth noting. In the case of humans, when we go about solving a problem, we can in effect retreat and start over. If it seems to our advantage, we can retreat one step in order to go forward two steps. This is not true of natural selection: all it can do is go forward. Thus, although natural selection is, like human intelligence, a problem solving mechanism, it is one that is not as efficient as human intelligence.

(f) Functions and Functional Explanations

We speak of functions in the context of biology. "The function of the heart beating is to circulate the blood." It is evident that what this means, simply enough, is that the circulation of the blood is an effect of the heart beating: the presence of the beating heart is, normally at least, sufficient for the circulation of the blood. Of course, this statement is not true in full generality. Block the arteries and the blood will not circulate even though the heart keeps beating — at least for a short while. And during various surgical operations, the heart may be isolated from the circulatory system, kept beating but not circulating the blood, which function is carried out temporarily by an artificial heart. So, one has to say that "Normally, the function of the heart beating is to circulate the blood."

The word 'normally' which usually has to be inserted in such a statement indicates that the causal knowledge is gappy. But for all that, it is still knowledge. And because it is knowledge of effects, that is, of laws, it has explanatory power. Of course, we would like to do better, finding out, for example, more precisely what the conditions are that fill out the ignorance that is encapsulated in the use of 'normally.' But there are other imperfections also. It is clear that the blood is not circulated by the beating of the heart save in the context of a lot of other bodily structures — blood vessels, lungs,

components of the nervous system, and so on. These details, too, would have to be filled in as we seek to replace the simple, or even simplistic, function statement with more detailed, less gappy knowledge.

But on the whole, talk of functions is not difficult to understand in scientific terms. More difficult, perhaps, are what have been called "functional explanations."

It is sometimes said that "the beating of the heart exists because of the need of the organism to have its blood circulated." This sort of statement attempts to *explain* the fact that an organism has a beating heart. It attempts to explain that fact by arguing that its function fulfills a need of the organism: *the fulfilment of the need is the cause of the existence of the fact.*

The inference here is from the fact that there is a need to be fulfilled to the existence of a mechanism, the beating of the heart, that has the function of fulfilling that need. To say that beating of the heart has the fulfilment of the need as its function, is to say that the former is causally sufficient for the latter. To say that the beating of the heart exists because there is this need to be fulfilled, is to say, conversely, that the existence of the need to be fulfilled is causally sufficient for the existence of the beating of the heart.

It is one thing to say that the beating of the heart is sufficient for the fulfilment of the need. That, in fact, is not controversial. It is quite another to say that the fact that the need is being fulfilled is sufficient for the beating of the heart. For, it is not straightforwardly evident that the need can be fulfilled in only one way. Perhaps there are other mechanisms that could circulate the blood other than the beating of the heart. Indeed, during certain surgical operations on the heart, it happens, as we have just noted, that the heart is separated from the circulatory system and its functions are carried out, temporarily at least, by an artificial heart. Given that there such alternative mechanisms, then the fulfilment of the need is not sufficient for the circulation of the blood, and the latter cannot be inferred from the former. The attempted functional explanation would after simply not be an explanation: the premises would not entail the conclusion. One could deduce the existence of *some mechanism or other* but not the specific mechanism that one set out to explain.

In fact, it is sometimes suggested that functional explanations are simply fallacious. The function statement

"Being A has the function of producing being B" can be understood, as we have seen, simply as

(*)   being A is sufficient for being B

We then get the following sort of explanation schema for such function statements:

> being A is sufficient for being B
> This is A
> _____
>
> so,   This is B

which explains something being B in terms of the presence of A, the circulation of the blood in terms of the beating of the heart. The functional explanation seems to be

> being A is sufficient for being B
> This is B
> _____
>
> so,   This is A

which is simply fallacious — it is the fallacy of affirming the consequent. The functional explanation would be satisfactory if it were of the form

> being B is sufficient for being A
> This is B
> _____
>
> so,   This is A

But the major premise

(**)   being B is sufficient for being A

unlike the function statement (*), seems to be false: as we have just seen, we seem to be able to say

(***)   being B is sufficient for the existence of some mechanism or other that fulfills the function that being A fulfills

but not (**).

It is likely that this sort of criticism has some force against many functional explanations that have been offered in sociology. But it seems to have less force in biology, provided that we consider the context. Consider once again the functional explanation of the beating of the heart. The inference here is from the fact that there is a need to be fulfilled to the existence of a mechanism, the beating of the heart, that has the function of fulfilling that need. To say that beating of the heart has the fulfilment of the need as its function, is to say that the former is causally sufficient for the latter. The criticism is that one could infer the existence of *some mechanism or other* but not the specific mechanism of the beating of the heart that one set out to explain. However, if one recognizes that we are talking about the circulation of the blood in the context of a mammalian bodily organization, then it is difficult to see, given this constraint, what other mechanism could *under normal conditions* perform the same function. The caveat about "normal conditions" is meant to exclude such things as artificial hearts used during surgery, but there is nothing wrong with that: it simply marks the fact that the knowledge is to a certain extent gappy. And surely we *can* grant that in a bodily system of mammals, we know that we can reasonably infer from the fact that blood is circulating that the heart is beating: the former *is* sufficient in that context for the latter. In short, not only is the function statement (*) true, but so too is the statement (**) true. And that means that the functional explanation is also satisfactory as a scientific explanation — or, at least, satisfactory as far as it goes: we do have to recognize its gappiness.

We above characterized a need a something that an organism must do or acquire if it is to survive and reproduce in its normal environment. Organisms act to satisfy their needs. Within the Aristotelian scheme, this activity was understood as due to simple unanalysable tendencies or dispositions. It was simply their nature so to act, and once one had grasped that nature in a rational intuition one understood: nothing more was to be said. These Aristotelian explanations were derided in the early modern period as vacuous. Still, it is a fact that organisms are generally well adapted to meet their needs. What became increasingly apparent during the early modern period is that organisms have within themselves mechanisms that ensure that, for the most part, those organisms satisfy their needs. Some, such as the Cambridge Platonist Henry More, argued that the science of the day could not explain how such mechanisms could exist, and they concluded that one still had to have recourse to Aristotelian "plastic natures" or forms to explain the observed tendencies of things to satisfy their needs. But as more and more mechanisms came to be discovered and understood, it became increasingly plausible to hold that in the case of needs where there are no clear mechanisms nonetheless there exist other, as yet undiscovered, mechanisms through which the organism ensures those needs will be fulfilled.

The argument was of course inductive: past successes in finding such mechanisms made it plausible to suppose that there would be continued success in discovering such mechanisms. That is, the past success in discovering that there are mechanisms that ensure certain needs are fulfilled justified with respect to other needs the proposition that we earlier labelled (G)(c):

> For each species of organism there are various anatomical and physiological mechanisms the working of which in organisms of that kind in normal environments is necessary and sufficient for the satisfaction of the needs of those organisms.

or, more briefly,

> (+)   There are mechanisms that ensure these needs are fulfilled.

This states that where there is a need, then there is some mechanism or other that ensures that need will be satisfied. This generates the research task: find those mechanisms.

To be sure, one was as yet ignorant of those undiscovered mechanisms. But, as we have just seen, that is merely a basis for continuing our research to fill in those gaps. This ignorance is certainly not grounds for having recourse, as Henry More did, to non-scientific explanations. More's "plastic natures" are simply a means to cover up ignorance. Although philosophically more sophisticated than the appeals to "creation" as explanatory by creation scientists, More's "plastic natures" are of a piece with the latter, in evading a problem in science rather than solving it.

It was the increased understanding of the nature of the mechanisms whose function it was to ensure that

needs are fulfilled that prompted Boyle and Paley to revise the argument from design for the existence of a deity, and to make the causal efforts of a God to be part of our scientific causal understanding of the workings of nature. On this view, functional explanations are gappy, with the gap to be filled by the causal activities of the deity.

Suppose that we have a functional explanation of, say, the eye: "The eye exists because we have to be able to see if we are to survive and reproduce." We have seen that this explanation is satisfactory as far as it goes, but that it is gappy. It calls out for a further explanation: Just why do we have animals with these particular mechanisms in these particular environments? The explanation that Boyle and Paley gave was in terms of a designing deity. In one sense, they were on the right track. The answer is in terms of solving the problem of adapting organisms to the environments in which they are normally found. But the actual cause of this design, the actual problem solver, is far from the traditional deity: it is, we have known since Darwin, natural selection working on small variations.

Functional explanations, then, are quite satisfactory as scientific explanations, at least in the biological context. The principle (+) asserts that we will be successful in discovering mechanisms though which organisms efficiently satisfy their needs. One want to know, however, why (+) is true. We would like a theory that would locate (+) as a piece of gappy knowledge within the context of a broader picture of the biological world. That theory is Darwin's theory of the origin of species by natural selection. With this theory we know, in outline, and in abstract generic terms at least, the causes for the mechanisms the existence of which is asserted by (+).

(g) Genetics and the Modern Synthesis

If natural selection provides one basis for understanding (+), there is another aspect which must be mentioned. This is the development of the individual organism from a fertilized ovum to adult. If one wants to explain how the individual comes to be, the laws that describe these processes must be uncovered. Included in this will be the mechanisms of heredity, by which characteristics of individuals are passed on from parents to their offspring.

Darwin's theory of evolution by natural selection presupposes that there are mechanisms of heredity. This is part of what is asserted by the principle

(T8) For any species, there are small variations that appear and these are heritable from a set of parents to individuals of the nth generation, for all n.

which we noted as one of the central claims of the theory.

These mechanisms of heredity are predicted by (+). At the same time, if these mechanisms do exist, then they will provide explanations for the development in the individual of all these mechanisms, including the development within the individual of the mechanisms through which its own offspring will inherit the parental characteristics. That is, given (+), or, what is the same, (G)(c), it follows from (T8) that

(T8') For any species, there are small variations and there is a physiological-embryological-chemical mechanism such that these variations are heritable from a set of parents to individuals of the nth generation, for all n.

But of course, to know on the basis of either (+) or Darwin's theoretical (T8) is not yet to know what those mechanisms are. The discovery and location of these mechanisms has been the work of the last half century.

For Darwin himself, there were serious problems in developing a theory of heredity. He held the view that characteristics of parents appear in the offspring in blended form. There was considerable evidence of this view. Think of breeding dogs or cats: the offspring often have a colour that is somewhere in between those of the parents. Call this a "blending account" of heredity.

If a blending account of heredity is accepted, then small variations will very quickly, over a few generations, be blended out of existence, with a return to the standard. This is inconsistent with the equally crucial Darwinian thesis that small variations accumulate over time to form new species. The latter requires that the variations remain, as (T8), asserts, to the nth generation. If one accepts the blending account, then if variations are to accumulate, then evolution must proceed at a fairly rapid rate, more rapidly in fact that is consistent with it proceeding solely in terms of natural selection.

A consequence of Darwin's accepting a blending account of heredity was that he was forced to look for mechanisms that could speed up the evolutionary process. One such mechanisms that he introduced in later editions of *The Origin of Species by Natural Selection* was the mechanism of use and disuse. The suggestion was that a characteristic that an individual organism acquired during the struggle for existence and which proved useful could be passed on to future generations. Conversely, an inherited characteristic that proved useless in the struggle for existence would not be passed on.

If this suggestion is to become acceptable, then there must be, as (+) requires, mechanisms that make it possible. Thus, there must be mechanisms which transmit information from the environment in which the struggle for existence occurs to the germ cell which carries to the next generation the information that is needed for the fertilized ovum to develop into a new individual. Darwin himself proposed hypotheses about such a mechanism. The difficulty was that in due course it became more and more evident that the germ matter was in fact impervious to any such information. Scientists were able to examine the matter in the germ cell with very high power optical microscopes and these revealed no mechanisms by which information about what happened to an organism in the environment in which natural selection occurred could penetrate to the germ cell to influence what characteristics were inherited by the next generation. Optical microscopes, at the limit of their capacity to magnify, revealed that body of an organism is made of cells, and that inside the part of the cell called the nucleus there are pairs of string-like entities which were called "chromosomes." It was further discovered that sex cells (the "gametes") contain just one of the usual pair of chromosomes, and that each individual is the result of the combination of two such sex cells, one from each parent. It was clear that the chromosomes were the entities which carried the information from one generation to the next, and also that these entities received no information about the utility or disutility of various parts or activities of the organism in its struggle to survive and reproduce. In fact, the cells which form the gametes are already in the woman's ovaries before birth.

Darwin's theory was therefore confronted by a genuine problem: the requirements of the theory, so very well grounded in factual evidence, seemed to conflict with a fact about heredity, the assumed fact that inherited characteristics were a blend of the characteristics of the parents.

In the end it turned out that the blending account of heredity was simply wrong. People, including Darwin, had taken some simple minded and largely anecdotal — but nonetheless persuasive — evidence for the blending account for granted, but had never subjected it to serious scientific investigation. When it was subjected to investigation, it was discovered to be quite wrong: heritable characteristics were after all preserved entire and as it were undiluted, just as the Darwinian theory required.

The required research was done by the Moravian monk Gregor Mendel, who studied the heredity of peas. These had either white flowers or pink flowers. Mendel observed that these colours did not blend into a uniform light pink flowers; rather the colours were passed along to future generations undiluted. That established that the blending account of heredity was incorrect. Mendel further noted that parents with white flowers could have offspring with pink flowers. He concluded that the information concerning heredity was carried by units that could be preserved from generation to generation, sometimes producing and sometimes not producing the observable characteristic for which they were responsible. These units came to be called "genes." These entities had not been observed; their existence was inferred from their effects. The effect was the observable characteristic that a gene produced. The term 'gene', or, more exactly, 'pink colour gene' (or whatever effect was being considered) was shorthand for

[d]   the sort of entity that causes the pink colour

We abbreviate this expression by the term

[d']   gene

How do we know that such an expression actually refers? After all, we have not observed the entities to which it purports to refer. The answer is, clearly, that our theory assures us that it does refer. We have a law statement in the theory to the effect that

[Hd]   For every individual of this species, there is a unique sort of entity which is necessary and sufficient for this plant to have pink flowers.

[Hd] asserts that from the observed characteristics we can infer the presence of the relevant gene, and from the presence of the relevant gene we can infer that the observable characteristic will occur. The success of the theory justifies accepting [Hd], and this law in turn assures us that the expression [d] does actually refer to a kind of entity which is found in all individual pea plants and which causes these plants to have pink flowers. Such expressions as [d] which theories justify us in using to refer to unobserved entities are what logicians call "definite descriptions."

What we have seen is that in order to account for the observed characteristics of things we make the assumption about genes that each parent contributes one such entity. Moreover, the observable effects, those characteristics that go to make up what is referred to as the "phenotype" of the organism, are not the only constraints on what it is to be a gene. Since a gene is the unit of heredity, it must fit with what had been discovered through optical microscopes about heredity. Genes clearly had to be located on chromosomes. Besides [Hd], then, the theory includes as an additional hypothesis about genes the *Law of Parental Contribution*, to the effect that

For each sexual individual, each parent contributes one and only one of the genes at every locus on the chromosome. These genes come from the corresponding loci in the parent.

Assume for the sake of simplicity that a single gene is responsible for the colour of the pea flowers. Since there are white flowers and pink flowers, there have to be two genes for the colour. A white flower will have it colour determined by the gene for whiteness, but since it can have a pink offspring even when it mates with a white flower, it follows that the gene for pink flowers must also be present, ready to be passed on. Both genes are therefore present in some individuals. It was therefore assumed, in conformity with the known fact of two chromosomes, that all genes come in pairs. The genes at a particular place or "locus" on a chromosome determine some one phenotypical characteristic. The genes at a locus may be the same (both white producers) or different (one white producer, one pink producer). The set of all genes that can occupy a locus are called "alleles"; there may be more than two members in a set of alleles, but at most two of the set can occupy the same locus on the two chromosomes. If the two genes are at a locus are the same, then the organism is said to be "homozygous"; if, however, the genes are different, then the organism is "heterozygous." If there are two pink genes, then the flower will be pink. If there are two white genes, then the flower will be white. However, if the organism is heterozygous with one white gene and one pink gene, then the flower will be white. When an organism is heterozygous, then the gene whose characteristic appears in the phenotype is said to be "dominant," the other "recessive." In our example, the white gene is dominant, the pink gene recessive. However, though the pink gene is recessive and the colour it determines does not appear, it nonetheless continues to exist, capable of being passed on to the offspring. If the offspring should receive this pink gene and also a pink gene from the other parent the result will a be homozygous organism which will have pink flowers.

There are four possibilities for gene pairs: white-white, white-pink, pink-white, pink-pink. Mendel, through a series of breeding experiments, was able to establish that in a large population the probability of getting a plant with pink flowers was 1/4, while the probability of getting white flowers was 3/4. This is precisely what would happen if the probability of a gene being transmitted from a parent is equal to the probability of transmission of the other gene at the same locus. This is *Mendel's First Law* (or the Law of Independent Segregation):

> The probability of a gene at a locus in a parent being transmitted to the offspring is equal to the probability of transmission of the other gene at the same locus.

If the set of alleles consists of two genes, as in our little example, then the probability of either gene being transmitted is ½. If we assume that the numbers of white genes and the numbers of pink genes in the population are equal, and that the members of the population randomly mate, then the probability of an offspring

receiving a white gene is ½ and the probability of receiving a pink gene is ½. It follows from this that the probability of getting a pair with a white, dominant, gene is ¾ and the probability of getting a pair of pink genes is ¼. Hence, the probability of a white flower is ¾, and the probability of a pink flower is ¼. This is exactly what is observed. But of course that does not surprise us. We build into our assumptions about genes precisely those hypotheses that are needed if we are to account for the distributions that we observe in the phenotypes of our population of organisms.

It also turns out that the genes of other characteristics turn up independently of the genes for colour. More generally, we have *Mendel's Second Law* (or the Law of Independent Assortment):

> The probability of an offspring receiving a particular gene from a particular parent is independent of the probability of the offspring receiving any other gene (at a different locus) from that parent.

Thus, the genes at a locus segregate independently of genes at other loci.

This simple picture was complicated over time as more and more breeding experiments were conducted. It was found that one had to assume that certain genes were dominant over others but recessive to still others. Moreover, there is no simple one-one relationship between the characteristics determined by the genotype and the characteristics observed in the phenotype. In the first place, of course, it is not just the genes but also the environment that determines the phenotype. Thus, height is determined in part genetically, but also by one's diet, that is, by the environment. But besides this, it turns out that one gene may have effects on several phenotypical characteristics. Conversely, several genes may be required to determine a single phenotypical characteristic. Various exceptions to Mendel's Laws were also discovered, and further mechanisms at the genetic level had to be postulated to account for these exceptions. In particular, one had to allow that chromosomes occasionally split during the reproduction process and re-unite with the opposite pair, affecting the probabilities of transmission. Note that, while chromosomes may split, genes may not. Genes are the units of heredity; in the postulated mechanisms it also turns out that they are the smallest units of crossing-over.

In general, genes remain identical during the processes in which they are passed from parents to offspring. But occasionally they change. Such a change is a **mutation**. *Mutations are the source of the small variations that Darwin notes are there and the slow accumulation of which under the pressures of natural selection account for the origin of species.*

The causes of mutations are such things as cosmic rays or radiation. These are factors which act independently of the forces of natural selection. Relative to the latter, they, and therefore mutations, are random. In particular, the change in a gene does not occur in response to the needs of the organism — contrary to what Darwin thought was needed. Mutations are not made to order; when one appears the organism must make of it what it can. As it turns out, most mutations, and certainly all large mutations, are not adaptive at all; but this should not surprise one, given that the origins of mutations are unrelated to its needs.[72]

Modern biological theory unites Darwin's theory of evolution by natural selection and Mendel's theory of genetics. It has been remarkably successful. Genetic theory led to a variety of predictions that were confirmed in further breeding experiments, just as failures of predictions led to revisions of the theory. In this respect, it is like other successful theories, e.g., that of Newton, in being on the one hand being confirmed through successful prediction, and on the other hand being revised to make it fit the facts where its initial predictions have failed.

One major difference from Newton's theory is worth noting, however. This is the fact that the predictions were statistical, about populations, rather than about individual systems. That makes the inferences sometimes more complicated in the case of genetic theory. But the basic logic of research is the same..

It is one thing to postulate genes, another to observe them. Mendel's genetic theory does not claim to identify genes save as the causes of certain results observed in breeding experiments. Genes are referred to by means of definite descriptions like [d]. The theory was predictively successful, but throughout one prediction which it made remained unconfirmed. This is the prediction that is involved in the use of definite descriptions like [d].

This is the prediction that

there are (chemical) entities on chromosomes that are the units of heredity which account for the phenotypical characteristics of organisms

where these entities fulfill Mendel's Laws of Independent Segregation and Independent Assortment. This creates a *research problem*, that of *identifying* these entities that the theory asserts are there.

The problem, of course, was that one reached the limits of optical microscopy with the chromosome. By hypothesis, genes are smaller units, which therefore cannot be observed through those instruments. At the time of Mendel, and much later into the 20th century, tools for observing such small entities as genes had to be supposed to be were not available.

Now, however, through a variety of techniques, including those of the electron microscope, molecular geneticists have actually succeeded in observing and identifying genes. These turn out to be parts of the DNA molecules within the nucleus of each cell. Not without surprise, these developments have led to further refinements in genetic theory.[73]

The logical structure of the Mendel's genetic theory is clear enough, as is the structure of the argument that justifies accepting it.

There are certain processes that seem to begin at the same point and yet end in different ways. The pea plants seem much the same, yet when some plants with white flowers give rise to offspring the flowers of the latter are white, while in other cases plants with white flowers give rise to offspring with pink flowers. If we assume that these differences have a cause, then we will have to infer that there is an unobserved cause for these differences. In fact, we have a principle on which to base this inference that is stronger than the simple principles that every difference must have a cause. This is the principle of Darwin's theory that we called (T8'):

For any species, there are small variations and there is a physiological-embryological-chemical mechanism such that these variations are heritable from a set of parents to individuals of the nth generation, for all n.

(T8) is justified by the fact on the one hand that there are heritable variations that are preserved through many generations and by the fact on the other hand that scientists had been very successful in finding mechanisms that function to satisfy the various needs of organisms.

The law [Hd] to which we earlier referred is just a special instantiation of the more generic (T8'). [Hd] justifies our assurance that the definite description [d] = 'gene' actually does refer to an entity. On the basis of [Hd] and [d] we can infer that

[Hd'] For every individual of this species, a specific sort of gene causes this plant to have pink flowers.

A principle such as [Hd'] can be referred to as a *bridge principle*, linking the observed characteristics of things to the hypothetical causes that account for these characteristics.

Besides bridge principles, the theory also assumes laws about the unobserved entities mentioned in the bridge principles. These are laws such as the Law of Parental Contribution, and Mendel's two Laws of Independent Assortment and Independent Segregation. We may call these the *theoretical principles*.

From the bridge principles and the theoretical principles, one can deduce the laws that connect the observable characteristics of things. The latter are gappy, allowing an apparently similar process to turn out one way in one instance and another way in another instance. The theory fills in the gaps in such a way that we can understand why it comes out one way on some occasions and another way on other occasions. The theory of genetics, in other words, introduces what we earlier referred to as *intervening variables* in order to help us eliminate the gaps in the laws about the observable characteristics of organisms.

What justifies our accepting this theory is its predictive success. But during the early stages of the history of this theory, this predictive success so far as observation is concerned was with respect to the observable characteristics of things. Observation did not extend as far as the hypothesized entities referred to by the definite description 'gene': entities of these sorts had never been observed.

This meant that when the theory encountered predictive failures and had to be modified to bring its predictions into line with what was actually observed at the phenotypical level, it was possible to modify it in two ways, either by altering the bridge principles or by altering the theoretical principles. There were no independent constraints on either.

This way of dealing with theoretical change altered when instruments were introduced by means of which the hypothesized entities could be identified.

By means of, let us say for the sake of simplicity, the electron microscope, it was possible to discover a law to the effect that

[Hd"] For every individual of this species, a specific piece of DNA is necessary and sufficient for this plant to have pink flowers.

Since, by [Hd] = [Hd'], a specific gene is precisely that entity that causes the plant to have pink flowers, it follows deductively that this gene just is that specific piece of DNA. We have, in other words, established the identificatory hypothesis

[id] the pink flower producing gene = this specific piece of DNA

Once we have confirmed this identificatory hypothesis by means of the theory of the electron microscope, we are no longer simply free to alter either the bridge principles or the theoretical principles of Mendel's theory when we run into predictive difficulties. The identificatory hypotheses such as [id] place constraints on the bridge principles, and since we can independently confirm or falsify, by means of the electron microscope, laws about the pieces of DNA that are the genes, we can alter the theoretical principles only so far as the alterations conform to the patterns observed to hold among the pieces of DNA that we observe.

So long as a theory refers to unobserved theoretical entities, there is a certain flexibility in the way we construct and re-construct our theory; once we observe those entities, and identify them, those observations, even though mediated by an instrument, place strong constraints on our ability to modify our theory.

The point is one of interlocking theoretical structures. We have on the one hand Mendel's theory T_M; and have, on the other hand, the theory of the instrument T_I.

We saw earlier during our discussion of Darwin's theory, that even in its earliest stages there were interlocking patterns of deduction and confirmation. These interlocking patterns served to make the theoretical structure more secure, more likely that it would not run into contrary data. We see the same sorts of interlocking patterns in the present case.

T_M makes two sorts of prediction. There is, first, the prediction of the unobserved entities, genes. There are, second, the predictions about patterns of observable characteristics. The latter justify our accepting the theory for purposes of explanation and prediction. Initially at least, the other prediction is not verified.

On the basis of the well confirmed theory of the instrument T_I, we can infer something roughly to the effect that such and such an image is observed if and only if a piece of DNA is present. We observe an image of that sort and infer the presence of the relevant piece of DNA, and further confirm the hypothesis [Hd"], or, what is the same, confirm the identificatory hypothesis [id]. But this is to confirm the prediction of T_M that genes exist. So, by means of T_I we are able to confirm the existence of the hitherto unobserved entities the existence of which has be predicted by the confirmed theory T_M.

Strictly speaking, of course, what we observe are, on the one hand, the observable phenotypical characteristics of organisms, and, on the other hand, images presented by the electron microscope. The former are the initial condtions IC, the latter is the prediction P made by the conjunction of the two theories. From IC and T_M we deduce that there are certain unobservable entities G. from G and T_I we deduce the prediction P.

This prediction P is deduced from IC and the conjoined theories (T_M & T_I). This successful prediction raises the probability of this conjoined theory. But since T_M and T_I are independent of each other, both logically and in point of having their own separate confirmations from other predictions, it follows that each of them also has its probability increased. In particular, of course, Mendel's theory T_M has its probability increased by virtue of this additional successful prediction. Indeed, since the existence of genes = pieces of DNA as the causes of phenotypical characteristics would be unexpected if T_M were false, the confirmation of T_M is quite strong.

We may conclude, then, that when a theory makes predictions about unobserved entities, even if those entities come to be observed only indirectly by means of an instrument, nonetheless those observations strengthen the confirmation of that theory.[74]

Another point about explanations using Mendel's theory of genetics is worth making. This theory can be used to explain the phenotypical characteristics of organisms. We can explain, for example,

> This flower has a pink colour by virtue of its having a pair of pink producing genes.

This explanation is justified by the law, or bridge principle, [Hd]. We observe the effect, the pink flower. From this, via [Hd], we deduce the presence of the cause. We can then appeal to this cause the presence of which we have thus discovered to explain the occurrence of the effect. To be sure, this explanation of the effect is only *ex post facto*: we could not have predicted the effect from the cause because we have not — as yet — found a way to identify the presence of the cause apart from its effect. This does indicate that our knowledge is gappy. But it does not, as we have previously seen, imply that we do not have an explanation. It is just that it is an explanation which we hope research will enable us to eliminate by finding a less gappy explanation.

This was done in due course. But Mendel's theory overcame this defect in part by through a means that we have seen used elsewhere in biology to overcome the same sort of defect due to gappiness, namely, statistics. The theory could not predict the colours of individual flowers, but it could use the methods of statistics to predict the percentages of different colours in the population. So the theory could predict percentages in mass events, but explain only *ex post facto* individual events.

Eventually, however, this defect was overcome. Less gappy explanations were achieved when, by means of instruments, scientists were able to identify genes with pieces of DNA. With a way now found to identify genes apart from their effects, we can, in principle at least, discover the presence of genes and from them predict the presence of the phenotype.

The progress of molecular genetics has been important here. It has allowed for the identification of genes which carry certain medical defects prior to those defects manifesting themselves, and has thus improved immensely not only medical diagnostics but also the possibilities for reasonable decision-making about such things as potential birth defects. It is one thing to say that a defect will appear in such and such a percentage in a population, and then to tell parents when the defect appeared in their offspring that they were carriers of the gene that causes it. It is quite another, and something better, to be able to predict on the basis of identifying that gene beforehand that a couple's offspring will have that defect. Of course, the couple may in fact choose not to know, even if the information is available. But they could also choose to know and choose to act as they deemed fit on the basis of that knowledge. Either option is possible, but either is better than acting on the basis of pure ignorance or even on the basis of statistical probabilities.

Endnotes

1 For accessible discussions of the historical issues, see T. Kuhn, *The Copernican Revolution* (Cambridge, Mass.: Harvard University Press, 1957); and I. B. Cohen, *The Birth of a New Physics* (Garden City, NY: Anchor, 1960).

2 From I. B. Cohen, *Birth of a New Physics*, p. 39.

3 From M. Hoskin, "Astronomy in Antiquity,' in M. Hoskin, ed., *Cambridge Illustrated History of Astronomy* (Cambridge: Cambridge University Press, 1997), p. 41

4 This and the next Figure are from T. Kuhn, *The Copernican Revolution*, p. 61.

5 From I. B. Cohen, *The Birth of a New Physics*, p. 41.

6 From I. B. Cohen, *The Birth of a New Physics*, p. 45.

7 Nicholas Copernicus, *On the Revolutions of the Heavenly Spheres*, in Nicholas Copernicus, *Complete Works*, 3 vols. (Warsaw: Polish Scientific Publishers, 1978), vol. II, p. 10.

8 From I. B. Cohen, *The Birth of a New Physics*, p. 50.

9 Ptolemy, *Almagest*, trans. G. J. Toomer (London: Duckworth, 1984), p. 45.

10 For an extended discussion of Aristotelian explanations, see F. Wilson, *Logic and the Methodology of Science in Early Modern Thought* (Toronto: University of Toronto Press, 1999), especially Study One.

11 See Stillman Drake, *History of Free Fall* (Toronto: Wall and Thompson, 1989).

12 Cf. Stillman Drake, *Galileo: Pioneer Scientist* (Toronto: University of Toronto Press, 1990).

13 Galileo Galilei, *Two New Sciences*, trans. S. Drake (Madison, Wisc.: University of Wisconsin Press, 1974).

14 Other theories of projectile motion deriving from the ancient world were equally metaphysical, or, to use a phrase of Michael Wolff, "speculative". This includes the influential theory of impetus deriving from Philoponus, as well as other ancient theories of motion such as Aristotle's. As Wolff emphasizes, none of them were empirical. There is in fact a discontinuity between the metaphysical tradition, and the new science. Wolff puts the point this way:

> ... it is now rather difficult to regard the theory [of impetus] as a connecting link between Aristotelian physics and classical mechanics. For the words 'connecting link' imply more than the mere temporal order of these theories. They are intended to point to the continuous evolution of one theory into another. But, if we keep in mind the non-empirical character of impetus theory, it is difficult actually to recognise such continuity.

See Michael Wolff, "Philoponous and the Rise of Preclassical Dynamics," in R. Sorabji, ed., *Philoponus and the Rejection of Aristotelian Science* (London: Duckworth, 1987), pp. 84-120. The passage cited occurs pp. 85-86.

15 We thus see that J. C. Pitt, "Galileo: Causation and the Use of Geometry," (in R. E. Butts and J. C. Pitt, eds., *New Perspectives on Galileo* [Dordrecht, The Netherlands: Kluwer Publishers, 1978], pp. 181-95), is just wrong to suggest that Galileo was "refining, for scientific purposes, the Aristotelian modes of causal analysis" (p. 187). Galileo was in fact eliminating the whole idea of Aristotelian causation in favour of a very different concept of science.

16 See also his "Impetus Theory Reappraised," *Journal of the History of Ideas*, 36 (1975), pp. 27-46.

17 Galileo Galilei, *Dialogue concerning the Two Chief World Systems*, trans. S. Drake (Berkeley: University of California Press, 1953).

18 In fact, there is in Newton's own exposition some residual Aristotelianism. It is this that (mis-)leads him into thinking that the second Law is not simply a definition of "force" but a Law parallel to the Law of Inertia. See F. Wilson,

"Explanation in Aristotle, Newton and Toulmin," *Philosophy of Science*, 36 (1969), pp. 291-310, pp. 400-428.

19 In actual fact, Newton's inference was more complicated than what follows, as we shall see in due course. But the inference that we are about to notice is, on the one hand, a good approximation to Newton's actual inference, and, on the other hand, the presentation that Newton's followers used to explain to a more general public the central points of Newton's physics.

20 Epigraph to Stillman Drake's translation of Galileo's *Dialogue concerning the Two Chief World Systems*.

21 P. Duhem, "Essai sur la notion de théorie physique," *Annales de philosophie chrétienne*, 79ᵉ année, t. 156, p. 588, pp. 591-2.

22 *Ibid.*, pp. 374-5.

23 More exactly, given the gravitational interaction among the planets that causes deviations from exact ellipses, Newton is able to demonstrate that the force that moves the planets is either inverse square or very close to inverse square.

24 Cf. Cecilia Payne-Gaposchkin, "Worlds in Collision," *Popular Astronomy*, 58 (1950), pp. 278-86, at p. 285.

25 Cf. *ibid.*, p. 279.

26 This terminology is due to J. L. Mackie, "Causes and Conditions," in E. Sosa, *Causation and Conditionals* (London: 1975).

27 E.g., M. Scriven, "Explanations, Predictions and Laws," in H. Feigl and G. Maxwell, eds., *Minnesota Studies in the Philosophy of Science*, vol. III (Minneapolis: University of Minnesota Press, 1962).

For a detailed discussion of those who misunderstand the nature of gappy knowledge, see F. Wilson, *Explanation, Causation and Deduction* (Dordrecht, The Netherlands: Reidel, 1985).

28 For greater detail on process knowledge, see G. Bergmann, *Philosophy of Science* (Madison, Wisc.: University of Wisconsin Press, 1956), Ch. Two. Bergmann also discusses gappy, or, as he calls it, "imperfect", knowledge. See also M. Brodbeck, "Explanation, Prediction, and 'Imperfect' Knowledge," in H. Feigl and G. Maxwell, eds., *Minnesota Studies in the Philosophy of Science*, vol. III.

29 This terminology derives from J. L. Mackie, "Causes and Conditions."

30 Francis Bacon, "Description of the Intellectual Globe," in *The Philosophical Works of Francis Bacon*, ed. J. H. Robertson (London: Routledge & Sons, 1905), p. 506.

31 Cf. T. Roszak, *The Making of a Counter Culture* (Garden City, NY: Doubleday, 1969), and *The Voice of the Earth* (New York: Simon and Schuster, 1992).

See also Leo Marx, *The Machine in the Garden* (New York: Oxford University Press, 1964).

32 Evelyn Fox Keller, "Feminism and Science," *Signs: Journal of Women in Culture and Society*, 7 (1982), pp. 589-602.

33 *Ibid.*, p. 601.

34 *Ibid.*, p. 601.

35 C. G. Hempel, "The Theoretician's Dilemma," in H. Feigl, M. Scriven and G. Maxwell, eds., *Minnesota Studies in the Philosophy of Science*, vol. II (Minneapolis: University of Minnesota Press, 1958).

36 Hempel should have taken the trouble to read G. Bergmann's *Philosophy of Science*, Ch. Two, where the point is fully explained.

37 Cf. S. Bromberger, "Why-Questions," in B. Brody, ed., *Readings in the Philosophy of Science* (Englewood Cliffs, NJ: Prentice Hall, 1970).

38 For a more extended discussion of these sorts of confusion, see F. Wilson, *Explanation, Causation and Deduction* (Dordrecht, The Netherlands: Reidel, 1985), Chapter Two.

39 From G. Bergmann, *Philosophy of Science*, p. 133.

40 For greater detail on this, see Fred Wilson, *Psychological Analysis and the Philosophy of John Stuart Mill* (Toronto: University of Toronto Press, 1992), Ch. 8.

41 G. Resch and V. Gutmann, *Scientific Foundations of Homoeopathy*, English trans. of the revised German text by G. Resch and V. Gutmann (Berg am Starnberger See, Germany: Barthel & Barthel Publishing, 1987), p. 355.

42 *Ibid.*, p. 349.

43 H. Dreisch, *The Science and Philosophy of the Organism*, 2 vols. (London: Unwin, 1908).

44 See for example, G. Resch and V. Gutmann, *Scientific Foundations of Homoeopathy*, Ch. 2, sec. 6.

45 Judge William J. Overton, "The Decision of the Court," in A. Montagu, ed., *Science and Creationism*, p. 380.

46 Evelyn Fox Keller, "Feminism and Science," p. 601.

47 *Ibid.*, p. 600.

48 There was a similar re-enchantment of science, on similar grounds, in Germany from the end of the 19th century through the 1930's, with disastrous results. See Anne Harrington, *Reenchanted Science: Holism in German Culture from Whilhelm II to Hitler* (Princeton: Princeton University Press, 1996).

49 Act 590 of 1981, General Acts, 73rd General Assembly, State of Arkansas.

For a clear account of how it came about that this statute was enacted, see Roger Lewin, "A Tale with Many Connections," *Science*, 212 (1982), pp. 484-488.

50 MacLean vs. Arkansas: Opinion of William R. Overton, U. S. District Judge, Eastern District of Arkansas, Western Division (Dated 5 January 1982).

51 See Theodosius Dobzhansky, "Nothing in Biology Makes Sense Except in the Light of Evolution," in J. Peter Zetterberg, ed., *Evolutionism vs. Creationism* (Phoenix, AZ: Oryx Press, 1983), pp. 18-28.

52 From the *General Catalogue of Christian Heritage College*, Christian Heritage College (2100 Greenfield Dr., El Cajon, Calif. 92021, U. S. A), 1977-78, pp. 17-20.

53 For a discussion of this argument, see F. Wilson, *Logic and the Methodology of Science in Early Modern Thought: Seven Studies*, Study Six.

54 Henry Morris, *Scientific Creationism*, p. 136.

55 For an excellent discussion of the methods of radioactive dating, see Kenneth R. Miller, "Scientific Creationism vs. Evolution: The Mislabeled Debate," in A. Montague, ed., *Science and Creationism*, pp. 18-63.

56 Morris, *Scientific Creationism*, pp. 157-58.

57 John C. Whitcomb and Henry M. Morris, *The Genesis Flood* (Grand Rapids: Baker Book House, 1961), pp. 341-2.

58 Whitcomb and Morris, *Genesis Flood*, pp. 343-44.

59 See Paul Thompson, "Some Punctuationists *Are* Wrong about the Modern Synthesis," *Philosophy of Science*, 14 (1983), pp. 74-85.

60 Duane Gish, *Science Digest*, October 1981, p. 84.

61 See Compton and Parker, *American Scientist*, 66 (1978), pp. 131-53.

62 Whitcomb and Morris, *Genesis Flood*, p. 327.

63 This is built into the Arkansas Act that mandated that the two models be taught in the schools of that state.

64 Charles Darwin, *The Origin of Species by Natural Selection*, Ch. vi, in P. Appleman, ed., *Darwin: Texts, Backgrounds, Contemporary Opinions, Critical Essays*, Second Edition (New York: Norton, 1979), p. 98.

65 See Dan-E. Nilsson and Susanne Pelger, "A Pessimistic Estimate of the Time Required for an Eye to Evolve," *Proceedings of the Royal Society: Biological Sciences*, 256 (April, 1994), pp. 53-58.

66 Robert Boyle, *A Free Enquiry into the Vulgarly Receiv'd Notion of Nature* (London: John Taylor, 1685/6).

67 Robert Boyle, *A Disquisition about the Final Causes of Natural Things* (London: John Taylor, 1688).

68 William Paley, *Natural Theology*, fourth edition (London: R. Foulder, 1803).

69 David Hume, *Dialogues concerning Natural Religion*, ed. Norman Kemp Smith, second edition (London: Nelson, 1947).

70 For more detail, see F. Wilson, *Empiricism and Darwin's Science* (Dordrecht: Kluwer, 1990).

71 Hume, *Dialogues concerning Natural Religion*, p. 227.

72 For a more extended discussion of modern genetic theory, see M. Ruse, *Philosophy of Biology* (London: Hutchison, 1973), Ch. 2 and 3.

73 For discussion of these changes, see D. Hull, *Philosophy of Biology* (Englewood Cliffs, NJ: Prentice-Hall, 1974).

74 We shall look further at some of these points below, Chapter Four, section (iv).

CHAPTER 4

METHODS
SCIENTIFIC
AND UNSCIENTIFIC

We are interested in drawing the distinction between science and pseudoscience. In order to do that we need to understand the *norms* or *standards* for accepting an explanation: under what conditions is a purported explanation *worthy* of acceptance? What *justifies* that acceptance as *rational* or *reasonable*?

We have begun to construct a picture of science. In outline, these norms of justification begin with observational data and proceed by means of *inferences* to justify accepting certain generalizations as worthy of use in explanation — and prediction. But pseudoscience often seems to proceed in the same manner, making inferences based on various data. Moreover, these inferences seem similar on the surface to those that are used in science. What we have to do is examine in greater detail the inferences that are made in science, and, further, to examine how they may be justified, so that we may establish which norms, which inferences, are rational or reasonable. With a knowledge of which standards are reasonable, we will be in a position to distinguish those that yield acceptable conclusions from those that *only apparently* yield acceptable — rationally acceptable — conclusions. The latter, of course, will be the conclusions of pseudoscience, the former the conclusions of science.

So far, our discussion of examples of scientific explanations yields a picture of science, and the inferences that it involves, that goes something like the following. One begins with observational data. One infers from these observations certain generalizations. These generalizations apply to a population even though, of course, all that one has observed is a sample. That means that the inference from observations to generalization is always hazardous, in the sense that there is no guarantee, even given the observational premises, that the conclusion, the generalization to which one infers, is true. To be sure, the observational data confer upon the generalization some likelihood. But they do not confer certainty. This inferential step, from observation to generalization, is referred to as an **inductive inference**, and, in contrast to *deductive inferences*, there is no guarantee that, if the premises are true, then the conclusion is true. *In inductive inference, the premises make it likely that the conclusion is true, if the premises are true, but they do not guarantee the truth of the conclusion.*

One thus arrives at generalizations as the conclusions of inductive inferences. The next step is *deductive*. From the inductively inferred generalizations one proceeds to discover what can be inferred from it deductively. If the propositions that are thus arrived at by deductive, and therefore certain, inference can be verified against observational experience, then the generalizations are accepted as worthy for use in explanation and prediction. This will provide further inductive support for the generalization. But again, the inference is not certain: there is still a gap between the sample examined, and the population described by the generalization.

The process of scientific inference that is the basis for the testing and acceptance of generalizations as worthy for use in explanations thus seems to go as follows.

The justification starts with a series of observations. One then proceeds by inductive inference to a generalization that at once both explains and is — at least somewhat — justified by those data. One has to say 'at least somewhat' here, since the observations will yield only a sample while the generalization is an assertion about a population. The assertion of the generalization can therefore be at best tentative: there is no guarantee. Inductive inference, unlike deductive inference, provides no guarantee that if the premises are true then the conclusion must be true. Given the logical gap between sample and population, it is always possible that what holds in the sample will not hold in the population, and that the conclusion of an inductive inference will be false even where it agrees with what holds in the observed sample. The classical example, of course, is that of swans. Until the 17th century all swans observed by Europeans had been white, and they felt cognitively safe in their assertion that all swans are white. But then some Dutch merchantmen sailing for the East Indies from around the Cape of Good Hope were blown off course and landed in what is now Western Australia. There they observed black swans. In the short run, all observed swans were white, and the inference to "All swans are white" was reasonable; but in the longer run, when one came to observe a larger proportion of the population, the further observations showed that that inductive inference in fact had a false conclusion, however justified it was on the basis of previous data, that is, on the basis of the previous, smaller, sample. What happened was that the further data enable the Europeans to *correct* their previous inferences.

It is this that is the crucial characteristic of the scientific method: it accepts generalizations only tentatively; checks those assertions against further data that it collects, deciding if the tentatively accepted generalization fits those data or is in conflict to them; re-affirming, or continuing to affirm the generalization previously accepted if it is supported by those data; and, if it conflicts with those data, it corrects its views by rejecting that generalization, adopting another in its place, one that fits not only the old data but also the increased sample that has come to be observed; and then proceeds to check this new hypothesis or the re-affirmed old hypothesis by deducing predictions from it and comparing these predictions to still further observational data, again revising if this is what is called for. *The scientific method consists of a constant adjustment of one's hypotheses to a constantly increasing set of observational data.*

We may contrast the scientific method for choosing which generalizations to affirm with eight other methods. Of these, the first three were identified by the philosopher Charles Sanders Peirce.[1]

(1) The first of these non-scientific or even anti-scientific methods is the **method of tenacity**: *continue to assert, no matter what, whatever generalization you have already asserted*. If you have learned to assert a generalization when seated at your mother's knee, if the beliefs of your fathers have been passed on to you, then this rule states that you should continue to affirm these generalizations, no matter what. In particular, it enjoins you to continue to affirm these generalizations no matter what further data come to hand.

But once this is seen, it is clear that this method will inevitably lead one to the affirmation of some generalizations which are false. No matter how wise your mother, no matter how informed your fathers, they, too, are only human and no more than any others can they ever observe anything more than a sample. Their affirmations, too, therefore, must be inductive and go beyond the observational data of a sample to make assertions about a population. These people have therefore inevitably given you beliefs that will fail to measure up as further data become available. Nonetheless, the method of tenacity enjoins you to continue to affirm these propositions that fail the predictive test.

There might be circumstances in which conforming to this norm might somehow be reasonable. But not for science. *Science is motivated by a cognitive interest in matter-of-fact truth*. What we have just seen is that the method of tenacity inevitably leads one to have beliefs that are false and therefore in conflict with what one would hope to attain if one has the scientist's cognitive interest in the truth.

Of course, if one is uninterested in the truth, the method of tenacity might be a good rule. But if one assumes, as one must for science, a cognitive interest in matter-of-fact truth, then conformity of one's inferences and beliefs to the method of tenacity will lead to beliefs that are incompatible with that interest. Given a cognitive interest in the truth, then, one can conclude that conformity of one's inferences and beliefs to the norms of the method of tenacity is *not a good, or efficient, means to achieve one's (cognitive) ends*. It should therefore be rejected by anyone who accepts the cognitive interest that moves science.

(2) The second method is the **method of authority**: *assert whatever belief is enjoined upon you by a recognized authority*.

The method of tenacity inevitably leads to conflict. What one person learns at his or her mother's knee is not what another person learns at his or her mother's knee. The method of tenacity thus leads people to have beliefs which conflict with each other. People found that they could resolve such conflicts if they were agreed upon some *common authority* that would determine which beliefs were to be accepted.

We can recognize the method of authority used by "creation scientists." The Creation Research Institute, the major centre for those who work to defend, and spread, "creation science," require applicants for membership to subscribe to the belief that the Book of Genesis is "historically and scientifically true in all of the original autographs. To the student of nature, this means that the account of origins in Genesis is a factual presentation of simple historical truths."[2] The authoritative account in the Bible becomes the basis for doing science. Thus, H. W. Morris states that

> ... it is ... quite impossible to determine anything about Creation through a study of present processes, because present processes are not creative in character. If man wishes to know anything about Creation (the time of Creation, the duration of Creation, the order of Creation, the methods of creation, or anything else) his sole source of true information is that of divine revelation. God was there when it happened. We were not there ... Therefore, we are completely limited to what God has seen fit to tell us, and this information is in His written Word. This is our textbook on the science of Creation![3]

Another defender of "creation science" states that

> Some object [to "creation science"] saying that the Bible is being placed on a scientific basis and being used as a textbook of science. To this objection it may be pointed out that the Bible is not a textbook, but that it does contain fundamental truths by which all scientific study must be oriented....The Bible must not be judged by men's ideas of science, for scientific theories came and go; on the other hand the Word of God abides forever, and human theories must be brought to the unerring standard of the Word. The Bible may not give details, they are left for man to discover; it does, however, lay down basic philosophical principles on which science is to be interpreted.[4]

Unfortunately, different groups, while each agrees within itself about the appropriate authority, turn out to disagree with each other on which authority. Some opted for the Bible. Fundamentalist Jewish groups agree with fundamentalist Christian Protestants on the authority of the Book of Genesis and the claim that, on the basis of this authority, we ought to reject the Darwinian account of evolution. Still, these two groups disagree about whether what Christians have called the "New Testament" should also be taken as authoritative. But there are authorities other than the Bible that have been accepted by various groups. Some have opted for the Pope in Council, some for the Koran, some for an Ayatollah, some for a Marxist dictator, and so on. And unfortunately, these different authorities disagree with each other. Worse, even a single authority was often ambiguous, admitting of various, conflicting interpretations. The different groups found various techniques for reconciling authorities, or, more exactly, for reconciling their chosen authority with itself. These included the techniques of scholastic theologians who developed procedures that made them adept at reconciling apparently conflicting authoritative texts. Other solutions were also developed to solve the problem of getting people to agree upon which authority ought to be accepted, and which interpretation of a given authority ought to be accepted. Do we use Jewish, Christian or Muslim scriptures? or Ron Hubbard's tracts on scientology? Unhappily, the methods used to induce agreement on an authority included exterminating those who accepted different authorities. St. Dominic developed this mode of securing a single authority during the Albigensian crusade; Lysenko was able to convince Stalin to use similar methods against Vavilov and the other defenders of the rational acceptability of Mendelian genetics.

In any case, it must be said once again that so far as concerns at least matter of fact generalities, authorities are no more infallible than anyone else. No one now takes seriously the defence of Michruinist genetics in terms of dialectical materialism: it is beyond doubt a matter of knowledge that Mendelian genetical theory is the true theory. Or again, consider the statement in the Bible, I Kings, 7: 23. This concerns the building of Solomon's palace by the skilled artisan Hiram of Tyre, who particularly excelled in bronze work. This verse tells how Hiram

> ...made a molten sea, ten cubits from the one brim to the other: it was round all about, and its height was five cubits: and a line of thirty cubits did compass it round about.

The diameter of this "molten sea" is given as 10 cubits and its circumference as 30 cubits, yielding a ratio of circumference to diameter of

$$3 : 1$$

In fact, this ratio of circumference to diameter, now called π ("pi"), is known since the time of the ancient Greek mathematicians to be

$$\pi = 3.1416...$$

So much for the inerrancy claimed for the Bible by fundamentalist Christians. Or yet again, consider the 1966 request of Sheik Abd el Aziz bin Baz to the King of Saudi Arabia that the latter suppress a heresy that was becoming widespread in his country:

> The Holy Koran, the Prophet's teachings, the majority of Islamic scientists, and the actual facts all prove that the sun is running in its orbit...and that the earth is fixed and stable, spread out by

God for his mankind .... Anyone who professed otherwise would utter a charge of falsehood toward God, the Koran, and the Prophet.[5]

Here there is a conflict between the authority of the Koran, or at least of the Koran as glossed by the Sheik, and the facts that have been gathered by science since the time of Copernicus. The Catholic church at one time accepted the same sort of argument as the Sheik, and condemned Galileo for defending the thesis that the earth moves about the sun. The church has revised its position, deciding that Galileo and Copernicus were correct about the facts after all and that a new authoritative reading of Scriptures was necessary, one that was consistent with the acceptance of the heliocentric viewpoint. For most besides the Sheik, the geocentric theory has taken its place alongside the flat earth theory, as being quite unacceptable.

The point here is that conflicts almost inevitably arise between the authority and matters of fact. In other words, some authoritatively accepted beliefs turn out to be false, given the empirical data. Thus, the Biblical claim that God miraculously made the sun stand still in order to enable Joshua to win the battle of Gibeon turned out to be false, given the empirical data that support the Copernican scheme for the motions of the planets, including the earth, about the sun: if it is the earth that moves and not the sun, the sun could not be commanded to stand still. The method of authority demands that if this happens one continue to accept the authority. The Catholic church thus came into conflict with Copernicanism, just as nowadays so does the Sheik quoted above, and just as nowadays certain Protestant fundamentalist groups come into conflict with Darwinian theory in biology. No doubt one could continue to adhere to the word of the authority in a way that is perfectly satisfying and yields no conflict — unless, that is, one is committed, as scientists are, to the discovery of matter of fact truth. Again, the point is that we know that conformity of our inferences and beliefs to the norm of the method of authority is not an efficient way to attain our cognitive end of discovering matter-of-fact truth. To the contrary, we know that conflict will arise, that we will be required to accept beliefs that turn out to be, as a matter of fact, false, and therefore to be rejected if we are moved by the cognitive interest that moves science, namely, to repeat, a desire to know matter-of-fact generalizations that can explain the events in which we are interested.

There is a close connection between the method of tenacity and the method of authority: to accept the method of tenacity is to accept that you, yourself, are the authority, the person who determines which beliefs are worthy of acceptance.

(3) The third method identified by Peirce is the **a priori method**: *accept only those beliefs that are such that it is impossible to imagine that the contrary is true.* Thus, if you cannot imagine that the planets move in orbits that are non-circular, then this method requires that you accept as true the proposition that planets move in circular orbits (or perhaps in orbits compounded of circles). Here the authority dictating one's beliefs is one's own power of imagination.

Von Däniken uses this method to justify the acceptance of the hypotheses that he proposes. Thus, at one point in *Chariots of the Gods?*, discussing a being in a wall painting in Tassili, in the Sahara, he states that, "Without overstretching my imagination, I got the impression that the great God Mars is depicted in a space or diving suit."[6]

The method clearly is capable of taking what, in terms of empirical evidence, are unsupported *mere* hypotheses, and transforming these into acceptable theories:

> *Let us suppose* that foreign astronauts visited the territory of the Sumerians thousands of years ago. *Let us assume* that they laid the foundations of the civilization and culture of the Sumerians and then returned to their own planet, after giving this stimulus to development. *Let us postulate* that curiosity drove them back to the scene of their pioneer work every hundred terrestrial years to check the results of their experiment. By the standards of our present-day expectation of life the same astronauts *could easily* have survived for 500 terrestrial years.[7]

It is clear how propositions that are apparently postulates suddenly become the authoritative basis for judging matters of fact. They become authoritative because they appeal to von Däniken's imagination —

and to that of his readers, or at least to the imaginations of those of his readers who are unwary enough to let his rhetoric seduce them!

The fairly close connection between this method and the method of authority is fairly clear. It is just that one's own imagination becomes the authority to which one appeals, rather than some external authority. Von Däniken recognizes the similarity of his method to those who use the method of religious authority in science, e.g., the "creation scientist":

> Admittedly this speculation is still full of holes. I shall be told that proofs are lacking. The future will show how many of those holes can be filled in. This book puts forward a hypothesis made up of many speculations, therefore the hypothesis must not be "true." Yet when I compare it with the theories enabling many religions to live unassailed in the shelter of their taboos, I should like to attribute a minimal percentage of probability to my hypothesis.[8]

Unfortunately, this method has the same problems as the previous two methods.

In the first place, there is no means for reconciling disagreements among authorities. How do we reconcile the Bible and the Koran, where they disagree? How do we reconcile Protestant Biblical authority with Papal authority, where these disagree? How do we reconcile von Däniken's imagination with that of the "creation scientists"? or with that of Velikovsky? *For these methods, there is no means of correction when conflict arises.* In the end, adherents of competing authorities must agree to disagree, since they can agree on no single authority that will enable them to decide which one is correct. In the end, there is no procedure for resolving conflicts, no means for correcting error and securing agreement — except by the unfortunate method of killing off your opponents, a method that has in the course of history, alas!, been adopted far too often.

In the second place, the *a priori* method, like the methods of tenacity and authority, comes into conflict with matters of fact that we come to know. The principle that the planets move in circles came into conflict with Kepler's account of planetary orbits. If, then, one has adopted the cognitive interest in matter-of-fact truth that moves science, then the method *a priori* requires one to accept beliefs that are in conflict with that interest.

Since each of these three methods leads one to hold beliefs that are incompatible with the cognitive goals of science. each of them must be rejected as providing norms for making inferences and accepting beliefs.

(4) Next let us look at the **method of resemblance**.[9] John Stuart Mill identified this as one of the most common and "deeply rooted" fallacies of belief. It is the rule that *if x resembles y then x causes y*, or, as Mill put it, it is the principle that "the conditions of a phenomenon must, or at least probably will, resemble the phenomenon itself."[10] According to this rule, since there is a similarity between the fiery appearance of red hair and the metaphorically fiery behaviour of hot-tempered people, we should therefore infer, and accept as a true regularity, the generalization that red-heads in general are hot-tempered.

Astrology is permeated by inferences that fit this pattern, which might be called a "principle of correspondence" or "law of analogies".[11] Thus, Mars is said by astrologers to be associated with blood, war, and aggression, while the lovely Venus is said to be associated with beauty and motherhood, while Saturn is associated with gloominess and scholarship. No one has ever been able to establish in a scientifically satisfactory way that these correlations hold. Certainly, when they were first formulated, no such testing was undertaken; they came to be in a period long before the notion of a scientific test was common, let alone a statistical test. These causal judgments are not scientific but seem to have come about by resemblance thinking. The reddish colour of Mars is associated with the colour of blood and therefore with war and aggression. Since Venus is such a lovely sight it is associated with beauty and therefore woman and motherhood. Since Saturn is dull and the slowest of the planets it implies gloom and depression and (somehow) scholarship. The same sort of thinking extends to the signs of the zodiac. Libra, represented by the scales, signifies justice and harmony, while a Scorpio resembles the namesake of the sign in being secretive and aggressive.

In fact, this way of thinking is very common. People often estimate that someone, say Jones, is, for example, a civil servant by seeing how similar Jones is to their stereotype of a civil servant. This may lead either to severely underestimating the number of civil servants in society at large or to severely overestimating the number, but in any case getting it wrong.

This method is, of course, like the others that we have examined, one that comes into conflict with the aims of science. It leads us to accept as matter-of-fact regularities patterns which are far from being regularities. It thus conflicts with the cognitive aim of science to come to know true patterns of matter of fact regularity.

(5) Next there is the **method of fantasy**: the rule in this case is that *the wish is the father of the belief*. We are all unable as children to distinguish fantasy from reality. It is a distinction that we must all learn. Some learn better than others. Thus, there are some people who very strongly want structure to their lives and patterns that will enable them to understand somehow what is happening to them. Astrology can often provide a framework that yields such a structure, and because people who become involved in this pseudoscience want it to be true, they accept it as true, using it for purposes of explanation and prediction, in complete disregard of the fact that not only is there no objective evidence for its truth but the fact that there is considerable evidence that it is false.

One suspects something like this also lies behind the reliance of Velikovsky and von Däniken on the myths of the Bible. Many people look on the Bible with a certain nostalgic fondness for the time when it was taken as literal truth. They *want* the Bible to be true. But they know that it is not; it is myth. So, when someone says that, if we take it as true, then we can provide a scientific explanation for what it reports, they are very receptive to the claims of that person, however weak are those claims with regard to scientific support.

So strong is the pull of this method that special steps must be taken in scientific research to protect against it. Scientists put to the test what are very often their favourite hypotheses. That means that they are liable sometimes at least consciously to cheat and more often unconsciously to distort their evidence so that it seems to support more strongly than it does objectively the acceptance of their hypothesis.[12] Experimental controls, e.g., blind testing, should be introduced wherever such bias is a serious possibility.

In any case, it is clear that this method, like the others just discussed, cannot lead to the satisfaction of the cognitive aims of science. What science aims to discover are true matter-of-fact regularities. The method of fantasy leads to the acceptance of many claims that are false. Just because someone wants something to be true is no guarantee that *is* true: in this vale of tears there is no guaranteed happy ending.

(6) Then there is the **method of superstition**: here the rule is that *coincidence = causation*. Consider a story such as the following. Aunt Bertha has a vivid dream of her father Abner whom she has not seen for may years. Later that day, and out of the blue, she receives a phone call from her father, who tells her that he has been thinking very much about her lately and that he would like to talk to her and reconcile. "How can you explain *that*?" Aunt Bertha asks, and concludes that she has psychic powers.

There is of course a considerable amount of drama and emotion in the story for Bertha, but in fact there seems to be nothing about the events that would indicate that the two events happening together is something more than chance, a matter of *coincidence*. There is a strong tendency to infer causation from coincidence, especially if one of the events is of great emotional significance for us.

Coincidence is not causation, however. The ancient Chinese regularly set off fireworks during eclipses of the sun, in order to frighten away the dragon whose flight, they believed, obscured the sun from view. As they saw it, their remedy was successful: the eclipse ended. Other peoples held rain dances well into the dry season in order to bring on rains: the wet season tended regularly to appear. Or consider the person who regularly performed rituals aimed at keeping tigers away from his Toronto neighbourhood. When asked about its efficacy, he responded, "Have you seen any tigers recently?"

In these cases, the events happen together but there is no causation. Alternative hypotheses about the causes of the ending of the eclipse, about the coming of the rainy season, or about the absence of tigers, are not considered. Attempts to put the hypotheses to the test by trying to discover data that would falsify the causal hypothesis are not undertaken. Aunt Bertha never thinks to try to figure out how many times she has dreamed or otherwise thought of Abner and it has *not* been followed by Abner getting in touch with her. She never thinks, in other words, to find out how many times her dreams

have *not* been fulfilled. Causation is all too easily inferred from the coincidence.

Here is another example of the same inference, backed up by pseudoscience about the "vital energy" of the planet on which we live. The example concerns the Vice President of a Canadian firm that deals with many stock portfolios worth hundreds of millions of dollars. Yvon — for reasons which are perhaps obvious we are given only his first name —

> ...est branché sur l'« énergie vitale » de la planete. « C'est très utile, dit-il. Par exemple, un matin d'octobre 1995, j'ai senti que la Bourse américaine irait très mal. Dès l'ouverture des marchés j'ai donc vendu. À la clôture, le Dow Jones avait perdu 135 points. Mes clients avaient sauvé une petite fortune! Des choses comme ça m'arrivent souvent. »[13]

And how many times has the feeling *not* been confirmed by events?

Here again, from the viewpoint of science, Yvon's inference, like that of Aunt Bertha, is simply fallacious. It even has a name, so common is this fallacy. It is the fallacy of *post hoc, ergo propter hoc* — after this, therefore because of this. It is a fallacy because it leads us to accept as statements of cause or regularity sequences that are not really causal, not really regularities, just coincidences.

In fact, emotionally significant coincidences are fairly common. Thus, as John Paulos has pointed out,

> ... it can be shown ... that if two strangers sit next to each other on an airplane, 99 times out 100 they will be linked to each other in some way by two or fewer acquaintances.... Maybe, for example, the cousin of one of the passengers will know the other's dentist. Most of the time people won't discover these links, since in casual conversation they don't usually run through all their 1,500 or so acquaintances as well as all their acquaintances' acquaintances.[14]

Given probabilities of this order, it is not surprising that we very often come across emotionally significant coincidences.

What really needs explaining is not the fact that such coincidences occur as the fact that so many feel the need to construe such coincidences as cases of causation. The contingencies of social conditioning and re-inforcement can explain such practices as using fireworks to end eclipses or dancing to end seasons of drought. More personal factors, but also involving re-inforcement, can explain why someone like Aunt Bertha would leap to the conclusion that there was a causal relation between her dream and her father's call. She wants to be able to predict emotionally significant events, and the satisfaction of this desire on one occasion gives her confidence — unwarranted scientifically — that she can do it on all occasions. Perhaps, too, it is socially satisfying, and even monetarily rewarding to be known as a psychic.

In any case, it is clear that it is this method of inferring causes from coincidence which gives rise to many superstitions, whether they concern the efficacy of fireworks, or rain dancing, or the psychic powers of Aunt Bertha, or the aphrodisiacal powers of bear gall and ground rhinoceros horn according to traditional Chinese medicine. In the case of rain dancing, fireworks or medicinal claims of various traditional pharmaceutics the method of superstition is re-inforced by the method of authority. In the case of Aunt Bertha, the method of superstition is re-inforced, very likely, by the method of fantasy: Aunt Bertha really wants to be able to predict emotionally significant events, anticipate their occurrence. Don't we all? Indeed, it is so. But not everyone makes it into a rule for belief. Certainly, anyone interested in causal relations, anyone with the cognitive interests of science, in contrast to Aunt Bertha, rejects it as a rule for belief formation.

The inference from coincidence to causation appears often in pseudoscience. Thus, Velikovsky refers to what are most easily understood as coincidences among different ancient mythologies to infer a causal connection which links the several statements as reports of the ejection of Venus from Jupiter about 5,000 years ago and its passing close to the Earth prior to assuming its present orbit. Von Däniken also relies on coincidences among mythological statements to infer a causal connection among them as effects of the visits of ancient astronauts.

Certainly, the method of superstition is not that of science. assuming that our aim is knowledge of matter-of-fact regularities, the method is certainly inefficacious to that cognitive aim. It is in fact a fallacious mode of reasoning that often leads us into error and provides no means for detecting such error and remedying it. This is in stark contrast to the method of science.

(7) There is also the **method of prejudice**: This is really a combination of the previous two methods. I want something to be true, and search for evidence that will support that belief — any coincidence will do — and ignore evidence that conflicts with it. The example that the philosopher David Hume once gave of such a belief was the judgment, then common, that Irishmen lack wit. Dull Irishmen could be cited in support of the judgment; any person from Ireland who exhibited intelligence was skipped over in the catalogue of evidence.

Another example is the practice of placing in churches plaques giving thanks and credit to the patron saint who had saved those who put up the plaques from storms at sea. The wish is that there be unseen forces, saints and so on, which or who can save one from danger. Wanting it to be true, one believes it to be true. The evidence of the truth of the belief is the fact that one was saved from the storm upon praying to the saint. But what of all those who prayed and were not saved? Oneself being saved upon praying is hardly conclusive evidence that the praying helped unless we have some sense of the numbers who prayed and were not saved. The absence of this very relevant evidence is ignored by those who put up the plaques.

Since there are in fact many intelligent Irishmen, these individuals show it to be false that all Irishmen lack wit. Yet people continued in the 18th century, and indeed into this century, to hold that judgment. Another of Hume's prejudices, not so clearly recognized as such, was that Negroes are dull fellows. This prejudice continues to be held by many, in spite, unfortunately, of compelling counterexamples that show the judgment to be false. Since our cognitive aim is to come to know true matter of fact regularities, and since the method of prejudice leads us straightway into false beliefs about regularities, it follows that this method, like the others, is not a good method for anyone to use who accepts the cognitive goals that define science.

(8) Finally, we should recall the **method of Aristotle**. This method presupposes the truth of a non-empirical metaphysics which asserts the existence of non-empirical

natures or unanalysed dispositions and powers in things. These natures explain the empirically observed behaviour of objects. The method consists in acquiring rational intuitions of these natures. For Aristotle himself, the method is one of abstraction; abstracting the nature of the thing from its observed appearances, we have a rational intuition of that nature. For others, such as Descartes, we have innate ideas of the natures of things, and the methodological task is to organize these systematically from the simpler to the more complex, to give ourselves a knowledge of the ontological/causal structure of the world.

Clearly, this method is not one of science since it aims at knowledge of entities — natures — which lie outside the realm of ordinary experience. In fact, as scientists such as Galileo and Boyle argue, it runs counter to the cognitive aims of science, leading us to accept as laws regularities that objectively are not such.

Explanations in terms of powers are everywhere in pseudoscience, however. Consider the coincidence of Bertha's dream of Abner her father and the telephone call which he made to her. Bertha attributes this to her psychic powers. In fact, to explain the coincidence in terms of psychic power is to say very little indeed. The psychic power Bertha ascribes to herself is simply this:

x has a psychic power = if x dreams that an A-type occurrence happens then an A-type occurrence happens

Either this is true by definition of 'psychic power' or it is not; if it is not, then we are owed some explanation of what a "psychic power" is apart from its exercise.

Or consider what is apparently a very careful scientific exposition of the evidence of the validity of the theories of homeopathic medicine, *Scientific Foundations of Homoeopathy* by G. Resch and V. Gutmann. One is given chapters with considerable scientific detail on such matters as "Molecular Interactions" (Ch. 6) and "Molecular Systems Organizations" (Ch. 7). Then eventually we come to Part III on "Biological and Medical Aspects" which begins with Ch. 14 on "The Nature of Man" where it is not just science that is now relevant to understanding human being, illnesses and medical practice, but also the Aristotelian view of human being. We are referred to the four Aristotelian causes (material, formal, efficient and final). We need not go into the details; it suffices to note that the crucial one is "finality": "the three causes outside finality are open to different finalities and governed by them." The final cause is of course the teleological goal of the unanalysable strivings of the object or person. The material cause is the object whose nature is being considered; the formal cause is the form which the striving or activity takes; the efficient cause is the actual exercise of the power which is the nature of the thing; and the final cause is the goal at which the striving aims. The point for us is that none of this has anything to do with science; causes of this sort are rooted in the *non-empirical* natures of things. We are told that "...man is part of nature," that "within nature, the four [Aristotelian] causes connect everything, including man," and that "all life is subject to them."

Since man is subordinated to the structure and order of nature, everything that can be said about nature is applicable to man. The four causes are the basis of all natural events and phenomena which *cannot be reduced, and consequently must always be considered with regard to questions about nature.*

Like other parts of nature, *man illustrates the principle of finality.* (p. 344; their emphasis)

All of a sudden, then, out of nowhere, we have a transition from scientific explanations, or at least what purport to be scientific explanations, to something which is beyond the realm of the empirical and therefore not part of science. Whatever explanatory power these concepts may have to the practitioners of homoeopathic medicine, that power is not scientific. From the viewpoint of the latter, these concepts add obscurity and confusion, and interfere with the careful scientific evaluation of empirical evidence. These concepts may promote the aims of the practitioners of homoeopathic medicine, but they certain conflict with the cognitive aims of science — and with the aim of acquiring sound empirical knowledge about the healing power of various pharmaceutical practices.

In contrast to these other methods, we have the **inductive method of science**. This we have already attempted to describe, in outline at least, and we already recognize that, unlike the other methods we have considered, it does not come into conflict with the cognitive goals of science. For, as soon as one discovers that a belief is in conflict with that goal, then the method of science directs one to give up that belief. In contrast, the other methods all require one to hold a belief on certain grounds — authority, fantasy, or what not — quite independently of the empirical evidence. In science, one can give up a belief about the facts because no belief is accepted dogmatically, to be affirmed no matter what is the empirical evidence; all beliefs are accepted only tentatively, to be rejected if further data require it. At the same time, it will always be true that, so far as one can tell on the basis of presently available data, the hypotheses that one accepts are all in fact true. For, they have all been adopted precisely because they do, so far as we can tell, conform to all the data that have hitherto been collected. Where the available data lead to the formation of conflicting or contrary hypothesis, then the method tells us to suspend judgment between them until we have acquired further information that will enable us to decide between them — or against both and in favour of some third hypothesis (which might even be able to show that the two earlier hypotheses were both partial expressions of what really is the case). The method of science is, as we said, self-correcting, always adjusting the beliefs that it requires us to accept in a way that requires them to fit our always increasing sample of observed data.

This not of course to deny that sometimes a scientific theory has been held dogmatically by some who accept it. Scientists are only human too. But such dogmatism in the face of empirical evidence and observational data, while appropriate to one of the other methods we have noted, is contrary to the norms of science: for, these norms recognize that all generalizations go beyond the evidence available and that they are therefore liable for revision in the light of further data. A scientist who accepts his or her theory dogmatically, however human he or she may be, is nonetheless acting contrary to the norms of science.

What we have just given is a *justification* for the use of the method of science. This justification consists in a practical argument. We have certain cognitive ends. We undertake research to satisfy these cognitive ends. Some rules to which we conform our research efforts lead to results which do not regularly satisfy those cognitive ends. Such ineffective practices can be found in the method of authority, for example, or the method of tenacity or the method of fantasy. In contrast, conforming our research efforts to the rules of the scientific method does yield results which tend much more reliably to satisfy our cognitive ends. We are therefore justified, relative to those cognitive ends, in conforming our research practices to the rules of the inductive method of science.

Here is the practical argument:

> *Given that we have a cognitive interest in true matter of fact generalities, then the inductive method of science is, so far as we can tell from past practice, the methodological rule that is most effective in leading us to accept or reject hypotheses that satisfy that cognitive interest.*[15]

This argument appeals to the notion that the *rational person will adjust means to ends*. We choose an axe, not a butter knife, to cut down a tree. We do so because the axe is much more efficient for that end; it is more appropriate as a means towards our ends. Similarly, if our cognitive ends are those of science, then the inductive method is, so far as we can tell, more efficient than the other methods at which we have looked, e.g., the method of authority or the method of tenacity or the *a priori* method. The justification that we are considering appeals to this practical standard of rationality: adopt those means that so far as you can tell most efficiently lead to the satisfaction of your ends.

The argument, then, is this: **if we take for granted the cognitive ends of science, then the reasonable person, one who adjusts means to ends, will adopt the method of science**. At the same time, such a person will reject the practices required by the other methods at which we have looked.

Two points about this justification of the method of science should be noted.

First, the inductive method of science cannot guarantee that propositions that are accepted in conformity with its standards really are true. The method requires one to accept as true, for purposes of explanation and prediction, those propositions for which there are data that testify to their truth and for which there are data that do not testify to its falsehood, or at least to accept those propositions for which the data more strongly testify to truth than they testify to falsehood. But we never achieve a complete set of data: there is always an inference, to put it a bit simply, from an observed sample to a population. There is never any guarantee, then, that we will not come across among the individuals we have not yet observed some that falsify a proposition that we have hitherto accepted as true. Thus, the method of science never yields beliefs which are incorrigibly certain: *its results are always tentative.*

Second, we can contrast this feature of the scientific method to what the other methods achieve. They do allow for beliefs which are incorrigibly certain. Unfortunately, many of these turn out to be as a matter of fact false. The method of authority, for example, or the method of tenacity, enjoin the continued acceptance of propositions even after contrary empirical data have been discovered. In contrast, the inductive method of science takes the notion of falsification seriously. What it insists upon is that, if data are found that show a belief to be as a matter of fact false, then that belief is rejected. The method of science cannot guarantee truth — no method can guarantee truth — nor can it infallibly eliminate error — we may fall into error simply because we have not yet been able to acquire sufficient data — but it at least allows, as the other methods do not, for the *correction of error* by requiring us to reject any proposition for which we acquire falsifying data.

It is this account of science that was used by Judge William R. Overton to rule that "creation science" is not science but religion and therefore something that, on American constitutional grounds of separation of church and state, cannot be taught alongside the theory of evolution in biology courses. "Creation science" has some of the trappings of science, but when examined closely with regard to its deeper structure, we can see, Judge Overton argued in his decision, that it in fact is contrary to the basic norms of science. Apparently science, but not really so: it is pseudoscience.

What was at issue was the attempt by Protestant fundamentalists to secure the teaching of "creation science" alongside evolution in the public schools of Arkansas. They managed to convince the state legislature to require this to be done in Arkansas public schools.

This law was challenged in court, and Judge Overton's decision ruled, as indicated, that "creation science" is really religion, and that the state law requiring it to be taught in schools was unconstitutional.

The characteristics of science are the following, according to Judge Overton's opinion:

(1)   It is guided by natural law.
(2)   It has to be explanatory by reference to natural law.
(3)   It is testable against the empirical world.
(4)   Its conclusions are tentative, i.e., are not necessarily the final word.
(5)   It is falsifiable.[16]

These features characterize science and its method as we have described it. But none of these characterize "creation science." The explanations offered by "creation science" refer to supernatural entities. They are therefore not testable against the empirical world. For the same reason they are not falsifiable by any empirical data. In fact, the major tenets of the "theories" propounded by "creation scientists" are held dogmatically, no matter what: no contrary evidence would ever be acceptable. For the same reason, the conclusions of "creation science," in contrast to those of science, are not tentative, but dogmatic. Since the explanatory principles of science are both testable and falsifiable, they state matter of fact regularities. In particular, they make no reference to supernatural entities. When science explains by appeal to certain causal relations, then, it explains by appeal to natural law. But "creation science" rejects such explanations as insufficient when it insists that the explanation of the origin of species must, in the end, make reference to God, an entity that lies outside the natural world investigated by science. Finally, "creation science" is not guided in its research by natural law; it is clearly guided by its religious commitment to the unerring truth of the Bible story in Genesis. It follows that "creation science," whatever its surface appearances, is not science: neither its conclusions nor its method conforms to the norms of science. It is a peculiar kind of dogmatic fundamentalist Protestant religion that is disguised as science: it is pseudoscience.

Some pseudoscience is advocated by those who mistakenly, but genuinely, believe it to be science. It is pretty clear that this is the case in Velikovsky. But in the case of "creation science," the attempt to disguise religion as science is done clearly for the political religious purposes of the American religious right. The attempt was made to present the Biblical story of creation in scientific terms in order to secure its being taught in schools. It could not be taught in public schools in the United States in its own right, since that would violate the constitutional separation of church and state. It had therefore to be disguised. There is undoubtedly a good amount of bad faith in those who attempt to defend "creation science" as a case of *science*.

In any case, we asked the following question about the *norms* or *standards* for accepting a generalization as worthy of use for purposes of explanation: Under what conditions is a purported explanation *worthy* of acceptance? What *justifies* that acceptance as *rational* or *reasonable*? We have already seen part of the answer that is to be given. The method of science, unlike other methods such as those of tenacity, authority, and *a priori*, is able to satisfy the cognitive goal of science, namely, that of acquiring knowledge of matter-of-fact generalizations capable of explaining facts in which we are interested. Of the methods that we have looked at, the method of science alone keeps us on track as we strive to attain that cognitive goal.

Of course, if one does not have that goal, then this justification will not work. But to seek some absolute or transcendental standard that would justify science no matter what, and independently of the contingencies of human cognitive aspirations, is itself contrary to the tentative and empirical nature of science. Descartes sought such an *a priori* justification for the norms that he proposed, but science limits itself to the empirical and must reject any attempt to find a transcendental justification for itself. Science, if it is to be justified, must be justified relative to certain human ends. Its justification must, in other words, be pragmatic. Here one does not mean pragmatic in any crude or simplistic sense, in which those who adopt pragmatic solutions seek only those that serve human material ends and ignore the higher aspects of human being. Science may well serve, especially through its technological applications, these material ends. Nor are these goods that derive from science to be denigrated. In this respect, science has indeed served humankind extremely well. But science also serves our higher ends. Specifically, it serves our simple cognitive interest in matter-of-empirical-fact generalizations — our idle curiosity, if you will. But this curiosity, this love of truth for its own sake, is after all one of our human passions. *The method of science is justified just because it serves this passion, this love of truth, better than any other method.*

All this, however, is only a start. We must see in greater detail precisely how the norms of science serve the cognitive interests that move, and define science as a noble — but also useful! — human activity. It is to the detail of the norms of inductive scientific inference, and to their justification, that we must turn.

(i) The Logic of Scientific Inquiry

a. The Process of Inquiry: Three Examples

Our aim is to come to understand phenomena that we do not yet understand or at least do not yet understand completely. The understanding we seek, of course, is scientific understanding, that is, understanding in terms of matter of fact generalities. In order to obtain that understanding, we undertake inquiry. The inquiry aims to discover the knowledge that will fulfill our cognitive interest in coming to understand.

In order to obtain a better grasp of the process of inquiry, we will begin by looking at three examples.

Example (i): Faraday and the Mystical Turning of Tables

Round about the middle of the 19th century, people in England discovered that spirits could manifest themselves by moving tables that people were touching. The physiologist/psychologist Carpenter describes a typical session of this sort of spirit manifestation in this way:

A number of individuals seat themselves round a table on which they place their hands, with the *idea* impressed on their minds that the table will move in a rotary direction; the direction of the movement, to the right or to the left, being generally arranged at the commencement of the experiment. The party sits, often for a considerable time, in a state of expectation, with the whole attention fixed upon the table, and looking eagerly for the first sign of the anticipated motion. Generally one to two slight changes in its place herald the approaching revolution; these tend still more to excite the eager attention of performers, and then the actual "turning" begins. If the parties retain their seats, the revolution only continues as far as the length of their arms will allow; but not unfrequently they all arise, feeling themselves obliged (as they assert) to *follow* the table; and from a walk, their pace may be accelerated to a run, until the table actually spines round so fast that they can no longer keep up with it. All this is done, not merely without the least consciousness on the part of the performers that they are exercising any force of their own, but for the most part under the full conviction that they are not.[17]

The great British physicist, Michael Faraday, explored the phenomenon and communicated his results in a letter to *The Athenæum*.[18] The editor notes, in his introductory remarks, that "The communication is of great importance in the present morbid condition of public thought," when

… the effect produced by table turning has, without due inquiry, been referred to electricity, to magnetism, to attraction, to some unknown or hitherto unrecognized physical power able to affect inanimate bodies, to the revolution of the earth, and even to diabolical or supernatural agency (p. 801).

Table turning is no longer popular. But the inferences are still with us. The problem for us, as it was in the 19th century, lies with those "who refer the results to electricity or magnetism, yet know nothing of the laws of these forces" or who refer them "to some unrecognized physical force": they do so

…without inquiring whether the known forces are not sufficient (p. 801).

Faraday began with the assumption that the latter were in fact sufficient: the principles of the argument were the principles of scientific inquiry.

… the proof which I sought for, and the method followed in the inquiry, were precisely of the same nature as those which I should adopt in any other physical investigation (p. 801).

He began, in other words, with the assumption that **there exists** a **naturalistic causal explanation** of table turning of the sort described by Carpenter. This is the same sort of assumption that Herodotus made with regard to the Nile or Hippocrates with regard to epilepsy: there is nothing mysterious about these, they are natural phenomena, and as such they have naturalistic causes. The task of the scientific inquirer is not to find mystery, not to establish the "existence of a peculiar power" (p. 801), but to illuminate darkness by finding that natural causes of things.

Faraday obtained subjects who were, as he put it, "very honourable," and also "successful table-movers." These subjects all believed that they did nothing to move the table, and they certainly did not intend to move the table "by ordinary mechanical means," i.e., by pushing it. "They say, the table draws their hands; that it moves first, and they have to follow it…" (p. 801). Faraday was careful to search for electric and magnetic forces, since these had been cited as possibly explanatory. But he detected none. "No form of experiment or mode of observation that I could devise gave me the slightest indication of any peculiar force. No attractions, or repulsions, … nor anything which could be referred to other than the mere mechanical pressure exerted inadvertently by the turner" (p. 801).

Faraday had formed the hypothesis that there were unconscious expectations at work, which caused the table movers to move the table without their being aware of it. In order to ensure that these expectations were not connected to any particular substance, Faraday tested a variety of substances on the table, including sand-paper, rubber, cardboard, wood, etc. A bundle of these were placed under the hands of various table turners in several varied tests, and in each case the table turned.

Faraday performed two experiments.

In the first, he took five pieces of smooth, slippery cardboard and put one on top of the other with little pellets of a soft cement between them. The bottom piece was attached to a piece of sandpaper whose rough surface was on the table top. The edges of the cardboard sheets overlapped slightly, and on the underside a pencil line was drawn over the laps, indicating the original positions. The top piece of cardboard was larger than the others, so that a subject perceived only the large piece covering the table top. The cement between the sheets was, on the one hand, sufficiently strong to offer resistance to mechanical motion, and also sufficiently strong to hold the sheets in any new position to which they might move or be moved. On the other hand, the cement was weak enough that it would give way if a fairly strong mechanical force were exerted on it. The table mover placed his hands on the top card, and then waited for the table to move. It did. The point of the experiment was to see which moved first, the table or the hands. If the former, then the bottom cards would show that they had moved first and dragged along the higher cards. In contrast, if the hands moved first, they would have moved the top card first and dragged along the lower cards and, finally, the table.

Faraday reports that

> When at last the tables, cards, and hands all moved to the left together, and so a true result was obtained, I took up the pack. On examination, it was easy to see by displacement of the parts of the line, that the hand had moved farther than the table, and that the latter had lagged behind; — that the hand, in fact, had pushed the upper card to the left and that the under cards and the table had followed and been dragged by it (p. 802).

Faraday then constructed a second, more elaborate experiment. This involved two cards, one on top of the table and the other on top of the first, together with a visual index which would indicate which card moved first. Assuming the moving of the table is to the left, if the upper card moved to the left before the lower, the indicator would move to the right; in contrast, if the table moved first and the hands followed, as the subjects claimed, then the indicator would move to the left. The experiment was performed with the indicator not visible to the subject. The table moved to the left. This established that the apparatus did not interfere with the phenomenon.

Faraday relates that the indicator revealed that the hands were moving first.

> It was soon seen, with the party that could will the motion in either direction (from whom the index was purposely hidden), that the hands were gradually creeping up in the direction before agreed upon, though the party certainly thought they were pressing downwards only.

The subject vehemently believed that he was *not* pushing but rather was being pulled. However,

> When shown that it was so [that the indicator revealed that the hands were moving the table], they were truly surprised; but when they lifted up their hands and immediately saw the index return to its normal position, they were convinced (p. 802).

In fact, Faraday pointed out, people do not realize how difficult it is to press directly downward upon a fixed surface such as a table, "or even *to know only* whether they are doing so or not" (p. 802). This is especially so if there has been prolonged pressure, and the nerves in the hand are numbed. The feedback from the senses as the position of the fingers and the angle of their pressure is after a period of prolonged pressure reduced to a minimum. To use Faraday's example, "If a finger be pressed constantly into the corner of a window frame for ten minutes or more, and then, continuing the pressure, the mind be directed to judge whether the force at a given moment is all horizontal, or all downward, or how much is in one direction and how much in the other, it will find great difficulty in deciding; and will at last become altogether uncertain…" (p. 802). The importance of the index was that it provided feedback to, as Faraday says, the mind, when the senses had ceased doing so.

> … the most valuable effect of this test-apparatus is the corrective power it possesses over the mind of the table-turner. As soon as the index is placed

before the most earnest, and they perceive — as in my presence they have always done — that it tells truly whether they are pressing downwards only or obliquely; then all effects of table-turning cease, even though the parties persevere, earnestly desiring motion, till they become weary and worn out. No prompting or checking of the hands is needed — *the power is gone*; and this only because the parties are made conscious of what they are really doing mechanically, and so are unable unwittingly to deceive themselves (p. 802).

Faraday came to the phenomenon of table turning with a determination to come to understand it, that is, to subsume it under laws, known regularities. From the beginning he excluded "peculiar forces", that is, forces unknown to physics. He was then careful to exclude the possibility of electrical or magnetic forces, though he must have known that there was never any possibility consistent with electric and magnetic theories that there were powerful enough to move the table. The only force that remained was mechanical. But what was it that was exerting a mechanical force on the table sufficient to move it? The only possibility was the hands of the table turner. However, the table turners vehemently affirmed that they were pushing down vertically on the table, and that they were definitely *not* pushing obliquely. They asserted that they pushed with no horizontal component, and without a horizontal component their pushing would have no mechanical effect in that direction, and therefore could not be moving the table.

Faraday's first experiment demonstrated conclusively that there was in fact a horizontal component to the pushing of the table turners. *Since all other alternative possible causes had been eliminated, the only conclusion was that the table's moving was the effect of the subjects' pushing obliquely.*

It followed from this that the assertions of the table turners that they were pushing at an oblique angle were mistaken. Their self-monitoring, in other words, was erroneous. What the second experiment showed was that when the subjects were able to monitor themselves using the external sign of the index, maintaining through such monitoring a constant purely vertical and non-oblique pressure, the movement of the table horizontally also

ceased. *In the absence of an oblique pressure, table moving is absent.*

Faraday has thus shown that the table moving occurs if and only if there is oblique pushing. He has shown, in other words, that oblique pushing is necessary and sufficient for the table moving. The method or pattern of experiment that he has used, we shall indicate below, is what John Stuart Mill was to call the "joint method of agreement and difference." In applying this method, Faraday was guided by the background theory which gave him a knowledge of the sorts of natural forces that could possibly explain the motion of the table. The theory delimited the range of possibilities, and it became his task to precisely which among these it was that was the cause.

As a consequence of his experimental work Faraday was "enabled to give a strong opinion, founded on facts" (p. 801) to the people who required it. The experimental tests he performed justified the strength of his belief that table turning was to be explained mechanically, in terms of the pressure applied by the table turners. That is, the tests raised the posterior probability of that belief to a level in which it was, in effect, morally certain.

But there was another consequence of the set of experiments. "I think," Faraday added,

> ... the apparatus I have described may be useful to many who really wish to know the truth of nature, and would prefer that truth to a mistaken conclusion: desired, perhaps, only because it seems to be new or strange (p. 802).

Among those deceived were the table turners themselves who deceived themselves about how they were pushing on the table.

> It is with me a clear point [Faraday writes] that the table moves when the parties, though they strongly wish it, do not intend, and do not believe that they move it by ordinary mechanical power. They say, the table draws their hands; that it moves first, and they have to follow it, — that sometimes it even moves from under their hands....Though I believe the parties do not intend to move the table, but obtain the result by a *quasi* involuntary action, — still I had no

doubt of the influence of expectation upon their minds, and through that upon the success or failure of their efforts (p. 802).

With the apparatus of the second experiment, the self deception of the table turners was ended and, as Faraday says, "*the power is gone*".

Faraday concludes, "I must bring this long description to a close. I am a little ashamed of it, for I think, in the present age, and in this part of the world, it ought not to have been required" (p. 802). But required it was. He comments that "I think the system of education that could leave the mental condition of the public body in the state in which this subject has been found it must have been greatly deficient in some very important principle" (p. 801). The system remained deficient in his century, and has continued deficient in ours. Faraday's kind of close experimental analysis is the sort of thing that continued to be required in his century and continues to be required in ours. His experiments showed that the leap from table turning to peculiar, spiritual causes was quite illegitimate. People were to continue, however, to make such inferences; spiritualism remained a Victorian pre-occupation.[19] But they did not end with the Victorian era: they are with us today, and still require careful refutation. The quick leap from observed results to peculiar non-natural or spiritual forces, without the check of careful and controlled experiments, is a sure sign of pseudoscience.

Example (ii): Clever Hans, the Intelligent Horse

Clever Hans was a horse who was exhibited in Berlin in 1904. He had been educated by a former schoolteacher named von Osten, then about 70 years of age. He has spent three years training Hans, using the methods of a schoolteacher — patience and occasional rewards of carrots — rather than the whip of the circus animal trainer. Hans could solve arithmetical problems including those involving compound fractions and decimals, spell and define words, identify persons and objects by name, identify musical notes and intervals, and even express like and dislike of various kinds of music. Hans of course did not do this vocally. Rather, he responded to questions by tapping with his hoof, shaking his head, or walking over and pointing to letters on a board or objects on a rack.

Hans was examined by a select committee which included a circus manager, several educators, a zoologist, a veterinarian, a physiologist, and the famous psychologist Carl Stumpf. The committee reported that "This is a case which appears in principle to differ from any hitherto discovered, and has nothing in common with training, in the usual sense of that word, and therefore it is worthy of a serious and incisive investigation."[20]

Some thought that there had been investigation enough to conclude that Hans settled an issue that had been debated at least since Descartes, namely, the issue of whether animals have consciousness. It seemed that Hans' performances indicated that they did.

In fact, it was not a sound inference. The "serious and incisive investigation" had yet to be made. It was a young psychologist Oskar Pfungst who undertook it. He reported the results in what is now a classic study, *Clever Hans*.

Pfungst took pains to make friends with Hans, and then discovered that, when von Osten was absent, Hans was able to answer his [Pfungst's] questions correctly. Others had had the same experience, and admitting their failure to find natural causes jumped to the conclusion that Hans had quasi-human abilities, powers that were in some way supernatural. Pfungst too was baffled; he seemed unable to locate natural causes for Hans behaviour. But he was not inclined to make the leap to the supernatural. Like Herodotus when faced with the question of the flooding of the Nile or like Hippocrates when faced with the question of the cause of epilepsy or like Faraday when faced with the question of table turning, Pfungst refused to conclude that extraordinary causes had to be assumed. He refused to jump to the conclusion that here we have natural events that can be understood only by reference to what is beyond nature, what is non-empirical. Pfungst kept firm to the scientific attitude that *there are* naturalistic causes for natural events. To put it in the words of Judge Overton, Pfungst held firm to the view that science is "explanatory by reference to natural law."[21] *There are* naturalistic causes. The fact that we have not yet found them does not falsify this claim. Here Popper is correct about the logic of "there are" or existence claims. All that is implied by the failure to find an instance of what is claimed to exist is that we have not looked hard enough, that our search has *not yet* been successful.

Pfungst designed a series of experiments in which Hans was questioned in the usual way. But the questions were so arranged that the questioners knew only half the answers. The result was clear. Hans could give the correct answer only when the questioner knew. When the questioner did not know the answer, Hans became confused. Where the questioner knew the answer, Hans knew it; where the questioner did not know the answer, Hans did not know it. These experiments established (by, as we shall see, what John Stuart Mill called the method of agreement) that the questioner knowing the answer was a *necessary condition* for Hans knowing the answer.

Pfungst concluded that it was obvious that *there are* cues as to the correct answer to which Hans was responding. That is, more exactly, Pfungst concluded that

(&)   for every correct answer there is a cue which is necessary and sufficient that answer being given

Pfungst could draw this conclusion from his background knowledge about horses. There are alternatives. Thus, for example, Hans could have somehow learned the answers on his own or have been taught them. However, given what we know about the learning capacities of horses, based on past experience, this has indeed a very low prior probability, so low that it is not worth considering. Again, it could be suggested that Hans has psychic powers and can read the mind of the questioner. While this sort of hypothesis might appeal to some, it is certainly excluded by Pfungst's determination to seek a natural cause for Hans' ability. We do know about horses, however, that they are very sensitive to cues; that is why, for example, they can be trained to respond as show jumpers to very slight cues on the part of their riders. It is far more reasonable than not to suppose that Hans was responding to unconscious cues on the part of the questioner. Since this piece of background knowledge alone has a high prior probability, it is this that Pfungst accepted as a guide to his research into the discovery of the causes of Hans giving the correct answers.

Pfungst was, therefore, convinced that *there are* cues to which Hans was responding. But what, more specifically, were these cues?

Pfungst was able to narrow the range of possibilities still further by further experiments. It was shown that Hans responded correctly even in the absence of either auditory or tactual cues. Neither of these, therefore, could be necessary conditions; hypotheses that they were thus eliminated. Pfungst then used blinders to prevent Hans from seeing the questioner. Not only was Hans unable to give correct answers, but he balked at the situation and insisted upon trying to turn his head so that he could see the questioner. In the absence of visual contact, Hans was unable to give correct answers. Since auditory and tactile cues were present, and only the visual cues absent, the fact that correct answers were not given shows that such cues cannot be sufficient conditions. As the cues had to be sensory — this is what a naturalistic explanation required — , and since auditory and tactile cues had been eliminated, visual cues alone remained uneliminated as possible sufficient conditions. (It was clear that background knowledge made it reasonable not to consider the possibility of olfactory cues.) The cues that were necessary and sufficient were therefore visual.

Pfungst thus knew that the cues that were necessary and sufficient to elicit the correct answers were there, there to be discovered, and that they were of a certain *sort*, then, namely visual. But this is still only a generic description. Pfungst had yet to make a specific determination of what the cues were.

Unfortunately, although Pfungst tried hard to detect the cues, he was unable to do so. However, he did not give up. He realized that his failure to notice the cue did not falsify his inference that there were such cues. And the background theory which enabled him to affirm (&) had sufficient prior probability that he knew he had grounds for thinking the claim to be true that there were visual cues and that he had just not looked hard enough.

Finally, he did discover what he previously and all others had failed to discover.

When the questioner asked his question, his or her attention was directed at the horse's hoof in anticipation of the response. The horse responded by beginning to tap. The cue to which the horse responded was a certain tenseness in posture that caused an almost imperceptible lowering of the head when the questioner began to attend to the hoof. So here was the sufficient cue for Clever Hans to begin tapping. Then, when the horse had tapped his hoof the correct number of times, the questioner, confident that the horse would give the right answer, would almost imperceptibly relax, raise the head and straighten up. This was the cue to stop tapping. With this cue, it was guaranteed that the horse would answer correctly.

Pfungst had been unable to detect these cues when he attended to his own behaviour. It was only through a very careful examination of other questioners that he was able to discover them

Once they were known, however, Pfungst found that, while it was difficult, they could in fact be brought under voluntary control. He was then able to teach himself, and even others, the cues that controlled the horse's behaviour. These further experiments repeatedly confirmed that he had discovered the cues that were necessary and sufficient to elicit answers from Hans, and to give the answer that the questioner wanted to hear.

In this research Pfungst eliminated a variety of hypotheses until he had secured data that made it clear that what remained had to be the cause. In this research he was guided in the formation of hypotheses by his background knowledge. This knowledge gave him information about the *generic* form of the hypotheses that he had to consider.

Above all, Pfungst was guided by the scientific ideal that natural events require natural causes. He did not stop his research until he had found them.

There is the issue of how Clever Hans was trained, for trained he was. Pfungst reviews all the evidence that was available, and arrives at the conclusion that von Osten was as much caught up in the network of unnoticed behaviour and actions as the other questioners and even as Pfungst himself initially was. Pfungst's review shows that even though von Osten clearly believed that he was teaching the horse in essentially the same way that he taught schoolboys, in fact the actual procedures were such that they would generate rewards for the horse if the latter responded by starting or stopping on certain cues. It turned out that von Osten's methods of training were very much like the procedures that B. F. Skinner called "operant conditioning" and made famous by training pigeons to do a variety of amazing feats.

Von Osten had himself to learn how to train his horse. It took him three years of patient effort. Others who

came later, and who knew the real secret of his success, were able quite easily to teach other horses Clever Hans' complete repertoire of tricks in just a few weeks.

Many people have been able to impress others of their psychic powers by exhibiting a capacity to tell what others are thinking.

Oskar Pfungst acquired this capacity.

Pfungst himself took the role of Hans. His right hand played the role that the right foreleg played for Hans: it tapped out answers. A questioner stood before Pfungst. He or she would merely *think* of a question that could be answered by means of a number of taps. Pfungst would wait for the cue, the almost imperceptible slouch, that was the signal to begin tapping. He would then tap. But also observe the questioner. At the point where he thought he noticed a slight relaxation in the latter, Pfungst would cease tapping.

Pfungst went through over twenty-five subjects. All but two of them gave the same involuntary movements of the head that had been the cues for Clever Hans. Pfungst was able to use these cues to "read the minds" of his questioners. Indeed, if they were told that Pfungst was responding to such cues, they denied any knowledge of the involuntary movements.

Pfungst continued the experiments to include a series in which subjects, like Hans, were trained to give cues of various kinds even while they did not know that they were being trained to give such cues.

In short, Pfungst the naturalistic experimenter could train himself, and his subjects, so that he had the ability to "read their minds."

A capacity to read another's mind, to sense what they are thinking or feeling, is often claimed. There is no reason to deny that some people do have this ability. It is one that Oskar Pfungst has shown can be acquired, if only we use a little diligence. The point is, however, that there is nothing supernatural or "psychic" about all this. As Pfungst showed, both the abilities and their acquisition can be explained naturalistically. They give no evidence for peculiar powers beyond those of everyday experience.

Example (iii): Another Clever Horse

The parapsychologist J. B. Rhine investigated another horse for which supernatural powers were claimed. This horse was Lady Wonder, owned by Mrs.

Claudia Fonda of Richmond, Virginia. This horse had attracted national attention in 1927 by successfully predicting that Dempsey would beat Sharkey. In 1952, the horse again made national headlines when, after being consulted by the police of Quincy, Mass., she was given credit for the discovery of the body of Danny Mason, a four-year-old child who had gone missing two years previously. A police officer questioned Lady Wonder and she spelled out "Pittsfield Water Wheel." A search of water wheels in the Quincy area turned up nothing. It was then suggested by a policeman that "Field and Wilde Water Pit" might be what the horse had tried to spell out. The body of the boy was found in that pit. Lady Wonder died in 1957, at the age to thirty three, after having been written up in *Life* magazine.[22]

Rhine and his wife report their investigation in their essay "An Investigation of a 'Mind-Reading' Horse."[23] It was said that Lady Wonder "could make predictions, solve simple arithmetical problems, answer questions aptly and intelligently, and do all this without verbal command." Mrs. Fonda was convinced that even as a colt her horse was "'reading her mind,' that is, obeying her commands before she had expressed them" (p. 452). Rhine and his wife investigated Lady Wonder over six days in later 1927 and early 1928.

The initial experimental set up involved the horse facing a table upon which were placed either letter blocks or number plates with one digit on each. Mrs. Fonda stood by the horse's head, and the onlookers were situated across the table. A symbol was chosen and the horse would go and pick one of ten letters or numbers. Rhine discovered that so long as Mrs. Fonda knew which symbol was chosen, Lady Wonder performed admirably, but so long as the symbol chosen was kept secret, so long as Mrs. Fonda did not know it, Lady Wonder was unsuccessful in her choices. Rhine, in other words, quickly found that what Pfungst had found, that knowledge of the chosen symbol by someone present was necessary for successful performance. The correct answer was not due to superior intelligence but to communication from a questioner to the horse.

When Pfungst was at this point in his investigation, he concluded that he had to find the behavioural cue to which Clever Hans was responding. Rhine, too, searched for such cues. Rhine had Mrs. Fonda stand motionless, turn her head away, and even wear a blindfold. Lady

Wonder still performed at a better than chance level. However, when Rhine separated the horse and her owner, the horse could not perform. The presence of the trainer was a necessary condition for a good performance. It was necessary, then, to determine the cues that Mrs. Fonda was giving. Rhine introduced restrictions designed to "eliminate certain possibilities of signalling" (p. 459). In particular, a "screen [was] interposed between horse and trainer" (p. 458). In a final series of tests, Mrs. Fonda was not told the symbol chosen "yet was retained as an aid in controlling the colt and making her work for us" (p. 459). The horse still performed at a level better than chance.

The Rhine's considered three hypotheses to account for the successes of Lady Wonder. The first was the conscious presentation of cues by Mrs. Fonda. The second was the unconscious presentation of cues. The third was

Telepathy, or the transmission of mental influence by a process that does not involve the known senses, but which does involve some special susceptibility or sense in the subject (p. 462).

The tests concerning the possibilities of signalling were held to eliminate the second of these three possible hypotheses, while the experiments in which Mrs. Fonda was not told which symbol was chosen were held to eliminate the first. The Rhines conclude that

There is left then, only the telepathic explanation, the transference of mental influence by an unknown process. Nothing was discovered that failed to accord with it, and no other hypothesis proposed seems tenable in view of the results (p. 463).

But does this follow? Were the alternatives eliminated?

It would seem that not.

Consider the claim that the possibility of unconscious cues had been eliminated. In fact, when Mrs. Fonda was shielded by a screen, the "screen" was nothing more than "a board 18 inches square" (p. 458). Given what Pfungst had discovered, this was hardly sufficient.

Or consider the claim that the possibility of fraud, that is, the possibility that consciously given cues, had been eliminated. It turns out from a later paper that during the tests in which Mrs. Fonda was not told which symbol was chosen, Rhine "stood behind F [Mrs. Fonda] and wrote the number on a pad..."[24] Now, there are in fact many ways in which a skilled magician can secretly discover information that has been written down. The Rhines controlled for none of these. In fact, a professional magician, Milbourne Christopher, later attended a performance of Lady Wonder without informing Mrs. Fonda of his occupation. He discovered that she was a skilled pencil reader; that is, she could read what someone was writing by watching the motions of the top of the pencil. In Christopher's visit to Lady Wonder's performance, Mrs. Fonda gave him a long pencil and a pad, told him to stand on the other side of the room and write a number on the pad. Christopher traced out the figure eight but let the pencil touch the paper only at such points as to write down the figure three. Lady Wonder guessed the number eight. This was clear evidence that Mrs. Fonda was "pencil reading" and then giving cues to her horse.[25] Rhine failed to control for possibilities such as this. He had, therefore, not eliminated the possibility of conscious signalling on the part of Mrs. Fonda.

The Rhines later performed further tests on Lady Wonder. During these, the horse failed to perform in a way that convinced them that she had telepathic powers. They concluded that "a profound change in the horse's capacities [had] occurred,"[26] that she had lost her telepathic powers. But they remained convinced that the horse did have such powers when they first investigated her.

However, given the looseness of their experiments, it is clear that there is no significant support for the claim that Lady Wonder ever had telepathic powers. When compared to the investigations of Pfungst, the investigations of the Rhines with regard to their clever horse were woefully inadequate. Pfungst spent much more time and effort with Clever Hans, and used much more effective controls. Even then it was difficult to isolate the cues to which the horse was responding. The Rhines, in contrast, leapt to the conclusion of telepathic powers and extrasensory communication even though during the experiments the trainer was *always* present and partially visible.

The unwarranted leap to a conclusion from data that are weak but are claimed to be strong is a sure mark of pseudoscience. One can explain the fact that the Rhines came to their conclusion not on the basis of data but only on the basis of a clear conviction that extrasensory perception alone could explain the performances of Lady Wonder. Compare Herodotus, Hippocrates and Pfungst: these thinkers were guided by the notion that natural phenomena are to be accounted for in terms of natural causes. As Judge Overton insisted, theories to be scientific must be "... testable against the empirical world."[27] If the Rhinean hypothesis of extrasensory perception is testable, then it clearly failed the test. The Rhines did not abandon the hypothesis, however, even when, during the later tests it became dubious. Theirs was more the *method of fantasy* than the *inductive method* of science.

b. The Process of Inquiry: The Role of Hypotheses

The examples at which we have looked have made this much clear: Science does not consist in merely "studying the facts" or in "performing experiments." Such a characterization of science is superficial since we seldom go out simply to study the facts without there being something that we do not know and wish to come to know. Nor is it often that we simply perform experiments for their own sake, just to see how they go. Rather, we perform experiments because there is some *hypothesis* we want to put to the test. We want through the experiment to obtain evidence that will tell us whether this hypothesis is to be accepted as confirmed or rejected as false. And we are interested in the hypothesis we test because we think that it will yield the understanding that we want. We are interested in it, in other words, because the hypothesis is a general statement which, if it turns out to be acceptable, will thereby provide us with an explanation of the facts we hope to come to understand.

Scientific inquiry is thus a systematic affair. There is a problem. We have a cognitive interest in something, some state of affairs which we do not understand but wish to understand. The problem arises because, on the one hand, we have a certain cognitive interest in knowing something, but, on the other hand, do not know it. The problem arises because there is a gap between what we want to know and what we do know. Specifically, what we aim at is understanding, and so the gap consists in our wanting to know but not knowing some law, some matter-of-fact regularity. To fill that gap, we search for relevant data. But, to repeat, this search for data is no random search. It is, rather, a search for data that will test something that we think might satisfy our cognitive interest. We search for data that will test something that we *guess* will provide the knowledge that we want. That guess is the hypothesis that is put to the test. If the hypothesis passes the test, and those data render it acceptable, then we have the knowledge that we sought: the problem is solved and our cognitive interest is satisfied. Our inquiry, on this point anyway, comes to an end. If the hypothesis fails the test, then our inquiry must go on. We search for another hypothesis that we guess will, if acceptable, satisfy our cognitive interest, and then put that hypothesis in turn to the test. Such hypothesis testing continues until we have found one that does satisfy our cognitive interest.

Of this more directly. We can notice here, however, our **first criterion for a good hypothesis**:

a good hypothesis must be such that, if data render it acceptable, then it will provide the answer to the problem that generated the inquiry.

Not every hypothesis can be put directly to a test. Thus, for example, Galileo wanted to explain the motion of falling bodies. He formed the hypothesis that their acceleration is constant. But he had no way by which to measure acceleration directly. So the hypothesis that he formed could not be put to the test directly. Galileo had to develop the logical implications of his hypothesis. In particular, he deduced that the proposition that acceleration is constant

(*) $a = g$

is logically equivalent to the expression

(**) $s = 1/2\, g\, t^2$

which asserts that the distance fallen by a freely falling body is proportional to the time of the fall.

This proposition can be put directly to the test since we can measure both distance and time. A body which freely falls for two seconds, falls four times as far as a body which falls only one second; and a body falling three seconds falls nine times as far as the one which falls one second. When Galileo examined the motion of falling bodies he confirmed these relationships. He thereby confirmed (**). But, since (*) is logically equivalent to (**), he had succeeded in confirming the former also.

We should note how the patterns of confirmation go. First, hypothesis (**) is used to make predictions. These predictions turn out to be successful. These successful predictions confirm the hypothesis (**). Second, (**) is logically equivalent to (*). The data that confirm (**) therefore also confirm (*). This last has been dubbed the **equivalence condition** for confirmation:

if h is logically equivalent to h', then data that confirm h also confirm h'

We shall have more to say about this condition below.

A more general principle with much the same import is the **consequence condition** for confirmation:

if h' is logically implied by h, then data that confirm h also confirm h'

We shall have more to say about this principle below.

In the Galileo example, the hypothesis (*) was elaborated into a logically equivalent hypothesis. Often this is not possible. Often, to obtain an hypothesis that can be put directly to a test, one can find only a deductive consequence. Thus, for example, consider Newton's law of gravity, which states that every object attracts every other object with a force which varies directly as the product of the masses and inversely as the square of the distance between them. Call this hypothesis "H". Newton did not put this hypothesis to the test directly, but deduced from it two other laws which could be put to the test. One was Kepler's Laws, that the planets move in ellipses with the sun at one focus. Call this "H^". The other was Galileo's Law for falling bodies. Or, more accurately, Newton was able to deduce that, while acceleration of falling objects varied with their distance from the centre of the earth, for objects close to the surface of the earth the variation is so small as to be for practical purposes indiscernible. Call this "H^^". The point is, of course, that Galileo's Law, while literally false, is a close approximation to the truth that Newton subsequently discovered. Thus, data that predictively confirm Galileo's Law also confirm the Newton correction $H^{\wedge\wedge}$ of that law.

Data, of course, predictively confirmed $H^{\wedge}$ and other data predictively confirmed $H^{\wedge\wedge}$. Since H logically implied $H^{\wedge}$ and also logically implied $H^{\wedge\wedge}$, it was concluded that these data also supported H.

Again we should note how that confirmation patterns go. First, data predictively confirm $H^{\wedge}$ and $H^{\wedge\wedge}$. Since $H^{\wedge}$ and $H^{\wedge\wedge}$ are logically implied by H, these data also confirm H. This has been dubbed the **converse consequence condition** of confirmation:

> if h' is logically implied by h, then data that confirm h' also confirm h

We shall have more to say about this principle below. In particular we shall discover that it cannot be taken to hold in all generality.

Another point about confirmation is also worth noting. The data that confirm Galileo's Law $H^{\wedge\wedge}$ are very different than the data that confirm Kepler's Law $H^{\wedge}$. Here the principle seems to be that **data of diverse sorts have greater confirming power than data which are repetitions of the same sort**.

We shall have more to say about this principle below.

We have, then, our **second criterion for a good hypothesis**:

> **a good hypothesis must be formulated in such a way that its deductive development reveals implications that can be directly tested in experience.**

In order to test an hypothesis against experience, it must be *empirical*. There are many hypotheses that one could think of but which have no empirical consequences that could be tested. For example, the hypothesis that a God created the universe out of nothing cannot be tested against experience. Or again, the claim that various traditional medical procedures — or, for that matter, various non-traditional procedures — restore the balance of energy or of forces in the body can have no testable consequences unless we can independently identify and measure the energies or forces mentioned in the hypothesis.

In order to make the hypothesis empirical, its non-logical concepts must either refer directly to observable individuals or the properties or and relations among such individuals, or else the concepts must be tied to such concepts. Without such hook-ups to observable entities, empirical tests would not be possible.

In a 1998 column in Toronto's national newspaper, *The Globe and Mail*, Pat Moffat, an advocate of "alternative medicine," describes the work of Tetauro (Ted) Saito of the Shiatsu Centre in Toronto.[28] He organizes his treatments of those who come to him on the basis of the traditional Chinese account of the body. On this view the vital energy of the body known as *chi* (also referred to as "Qi") flows along various "meridians" in the body. Moffat tells us that Mr. Saito "envisions the meridians as organizing the body internally and also swirling in layers externally to connect us to the cosmos, to universal energy." It is of course a good thing to be hooked up to the cosmos. No doubt the ancient Greeks thought something of the same sort when they proposed that all springs and rivers, including the Nile, were somehow hooked up to the encircling river Ocean. But as Herodotus saw, notions like those of the cosmos or universal energy or the river Ocean are mythological and metaphysical and have no place in science. There is no way a theory involving such a concept can be put to the test.

Still, not one to readily notice difficulties, Ms. Moffat ignores the weirdness of Mr. Saito's views and bravely soldiers on in her presentation:

> "This is so difficult to explain," he [Mr. Saito] said, "because scientifically we can't prove meridians at all. And the *chi*, it is partially electromagnetic energy, partially ultraviolet light. I can sense it, I can draw the meridians exactly,. But other people can't detect it, so this is a problem."

It is indeed. If it cannot be detected, if the concept has no empirical content that enables the hypotheses in which it appears to be put to the test and either confirmed or falsified, then those hypotheses are simply not scientific.

In fact, Mr. Saito is simply inconsistent here. On the one hand he asserts that it is partially ultraviolet light and partially electromagnetic energy. Note that Mr. Saito does not seem to know that ultraviolet radiation is itself a form of electromagnetic radiation. Note also that others have suggested that this special sort of energy is a form of infrared radiation.[29] Infrared and ultraviolet are of course at opposites ends of the visible light spectrum. In any case, these are forms of energy or radiation that can easily be detected, and their changes and flows measured with considerable accuracy. To the extent that the force that Mr. Saito claims to be present involves those things then it can be detected and he should be able to explain to others how this can be done. On the other hand, he also insists that the force is beyond our means to detect, and holds therefore that his theories cannot be put empirically to the test. The claim is, then, that this new concept of force is somehow empirically grounded just like concepts of other well known forces, and so is a legitimate notion, while also being non-empirical, so that he does not have to specify how the theories about it can be tested. To have it both ways would be nice indeed, but logically impossible.

Moffat proposes a question for future research to tackle.

> Does an organized energetic system exist in the human body that has clinical applications?

But really, how does she propose that this research be undertaken? Just how are we to determine whether or not such an energetic system exists.? She in fact does not have any idea of how to go about it. Nor *could* she have any such idea: a concept that does not hook up to reality as we experience it cannot be counted among the concepts that it is legitimate to admit into the language of science. If the relevant concept of energy lacks empirical content, then we simply have no basis on which to proceed in our investigation: we don't know what we are looking for. We are no better off than we would be if we proposed that we need research into whether bumbies are the causes of various diseases. Without knowing what bumbies are and without knowing what *chi* is we have no hypothesis which could ever form the basis for any scientific research programme.

Moffat indicates that she agrees with Mr. Saito's judgment that "this [the fact that the *chi* cannot be detected] is a problem." It is, she says, "certainly a problem for conservative elements within the medical community." Conservative, that is, old fashioned and perhaps even dogmatic, certainly people who have their own vested interests to protect. Good rhetoric, but she manages to miss the point completely. But that is the way the defenders of alternative medicine so often proceed: they prefer rhetoric to rational discourse, and the scoring of the debating points to the search after truth.

It is precisely the latter that is important. There is no way in which we can decide if an hypothesis is true or false unless we can figure out how to put it to the test. Mr. Saito may somehow and in his own way believe that his theories are true. But unless those theories are given some sort empirical content they simply cannot be put to the test. They are, therefore, worthless so far as concerns anyone who has a cognitive interest in matter-of-fact truth. This includes those who have the pragmatic interests of a physician, in coming to know the empirical causes of physical dis-ease. Just as Hippocrates rejected the notion that epilepsy was the "sacred disease" and somehow special and beyond empirical investigation, so any other physician concerned to eliminate the dis-ease of his patients will reject theories that are so stated that they cannot be subjected to tests in the ordinary way.

Moffat simply ignores what it is to be a good scientific hypothesis, and in particular what we may label as our **third criterion for a good hypothesis**:

> **a good hypothesis must be such that its non-logical concepts refer to observable entities or are tied to such concepts in a way that permits the hypothesis or its deductive consequences to be tested against observable reality**.

While this criterion is necessary, it is not sufficient. Consider the hypothesis that

> Opium, if ingested, puts one to sleep because it has a soporific power.

or, more perspicuously,

(@)   For any person y, since opium has a soporific power then if y ingests opium then y goes to sleep

The various non-logical concepts — opium, sleep, ingestion — are all empirical. Yet the hypothesis, as the early modern scientists such as Galileo argued, is not worthy of consideration. For, it is true *ex vi terminorum*, and therefore tautological. In fact, "soporific power" is defined in terms of putting people to sleep:

x has a soporific power = $_{Df}$ if x is ingested by y then y goes to sleep

In particular, this applies to opium:

opium has a soporific power = $_{Df}$ if opium is ingested by y then y goes to sleep

Thus, the hypothesis (@) states no more than

(@@)   For any person y, since if opium is ingested by y then y goes to sleep then if y ingests opium then y goes to sleep

This has the form

For any person y, if y is F then y is F

which clearly explains nothing. Just as one cannot explain why someone is an unmarried male by citing the fact that he is a bachelor.

The hypothesis (@) = (@@) thus satisfied the third criterion for a good hypothesis while yet not being one. The problem with (@), as its expansion into (@@) reveals, is that it is *empirically vacuous*: while true, its truth is compatible with the world being way whatsoever. To be a good hypothesis, in short, a proposition must be non-vacuous in the sense that there must be ways in which the world could be which are incompatible with the truth of the proposition. The hypothesis must not only be empirical but it must also be possible that there are states of affairs that would show it to be false, or, in short, it must be *falsifiable*.

This does not mean that the hypothesis must be falsifiable by a single counter-example. Thus, because

(&)   There are unicorns

makes an *existential claim* it cannot be falsified by a single observation. If, having searched through the zoo, we declare that there are no unicorns, that conclusion simply does not follow: all that follows is that we have not looked everywhere, and it may be that there are as yet unobserved unicorns elsewhere. However, there are still data that could lead us to conclude that (&) is false. Thus, we know that a unicorn is a horse with a single horn growing out of its forehead. A unicorn has the same physiology as a horse, and in particular the same structure of bones. The bone structure of the horse's forehead is weak compared to the bone structures of animals with horns. If a horn is to be supported and to serve the functions that horns serve, e.g., in struggles with other animals, then a rather massive bone structure is necessary to prevent the skull from breaking. The existence of a horn is thus incompatible with the bone structure of the horse's skull. What we are appealing to are data which support the lawful claim that

(&&)   Horns do not exist in animals unless supported by a massive bone structure

Since we know about a unicorn that it has the physiology of a horse, we know that

Unicorns do not have a massive bone structure

It follows that unicorns lack horns. Or rather that, since by definition unicorns have horns, it follows that unicorns do not exist. Thus, the claim (&) is incompatible with the lawful generalization (&&). So the data which support (&&) support the claim that (&) is false, and lead us to *reject* the latter.

Thus, there are data which tend to support the claim that (&) is false and in that sense the proposition is *falsifiable*, even though simply searching about for a counter-example would never show it to be false.

This yields our **fourth criterion for a good hypothesis**:

a good hypothesis must be falsifiable in the sense that it must be possible that there be observational data which, if

observed, would provide evidence that the hypothesis is false.

It should be pointed out that it is **not** laid down as a requirement for a good hypothesis that it be *true*. In the first place, if we have the evidence that it is true, then there is no point in undertaking the inquiry: the proposition would no longer be simply an hypothesis but to the contrary a proposition for which there was evidence for its truth and which would justify its acceptance. But in the second place, the experience that we have had in conducting inquiry is that an hypothesis that turns out to be false may nonetheless have yielded, in the process of being tested, information that turns out to be important to the inquiry. Thus, for example, the phlogiston theory of combustion in chemistry in the end turned out to be false. For a while it successfully guided research, but in the end it was rejected when evidence was shown to provide conclusive support for the explanation of combustion in terms of the oxygen theory, a theory that is logically incompatible with the phlogiston theory. As the great English logician Augustus De Morgan stated, "Wrong hypotheses, rightly worked, have produced more useful results than unguided observation."[30]

Of course, we would not deliberately choose as an hypothesis one that we already knew to be false. Since we wish to explain some phenomenon, what we need is an *hypothesis* for which, at the end of the inquiry, there are data that *support its acceptance as true*. The point is that an hypothesis may turn out at the end of the inquiry to be false and yet have proved useful in the overall inquiry by providing information useful in subsequent developments in the process of inquiry.

We want, therefore, to exclude hypotheses which we have reason to believe are false. Such reasons might be data that demonstrate such falsehood, or at least tend to disconfirm the hypothesis. Or they might include a body of acceptable laws and theories with which the proposed hypothesis is inconsistent.

This yields the **fifth criterion for a good hypothesis:**

a good hypothesis must be consistent with background knowledge in the sense that there are no observational data which require the rejection of the hypothesis nor any acceptable body of laws or theory which imply its falsehood.

There remains one final criterion for a good hypothesis that we should note.

Suppose there is a bench in a park in North Bay and suppose that, just as a matter of fact, by the time it was worn out and replaced, only Letts ever sat on it. Suppose we wonder why only Letts sat on the bench. This generates an inquiry. We formulate the hypothesis

Everyone who sits on this bench is a Lett

But this, surely, is hardly a good hypothesis. It is a generalization, it is empirical, it is falsifiable, it can be verified in experience. Why, then, is it not a good hypothesis? The problem is that its scope is effectively limited to a set of facts that are *already known* and that we therefore *cannot put it to a test*. Although it is in principle falsifiable, the fact that its scope is limited to what is already known means that in fact we will never be able to find data that would falsify it. For an hypothesis to be acceptable, it must be possible to put it to a test. That is, it must not only account for what is already known, but it must also *predict* certain observable states of affairs the observation of which would reveal the truth of propositions which we not known or even suspected at the time the hypothesis was formulated and the prediction made. Galileo's hypothesis explained not only the falling bodies that he had observed, but all falling bodies, including the unobserved ones. It thus yielded predictions about things unobserved, so that the prediction of the latter did — and still does — put the hypothesis to a test. It tested the hypothesis in the sense that, for all one knew about the unobserved states of affairs that were predicted, those states of affairs could have turned out otherwise, resulting in the falsification of the hypothesis. If an hypothesis yields a prediction, then the hypothesis *risks falsification*. An hypothesis is tested only if it risks *elimination as false*.

This yields the **sixth criterion for a good hypothesis:**

a good hypothesis must yield *predictions*; these predictions must be such that, if they turn out to be false, then that fact testifies to the falsity of the hypothesis.

Of course, an hypothesis, being a generalization that goes beyond the known facts, can never be known for certain to be true. To be sure, it can be confirmed if its predictions are successful, but since not all its instances are observed, it will never be beyond all possible doubt. In that sense, as we have emphasized, the acceptance of an hypothesis at the end of inquiry must always be tentative.

These criteria are clearly related to the criteria for distinguishing science from pseudoscience that were developed by Judge Overton, in his judgment concerning the teaching of "creation science" in Arkansas schools.

The characteristics of science are, as we have seen, the following, according to Judge Overton's opinion:

(1)   It is guided by natural law.
(2)   It has to be explanatory by reference to natural law.
(3)   It is testable against the empirical world.
(4)   Its conclusions are tentative, i.e., are not necessarily the final word.
(5)   It is falsifiable.

Criterion (1) corresponds to our requirement that a good hypothesis be consistent with background knowledge of laws and theory: that background knowledge guides us in our choice of hypotheses. Criterion (2) is the requirement the hypothesis must be a generalization which, if data render it acceptable, will solve the problem that initiated inquiry. Criterion (3) is the pair of requirements that the hypothesis be empirical and that it yield predictions of hitherto unknown facts. Criterion (5) is, of course, the requirement that the hypothesis is falsifiable. Criterion (4) is simply the requirement that we accept hypotheses on the basis of empirical evidence and not on the basis of, say, authority. Since the empirical data available through observation that might support an hypothesis is always incomplete, the acceptance of an hypothesis within the context of scientific inquiry must always be tentative.

c. The Process of Inquiry: The Role of Theory

It has sometimes been said that there is no formula for coming up with an hypothesis that will satisfy the above criteria, let alone an hypothesis which will, after experimental data have been obtained, turn out to be true. The point was put nicely by Augustus De Morgan in the following way:

> A hypothesis must have been started not by rule, but by that sagacity of which no description can be given, precisely because the very owners of it do not act under laws perceptible to themselves. The inventor of hypotheses, if pressed to explain his method, must answer as did Zerah Colburn [a Vermont calculating boy of the early 19th century] when asked for his mode of instantaneous calculation. When the poor boy had been bothered for some time in this manner, he cried out in a huff, "God put it into my head, and I can't put it into yours."[31]

Morris Cohen and Ernest Nagel, however, suggested that there is in fact a source that provides us with a guide to good hypotheses, namely, analogies.[32] This line of thought has been more recently pursued in detail by N. R. Hanson and Thomas Kuhn.

Hanson, in his essay, "Is There a Logic of Scientific Discovery?",[33] distinguishes

(1)   reasons for accepting an hypothesis H

from

(2)   reasons for suggesting H in the first place

But he then re-phrases this in a more guarded way, as a distinction between

(1')   reasons for accepting a particular, minutely specified hypothesis H

from

(2')   reasons for suggesting that, whatever specific claim the successful H will make, it will, nonetheless, be a hypothesis of one *kind* rather than another

With this notion of *kind* we notice the introduction of the idea of Cohen and Nagel, that what is important is analogy.

Hanson uses the example of the *form* of Jupiter's orbit, as anticipated by Kepler. What Kepler did first was to discover a law for Mars' orbit, a law which goes something like this:

(*)   Any place p such that Mars is located at p is also such that it is situated on a path the form of which is an ellipse with eccentricity e_1, and with the sun at one focus

This is a *generalization* about the places at which Mars is located, a *law* about the positions of that particular planet. Kepler immediately generalized. If the orbit of Mars were an ellipse, then, since Mars is a typical planet, other planets, too, must have elliptical orbits:

(**)   For any planet there is a form f which is an ellipse with the sun at one focus such that, for any place p such that planet is located at p is also such that it is situated on a path the form of which is f.

This latter law states that *any* planetary orbit will have a certain *form*, specifically it will have the form of an ellipse. It asserts that there is a common *logical form* that is shared by the laws for planetary orbits for all planets. This shared logical form makes them analogous to each other, analogous in respect of this form. This law enables Kepler to predict that the law for Jupiter's orbit will be of this *kind*. That is, from (**) and the assumption that Jupiter is a planet we can deduce that

(***)   There is a form f which is an ellipse with the sun at one focus such that, for any place p such that Jupiter is located at p is also such that it is situated on a path the form of which is f.

The law (***) asserts that *there is* an elliptical orbit for the planet Jupiter; it states that there is a law of a certain *kind* or *generic form*, though it does not say *specifically* what ellipse it is, that is, what specifically the law is that it asserts is there. (***) asserts that there is for Jupiter a law like that of (*) for Mars, a law of the same generic form. It is clear that the law (*) is the minutely specified law that Hanson mentions in his (1'), while (***) provides

a reason for supposing, as Hanson says in his (2'), for supposing that there is a law of a certain *kind*. But, of course, we have reason to suppose that (***) is true only because we deduced it from the generic theory (**). Thus, laws of the sort (**), that is, laws about laws, or, more exactly, laws about the forms that laws exemplify, provide the sorts of reasons that Hanson mentions in his (2'). A law such as (*) is his minutely specified or specific hypothesis H, and the reasons for accepting (*) have to do with its confirmation by observational data. Why suggest a *similar* law for Jupiter, one that is *analogous* or *of the same kind*? The suggestion comes from the law (**), which is a generalization from (*). The reasons that justify the inference that Jupiter's orbit will have an elliptical *form* thus lie in the theoretical and generic generalization (**), and, beyond that, the observational evidence that supports the generalization (**).

The point to be made here, in agreement with Hanson, and against De Morgan's remark, is that the discovery of hypotheses at the beginning of inquiry is not simply a mysterious psychological process. While there may not be reasons, right at the beginning, for proposing some very specific hypothesis, there often are *good* theoretical reasons, rooted in a background *abstract generic theory*, for proposing an hypothesis that is of a certain *kind*.

We should again note how the confirmation patterns go. The data that support the acceptance of the law (*) in turn support the generalization (**). Why? Well, the law (*) entails

(*')   There is a form f which is an ellipse with the sun at one focus such that, for any place p such that Mars is located at p is also such that it is situated on a path the form of which is f.

(*') is entailed by (*), and so the data that support the latter also, by the consequence condition, support (*'). But (*') is entailed by (**). Hence, by the converse consequence condition, what supports (*') also supports (**). Moreover, (**) entails (***), so by the consequence condition, what supports the former also supports the latter.

Furthermore, when a specific ellipse is found to describe the orbit of Jupiter, the observational data that support this specific law will, for reasons now clear, also support (***). But this means that these data will also support (**) which entails (***). And if they support (**), then they also support (*). So the latter receives support not only from the data that support the specific law (*) about Mars' orbit but also from the data that support the specific law about Jupiter's orbit.

Once again we note how the deductive interconnections enable data to provide a network of supports for the various hypotheses that are united by the generic law (**).

Kuhn also develops the points of Cohen and Nagel about the role of analogy in selecting from among all possible hypotheses those that it is reasonable to put to the test.

The picture that Kuhn gives of normal scientific research is something about which we shall more to say later, but for now we can simply note that on this account research goes roughly like this.[34] Most research is guided by a theory or paradigm. This theory applies to a variety of specific areas (pp. 17-18). It is therefore a generic theory. Past successes in guiding research testify to its utility (p. 23). The successes consist in the discovery of laws that apply in specific areas to which the generic theory or paradigm applies. The data that confirm these specific laws, testify to the utility of the paradigm, that is, make it reasonable to accept — tentatively, of course — the paradigm as true. When the generic theory or paradigm is applied to a new, as yet unexamined, specific area, it asserts that *there is* a law of a certain sort there to be discovered, while not, on the other hand, asserting *specifically* what that law is (p. 25ff). The research task, or puzzle, is to find the law that the theory asserts to be there. Failure to find the law does not falsify the theory or paradigm; it testifies more to the lack of skill on the part of the researcher (p. 35ff, p. 147). The Kuhnian puzzle is thus very much of a piece with what we have been calling a problem. The problem or puzzle involves a gap between what we do know and what we want to know, and we undertake inquiry, research, to find a solution to our puzzle. In this search, we are guided by a paradigm or theory. This theory asserts that there is a solution to our puzzle, and that this solution is, as Hanson also argued, of a certain sort or kind. Kuhn, then, is defending with Cohen and Nagel the role of analogy between laws as a guide in locating hypotheses that are

potential solutions to our problem and for which it is reasonable that we put them to the test of experiment.

We can provide a simple model of the sort of theory that guides inquiry.

Suppose that we have a series of diseases D_1, D_2, D_3, of genus **D**, and that these are caused, respectively, by the presence of germs of species G_1, G_2, G_3, all within the genus **G**. We thus have the laws

(a)   For all x, x is G_1 if and only if x is D_1
        For all x, x is G_2 if and only if x is D_2
        For all x, x is G_3 if and only if x is D_3

Together with the data that eliminate other possible causes, these entail respectively

(b)   There is a unique species g such that g is of genus **G** and such that for all x, x is g if and only if x is D_1
        There is a unique species g such that g is of genus **G** and such that for all x, x is g if and only if x is D_2
        There is a unique species g such that g is of genus **G** and such that for all x, x is g if and only if x is D_3

Given that the D_i are all **D**, we can now generalize again to a *law about laws* — as John Stuart Mill put it, it states that "it is a law that there is a law for everything"[35] — to the effect that

(c)   For any species f, if f is of genus **D** then there is a unique species g such that g is of genus **G** and such that for all x, x is g if and only if x is f

We now run across a new species of disease of the same sort, e.g., D_4. Since

D_4 is **D**

we can deduce from (c) the *existential hypothesis*

(d)   There is a unique species g such that g is of genus **G** and such that for all x, x is g if and only if x is D_4

This asserts that *there is* a species of germs that causes disease D_4.

We have observational data that justify the law-assertion of each of (a). These law-assertions justify the law-assertion of the generalities (b). Since we have in the past succeeded in finding data that justify these laws, or, rather, law-assertions, we may reasonably infer that in the future we will find data that confirm *analogous laws* in *analogous cases*; that is, our successes in confirming the laws (b) justify our law-asserting a generalization about such generalizations, namely, (c). This law-assertion in turn justifies our law-asserting the existential hypothesis (d). The inference proceeds from what we have found in *parallel* cases to what we may reasonably expect in the present case. The result is a justified assertion of an existential hypothesis which asserts that a certain sort of cause *is there to be discovered*, or, what is the same, that a certain sort or form of law obtains, without, however, asserting specifically what that cause or law it. The hypothesis thus presents a *research problem*; we know that there is a law there to be discovered, and the task is to discover it.

In this example, the principle (c) functions as a *principle of determinism* — it asserts that for any causal factor of the relevant sort then *there is* another factor which explains its presence — and as a *principle of limited variety* — it asserts that the possible conditioning factors are limited to the range determined by the genus **G**. It is, as we said, a *law about laws*. It is achieves this status because it *abstracts* from the specific laws a certain *generic form*. The fact that theories can take on an abstractive, generic form is of crucial importance to any adequate empiricist account of science.

Once again we should note how evidence travels up and down the hierarchy of laws in apparent conformity with the consequence and converse consequence conditions for confirmation.

Once we have the law (d) to guide us, we know what *kind* of hypothesis to propose; it must be of the form

(f) g is **G** and for all x, x is g if and only if x is D_4

a form which it shares with other laws unified by the same generic theory (c). Given that the theory (c) is empirical, then so is any hypothesis of the form (f). Any such hypothesis will, moreover, be falsifiable. As well, it

will be consistent with the background theory. And it will yield predictions that could put it to the test. So if we propose as a solution to our problem some hypothesis of the form (f), we will have proposed a solution to our problem that satisfies the criteria we laid down for being a good hypothesis. And even if the hypothesis that we proposed fails to pass the test and thus turns out to be false, we still *know* that at some other hypothesis of the same form will turn out to be correct: after all, that is what is asserted by the law (d) that we are reasonably using to guide our research.

Let us suppose that we have a relatively narrow range of alternatives, say three. Let us say that we have

(g) g is of genus **G** if and only if either $g = G'$ or $g = G''$ or $g = G'''$

If this is so, then what (d) asserts is that exactly one of the following possibilities is true:

(h)
for all x, x is G' if and only if x is D_4
for all x, x is G'' if and only if x is D_4
for all x, x is G''' if and only if x is D_4

Until these are put to the test, we do not know which of them is true. But, given (g) and (d), then we know that one of them is true. The task of the inquiry, the research task, is to find out which one, to find the observational data that will enable us to decide.

These points can be put in another way. The law (d)

There is a unique species g such that g is of genus **G** and such that for all x, x is g if and only if x is D_4

states that *there is exactly one* **G** such that it is necessary and sufficient for D_4. Since it asserts both existence and uniqueness, we can form the *definite description*

(r)   *the* species of genus **G** that is necessary and sufficient for D_4

If we think of disease D_4 as "the flu" of a certain specific sort, then the definite description (r) corresponds to the phrase "the flu bug". And just as we can say that the flu bug causes the flu, so we can say that the property denoted by (r) is necessary and sufficient for D_4. If we abbreviate (r) to

(r') *the* species of genus **G** that is necessary and sufficient for $D_4 =_{Df} G^\wedge$

then we can infer that (d) is logically equivalent to

(j) for all x, x is $G^\wedge$ if and only if D_4

Just as (d) is an empirical generalization, so is (j).[36]

We are now in a position where we can explain the presence of D_4 in some individual, say a, by appeal to the presence of $G^\wedge$.

Of course, we cannot know beforehand that $G^\wedge$ is present in an individual, since we have no way of identifying the presence of $G^\wedge$ save through its effect. But suppose we know that

a is D_4

i.e., suppose that we know that the effect has occurred. Then we can infer the presence of the cause. Which is to say that from (j) we can deductively infer that

a is $G^\wedge$

which in turn yields the causal explanation

for all x, x is $G^\wedge$ if and only if x is D_4
(E) a is $G^\wedge$
_____

so, a is D_4

This asserts something to the effect that Jones has the flu because he has been infected by the flu bug.

Notice the pattern with which we are by now familiar: owing to the existential quantifier in both the major and minor premises in (E), this argument form could not have been used for prediction. But we *can* use it to explain. The point is that the explanation is *ex post facto*.

This is a mark of the gappiness of the explanation. Which is another way of saying that (d) = (j) is a gappy law.

But this should be evident. It asserts that *there is* a generalization which is a solution to our problem, and describes that generalization in *generic* terms, but it does

not say *specifically* what the law is. The gappiness of our knowledge will be remedied when we have discovered precisely which among the possible alternatives which it allows really is the true statement of law.

The cognitive interest that motivates the inquiry is to discover the specific law that describes a certain area. We are guided by an abstract generic theory that gives a generic description of the specific law without stating that law in its specific detail. Since we know that there is a specific law that holds but do not know specifically what it is, our knowledge is gappy. Thus, in the context of theory-guided research aimed at discovering specific laws, the cognitive interest that moves us is the elimination of gaps in our knowledge of laws.

Inquiry, then, aims to discover which of the three possible hypotheses (h)

for all x, x is G' if and only if x is D_4
for all x, x is G" if and only if x is D_4
for all x, x is G"' if and only if x is D_4

is true. Another way of putting the same point is this. What inquiry aims to do is identify the **G**-ish cause of D_4. This cause is denoted by the definite description "$G^\wedge$". This cause we know to be one of G', G", or G"'. Thus, in effect what we want to do is discover which among the following *identificatory hypotheses* is true:

$G^\wedge$ = G'
(h*) $G^\wedge$ = G"
$G^\wedge$ = G"'

Indeed, if we expand the definite description according to the Russellian rule, then the hypotheses (h*) turns out to be logically equivalent to the corresponding hypotheses in (h) together with the information that the relevant G^i is a **G**.

If we are thinking of D_4 as "the flu", then what (h*) makes clear is that the task of the researcher, the aim of the inquiry, is to *identify* the flu bug.

One should note that the paradigm or generic theory (c)

For any species f, if f is of genus **D** then there is a unique species g such that g is of genus **G** and such that for all x, x is *g* if and only if x is *f*

makes only *determinable predictions*.

To see what this means, we must first introduce some further terminology.

Consider the predicate

x is coloured

To assert of something that it is coloured is to assert that

there is an f such that f is a colour and x is f

If we let 'f is **C**' mean 'f is a colour', then we may define 'x is coloured' thus:

x is **C**-ed = ₚₑ there is an f such that f is **C** and x is f

The point is that, where 'red' is a *determinate* predicate, referring to a specific colour, 'coloured', in contrast, is a *determinable* predicate. When it is predicated of an individual, the specific colour that individual has remains to be determined; all that we know is the existential claim that *there is* some determinate property or other of the genus colour which is present in the individual.

We may similarly define the determinable predicates

x ix **D**-ed = ₚₑ there is an *f* such that *f* is **D** and x is *f*
x is **G**-ed = ₚₑ there is an *f* such that *f* is **G** and x is *f*

It is easy to show that the following deductive argument is valid:

> For any species *f*, if *f* is of genus **D** then there is a unique species *g* such that *g* is of genus **G** and such that for all x, x is *g* if and only if x is *f*
>
> *a* is **D**-ed

─────────────────────────────

so, *a* is **G**-ed [37]

Thus, while (c) cannot be used to make *determinate predictions*, it does yield *determinable predictions*.

However, since the conclusion of this argument is

(p)   *a* is**G**-ed = ₚₑ there is an *f* such that *f* is **G** and *a* is *f*

it is clear that the conclusion contains an existential quantifier. *It is therefore not falsified by the failure to find an instance of the genus* **G** *present in the individual a.* This, of course, is simply another way of making once again the logical point that, since (c) contains an existential quantifier, it is not falsifiable by a single counter-example.

Still, while (c) is not falsifiable by a single counter-example, it might well turn out that the prediction (p) is false. There are states of affairs which are incompatible with the determinable claim that

so and so is coloured

and similarly there are states of affairs which are incompatible with the determinable claim (p). In this sense, (c) remains falsifiable, inconsistent with certain possible states of affairs.

Laws of this sort are common enough in science. To recall but one example, consider the Principle of the Origin of Species by Natural Selection, which states, roughly, that

> For any species of organism S there is an earlier species S', there are variations in S' and there are forces of natural selection such that these variations and forces are necessary and sufficient for the divergence of S from S'

On the basis of the premises — the determinable premises — that there are variations and there are forces of natural selection, then this Principle yields the determinable prediction that species will evolve and diverge. For this reason, the Darwin's Principle of the Origin of Species by Natural Selection is, as Judge Overton indicated, falsifiable.

For Popper, of course, a law about laws such as (c) or the Principle of the Origin of Species by Natural Selection is not scientific because it contains an existential quantifier. It is, rather, in his terminology, metaphysical. He is therefore inclined to reject Darwin's theory as non-scientific. With this he provides arguments that the creation scientists can use to attack the theory of evolution by natural selection. They argue that, just as their theory — creation science — has many aspects which are non-scientific, so does the standard theory of Darwin contain many such components. They can cite Popper's authority on this. Unfortunately, Popper's authority is not so good. Creation science is not science because so many of its statements — those which are not empirically false — are non-empirical, about a God who creates the universe from nothing. Darwin's science, in contrast, is empirical and even falsifiable. Popper misleadingly classifies it as metaphysical, simply because he spots the existential quantifier. But an existential quantifier does not make a statement non-empirical. The proposition that

> There are dogs

is clearly empirical, in spite of Popper, and to call it otherwise is clearly arbitrary. Similarly, the Principle of the Origin of Species by Natural Selection is clearly empirical, and to call it otherwise simply because it contains some existential quantifiers is clearly arbitrary.

Popper is, however, prepared to allow that Darwin's theory is not so metaphysical as not to have some predictive power. He tells us about Darwin's theory that

> It allows us to study adaptation to a new environment (such as a penicillin-infested environment) in a rational way: it suggests the existence of a mechanisms of adaptation, and it allows us even to study in detail the mechanism at work.[38]

Popper is here pointing to the fact that Darwin's theory is one that asserts the existence of certain things (mechanisms of adaptation) — "suggests" is the way he puts it — without stating specifically what they are. It is the same feature of paradigms to which Kuhn has directed our attention; and it is that feature which makes theories of this sort able to guide research.

Popper contrasts theories of this sort — abstract generic theories — with specific laws. An abstract generic theory such as Darwin's, he correctly points out, asserts that *there are specific laws of certain generic sorts*, so that it allows that "...changes in the genetic base of the living forms ... are — at least 'in principle' — explained by the theory...", that is, they are explained "in principle"

because the theory as such, that is, the generic theory, provides only a gappy explanation, but in so doing asserts that *there is* a less gappy, specific explanation there to be discovered. Popper goes on to make this contrast more explicit:

> ... "explanation in principle" is something very different from the type of explanation which we demand in physics. While we can explain a particular eclipse by predicting it, we cannot predict or explain any particular evolutionary change...; all we can say is that if it is not a small change, there must have been some intermediate steps — an important suggestion for research: a research programme.[39]

In physics we have specific, non-gappy, laws. These yield specific predictions, and are such that explanation and prediction are symmetric. In contrast, Darwin's theory contains gappy laws which do not yield specific predictions, and where explanation is at best *ex post facto*. These are indeed two different things. But Popper is wrong to suggest that laws of the logical sort that appear in Darwin's theory do not appear in physics. We have in fact seen that such axioms in physics as the Law of Inertia have just the same sort of abstract generic form that one finds in the laws of Darwin's theory. They have the same gappy nature that yields only imperfect, *ex post facto*, explanations at the specific level. But precisely that abstract generic nature which makes explanations in terms of them at the specific level less than ideal is that logical feature which enables them to guide research. It is that logical feature that allows them to provide "suggestions for research" and to provide scientists with "a research programme," as Popper says.

However, since the theory contains existence claims, it cannot be conclusively falsified by a single counterexample. Only strictly specific laws can be conclusively falsified by a single counterexample. Because such falsification is not possible, Popper refuses to call Darwin's theory "scientific"; it is, rather, in his idiosyncratic terminology, "metaphysical."

> ...Darwinism is not a scientific theory, but metaphysical. But its value for science as a metaphysical research programme is very great,

especially if it is admitted that it may be criticized and improved upon.[40]

The terminology is idiosyncratic since it would result in such laws as the Law of Inertia or the Law of Conservation of Energy being counted not as scientific laws but as metaphysics. At the same time, it would link these laws, which are empirical, with such claims as

God created the universe from nothing

which are metaphysical precisely because they are in some sense factual claims but certainly *not empirical* claims. Popper's terminology mixes together what rationally ought to be distinguished.

Darwin's programme is valuable for science just because it leads to the discovery of new laws at the specific level. These discoveries have shown the value of the theory. In fact, as we have been emphasizing, such discoveries of new laws serve to *confirm* the theory. At the same time, the theory serves to provide an interlocking structure that serves to transmit confirmation from one set of specific laws to another via the consequence and the converse consequence relations. This is the role of the abstract generic theory in creating patterns of *coherence* among the laws in different specific areas.

Popper indicates, not incorrectly, that there are three basic propositions to Darwin's theory. These are: (1) that organisms reproduce in kind fairly faithfully; (2) that there are small, accidental, hereditary mutations; and (3) that there are forces of natural selection at work.[41] Popper suggests that if the first two claims are true, then the forces of selection begin to operate and variety in specific forms will be the result. That is, the theory *predicts* that there will be an increase in the number of different species. Popper argues, however, that Darwinian theory does not really make such a claim about a variety of forms.

For [he says] assume that we find life on Mars consisting of exactly three species of bacteria with a genetic outfit similar to that of terrestrial species. Is Darwinism refuted? By no means. We shall say that these three species were the only forms among the many mutants which were sufficiently well adjusted to survive. And we shall

say the same if there is only one species (or none). Thus Darwinism does not really *predict* the evolution of variety. It therefore cannot really *explain* it.[42]

There is a point to what Popper says, but what he says about the theory is not exactly correct. In the first place, the theory does not predict that every species will develop into something else. The crocodile has proved well adapted to its environment, and the basic form that exists now is the same as the form that existed before the time of the dinosaurs. This species has been stable and the forces of natural selection have eliminated all new mutations. And in the second place, the theory predicts that the forces of natural selection will begin to work in adapting forms only if there is some advantage to be achieved, for example if the environment changes and there are new ecological niches to be occupied. This is why Popper's appeal to Mars is misleading. For, on Mars, the environment is at once inhospitable to life and also relatively unchanging.

However, suppose we had a planet with a wide variety of environmental conditions — wet, cold, dry, warm, hot, etc. — and that there was a large population of organisms which had a high degree of genetic variation manifesting itself in the observable characteristics (phenotypes). Suppose, further, that there existed situations in which small groups of organisms could become isolated from the main body of the organisms, existing in a somewhat different ecological niche. Suppose, finally, that we know, perhaps through a fossil record, that life had existed on this planet for a very long time. *Then*, as Ruse has pointed out,[43] these data would count against the theory.

To be sure, they would not absolutely refute the theory. The theory asserts that *there are* forces of natural selection. Since form has been preserved, there must be forces at work which maintain that form — as in the case of crocodiles. Given the environmental variations and the variations in the organisms, we would expect whatever forces of natural selection that there are to produce variety in species. If we have not discovered those that maintain stability, all that follows is that we have not looked far enough. In this sense, Popper is correct: the theory is not refuted.

Nonetheless, Ruse is quite correct. What is the task of an abstract generic theory for the scientific researcher?

It is to guide the research in the discovery of laws. If the Darwinian theory ceases to this, and we cannot discover the forces of natural selection that are maintaining stability, then it is ceasing to do the job expected of it. Researchers will begin to consider it time to search for a new theory which *does* do the job expected, that is, does lead to the discovery of new laws. Under the conditions cited, the failure to find the forces of natural selection that maintain stability would lead the scientists to search for an alternative theory, one that is contrary to Darwin's in certain respects, but also one that leads, unlike Darwin's in the hypothesized circumstances, to the discovery of the specific laws that account for the observed phenomena. If scientists begin to search for such a contrary theory — in what Kuhn calls periods of "revolutionary science" — , they are searching for a theory that, if accepted, would require the rejection of Darwin's theory as false. The failure to find the forces of natural selection required by Darwin's theory thus tells in favour of rejection of that theory. In this sense, Ruse is quite correct: in the situation hypothesized by Popper, the data would point towards rejection of Darwin's theory. And, of course, if the facts can count against Darwin's theory, then it is an empirical claim. Perhaps, in Popper's sense, it is metaphysical. For all that, it is also empirical.

This points toward the answer to the question that arises when we allow, contrary to Popper, that abstract generic theories which cannot be falsified are part of science, namely, how do we actually come to reject a theory such as (c) as false, if not by appeal to single experiments?

Kuhn makes an important suggestion here. Where Popper holds that a theory is rejected in effect by a confrontation with observational data, Kuhn argues that a theory is rejected not so much through a confrontation with observable data as through a confrontation with another theory.

We have suggested just this in connection with our discussion of Popper on Darwinian theory. We can see the same point by means of an example drawn from the history of chemistry.

Towards then end of the 18th century, a major theory concerning combustion was the "phlogiston theory." This theory asserted that combustion is a process in which a substance changes by giving off heat and another substance called phlogiston. Here the relevant law is something like

(PT) There is a sort of substance P such that for any sample of any other sort of substance Q, if that sample undergoes combustion then heat and a sample of P are given off.

Chemists then introduced the definite description

the substance P such that for any sample of any other sort of substance Q, if that sample undergoes combustion then heat and a sample of P are given off.

They then abbreviated this definite description as

phlogiston

This enabled them to re-state their theory as

For any sample of any substance other than phlogiston, if that sample undergoes combustion then heat and a sample of phlogiston are given off.

This law is gappy, but they could use it to explain *ex post facto* processes of combustion. Of course, they had not yet identified phlogiston, but many chemists tried hard and believed, given their theory, that it was only a matter of time until they succeeded. Thus, it was, for example, suggested that

hydrogen = phlogiston

But it was possible to demonstrate that there are occurrences of combustion where hydrogen is not given off.

Late in the history of the phlogiston theory, a rival developed. This was the oxygen theory. According to this theory, the law for combustion goes something like this:

(TO) There is a sort of substance P' such that for any sample of any other sort of substance Q, if that sample undergoes combustion then heat is given off and a sample of P' is absorbed.

Chemists then introduced the definite description

the substance P' such that for any sample of any other sort of substance Q, if that sample undergoes combustion then heat is given off and a sample of P' is absorbed.

They then abbreviated this definite description as

oxygen

This enabled them to re-state their theory as

For any sample of any substance other than oxygen, if that sample undergoes combustion then heat is given off and a sample of oxygen is absorbed.

This law is gappy, but they could use it to explain *ex post facto* processes of combustion. Unlike phlogiston, which was never identified, oxygen was.

Joseph Priestly, a defender of the phlogiston theory, discovered that when one heated red calx of mercury (now called mercuric oxide), a gas was given off. Since things burned more brightly in this gas, Priestly agued that it lacked phlogiston and therefore was able to absorb more of it, permitting things to burn more quickly in it. Given this further hypothesis, he referred to this gas as "dephlogisticated air", that is, he made the theoretical identification:

the gas given off when red calx of mercury is heated = dephlogisticated air

But this identification required the theory (PT) to be acceptable, where in fact no one had ever successfully identified phlogiston.

In contrast, Lavoisier was able to establish by very careful measurements of the weights of the products of processes of combustion that in that during such a process a substance was indeed absorbed. For, the weight of the product of the process was greater than the weight of the sample initially. He was further able to confirm the identificatory hypothesis that

oxygen = the gas given off when red calx of mercury is heated

This definitely confirmed the oxygen theory (TO).

At the same time, this led to the rejection of the phlogiston theory (PT). For, (TO) and (PT) are contraries; they cannot both be true (assuming of course that there are actual processes to which they apply — but this assumption is true). Lavoisier's measurements confirmed that during combustion, the sample gained weight. That meant that a substance had been absorbed. This is contrary to the prediction of (PT) that a substance is given off. Lavoisier's measurements therefore confirmed the theory (TO), and rendered it acceptable. Then, since (TO) is contrary to (PT), the acceptance of the former required the rejection of the latter.

The only reply that the defenders of the phlogiston theory had was that phlogiston was indeed given off, but the sample gained weight because phlogiston had negative mass, or, what amounts to the same, levity rather than gravity. However, Newton's physics lay in the background. It took for granted that *every* particle of matter had mass, and that *every* particle of matter gravitated towards every other particle. I.e., there was no place in Newton's physics for negative mass or for a levity that opposed gravity. The revised phlogiston hypothesis therefore conflicted with Newtonian physics. Since the latter was highly confirmed, it followed that the revised phlogiston theory could be rejected on the grounds that it was contradicted by a highly confirmed theory. It was simply inconsistent with background knowledge and therefore failed to meet one of the criteria for being a good hypothesis.

No one succeeded in refuting the phlogiston theory by means of a direct confrontation with observational data. It was always open for the phlogistonists to argue that phlogiston existed but that no one had yet succeeded in identifying it. It was not that the theory was wrong, it was just that the researchers had not searched hard enough. Lavoisier, however, was able to confirm observationally an alternative, *and contrary* theory. The observation data that confirmed this theory rendered it acceptable. Once it was accepted, then one had therein to reject the contrary phlogiston theory.

Data did lead to the rejection of the phlogiston theory. But not through a direct confrontation. Rather, the data led to the rejection of the theory because they supported a contrary theory and acceptance of the latter required the rejection of the phlogiston theory.

This, of course, is Kuhn's point that falsification and rejection come about not through a Popperian confrontation with observable data, but rather through a confrontation with another, contrary, theory for which observational data provide strong support.

We can provide a model for this in terms of our schematic theory (c):

> For any species f, if f is of genus **D** then there is a unique species g such that g is of genus **G** and such that for all x, x is g if and only if x is f

We might suppose that we have a second theory

(c*)   For any species f, if f is of genus **D** then there is a unique species g such that g is of genus **H** and such that for all x, x is g if and only if x is f

which is contrary to (c) by virtue of the fact that being **H**-ed and being **G**-ed are incompatible:

> for all x, there is a g of genus **G** which x has if and only if it is not the case that there is a g of genus **H** which x has

(c*) yields only determinable predictions, and therefore is not falsifiable by a direct Popperian confrontation with observable data. But now suppose that (c) is strongly supported by the observations; suppose that it has led to the confirmation of specific laws in areas to which it applies. These data require that we must accept (c). But (c) is incompatible with (c*). We must therefore reject (c*).

This proposal of Kuhn is a plausible account of the logic of the process through which paradigms, that is, abstract generic theories, come to be rejected. We shall more to say about it later.

There remains one issue that should be mentioned at this point, though we will not discuss it here.

As we have seen, Kuhn argues that scientific practice is this: if repeated efforts to solve a problem posed by a paradigm, that is, to find a law it asserts to be there to be discovered, fail, then that, while not falsifying the paradigm, begins to call into question whether there is,

after all, a law of the sort that the paradigm claims to exist.[44] Such repeated efforts by skilled scientists begin to make it reasonable to assume that, as Mill puts, "a real [instance] could scarcely have escaped notice."[45] At that point, one begins to search for alternative guides for research, and we enter a period of what Kuhn calls "revolutionary science"[46] where research is no longer guided by a *tested generic hypothesis*,[47] i.e., a paradigm. How, then, is it guided?

Cohen and Nagel were followed by Hanson and Kuhn in arguing that analogies with known cases could guide researchers during the process of inquiry in formulating hypotheses that could reasonably be put to the test. But if the theory that establishes what those analogies are breaks down, no longer provides hypotheses that pass experimental tests, no longer leads to the discovery of laws, then how is the scientist to be guided during the inquiry? Kuhn does in fact, as we shall see, provide a clear answer to this question of how research is guided in periods of revolutionary science. But of that, more later.

One final point.

In our sketch of the process of inquiry, we pointed out that, when the research is guided by a theory with the logical form (c)

> For any species f, if f is of genus **D** then there is a unique species g such that g is of genus **G** and such that for all x, x is g if and only if x is f

it is necessary to apply the theory to a previously unexamined area, D_4 we supposed. This required the scientist to look in this area for hypotheses that exemplify the form (f)

> g is of genus **G** and such that for all x, x is g if and only if x is f

Now, in our simple example, this form is fairly obvious. In the terms of Cohen and Nagel, it is clear what is the relevant analogy. But, as one would expect, in the real world of inquiry things are hardly so perspicuous. It is in fact often difficult indeed to see in the regularities of the new area exactly which ones exemplify the logical form that the background generic theory asserts is present in those regularities.

Cohen and Nagel put it this way:

It is a mistake ... to suppose that we always explicitly notice precise analogies and then rationally develop their consequences. We generally begin with an unanalysed feeling of vague resemblance, which is discovered to involve an explicit analogy in structure or function *only by a careful inquiry.* We do not *start* by noting structural identity in the bend of a human arm and the bend of a pipe, and then go on to characterize the latter as an "elbow." ... Usually it is rather the other way around.[48]

The forms of instruction in the natural sciences have acquired much of their structure precisely in order to increase the capacity to recognize, largely unconsciously, the generic pattern in the specific regularities in new areas. Students in the natural sciences are presented with textbook exercises in applying theories. Varieties of problems are posed to them which involve applying a theory to new situations. This is as true of the textbook examples in mechanics as it is in the standardized routines of Physics 100 lab exercises. As the students acquire skills in recognizing how apply the theories with which they must deal they progress to more and more difficult problems, until they reach the stage of skilled researchers where they can take off on their own in investigating problems that are really new and not just contrived exercises.

(ii) Elimination and the Experimental Methods

What is characteristic of the new science is the *method of experiment.* This is clearly exemplified in the work of William Harvey on the circulation of the blood. Harvey's book on this subject is crammed full of experiments. Thus, to take only one, Harvey discusses the issue whether the blood found in the veins is pure or, as held by his opponents who followed Galen, contains vapours. He points out that the quantity of vapours decreases in the presence of cold. He therefore proposes the experiment of ligating an arm, letting the veins swell up with blood, and then immersing it in cold water. If the

swelling decreases then there are vapours present, but if not then the blood is pure. When the experiment is performed, there is no decrease in swelling. Harvey concludes that the blood is pure: "this experiment shows that, below the ligature, the veins swell up not with thinned blood and inflated spirit or vapours (for the immersion into the cold would have depressed such bubbling out)...".[49]

The logic of the situation is clear. There are two contrary or mutually exclusive hypotheses which together exhaust the alternatives. These are:

The blood in the veins is pure [that is, a necessary condition for blood in the veins is that it is pure].

and

The blood in the veins contains vapours [that is, a necessary condition for blood in the veins is that it contains vapours].

We might represent these by:

If V then P
If V then not-P

Harvey presents data that falsify the second. The data themselves are derived by inference from other observations. He *takes it on the basis of other data to be true* that

If the blood in the veins is mixed with vapours then if it is immersed in cold then it decreases in volume.

or, in symbols,

If V then [if not-P then (if C then D)]

Since this is true, and since it is given that it is blood in the veins, i.e., V is true, it follows that [if not P then (if C then D)] is true. Since C is true and D is observed to be false, it follows that (if C then D) is false. From this it follows that not-P is also false. Hence it is false that

        If V then not-P

He then concludes that the first is true.

The logic of the situation is clear: we can see the mechanisms of eliminative induction at work. We are presented with a range of jointly exhaustive and mutually exclusive hypotheses; data are presented which eliminate all but one of the range; it is concluded that the remaining hypothesis is to be affirmed as true. *It is this method that is characteristic of the new science.*

The logic of experiment turns crucially, as we have just seen and as we shall see in detail below, on the use of data to *eliminate* all but one of several competing hypotheses. Elimination means, of course, falsification. We therefore begin our discussion of the logic of the experimental methods of science by looking at the most well known discussion of falsification and its role in science. This philosopher of science is Karl Popper.

a. Popper and Falsification

Popper has emphasized the role of falsification and of falsifiability in the logic of science. It is his fundamental thesis that falsifiability is the mark by which statements of empirical science are distinguished from metaphysical statements and from tautologies. His point is similar to that which is embodied in the old saw that "The exception proves the rule." The "rule" is a universal generalization, and "to prove" is, on an older meaning, "to test." The exception (or negative or falsifying instance) proves (that is, tests) the rule (the universal generalization). It does so, not by showing it to be true, but by showing it to be false. Positive instances, non-exceptions, do not test the rule since it is impossible to conclusively verify a universal generalization with any finite number of positive or confirming instances. But one exception, one negative instance, can conclusively falsify a universal generalization.

It is Popper's claim that science proposes theories, sets of universal generalizations, that form a body of hypotheses or conjectures. The aim is not to confirm or verify these but rather to test them through attempts at falsification. The function of the experimentalist is to devise ways in which they theories might be falsified.

I think that we shall have to get accustomed to the idea that we must not look upon science as

a "body of knowledge," but rather as a system of hypotheses; that is to say, a system of guesses or anticipations which in principle cannot be justified, but with which we work as long as they stand up to tests, and of which we are never justified in saying that we know that they are "true" or "more or less certain" or even "probable."[50]

Popper is certainly correct that a negative instance eliminates an hypothesis from consideration as a candidate for explaining and predicting. A universal generalization together with initial conditions yields a certain prediction. If the prediction turns out to be false, then at least one of the premises must be false; that is the notion of deductive validity, the argument form is that of *modus tollens*:

| If p then q | = If hypothesis then prediction |
not q	= but not as predicted
so, not p	= hypothesis is false

Taking the initial conditions as given, it follows that we must reject the universal hypothesis. None of this is controversial. What is harder is Popper's further claim that the *modus tollens* inference is the only principle available for the acceptance and rejection of hypotheses. Since it is useful only for rejection, it follows, as Popper recognizes, that acceptance has no place in science. This latter Popperian claim is, of course, one that we are disputing: we are trying to show that probabilistic reasoning does take place in science. But if we are to do this, then we must locate the notion of falsifiability within the context of probabilistic reasoning. This we shall do shortly. But there are some further distinctions that must first to be dealt with if we are to give a fair evaluation of Popper's claim that falsifiability provides a good, single criterion for separating science form pseudoscience.

The problem with Popper's proposal is, of course, that it ignores crucial differences — propositions can fail to satisfy the falsifiability criterion for a variety of reasons, all of which should be kept distinct if we are to be effective in our critical evaluations.

We can in fact distinguish a number of different ways in which a proposition may fail to be falsifiable.

(1) *Empirically Meaningless Propositions*

Consider the proposition that

All bumbies are gooches

This is not falsifiable by observational data because it is *not empirical*, and it is not empirical because *its concepts do not refer to entities in the world of ordinary experience*. If the concepts that appear in a proposition have not been tied to characteristics of things that can be observed in ordinary sense experience, then the proposition simply lacks the empirical content which alone would establish conditions under which it is true or false of the empirical world. Since it lacks that content that would determine it to have a truth value, since it is neither true nor false, it is could not be false, nor therefore could it be falsified.

Here is another proposition that lacks empirical content:

The health of the body is regulated by its vital force.

Here we have the concept of a 'vital force.' This is another concept that has not been defined with reference to observable characteristics of things. So the proposition lacks empirical truth value, and therefore cannot be falsified. The same is true of all propositions that make reference to Aristotelian powers or Natures of things, unanalysable dispositions to behave in one way or another. It is precisely this feature — that their truth value cannot be determined with reference to our ordinary experience of the world — that led Galileo to reject appeals to them as explanatory of the behaviour of things, and to substitute for such appeals the scientific concept of explanation in terms of matter-of-fact regularities.

And here is yet another proposition that lacks empirical content:

(!)   God has created the universe from nothing.

Creating things from nothing is not an activity that occurs in the world of ordinary experience: the notion of a relation in the everyday world requires that both terms of the relation be *something*. For example, consider the relation of kicking. It relates say Jones to a football, but whomever it relates to whatever, the whomever and the whatever are not nothing: the things related by relations in the ordinary world are always something, never nothing. Creation by God, however, has *nothing* as one of the things that it relates. Moreover, the concept God is also one that has no reference in empirical reality. Since this proposition is not empirical, it could not be falsified by any facts that could be discovered in ordinary experience.

Essentially, the problem with (!) is that we have never seen a deity creating things from nothing. It might be objected on the same sort of ground that neither have we seen evolution at work. This is an objection that is often raised by creation scientists, who argue that the religious hypothesis (!) is in this respect no different from the Darwinian hypothesis, and that therefore the two theories deserve equal treatment, either both being represented in the classroom or both being excluded.

But contrast the hypothesis, considered by Herodotus, that

(!!)   Every flood of the Nile at the summer solstice is caused by snow melting in inland Africa from whence the Nile flows.

This refers to *places* where Herodotus had never been. But drawing on knowledge established in places where he had been, and in particular the knowledge of causes that he has observed working in that part of the world he had observed, Herodotus had quite reasonably attempted to draw inferences about the world he had not seen. A proposition making assertions about places we have not seen does not cease to be empirical simply because it makes an assertion about the unseen. In similar way contemporary biologists use knowledge of causes derived from present experience to draw inferences about times past where they, and indeed no person, has ever been. A proposition which makes an assertion about the past we have not seen does not cease to be empirical simply because it makes an assertion about the unseen.

The point about (!!), one which applies equally to the evolutionary hypothesis, is that *concepts* in (!!) refer to properties and relations instances of which can be found in the world that Herodotus has visited. He had, for example, seen snow melting. Or again, he knew what it is for someplace to be inland from other places. The same is true of the biologists' hypothesis about the evolutionary past: the concepts that they use to describe that past all are such that they either have instances in the world the biologists presently experience, e.g., sedimentation, eating prey, and so on, or, if there are no such instances, then the concepts are defined in terms of others which themselves have instances in the world as presently experienced, e.g., the notion of the skin of a dinosaur, no example of which has been preserved for us to observe but which can be explained by reference to other things under genus of skin such as the skin of alligators.

What is significant in this respect about (!) is not simply that we have never observed a deity creating but that we not only have never observed in present experience any instances of the concepts involved. Nor are the concepts in the proposition (!) definitionally tied to concepts that have instances in present experience. If only the one or the other of these conditions were fulfilled would (!) be an empirical proposition, one such that empirical, matter of fact, considerations, or empirical inferences drawn form such considerations, could determine whether it was true or false. (!) is not like (!!), but rather like the proposition

> The Nile floods because it is connected to the River Ocean

which Herodotus rejected as an attempt at a *scientific* explanation because it introduced the non-empirical and mythological entity the River Ocean.

(2) Tautologies

Consider the proposition that

> Everything is either red or non-red

This is not falsifiable by observational data because it is not empirical. But it is different from the previous case because the non-logical concepts do refer to entities in the world of ordinary experience: we are all acquainted in ordinary sense experience with the property of being red. This proposition is not falsifiable, instead, because it is *tautological*, a *necessary truth*. It is, to be sure, a truth about the world of ordinary experience, unlike the previous proposition — it is a truth about the world of

experience because the non-logical concepts that occur in it are empirical. But it is a necessary truth about this world. It has the logical form

> for any x, x is R or x is not R

Since the matrix (x is R or x is not R) exemplifies the form of a tautology of sentential logic, "p or not p", it follows not only that the matrix is tautological but that its universal generalization is also.

Another example of a necessary truth is

> Bachelors are unmarried males

The non-logical concepts in this proposition also refer to entities in the world of ordinary experience. So this proposition is a truth about this world. But the one concept is *defined* in terms of the others:

> x is a bachelor = $_{Df}$ x is unmarried and x is male[51]

This statement of definition asserts that the left hand side (the definiendum) is *short for* the right hand side (the definiens). This means that the statement

> Bachelors are unmarried males
> = for all x, x is a bachelor if and only if both x is unmarried and x is male

is short for

> for all x, x is both unmarried and male if and only if x is both unmarried and male

which is an ordinary tautology: the matrix to which the quantifier is attached has the form "(p & q) if and only if (p & q)", which is the form of a tautology in sentential logic. It, too, is a necessary truth.

What we earlier referred to as "elastic hypotheses" when we discussed Lysenko are another kind of proposition which is not falsifiable by virtue of its being a tautology. These propositions are asserted with reference to a rule such as

(*)   All F are G

This is falsified by an experiment c in which

c is F and c is not G

It is proposed to as it were explain away this apparent counterexample by appeal to unknown forces that make an exception in the case of c. We have not so much (*) as

(**)   There is a factor f such that everything which is F is G if and only if it is not f

Now, any existence claim can be verified if but one instance of it is discovered. Thus, the existence claim that

There are dogs

can be verified if we observe that

Fido is a dog

Similarly, we can verify (**) if we find one factor that will do the job. We have so far found one counterexample to (*), namely c. If this is the only such case then we have

(&)   Everything which if F is G if and only if it is not c

Assume, as we said, that c is the only exception, we have in (&) an instance which verifies (**). This illustrates that it is trivially true that one can always in this way explain away exceptions; just mention the exceptions in the exceptive clause. The problem is that without limitation on the exceptive conditions mentioned in (**), it is tautologous that there are such exceptive conditions. The proposition (**) is elastic enough to fit every situation so as to explain away all exceptions to (*) precisely because it is a tautology: it explains nothing. And such elastic propositions cannot be falsified for the same reason: tautologies cannot be falsified by empirical facts.

As we noted previously, a proposition such as (**) becomes empirical, matter-of-factly either true or false, only if it conforms to the *Principle of Limited Variety*. Another way of saying that (**) is tautologous is to say that it places no restrictions on the conditions which will

count as exceptive conditions. So long as it allows for an infinite variety of possibilities for making it true, it is trivial, tautologous, that such possibilities exist. The proposition makes a statement of fact only if there are some generic constraints that allow for only a *limited variety* of conditions. Then, it is as a matter of fact true just in case one of the conditions under that constraint is a genuine exceptive condition; if none of the conditions under the genus constitutes a verifying instance of (**), then it is as a matter of fact false.

The move from (*) to (**) as a way of ensuring that one continues to have a true proposition achieves that end only at the cost of transforming an empirical proposition into a tautology. The proposition is as it were saved from falsification only by making it unfalsifiable.

There are other ways in which they same end might be achieved. Suppose that a member of the Institute for Creation Research is told of a Christian, e.g., Pope Paul, who has declared the Darwin's theory of evolution is empirically true. The member of the Institute for Creation Research might well reply, "Oh, but he is not a *true* Christian." This use of the term 'true' turns it into a tautology that a Christian, that is, a *true* Christian, rejects the theory of evolution and accepts as literally true the story told at the beginning of the Book of Genesis.

Similarly, suppose that Lysenko is presented with results that show that the process of vernalization does not work. If he replies that these cases did not involve "*true* vernalization" then he has transformed his concept of vernalization, that is *true* vernalization, into one that always works: it becomes trivial because tautologous, true by definition, that vernalization has the effects that Lysenko ascribes to it.

These points are relevant to the evaluation of results in parapsychology. These results are often unscientific for not being falsifiable, where the falsifiability derives from the fact that the purported explanatory hypotheses have become, not statements of matter of fact regularity, but rather *tautologies*, statements that are *true by definition*.

Suppose that a parapsychologist named Danube runs an experiment on Jones and finds that Jones can in a long run of draws call the cards correctly at a rate greater than that which would be expected on the hypothesis of chance. A sceptical psychologist Dr. Cam now comes along and runs a similar set of tests on Jones, with more rigid controls. Jones' success rate is about what would be expected on the hypothesis of chance, on the hypothesis that he was guessing. Dr. Danube and his colleagues explain the failure by citing the fact that when Jones gets bored he does not perform well, and the rigid controls, alas, bore him.

Previously what was being tested was a statistical hypothesis about a mass event. The mass event consists in Jones calling repeatedly the sort of card that has been drawn. This mass event has a certain property, namely, that the run of successes is more than would be probable under the assumption of chance. The hypothesis being tested, in other words, is this:

> Whenever Jones is subjected to a long sequence of tests, then his success rate is greater than that expected by chance.

Let us abbreviate this by

> Whenever Jones is F then Jones is G

But it turns out to be false, or, at least, when Dr. Cam comes along, he obtains a result that, within the limits of the statistical test, leads us to reject this hypothesis. He finds that he cannot accept that the mass event has the property G. Dr. Danube retorts, however, that we have not taken into account all the relevant variables. We also have to include the variable of boredom, call it "B":

> Whenever Jones is F then Jones is G if and only if Jones is not B

And so, if Jones is F and it turns out that he is also not G — in other words, doesn't get enough right — we can explain that negative result *ex post facto* by appeal to B. However, *is there any evidence that this new hypothesis is true?* After all, *we can rely upon this to provide* ex post facto *explanations that are acceptable only if it itself has a reasonable level of probability.* If there is no independent evidence for the role of the factor B, no background theory with some prior probability, then we are not entitled to introduce this hypothesis as a way to explain away negative results.

What is the criterion for boredom B? If it is merely Jones' say-so, then the possibility of cheating is clear.

Especially if Jones has some interest in achieving a better than chance result. All that he has to do, when faced with a result no better than that expected by chance, in order to preserve his reputation as someone with psychic powers, is to claim that he was bored. A criterion of boredom independent of Jones' say-so is needed if we are to eliminate the hypothesis of cheating.

Even worse are attempts to explain away negative results with hypotheses such as

> Whenever Jones is F then Jones is G if and only if Jones is not rigidly tested

or

> Whenever Jones is F then Jones is G if and only if Jones is not subject to the bad vibes of a sceptical tester

The problem is that neither of these hypotheses remains scientific: they are in fact both tautological. Without any criterion of a rigid test other than its being one that prevents the result C, then of course the first hypothesis is true. Except that it is no longer a scientific hypothesis: every test that is sufficiently unrigid is going to confirm the hypothesis of necessity. Similarly, the second hypothesis is tautological, since every negative result, every result that purports to falsify the hypothesis, will be one where bad vibes are present. Again, every test will confirm the hypothesis: negative results will simply not apply. But we should not be surprised that tautological hypotheses are confirmed by the data. At the same time, neither do they explain anything, nor, especially, do they lead to any astounding predictions or any astounding ascriptions of special ESP capacities.

In principle, if the notions of "rigid test" or of "bad sceptical vibes" were independently defined then even these hypotheses could be subjected to a variety of tests, And these tests would enable one in principle to begin to put the real question to the test, of whether Jones has powers that, even in special circumstances — "no bad vibes" — , would permit him to call the cards at a rate better than that which can be expected if he were only guessing.

Such independent definitions of these notions are never given by the parapsychologists. Nor, really, should that surprise us. It is clear that they are introduced more to explain away adverse criticism than to try to advance scientific research. It is evident that parapsychologists are moved more by the desire that their hypotheses regarding psychic powers are true than they are moved by the desire to subject them to the tests of the inductive method.

One of the charges often made against Darwin's theory of evolution by natural selection is that it is tautological. As one Evangelical Christian put it, "A chief reason that some scientists are unhappy with regarding evolution as a theory like other scientific theories is that it cannot be tested."[52] This particular charge, that the theory is untestable because it is tautological, has been made by Popper.

> ... we use the terms 'adaptation' and 'selection' in such a way that we can say that, if the species were not adapted, it would have been eliminated by natural selection. Similarly, if a species has been eliminated it must have been ill adapted to the conditions. Adaptation or fitness is *defined* by modern evolutionists as survival value, and can be measured by actual success in survival: there is hardly any possibility of testing a theory as feeble as this one.[53]

Popper's argument is that the theory is not testable because it is tautological. However, at the same time, as we saw earlier, he holds that the theory has sufficient content to be able to make the "suggestion" that *there is* a "mechanism of adaptation." It surely cannot both have empirical content and also be tautological. We can see here that Popper is confusing non-falsifiability by virtue of containing an existence claim, on the one had, and, on the other, non-falsifiability by virtue of being a tautology.

The important point to be made is that Popper's argument that the theory is tautological is simply mistaken. Popper simply fails to note that *fitness* is defined in terms of *probabilities*: a characteristic is fit-making just in case that its presence increases the probability that organisms with it will achieve their goals of survival and reproduction. Since this is a probabilistic notion, it is compatible with it to say that some of the non-fit survive and reproduce while some of the fit do

not survive and reproduce. It is, therefore, simply false to say that it is tautological that the fit are those that survive.

Popper's charge that the theory of evolution by natural selection is not explanatory and cannot be tested by virtue of its being tautological is not justified. He is thus yet another of the spurious authorities so often cited by creation scientists.

(3) Falsifiability and the method of authority

The next sort of non-falsifiability can be explained if we turn to the case of Duane Gish, the defender of creation science. Gish, like other creation scientists, is concerned to argue that there are no transitional forms in the fossil record, thus challenging the claim that more complex species evolved from simpler ones. One of the cases with which Gish deals is that of "Lucy", the 40% complete skeleton of an *Australopithecine* (early hominid) discovered in Ethiopia in 1973. Lucy is considered by anthropologists to be strong evidence for human evolution because she shows both ape-like and human characteristics, including among the latter the capacity to walk upright. Gish, in his debates,[54] attacks the claim that Lucy is transitional by citing the authority of the anthropologist Lord Solly Zuckerman, who denied that the Australopithecines could walk upright. Gish regularly implies that Zuckerman examined the Lucy fossil. In fact, Zuckerman never examined Lucy, and came to his conclusion three years before Lucy was discovered. Nor did Zuckerman work with any of the original Australopithecine fossils. All he ever examined was the cast of one half a pelvis of a single fossil.

Here is something of Gish's record:

> In 1982, at a high school in Lion's Head, Ontario, Gish debated Chris McGowan, a zoologist from the University of Toronto. A member of the audience, Jay Ingram (former host of the national Canadian radio program *Quirks and Quarks*), heard Gish's Lucy story, which clearly implied that Zuckerman had studied Lucy herself and concluded that she, along with *other Australopithecines*, did not walk upright. Knowing this was not true, Ingram asked Gish in the Q & A period why he had misled the audience. A show of hands indicated that about 90% of the audience had assumed from what

Gish had said that Zuckerman had studied Lucy. Gish became very upset, lost his temper, and railed that he wasn't responsible for people misinterpreting his remarks....

Gish has never bothered to change his misleading story; in fact, he went on to increase its inaccuracy. In a 1991 debate with biologist Fred Parrish, Gish stated outright that Zuckerman had examined the Lucy skeleton itself: {he asserted that} "For 15 years ...[Zuckerman] studied fossils of Lucy and fossils of 1-2 million years younger than Lucy [sic]."[55]

Another example of Gish's way of thinking and debating about these things is his description of the process of evolution. In his book, *The Amazing Story of Creation from Science and the Bible*, he argues that

Mutation is the commonly accepted mechanism required, by evolution, to change the first form of life into all other living creatures.[56]

Now, mutations occur at random relative to the evolutionary process. So Gish's statement implies that evolution proceeds solely by chance. And built onto this is the conclusion that, of course, something like the eyeball could not arise solely by chance, so we need to introduce an intelligent creator if we are to understand the existence of an organism with a mechanism as intricate and as adapted as the eyeball.

But of course, mutations are not the whole story of evolution. There is also natural selection. Once this is introduced, one has something more than chance, something that drives the process of increasing levels of adaptation. Without natural selection, as many have emphasized, there is no hope of understanding the fact of evolution. Rather than presenting an accurate picture of the theory he opposes, Gish omits an essential ingredient, precisely that ingredient that accounts for, that is, explains causally and naturalistically, the adaptation, the element of "design", that Gish wishes to attribute to a creating deity.

If Gish were genuinely interested in arguing the issues about facts and evidence, moved only by a cognitive interest in matter of fact truth, then he would have the responsibility to present the opposing view honestly and with a certain degree of accuracy. This he does not do. Nor does he present the evidence about Zuckerman in a way that is at all accurate. In fact, when inaccuracies are pointed out, he continues to adhere to the same position in spite of the evidence against it.

This suggests that Gish is not really concerned with the truth of things in the way in which those moved by the scientific motive of curiosity are interested. Scientists need evidence, and base their explanations on hypotheses tested against experience. Moreover, these explanations are, as Judge Overton insisted, only tentative, and if further evidence comes in and requires revisions in one's explanations, then those explanations are indeed revised. It is precisely this that is lacking in Gish. It is clear that he is in no way committed to the inductive method of science. Rather, it is evident, his commitment is to the **method of authority**: *assert whatever belief is enjoined upon you by a recognized authority.* Gish's authority is the Book of Genesis in the Bible, as that Book is understood by Christian Protestant fundamentalists.

In particular, the method of authority enjoins you to continue to affirm the authoritatively recognized beliefs no matter what further data come to hand. The acceptance of this method thus requires that you give up the idea of falsification: evidence that, in terms of the logic of the situation, leads to the falsification of a belief, and therefore, in terms of scientific rationality, to the rejection of the belief, is simply ignored. In this sense, *a belief based on the method of authority is not falsifiable.*

Popper recognizes this sort of non-falsifiability as non-scientific. His favourite example is that of Marxism which, he argues, in its original version was falsifiable. It was also subsequently falsified. But many Marxists continue to adhere to it in spite of the fact of falsification. It is the method of authority at work in a different area from that of Gish, but the upshot is the same: unfalsifiability. Popper writes that

Marxism was once a science, but one which was refuted by some of the facts which happened to clash with its predictions.

However, Marxism is no longer a science; for it broke the methodological rule that we must accept falsification, and it immunized itself against the most blatant refutations of its predictions.[57]

Popper is correct here. We have seen this pattern at work in Lysenko's appeal to the authority of dialectical materialism as glossed by Lenin and Stalin. Here, too, propositions are rendered impervious to empirical evidence and to falsification by reference to an authority which declares them, incorrigibly, to be true. But it is important to notice that this has nothing to do with the *logical form* of the propositions that are protected from falsification. In fact, those propositions *are* falsified; Lysenko's theories were false because they conflicted with the well confirmed theory of Mendelian genetics. It is just that propositions based on authority continue to be accepted in spite of being falsified for the simple reason that acceptance is not a matter of empirical evidence but of the *authority*. Popper's blanket use of the term 'falsifiable' blurs the difference between "not falsifiable because of logical form" and "not falsifiable because accepted on the basis of authority (or some other non-scientific method)". Effective criticism of pseudoscience requires one to keep clear what Popper in this way suggests are the same sort of thing.

(4) Falsifiability and Abstract Generic Laws

Finally, we should emphasize once more propositions like those involved in the theory of evolution by natural selection, say the law

> For any species of organism S there is an earlier species S', there are variations in S' and there are forces of natural selection such that these variations and forces are necessary and sufficient for the divergence of S from S'

which, roughly speaking, fits our model (c)

> For any species f, if f is of genus **D** then there is a unique species g such that g is of genus **G** and such that for all x, x is g if and only if x is f

Any law of this sort is not falsifiable by a single counter-example, and therefore does not fit the simple logical model of falsification that Popper presents. The latter is based upon the simplistic view that all laws in science have the simple specific form

> All F are G = for all x, if x is F then x is G

which is falsified by an instance which is both F and not G:

> a is F and a is not G

A law of the form (c) is not falsifiable in this simple minded way, though it is to be sure falsifiable in the sense that there are possible observational data which can lead to its rejection.

The point is that a law of the form (c) resists falsification by a single counter-example because of its logical form: *it contains an existential quantifier*. In proclaiming that falsifiability by a single counter-example is the criterion for distinguishing science from both non-science and pseudoscience Popper excludes form science things such as (c) or the theory of the origin of species by natural selection. These become "metaphysics."

What is characteristic of science, Popper argues, is that its propositions are universal propositions of a logical form that renders then capable of conclusive refutation by singular observation statements. A singular observation statement is of the form "a is F and a is not G" and the universal proposition therefore has to be of the form

> All F are G = for any individual x, if x is F then x is G

An individual that we know to be F but which we do not know to be G nor know to be not-G is a potential falsifier. In criticizing a scientific theory we search out such potential falsifiers. If the potential falsifier turns out to be not-G then the theory is eliminated. But if the potential falsifier turns out to be G, then the theory has survived the attempt to find a falsifier. In Popper's idiosyncratic terminology, the universal proposition that has survived such attempts at criticism is said to be "corroborated." This is a variation on what others meant by "confirmation," but in any case it is Popper's claim that the body of propositions acceptable as scientific are those that have been corroborated. Nor is this wrong: it is generally accepted that, when a generalization passes a test in which it has risked falsification, it is strongly confirmed.

The problem is that there is always a vast range of possible hypotheses. What we have to do is single out a subset as those that are reasonable for criticism or testing, that is, reasonable candidates for acceptance, if they survive testing, into the body of scientific propositions. In his essay on the *Poverty of Historicism*,[58] Popper allowed that such a choice of potential explainers, potential entrants into the body of scientific knowledge, was delimited by "points of view" or "interpretations" — in the case of history, "historical interpretations" (p. 150ff). These points of view are like theories in being universal — they assert that for all systems of a certain generic sort... — but, as Popper also emphasizes, they also make existence claims — they assert that for all systems of a certain generic sort *there are* (specific) laws of such a such a generic form. That is, they have something like the form of our (c). Now, while a purely universal proposition has a logical form that permits it to be conclusively falsified by a singular observation statement, a purely existential statement cannot be falsified. Observing a non-dog does not show that it is false that there are dogs; and failure to find a mermaid does not falsify the proposition that there are mermaids. So, according to Popper, such propositions are not of a logical form proper to propositions that are candidates for the status of scientific. But purely existential claims can, of course, be conclusively verified by a singular observation statement: if I observe that the individual thing Fido is a dog then that conclusively verifies the claim that there are dogs. Thus, where purely universal propositions can be conclusively falsified but not conclusively verified, purely existential propositions can be conclusively verified but not conclusively falsified. Since Popperian "points of view" are propositions that are universal, they cannot be conclusively verified, but at the same time they make existence claims and therefore cannot be conclusively falsified. They are therefore not part of science, according to Popper. They are, rather, "metaphysical."[59] Popper has argued that such metaphysics is of central importance to science, and even suggests that it has a central place in the methodology of science insofar as it can provide guidance about which hypotheses it is reasonable to put to the experimental test. Indeed, as we have seen, he makes just this point

about Darwin's theory of evolution by natural selection, that "although it is metaphysical ..., it suggests the existence of a mechanism of adaptation."[60] Darwin's theory, in other words, makes an empirical claim (since it has the logical capacity to suggest the existence of certain empirical mechanisms), is universal (so that it cannot be conclusively verified), and also involves an existential claim (so that it is not conclusively falsifiable by a single counterexample). Or, to put it our way, it has the logical structure of an abstract generic theory.

But such propositions, though to be sure ignored by those logical positivists who followed Carnap,[61] have long been well known to empiricists. Thus, Newton's axioms state, for example, that "for all mechanical systems, there are laws regarding forces that satisfy certain generic conditions." Such laws do indeed pose research problems, as Popper asserts; these laws are about laws, asserting about systems not yet investigated that certain specific (and therefore purely universal) laws do obtain in those systems, and that these laws will be of a certain generic form. The "point of view" does not state specifically what those laws are, but only asserts that they have a certain generic form and *are there* to be discovered. The research problem is to find that law. One goes about investigating the hypotheses in the delimited range, that is, the hypotheses which have the generic form described in the law about laws or "point of view." In particular, one sets about finding instances of those hypotheses to eliminate the false and accept the one corroborated hypothesis into the body of science.

If one succeeds in corroborating one hypothesis picked out by the "point of view," then traditional empiricists such as John Stuart Mill would say that the proposition expressing the "point of view" has been confirmed. But Popper has excluded it from the status of being scientific since it makes an existential assertion and is therefore not falsifiable. He therefore speaks of "useful" metaphysics. It amounts to much the same, however.

The point is, of course, that what Popper has described when he speaks of falsification guided by "points of view" is little different in logical form from the eliminative methods of experimental science first described by Francis Bacon and elaborated by John Stuart Mill. Popper first excludes such abstract generic laws from the body of science because they are not falsifiable, and then re-describes them not as laws but as metaphysical points of view. He then re-introduces them into his methodology to do the job that others have abstract generic theories do, namely, that of guiding research by delimiting a range of hypotheses that are deemed worthy of being put to do the test. By thus re-naming abstract generic laws, Popper has disguised his affinities to do the earlier empiricists as well as his affinity to do more recent thinkers such as Hanson and Kuhn. Popper's "falsificationist" methodology therefore seems to himself to be an original contribution. To others, who are not misled by the terminological innovations, it amounts to re-stating or re-emphasizing certain aspects of traditional empiricism.

Popper is misleading, then, when he takes his criterion of falsifiability to exclude from the body of science propositions of the form (c): his too narrow a criterion excludes from science what rightly belongs there and what Popper himself recognizes rightly belongs there when he himself smuggles it back in with a terminological shift.

(5) Falsifiability and the Role of Auxiliary Hypotheses

Popper suggests yet another way in which we can keep a theory from being falsified. This is by introducing auxiliary hypotheses that will save it. This point has been elaborated by Popper's student, I. Lakatos.

The story is about an imaginary case of planetary misbehavior. A physicist of the pre-Einsteinian era takes Newton's mechanics and his law of gravitation (N), the accepted initial conditions, I, and calculates, with their help, the path of a newly discovered small planet, p. But the planet deviates from the calculated path. Does our Newtonian physicist consider that the deviation was forbidden by Newton's theory and therefore that, once established, it refutes the theory N? No. He suggests that there must be a hitherto unknown planet p' which perturbs the path of p. He calculates the mass, orbit, etc., of this hypothetical planet and then asks an experimental astronomer to test his hypothesis. The planet p' is so small that even the biggest available telescopes cannot possibly observe it:

the experimental astronomer applies for a research grant to build yet a bigger one. In three years' time the new telescope is ready. Were the unknown planet p' to be discovered, it would be hailed as a new victory of Newtonian science. But it is not. Does our scientist abandon Newton's theory and his idea of the perturbing planet? No. He suggests that a cloud of cosmic dust hides the planet from us. He calculates the location and properties of this cloud and asks for a research grant to send up a satellite to test his calculations. Were the satellite's instruments (possibly new ones, based on a little-tested theory) to record the existence of the conjectural cloud, the results would be hailed as an outstanding victory for Newtonian science. But the cloud is not found. Does our scientist abandon Newton's theory, together with the idea of the perturbing planet and the idea of the cloud which hides it? No. He suggests that there is some magnetic field in that region of the universe which disturbed the instruments of the satellite. A new satellite is sent up. Were the magnetic field to be found, Newtonians would celebrate a sensational victory. But it is not. Is this regarded as a refutation of Newtonian science? No. Either yet another ingenious auxiliary hypothesis is proposed or ... the whole story is buried in the dusty volumes of periodicals and the story never mentioned again.[62]

The idea is this. In any interesting scientific theory, the logical structure is always complex. It always involves not only a theory but also auxiliary hypotheses, e.g, hypotheses about how instruments work. That is, the evidence never tests just T but always some theory

$$T \& A$$

Suppose (T & A) logically implies evidence E:

$$(T \& A) \text{ entails } E$$

Suppose further that it turns out that E is false. In that case we have

$$\text{not both } T \& A$$

But this is logically equivalent to

either not T or not A[63]

This means that given the observational data that we have, we must give up one of T or A as false. But the data do not force us to give up both. We can therefore continue to accept T as true. Or rather, we can do so provided that there is a way to explain the new data not-E. (T & A) would have been able to explain E had it occurred, since they logically imply E. But E did not occur; we observed not-E instead. Our theory, if it is to do the job of explaining the observed facts, must be modified so as to explain, or, what is the same, logically imply not-E. To do this, we need to introduce an new auxiliary hypothesis A* such that

(T & A*) entails not-E

Popper and, following him, Lakatos claim that any theory can be saved from falsification by introducing some suitably revised auxiliary hypothesis. A research strategy of *always* modifying one's auxiliary assumptions, never one's theory, is unscientific, they argue, because it in effect renders the theory unfalsifiable.

This, however, is *not* a point about the logical structure of theories. In fact, it allows that theories can be falsified, that is, that their logical structure is such that confrontation with the facts will refute them and lead to their rejection. The theories, therefore, *are* falsifiable in the sense that they can be refuted by a single counter-example. The point concerns, rather, *part of the theory*. It is a point about the *settled determination* to always modify one's *total theory* (T & A) in such a way that *part* of this theory, the part T, what may balled the "hard core",[64] will never be modified, only the auxiliary hypotheses. This *settled determination* becomes an **authority** that settles something about the logical form of the revisions that will be made in a theory after it has been falsified: these revisions will always be made in such a way that hard core T will remain unmodified.

This makes the role of a "hard core" theory sound much like an abstract generic law. After all, both are supposed to play the role remaining essentially the same through a process of inquiry, guiding the inquirer as to what hypotheses to use during the process. Except, of course, they remain the same for very different reasons. An abstract generic law is not falsifiable because of its logical form: it contain an existential quantifier. A "hard core" theory, in contrast, is not an instance of a "point of view" or of Popperian "metaphysics," and so it must, one presumes, be reckoned to be a law at the specific level. That makes its logical form such that it can be falsified by a single counter-example. The fact that it is never rejected on grounds that a counter-example has been found is due to the *settled determination* to find a new auxiliary hypothesis that will explain away the alleged counter-example. So the abstract generic theory of Hanson and Kuhn remains the same during the process of inquiry for very different reasons than the Popper-Lakatos "hard core."

If the new auxiliary hypothesis A* is designed so that it does nothing more, when it is conjoined to the hard core T, than explain the facts not-E that refuted the previous version (T & A), then the changes will be simply *ad hoc*. That, after all, is the meaning of "*ad hoc*". If, however, the new theory (T & A*) makes predictions that go beyond the data not-E, predicting in addition say E', that was not predicted by the previous version, then it may be said to be *theoretically progressive*[65]; if this prediction is subsequently confirmed, the new theory (T & A*) is confirmed, passing the test of successful prediction, and may be said to be *empirically progressive*.

If a hard core T is saved by an *ad hoc* manoeuvre, then there has been no progress in the process of inquiry. For no new test will be possible. It would seem, therefore, that one would have created a theory by this manoeuvre for which the acceptability could never be increased through passing (further) tests. This is not so, however. In general, a theory and some auxiliary assumptions are logically independent. This means that if the auxiliary assumption is less well confirmed than the theory itself, then that is what must be rejected, not the theory. Or if the auxiliary assumption is strongly confirmed, more strongly confirmed than the theory, then it would be the theory that is rejected.

But Popper and Lakatos not only suggest that a hard core can be saved by an *ad hoc* auxiliary hypothesis, they make the stronger suggestion that these manoeuvres to save a hard core T can *always* be made in the sense that one can always find an auxiliary hypothesis that, when conjoined to the hard core, accounts for the data that refuted the previous theory, and further is such that it is not *ad hoc* but is rather theoretically progressive. Briefly, they claim that if one forms the settled determination to save a hard core from refutation by finding a theoretically progressive auxiliary hypothesis then one can do so. This they do not establish, however. Clearly, such moves to save a hard core are *often* possible. The story from Lakatos makes this clear. But the claim that is made by Popper and Lakatos is that this sort of thing can be done in *every* case, a much stronger claim. That is, what they need to establish is that *for **any** hard core* T, if data not-E refute it, then *there is* an auxiliary hypothesis A* such that the conjunction (T & A*) of theory and auxiliary assumptions logically entails not-E. This stronger claim is nowhere established.

However, again as we shall see below, there are in fact circumstances in which data will force one to treat the hard core and the auxiliary hypotheses differently. This means that it will be possible in many cases to find data that require one to reject, perhaps not as false, but as improbable, the auxiliary assumptions while at other times there will be data that require one to reject as improbable the hard core.

We see, then, that on this matter there are four important points to be said with regard to the Popper-Lakatos thesis that a "hard core" specific theory can always be saved from falsification by finding the appropriate auxiliary hypothesis.

One. In their discussion, Popper and Lakatos confuse the grounds for a proposition being non-falsifiable. It is one thing to be non-falsifiable because of logical form and another to be non-falsifiable because of a settled determination to find appropriate auxiliary hypotheses.

Two. Beyond the blurring of grounds for a proposition being preserved from rejection, they also claim that anyone can anytime render a proposition immune from rejection by adopting as authoritative a settled determination to save a hard core from falsification. We now see that this is not so, *provided that one accepts the cognitive standards of empirical science.* Of course, if one proceeds in the fashion of Duane Gish, and accepts a hard core on the basis of some **authority**, then of course the result will often be the acceptance of hypotheses that are contrary to what is required by sound scientific practice.

Three. Popper and Lakatos never establish that it *always* possible to devise an auxiliary hypothesis that will save a hard core theory from falsification by explaining away an alleged counter-example.

Four. In practice, we often rely upon data to differentially distinguish between theories and auxiliary hypotheses. This implies that data, even if they do not render an hypothesis rejectable by straight off falsifying it, can nonetheless often render an auxiliary hypothesis, or, conversely, a theory, *more improbable than not*, and therefore, in that somewhat weaker sense, rejectable. Popper and Lakatos in their discussion ignore the fact that **data which do not falsify an hypothesis can nonetheless render it improbable**.

If Popper is confused about falsifiability, it is also true that the simple sort of falsification upon which he bases his argument is very important for science. In particular, it applies to laws at the *specific level*. Further, as we shall now argue, falsification has an important role to play in the scientific method and in the confirmation of hypotheses.

b. Some Paradoxes of Confirmation

There are various apparently paradoxical results association with the confirmation of laws and theories. It will help if we look at a few of these, and try to decide whether they really are paradoxical.

(i) The Paradox of the Ravens

Consider generalizations of the form

(H^)   All F are G ,

that is,

For all individuals x if x is F then x is G.

This is true if and only if all instances of the propositional function

(@)   if x if F then x is G

are true. An instance of this is obtained by putting the name of an individual for the variable 'x'. Thus, if Joe is F then Joe is G

is an instance of the propositional function. This instance will be false if Joe is both F and not G. For example, if (H^) is the proposition that

All swans are white

then the instance

if Joe is a swan then Joe is white

will be false if Joe is a black swan. But if (H^) is

All dogs are mammals

then the instance

if Joe is a dog then Joe is a mammal

is true since Joe is a dog and Joe is a mammal. Thus, an instance of (@) is true if, in that instance, we have

x is F and x is G

while it is false if, in that instance, we have

x is F and x is not-G

At the same time, since the statement

if Joe is a dog then Joe is a mammal

says that "**if** Joe is a dog …", it is also true when Joe is a turtle, and when Joe is a cat. That is, it is true if either

Joe is a cat and Joe is a mammal

or

Joe is a turtle and Joe is not a mammal.

Thus, to summarize, if (H^) is to be true, then all instances of the three forms

x is F & x is G

x is not-F & x is G
x is not-F & x is not-G

must be true while no instances of the form

x is F & x is not-G

can be true. The latter sorts of instance are falsifying instances of the generalization, while the former sorts of instance we may refer to as verifying instances.

The so-called *paradox of the ravens* arises because it seems odd to say that instances of the sorts

x is not-F & x is G
x is not-F & x is not-G

tend to confirm the generalization, tend to make it more probable. Consider the claim that

(h)   All ravens are black
      = For any individual x, if x is a raven then x is black

Now consider my chair (call it "c"). My chair is not a raven, and not black. We therefore have

c is not a raven and c is not black

We therefore have in my chair an instance of the hypothesis (h) which is true. Since it is one of the conditions that must obtain if (h) is to be true, as we have just seen, it follows that this instance should be taken as confirming (h). But how could observing my chair confirm the hypothesis that all ravens are black. How is indoor ornithology possible? It would seem, then, that instances of (H^) of the form

x is not-F & x is not-G,

while they may be necessary for the truth of (H^), can hardly be counted as confirming instances. Similar considerations establish with regard to instances of (H^) of the form

x is not-F & x is G

while also necessary for the truth of (H^), can hardly be counted as confirming instances.

In order to avoid this paradox, Jean Nicod declared[66] that the generalization (H^)

> All F are G
> = For any individual x, if x is F then x is G

is to be considered as confirmed by only some of its verifying instances. In particular, he proposed that we count this generalization to be confirmed by instances of the sort

> x is F & x is G

and falsified by instances of the sort

> x is F & x is not-G

but that instances such as

> x is not-F & x is G

and

> x is not-F & x is not-G

neither confirm nor disconfirm the hypothesis. This has been called the "Nicod Criterion" for being a confirming instance.

There is something intuitively appealing about the Nicod Criterion. Certainly, it seems that it is the observation of black ravens that confirms, makes more probable, the hypothesis that all ravens are black, and not the observation of typewriters, whether black or red. But the proposal lacks any justification beyond the intuition that we ought to avoid the cases that give rise to the apparent paradoxes. If it is to be used without the appearance of being simply *ad hoc*, then some further rationale is needed, something that shows that our intuition is in fact reasonable and not just some unfounded prejudice. But in any case, the Nicod Criterion, for all its intuitive appeal, is not without its problems. For it comes into conflict with some other intuitions that seem to be equally clear.

C. G. Hempel suggested[67] another criterion, the "Equivalence Condition": if two hypotheses are logically equivalent, then data that confirm one of them confirms the other to the same degree. Under this condition, data which confirm (H^)

> All F are G

that is,

> For any individual x, if x is F then x is G

must equally confirm its logically equivalent contrapositive

> (H^^) All non-G are non-F,

that is,

> For any individual x, if x is non-G then x is non-F

But, by the Nicod Criterion, while an instance of the form

> x is F & x is G

confirms the (H^), it does not confirm (H^^). The Nicod Criterion and the Equivalence Conditions thus conflict.

The same conflict has been expressed in other ways. Take "All F are G" to be

> All ravens are black,

and consider the observational datum that

> This magnifying glass is a non-black non-raven.

Then this datum, by the Nicod Criterion, confirms the hypothesis that

> All non-black things are non-ravens,

and in turn, by the Equivalence Condition, confirms the hypothesis that

> All ravens are black.

Most have found something odd, if not paradoxical, about the idea of indoor ornithology, the idea of confirming that all ravens are black by looking over the items on one's desk. We are back to the paradox of the ravens.

There have been a number of suggestions about how this "paradox," if it be such, is to be resolved. A number of them turn on the same sort of suggestion that Johnson made in the case of material implication, namely, to look at the role of the apparently paradoxical proposition in inferences.

An hypothesis in science must be one that yields a *prediction*, or, more generally, and what amounts epistemically to the same, an inference from the known to the unknown. It is precisely this that is important in confirmation: success in predictions is what renders an hypothesis acceptable. The important point to notice is that not *every* instance can yield a prediction. Consider again the case of hypotheses like (H^)

All F are G

What we must do is distinguish between instances of this hypothesis which are among its truth conditions and those which confirm it: not all the former are among the latter. Instances of the form

x is not-F & x is G

and of the form

x is not-F & x is not-G

are verifying instances, that is, instances that are among its truth conditions, but, since neither can be used in inferences which are *predictions*, it would seem that neither sort of instance can be a confirming instance. They cannot be used in predictions because arguments of the form

For any individual x, if x is F then x is G
x is not-F
_____
∴ x is G

and of the form

For any individual x, if x is F then x is G
x is not-F
_____
∴ x is not-G

are both invalid, both cases of the fallacious form of denying the antecedent.

The only verifying instances which could be used predictively, and therefore be counted as confirming instances, are instances of the form

x is F & x is G

For, only verifying instances of *this* sort can appear in valid argument forms:

For any individual x, if x is F then x is G
x is F
_____
∴ x is G

We can now see why Nicod would have concentrated on instances of this form, and argued that only verifying instances of this form are confirming instances.

We must note, however, that what makes verifying instances of this form confirming instances is an *epistemic condition*. An instance of the form

x is F & x is G

is a confirming instance only if it can be *used predictively*. If we are to use

All F are G
= For any individual x, if x is F then x is G

predictively, in the argument

For any individual x, if x is F then x is G
x is F
_____
∴ x is G

then we must (1) *know* that the minor premise 'x is F' is true and (2) *not know* that the conclusion 'x is G' is true.

It follows that although some verifying instances of the form

$$x \text{ is } F \ \& \ x \text{ is } G$$

are confirming instances — those that are used in successful predictions — other instances of that form are *not* confirming instances. If 'x is F' and 'x is G' are *both* antecedently known, so that the epistemic conditions (1) and (2) are not simultaneously fulfilled, then we have an instance which is verifying but not confirming.

A parallel argument shows that some, but not all instances of the form

$$x \text{ is not-G} \ \& \ x \text{ is not-F}$$

are confirming instances of the hypothesis that "All F are G."

This fits in with the suggestion of the Popperians that what is crucial is the *possibility* of falsification: an hypothesis H is confirmed, or, in their technical terminology, "corroborated," just in case that it is used in an *inference* from the known to the unknown, an inference that *puts it at risk*, or, rather puts its affirmation at risk, leaves it open that it might turn out to be false.[68] An hypothesis that successfully passes such a *test* is said to be corroborated. The epistemic conditions on an argument being useful in an inference can thus be used to distinguish those cases in which a verifying instance is also a confirming (or corroborating) instance.[69]

Moreover, unlike the proposal of Nicod, this proposal of the Popperians is not merely *ad hoc*, introduced solely to avoid the paradoxes. This proposal has a clear rationale, since it depends upon taking seriously the notion of a *prediction*, something which has an independent justification.

We may therefore conclude that the so-called "paradox of the ravens" and the possibility of indoor ornithology disappears once one invokes the notion of a prediction and the epistemic conditions which that implies.

(ii) The Transitivity Paradoxes

We have noted that observational data that confirm a specific hypothesis within the context of research that is guided by an abstract generic theory will provide additional confirmation for the background abstractive generic theory that guides the research and renders hypotheses worthy of test.

It has been suggested by C. G. Hempel that the confirming power of these data is carried from the specific level to the more generic level by two principles, referred to as the "Consequence Condition" and the "Converse Consequence Condition" of confirmation.[70] The Consequence Condition states that

> If evidence E confirms H and H entails H', then E confirms H'

The Converse Consequence Condition states that

> If evidence E confirms H and H' entails H, then E confirms H

Kepler's Laws form an example of the former. These laws are entailed by Newton's Theory of the solar system. The data that supports Newton's theory therefore supports the hypothesis that the orbit of an hitherto unobserved planet will be elliptical, as required by Kepler's Laws. This support for the hypothesis about the orbit of the unobserved planet is the support that is there for Newton's theory. This support is transmitted by the Consequence Condition to the hypothesis.

As for the Converse Consequence Condition, if we do observe the positions of the new planet when we come to observe it, and those data support the claim that the orbit is elliptical, then those same data lend further support to Newton's theory which entails that the orbit must have the Keplerian shape. This support is transmitted by the Converse Consequence Condition to the hypothesis.

The examples give some plausibility to the two conditions. Unfortunately, if these conditions are taken to hold in full generality, then the so-called "Paradox of Transitivity" results.

Let suppose that T entails H:

$$T ==> H^{71}$$

Suppose, further, that E confirms T. Then E, by the Consequence Condition, confirms H. Now let K be any other hypothesis. Now, it is also true that

$$H ==> (H \text{ or } K)$$

and that

$$K ==> (H \text{ or } K)$$

Since E confirms H, it also, by the Consequence Condition, confirms (H or K). But then, by the Converse Consequence Condition, it also confirms K. However, K is any hypothesis whatsoever. Hence, evidence E, if it confirms any hypothesis confirms every hypothesis. This is indeed paradoxical.

Something has to go. The consequence or entailment relation and its converse cannot be used indiscriminately to transitively transmit confirmational support, however plausible both the Consequent and Converse Consequence Conditions may be made by means of examples.

The notion of entailment is such that, if T entails H, then the truth of T guarantees the truth of H. Hence any data that testify to the truth of T at least as strongly testify to the truth of H. Confirming power is thus clearly transmitted by the Consequence Condition. It follows that if we are to avoid the paradox at which we just arrived, then it is the Converse Consequence Condition that must go.

Nonetheless, by examples such as those that we have given, it would seem that there is some plausibility to that Condition. In some cases, at least, it would seem that we ought to accept the Converse Consequence Condition. Or rather, we ought not so much accept the condition as recognize that *in some cases* where E supports an hypothesis H it also supports a theory T which entails H. What we have to discover are the special conditions where this holds. Perhaps not surprisingly, it turns out that were the relationship is that which obtains in the context of the mechanisms of elimination, then data which support the specific hypotheses also support the more generic hypotheses.

The important point is one that we have already noted on several occasions, that a generic theory or paradigm which guides research — unlike the specific laws and other hypotheses which it testifies are worthy of being put to the test — cannot, on the one hand, be falsified by particular data, it nonetheless, on the other hand, *can be confirmed* by such data.

This evidential support which it derives from guiding us to discover laws which are then confirmed in experience is transmitted, as it seems, via the consequence and converse consequence relations. The data that support the specific hypothesis

(a)   For any individual x, if x is F then x is G

also support the gappy law

(b)   There is an f such that f is **F** and such that for any individual x, if x is f then x is G

since the latter, assuming that

F is **F**

is entailed by the former. We thus have confirmation transmitted via the consequence relation. But (b) is predicted by the abstract generic theory

(d)   For all g such that g is **G**, here is an f such that f is **F** and such that for any individual x, if x is f then x is g

(b) is predicted by (d) because the latter entails the former — given the assumption that

G is **G**

Since (b) is predicted successfully, that prediction supports (d). Thus, the data that support (b) also supports (d). We thus have confirmation transmitted via the converse consequence relation. It is clear that it is the specific forms of these relations that obtain in these cases that provides the rationale that enables these relations to transmit confirming evidence, something which they cannot in general do.

There is another way in which these confirmation relations work. The determinate prediction that "x is F" or "x is G" entails the determinable prediction that "x is **F**-ed" or "x is **G**-ed." Thus, if one has the initial condition

a is F

then one also has

a is **F**-ed

But if one has the latter, then this and (d), assuming alternatives to (a) have been eliminated, yields the prediction

a is **G**-ed

If the prediction using the specific law (a) is successful, then there is automatically also a successful prediction using the generic law (d). *A confirmation of the law is eo ipso a confirmation of the generic law.*

These specific forms of the converse consequence relation have to do with the specific/generic relations that obtain among (a), (b), and (c). It is also these relations that, as we have seen, provide the element of coherence by which evidence for one specific law if transformed by a shared generic theory into evidence that supports other, different specific laws. The point here is that the paradox of transitivity occurs only because the converse consequence relation is allowed *unrestrictedly* to transmit evidential support. Once it is restricted to the mentioned cases, where the relation is one of specific to generic, then the problem does not arise. But we also see that it is not the converse consequence relation as such that transmit evidential support but rather the fact that laws and theories are connected by species/genus relations.

c. The Experimental Methods[72]

In this section, we examine in greater detail than hitherto the role of experiment in providing data that confirm hypotheses and render them acceptable for purposes of explanation and prediction. As it turns out — and in this we are likely not surprised — falsification, that is, elimination plays a crucial role. But it can play a crucial role, we shall see, only if research is guided by the sorts of theoretical considerations advanced earlier by Bacon and John Stuart Mill and more recently by Hanson and Kuhn, and which we examined above. The patterns were shown to fit with 20th century developments in formal logic by three Cambridge philosophers, W.E. Johnson, John Maynard Keynes and C.D. Broad. In particular, we shall see that the confirmation of specific laws leads to the confirmation of the theories that guided the research.

We also discussed above some of the patterns of confirmation which occur when research is theory-guided. These we referred to as the consequence and converse consequence conditions for confirmation. We shall see in the present section that these principles cannot be accepted in full generality if we wish to avoid paradox. But in coming to that conclusion, we shall also see that they do have a definite, if limited, range of application. The relevant restrictions turn out to hinge on the logical structure of abstract generic theories or what Kuhn called paradigms.

(α) Method of Difference

Suppose that we have a certain property G where we have a cognitive interest in discovering its **sufficient condition**. That is, we hope to discover some law of the form

for all x, if x is F then x is G [= All F are G]

G is said to be the *conditioned property*. The property F is the *conditioning property*. The research problem is to find the conditioning property from among the set of *possible conditioning properties*. Suppose that the set of possible conditioning properties consists of the following:

$$F_1, F_2, F_3, F_4$$

This means, in effect, that we have the following four *hypotheses* about the sufficient condition of G:

for all x, if x is F_1 then x is G
for all x, if x is F_2 then x is G
for all x, if x is F_3 then x is G
for all x, if x is F_4 then x is G

What we have to do is obtain *data* that will *decide among* these *competing hypotheses*. More specifically, what we have to do is find data that will, on the one hand, *confirm one of the hypotheses as that which correctly locates the property that is the sufficient condition* for G, and, on the other hand, *eliminate those hypotheses which are false*.

To eliminate an hypothesis as false, one must, of course, find an instance which falsifies it. If we consider the first of the above alternatives,

for all x, if x is F_1 then x is G

then an instance

b

such that

b is F_1

is true and

b is G

is false would eliminate this alternative. The task is to find instances that eliminate all but one of the competing alternatives. At the same time, we must also establish that the uneliminated hypothesis is not vacuous: we must find at least one positive confirming instance.

The following set of data would do this job, eliminating all but the fourth hypothesis and confirming its truth.

	G	F_1	F_2	F_3	F_4
Table 4.1					
a	t	f	t	t	t
b	f	t	t	f	f
c	f	f	t	f	f
d	f	f	f	t	f

Here, we have represented in the first row the instance or event "a". When a "t" occurs in the "a" row under "G", it indicates that G is present in a, that is, that 'a is G' is true. Similarly, when an "f" occurs in the "d" row under "F_1", it indicates that F_1 is absent from the individual or event d, that is, that 'd is F_1' is false.

Since the second row indicates that we have

b is F_1 and b is not G

it follows that the first of the alternative hypotheses is eliminated as false. It also indicates that we have

b is F_2 and b is not G

which falsifies the second of the alternative hypotheses; so this hypothesis is also eliminated as false. Similarly, event c eliminates the second hypothesis. Finally, event d eliminates the third hypothesis. The only hypothesis that is uneliminated is the fourth.

But event a has

a is F_4 and a is G

which confirms the fourth hypothesis. Thus, the hypothesis

for all x, if x is F_4 then x is G

alone among the possible alternatives is both uneliminated and confirmed. We therefore accept this hypothesis as stating correctly, so far at least as we know, what is the sufficient condition for G. We conclude, in other words, that this hypothesis solves our puzzle about the sufficient condition for G, that it satisfies our cognitive interest.

John Stuart Mill describes this method as follows:

If an instance in which the phenomenon under investigation occurs, and an instance in which it does not occur, have every circumstance in common save one, that one occurring only in the former; the circumstance in which alone the two differ is the effect, or the cause, or an indispensable part of the cause, of the phenomenon.[73]

This makes clear why he referred to this method as the "method of difference."

Mill himself did not use the terminology of "sufficient conditions." This appears in his mode of expression. Depending upon the context, a sufficient condition may be either a cause, in the ordinary usage of that term, or an effect, again in the ordinary usage of that term. So, depending upon the context, the method of difference will detect a cause or an effect, as Mill says. As for the possibility of being an essential part of a cause, Mill is here referring to complex causes, that is, sufficient conditions that are conjunctions of simpler conditions. We shall discuss the methods of elimination with regard to complex conditions in greater detail below.

Here is an example of the method of difference.

Scurvy was a disease that created serious problems in the 18th century for sailors in the British navy on long voyages. It was discovered that among a variety of possible remedies that were tried, drinking lime juice alone succeeded in preventing scurvy. In a population of sailors on long voyages, the difference between those who were victims of scurvy and those who were not lay in the fact that the latter drank lime juice. The method of difference therefore leads to the conclusion that drinking lime juice is a *sufficient condition* for the absence of scurvy.

In the light of this information about how scurvy could be prevented, the British navy made it policy for its sailors to be provided with limes on long voyages. The scourge of scurvy was eliminated.[74]

It should be noted that in certain cases a single experiment can detect a sufficient condition. To use John Stuart Mill's example, if a person is shot through the heart and dies, then, since all other conditions remain unchanged, we can immediately infer that the cause of his death, the sufficient condition for it, was his being shot. If we let "G" be the property of being alive, and "F" the condition of being shot through the heart, then the following table shows how this inference goes:

	G	F	other conditions
before shot	t	f	t
after shot	f	t	t

Table 4.2

This makes clear that F is the only way in which the person differs between the item when he is alive and the time when he is not alive. We conclude that F is sufficient for G: for all x, if x is F then x is G.

The use of **control groups** is an instance of the use of the method of difference.

In a brief history of the notion of a control group,[75] Richard Solomon notes that

The earliest use of this kind of experimental design [one involving a control group] in the study of transfer of training is found in the work of Thorndyke and Woodworth,[76] published in 1901. These investigators used independent educational groups as controls on other groups. They were studying the influence of various educational sequences on proficiency in various subjects. In their observations a sequence of training such as **ABC** could be evaluated against a control group with sequence **ADC**. They attributed any differences in performance in **C** in terms of differential effects of **B** and **D**.[77]

Here we consider a population **A**, and two different **C**-type outcomes. These are, say, "high score" (C_1) and "low score" (C_2), and, since the two are mutually exclusive we have in effect that C_1 = not-C_2. Some of **A** are given

training **B**, while others are not, so that **D** = not-**B**. If we have the following sorts of result then we can conclude that the training **B** was in fact efficacious in bringing about learning, that is, the high score C_1.

	B	C_2	other conditions
tested	t	f	t
control	f	t	t

Table 4.3

In the control group, without the training **B**, we find the result C_2, that is, low scores. The tested group, in contrast, were given **B**, and here we find that C_2 is absent, or, what is the same, that not-C_1 is absent, or, again what is the same, that C_1 is present, i.e., high scores. From this experiment, by means of the method of difference, we can conclude that administering the training **B** is sufficient to produce high scores.

The logic of the **method of difference** for the discovery of **sufficient conditions** consists primarily of the elimination of all but one of several possible hypotheses. It proceeds on the basis of the principle that

a property that is present when the conditioned property is absent cannot be a sufficient condition for that property.

What this rule makes clear is that the elimination proceeds by falsification. It thus embodies Popper's point that falsification by counter-examples plays a crucial role in the method of scientific inquiry. But in providing us with this insight, Popper was following in the footsteps of Bacon and John Stuart Mill.

The logic of falsification is clear. We have an *hypothesis* at the *specific level*, that is, of the logical form

All F are G = for all x, if x is F then x is G

This is falsified by a counter-example in which the antecedent is present and the consequent absent:

a is F and a is not G

This logical point about the mechanisms of elimination apply equally to the other methods of eliminative induction to which we shall now turn.

(β) *Method of Agreement*

Suppose that we have a certain property G where we have a cognitive interest in discovering its **necessary condition**. That is, we hope to discover some law of the form

for all x, if x is G then x is F [= All G are F]

Again, G is said to be the *conditioned property*, while the property F is the *conditioning property*; the research problem is to find the conditioning property from among the set of *possible conditioning properties*. Suppose that the set of possible conditioning properties consists of the following:

F_1, F_2, F_3, F_4

This means, in effect, that we have the following four *hypotheses* about the necessary condition of G:

for all x, if x is G then x is F_1
for all x, if x is G then x is F_2
for all x, if x is G then x is F_3
for all x, if x is G then x is F_4

What we have to do is obtain *data* that will *decide among* these *competing hypotheses*. More specifically, what we have to do is find data that will, on the one hand, *confirm one of the hypotheses as that which correctly locates the property that is the necessary condition for G*, and, on the other hand, *eliminate those hypotheses which are false*.

To eliminate an hypothesis as false, one must, of course, find an instance which falsifies it. If we consider the first of the above alternatives,

for all x, if x is G then x is F_1

then an instance

b

such that

b is G

is true and

b is F_1

is false would eliminate this alternative. The task is to find instances that eliminate all but one of the competing alternatives. At the same time, we must also establish that the uneliminated hypothesis is not vacuous: we must find at least one positive confirming instance.

The following set of data would do this job, eliminating all but the fourth hypothesis and confirming its truth.

Table 4.4					
	G	F_1	F_2	F_3	F_4
a	t	f	t	t	t
		eliminates 1st alternative hypothesis			
b	t	f	t	f	t
		eliminates 1st and 3rd alternative hypotheses			
c	t	t	t	f	t
		eliminates 3rd alternative hypothesis			
d	t	f	f	t	t
		eliminates 1st and 2nd alternative hypotheses			

Here, we have represented in the first row the instance or event "a". As before, when a "t" occurs in the "a" row under "G", it indicates that G is present in a, that is, that 'Ga' is true. Similarly, when an "f" occurs in the "d" row under "F_1", it indicates that F_1 is absent in individual or event d, that is, that 'd is F_1' is false.

Since the first row indicates that we have

a is G a is not F_1

it follows that the first of the alternative hypotheses is eliminated.

Since the second row indicates that we have

[b is G & b is not F_1] & [b is G & b is not F_3]

it follows that the first and third of the alternative hypotheses are eliminated as false. Similarly, event c also eliminates the third hypothesis. Finally, event d eliminates the third hypothesis. The only hypothesis that is uneliminated is the fourth.

But event a has

a is G and a is not F_1

and similarly for all the other events. Each of the events therefore confirms the fourth hypothesis. Thus, the hypothesis

for all x, if x is G then x is F_4

alone among the possible alternatives is both uneliminated and confirmed. We therefore accept this hypothesis as stating correctly, so far at least as we know, what is the necessary condition for G. We conclude, in other words, that this hypothesis solves our puzzle about the necessary condition for G, that it satisfies our cognitive interest.

John Stuart Mill describes this method as follows:

If two or more instances of the phenomenon under investigation have only one circumstance in common, the circumstance in which alone all the instances agree is the cause (or effect) of the given phenomenon.[78]

Again, we note that Mill does not use the language of necessary, sufficient and necessary and sufficient conditions. The method of agreement picks out necessary conditions, but these are sometimes in the ordinary discourse of causation characterized as effects, sometimes as causes.

At the same time, however, Mill does tend to think of causes as for the most part sufficient conditions. He is therefore led to make the suggestion that the method of agreement is not as powerful as the method of difference. In this, there is a sense in which he is correct: the method of agreement will not pick out sufficient conditions where the method of difference will. At the same time, however, he is strictly speaking wrong. For, the method of agreement is just as conclusive in discovering necessary conditions as the method of difference is in discovering sufficient conditions.

The logic of the **method of agreement** for the discovery of **necessary conditions** consists primarily of the elimination of all but one of several possible hypotheses. It proceeds on the basis of the principle that

a property that is absent when the conditioned property is present cannot be a necessary condition for that property.

What this rule makes clear is that the elimination proceeds by falsification. This method thus also embodies Popper's point that falsification by counter-examples.

Here is an example of the method of agreement, adapted from John Stuart Mill. It is observed that all objects which have a crystalline structure have one, and only one factor in common, to wit the fact that they have solidified from a fluid state. Solidifying from a fluid state is thus a necessary condition for having a crystalline structure.

Note, however, that there are substances which solidify from a fluid state but which do not have a crystalline structure. Glass is such a substance. Thus, while solidifying from a fluid state is a necessary condition for having a crystalline structure, it is not a sufficient condition.

(γ) The Role of Theory

The methods of elimination have been criticized by methodologists such as Cohen and Nagel as incapable of discovering any scientific laws. Thus, they object to the method of agreement, arguing that it cannot effect a conclusion that some factor among those investigated is a necessary condition unless the cause of the phenomenon is in fact included in the list of alternatives. Unless the list of conditioning properties is *exhaustive of all alternatives*, the eliminative mechanisms will not yield a conclusion with regard to necessary conditions. They make a similar objection to the method of difference. The latter method requires that the phenomena be alike in respect of all possible conditioning properties save one, which is then reckoned to be the sufficient condition of the conditioned property. But no two phenomena are ever exactly alike, so the list of characteristics which the phenomena have in common must be restricted to those which are *relevant*. Moreover, one could never examine all characteristics, given that there are infinitely many; the set of relevant characteristics could never be exhausted and must therefore be *limited*.

Cohen and Nagel comment that

> ...[the methods of agreement and of difference require] the antecedent formulation of a hypothesis concerning the possible relevant factors. [They] cannot tell us what factors should be selected for study from the innumerable circumstances present. And [they require] that the circumstances shall have been properly analyzed and separated. We must conclude that [they are] not a method of the discovery.[79]

In this they are quite correct. The logic of elimination by itself does not establish the set of relevant conditioning properties on which the mechanisms of elimination work. But no one ever claimed that they did. Not even Popper so claimed, in spite of the fact that he emphasized the mechanisms of elimination, of falsification, for, as we know, he insists that the use of elimination must be guided by metaphysical "points of view." The issue is rather precisely how does science delimit the set of possible relevant conditioning properties. But the answer is clear: they are delimited by Popperian "points of view," or what we have seen can more adequately characterized as *abstract generic theories*, or what Kuhn has called "paradigms."

Consider the example of the method of difference given above. There we considered the conditioned property G and four conditioning properties

$$F_1, F_2, F_3, F_4$$

This meant that we had the following four *hypotheses* about the sufficient condition of G:

(!)
for all x, if x is F_1 then x is G
for all x, if x is F_2 then x is G
for all x, if x is F_3 then x is G
for all x, if x is F_4 then x is G

The research task was to obtain data that would decide among these *competing hypotheses*. More specifically, we had to find data that would, on the one hand, *confirm one of the hypotheses as that which correctly locates the property that is the sufficient condition for G*, and, on the other hand, *eliminate those hypotheses which are false*.

Cohen and Nagel are quite correct: unless we know something besides observational data of the sort just described, it is not possible to conclude that the uneliminated hypothesis does in fact describe a sufficient condition for the conditioned property G. Specifically, we can *conclude deductively* that the uneliminated hypothesis is true if and only if we know that

(P1)   there is exactly one factor which is a sufficient condition for the conditioned property G

and that

(P2)   the one factor which is the sufficient condition is among the set of conditioning properties F_1, F_2, F_3, F_4

Principle (P1) asserts that **there is a cause**, and has been called a **Principle of Determinism**. Principle (P2) asserts that **the cause is one factor out of a delimited set of possible factors**, and has been called a **Principle of Limited Variety**. If (P1) is false, then none of (!) will be true, and to eliminate all but one will not ensure that the one that remains is true. Similarly, if (P2) is false, then again to eliminate all but one will not ensure that the one that remains is true.

The point is similar to that which applies to the use of disjunctive syllogism. If we have

H or K
not H
_____
so,   K

then the conclusion will follow only if the premises are true, and in particular only if the disjunctive premise is true. It may be false because no alternative is true. This corresponds to our needing (P1) to justify the inference in the case of the method of difference. Or, the disjunctive premise may be false because there is in fact an alternative that has been omitted: instead of

H or K

we should have instead

H or K or J

so that we cannot conclude K unless we have eliminated not only H but also the unmentioned alternative J. This corresponds to our needing (P2) to justify the inference in the case of the method of difference. We need to know that the range of alternatives is given [(P2)] and that one from among this range is true [(P1)].

It is evident that the two principles of Determinism and Limited Variety are *matter-of-fact generalizations*. If we take them together, then they jointly assert what we earlier represented by

(d)   there is a unique species f such that f is of genus **F** and such that for all x, x is f if and only if x is G

where we take it that "**F**" provides a generic characterization of the set of relevant conditioning properties.

f is **F** if and only [f is F_1 or f is F_2 or f is F_3 or f is F_4]

(d) is of course an *existential hypothesis* of a sort that we have seen is used to guide research in new areas of investigation during the process of scientific inquiry. As Kuhn and Hanson and Popper have all indicated, research proceeds on the basis of a knowledge or belief that *there is* an hypothesis that will eventually be discovered to be true and that this hypothesis will be of a certain *generic form*. What (d) asserts is precisely this: there is a sufficient condition, there to be discovered, and that this sufficient condition occurs within a delimited range.

Where does a law such as (d) come from? As we have suggested in our discussion of the way in which theory guides research during the process of inquiry, following Hanson and Kuhn, a law such as (d) is usually derived from some background abstract generic theory such as

(c)   For any species f, if f is of genus **F** then there is a unique species g such that g is of genus

G and such that for all x, x is g if and only if x is f

where we have the genus-species relationship

G is **G**

Background theory thus provides the relevant Principle of Determinism and relevant Principle of Limited Variety that must be assumed if the eliminative method of difference is to lead to the discovery of laws concerning sufficient conditions.

It is evident that a similar point holds with respect to the eliminative method of agreement and the discovery of necessary conditions.

In general, then, we see that

there are abstract generic theories that guide research and which can provide during the process of inquiry the Principles of Determinism and Limited Variety that are necessary if the methods of elimination are to succeed in locating or discovering sufficient conditions or necessary conditions (or necessary and sufficient conditions).

Cohen and Nagel do notice that analogies are relevant to delimiting a range of acceptable hypotheses from which research or inquiry is to eliminate all but the one that is true. They also recognize that the knowledge of the relevant analogies can derive from background theories. What they do not recognize is that precisely these analogical constraints on hypotheses deriving from background theory can also provide the principles that are needed if the methods of elimination are to be useful tools in the process of scientific discovery. As a result, they disparage the utility of the methods. We now see that their criticism is wrong, as, indeed, John Stuart Mill already knew.

NOTE: *We shall now turn to some further, more complicated uses of the methods of elimination. For each such method, if it is to lead to the discovery of a law, then an appropriate Principle of Determinism and appropriate Principle of Limited Variety needs to be assumed, just as for the simpler case of the* *methods of difference and of agreement at which we have just looked.*

Throughout the subsequent discussion of the different methods of elimination, we shall take for granted that we have ready for use the relevant Principle of Determinism and the relevant Principle of Limited Variety.

(δ) The Fallacious Method of Simplistic Induction

We have encountered the argument in Lysenko's biology and in alternative forms of medicine that this or that theory should be accepted because "**it works**." We have seen that this methodological rule is not one that can safely be relied upon. There is, of course, no guarantee in science. As Judge Overton emphasized, science is always tentative — unlike theology, whether of the creation science variety or of the dialectical materialist variety. What we should do is try to locate more carefully why it is an unsafe method.

The method of simplistic induction is often used to support what is in fact mere prejudice. Prejudice consists of a general belief, such as the belief that all Irishmen lack wit, which one will affirm even if one happens to know a few intelligent Irishmen. Such beliefs are rooted in the same natural tendencies of the human mind that give rise to justified casual belief; specifically, they are habits of inference caused by observed constant conjunctions. Habits sufficiently strong to maintain themselves in the face of counter-examples arise in certain specific sorts of situations. The crucial point is that phenomena are complex. We have cases where, one, (events of type) A cause(s) (events of type) B, but where also, two, most usually, though not invariably, or even always, but not essentially, A is accompanied by A'. In that case a habit of inferring B from A' will arise, even though, objectively, A' is not invariably connected with B. Such a habit will be strong enough to maintain itself in the face of — at least a few — contrary examples.

The rules of prejudice clearly cannot satisfy the passion of curiosity; the latter, aiming at knowledge of matter-of-fact regularities, cannot rest content with inference habits that are contradicted by experience.

However, the same sort of experience that gives rise to the law-assertions of prejudice may also bring about law-assertions that are not, in the normal course of events, contradicted by experience. The observations and inferences are like this:

(I)   all observed A are B,

which causes us to law-assert

(i)   Whenever A then B

like this:

(II)   all observed A' are B,

which causes us to law-assert

(ii)   Whenever A' then B

and like this:

(III)   all observed A have been accompanied by A' and conversely.

which causes us to law-assert

(iii)   A if and only if A'

Now, it might well turn out that instances unobserved by us make it objectively the case that only (i) is true; (ii) and (iii) are both, objectively, false. Only, in order to discover this one must go outside the narrow confines of the experiences of our quotidian life. That is, in our normal and everyday experience, (ii) and (iii) do hold. But if the researcher sets out deliberately to put these things to the test, then he or she will discover that one can indeed obtain objects that are A' but neither A nor B.

The person who asserts (i), (ii), and (iii) as laws worthy for use in explanation and prediction on the basis of the data (I), (II), and (III) that he or she has available is inferring in accordance with a "rule by which to judge of causes and effects," to use the phrase of David Hume, though it is not, of course, one of the rules that Hume was proposing to defend. The rule is what Bacon called the "childish" rule of *induction by simple enumeration*: assert as a law worthy for use in explanation and prediction any generality that holds in casual experience.

However, this rule will not lead to law-assertions that satisfy our curiosity. For, the regularities that it leads us to accept for purposes of explanation and prediction regularly fail to withstand critical experience. That is, we know by experience that where there are many factors involved, where the phenomena are complex, this rule leads to beliefs that will not withstand the test of a broader experience. It leads to judgments that are simply unsafe, easily liable to turn out to be false. These complicating circumstances must be eliminated before any regularity can safely be accepted. (And even then there is no guarantee.) In contrast to the method of simplistic induction, the rules of the experimental method, and specifically the Methods of Agreement and of Difference, require us to use, in such cases, not only data that confirm the hypothesis in question, but also data that *eliminate* or falsify, possible competitors. (Popper has, of course, more recently emphasized this Baconian point[80] that sound scientific inference requires not only confirmation but also falsification.) These rules, we find, lead to judgments that are both more general and less liable to being falsified in the future than those to which the method of simplistic induction leads. Thus, on the basis of experience we come to the conclusion that the rule of induction by simple enumeration cannot lead, save accidentally, to the acceptance of a regularity that might stand the test of the most critical examination, and which therefore cannot be acceptable to one committed to the cognitive standards of science.

In effect, the method of simplistic induction is what earlier logicians had referred to as the fallacy of *post hoc, ergo propter hoc*. It is the fallacy of using the rule of induction by enumeration, and it is a *fallacy* because we know that it leads to beliefs not of the sort that we desire, beliefs that will not stand the test of experience.

But then once again, the question arises: if experience tells us that such a rule does not lead to beliefs satisfactory to the natural passion of curiosity, then why do people still, as they clearly do, make inferences in conformity to it?

Return to our researcher, and suppose that she has accumulated data that eliminate (ii). With this data available, she can use the Method of Agreement to justify her acceptance of (i) for purposes of explanation of prediction. But this conclusion follows only if we also accept a Principle of Determinism and a Principle of Limited Variety in the area. We thus use this general rule to correct another general rule, namely, (ii) which is, given the data, contrary to it. Nonetheless, we *have* observed data that, by the natural workings of the mind, tend to cause us to accept (ii). Even after its correction, that tendency is there. There is, therefore, an opposition between the sound inference, and the fallacious one.

The point to be emphasized is that it is *natural to drift* into committing the fallacy of *post hoc, ergo propter hoc*. Moreover, there are passions other than the cognitive interest of curiosity that are relevant. These can function to generate and maintain beliefs that fail to satisfy the passion of curiosity. Anger, contempt, envy, flattery, satire, honour, etc. are all relevant. And in fact, "there are many things, in which the world wishes to be deceiv'd."[81] These other passions account for why we draw inferences by methods incapable of satisfying our natural curiosity, and for why we continue to accept the prejudiced beliefs we have even in the face of evident counter-examples. The mind indeed cannot rest content to stay within the realm of our quotidian experience, but it is not curiosity alone that leads him outside that circle. If it were, then means, i.e., rules of inference, would be adopted that would best serve the end of curiosity. Other passions are at work also, and it is these that generate the bold systems of superstition, such as those of Lysenko, creation scientists, Velikovsky, von Däniken, and so on, that subvert the aim of curiosity, that is, our cognitive interest in matter-of-fact truth, and ensure that we can never have any steady principles that will have the reasonable prospect of standing the test of future experience and careful scrutiny.

The same remarks extend to Pat Moffat who has written in Toronto's national newspaper, *The Globe and Mail*, about the form of alternative medicine performed at the Toronto Siatsu Centre by Tetsuro (Ted) Saito. Mr. Saito bases his practice on the theories of traditional Chinese medicine. These all involve such non-empirical concepts as *chi* ("Qi") and so cannot be put to the test; they fail the criterion for good scientific hypotheses. No matter.

…since they get results at the centre, I assume there's valdity in their underlying philosophy. It has worked for me.[82]

Very well. Let us grant that Ms. Moffat has had things done to her at the centre and that in each case the doing of these things has been followed by the disappearance of some dis-ease that was afflicting her. The question is, the question that she does not ask, but which would be asked by anyone with an interest in the truth, *why* has it worked?

From the fact, if it be such, that the procedures worked in the sense of being followed by relief from dis-ease, she simply cannot conclude that there is "validity in the underlying philosophy." Given that the underlying philosophy involves theories that cannot be put to any empirical test, it is not clear how it every could be discovered to have any validity of any sort. That is not a small issue, but put it to one side. The point is that "it works" does not justify an hypothesis concerning causal mechanisms unless alternative hypotheses have been eliminated. To mention only the most obvious one in this case, since Ms. Moffat seems to have a fairly robust capacity to believe, maybe it is in fact nothing else than the placebo effect. It may be more than that, maybe she is correct in so supposing. But she has not provided any evidence that it is not such an effect. And until she has done so, and until she has eliminated other alternatives, she is totally wrong in supposing that the fact that the treatments at the Shiatsu Centre relieve the dis-ease of patients provides support for the theories upon which they are based. Her unquestioned confidence that "there's validity in [the Centre's and Mr. Saito's] underlying philosophy" is totally misplaced. That is, it is so if she is a rational person, one with a genuine interest in the truth.

(ζ) The Joint Method of Agreement and Difference

The method of difference yields sufficient conditions. The method of agreement yields necessary conditions. To obtain conditions which are both necessary and sufficient, it is necessary to apply both methods simultaneously. The result is **the joint method of agreement and difference**.

Again suppose that G is the conditioned property, that property for which one is searching for a law which connects it to another condition which is both necessary and sufficient. Let the set of possible conditioning properties be, as before,

$$F_1, F_2, F_3, F_4$$

This means that we are concerned to discover which among the following hypotheses is true.

for all x, x is F_1 if and only if x is G
for all x, x is F_2 if and only if x is G
for all x, x is F_3 if and only if x is G
for all x, x is F_4 if and only if x is G

Assuming a Principle of Determinism and a Principle of Limited Variety, what we want to do is obtain data that will eliminate all but one of these hypotheses. This process will lead to the discovery of a condition that is necessary and sufficient for G. The data in the following table do this.

	G	F_1	F_2	F_3	F_4
a	t	t	f	t	f
b	f	t	f	f	f
c	t	f	t	t	t

Table 4.5

Consider events a and b. Event a shows that only F_1 and F_3 could be sufficient for G, since a sufficient condition for G must be present when G is present. But it cannot be present when G is absent. In event b, F_1 is present when G is absent. It follows that event a has eliminated this as a sufficient condition. It follows that since F_3 is uneliminated, it is the sufficient condition for G. In coming to this conclusion, we have used the method of difference. Now apply the method of agreement. Event a eliminates F_2 and F_4 as necessary conditions, since they are absent when G is present. And event c eliminates F_1 as a necessary condition for G. So all events save F_3 are eliminated as necessary conditions for G. It follows by the method of agreement that F_3 is the necessary condition for G. From these two conclusions it follows that F_3 is both necessary and sufficient for G.

Here is John Stuart Mill's formulation of the joint method of agreement and difference:

If two or more instances in which the phenomenon occurs have only one circumstance in common, while two or more instances in which it does not occur have nothing in common save the absence of that circumstance, the circumstance in which alone the two sets of instances differ is the effect, or the cause, of an indispensable part of the cause, of the phenomenon.[83]

It is sometimes objected that this method is not a separate method since it merely combines the methods of agreement and difference.[84] That it combines the two methods is certainly correct, but it does differ in that it picks out conditions that are both necessary and sufficient. Since the logical form of the conclusion is different from the logical forms of the other two methods, it ought to be recognized as a distinct method.

An example of the joint method of agreement and difference is the following. The Hungarian physician Ignaz Semmelweis noted that the number of women who, after giving birth, died from puerperal fever in the First Division ward of his hospital in Vienna was far higher than the number of women who, after giving birth, died from puerperal fever in the Second Division ward. What was the cause of the greater number of deaths? From these facts he could infer that, whatever was the cause, it was present in the First Division and absent from the Second Division. Semmelweis then proceeded to alter various factors in the First Division ward. He first altered those factors in the First Division which he knew to different from those in the Second, e.g., the mode of the priest's arrival, the position of delivery, etc. Semmelweis concluded that these were not necessary conditions, since the high rate of puerperal fever continued whether they were present or absent, and necessary conditions must be present when the conditioned property (high rate of death from puerperal fever) is present. The death of a friend from fever caught from the infection of a wound that occurred during a dissection drew his attention to a factor that he had neglected. He recognized that all his cases did contain one thing in common, namely, the medical students who examined the women had all come directly from the dissecting room with contaminated hands. The was the circumstance in which all instances of the conditioned property agreed even where they differed in all other possibly relevant factors. This, therefore, by the method of agreement, was the necessary condition for the high rate of death from puerperal fever.

Semmelweis then went on to show that this necessary condition was also a sufficient condition. He insisted that the students wash there hands not only with soap and water as they had hitherto done but also in chlorinated lime-water solution, and demanded further that all instruments and dressings be sterilized. The death rate immediately dropped from between 5% and 16% to 3% and then in two years to about 1.25%. Since the change to antiseptic conditions was the only difference between having a high death rate and having a low death rate, it followed by the method of difference that this was the sufficient condition for the high rate of death from puerperal fever.

Semmelweis' inference that the lack of antiseptic procedures was the cause of the high rate of death in the sense of being both necessary and sufficient was an application of the joint method of agreement and difference.[85]

Replication is an important notion in science. If an experiment which is claimed to confirm some hypothesis cannot be repeated in other laboratories — if, as one says, it cannot be *replicated* — then the hypothesis fails to receive acceptance in the scientific community as one that is worthy for use in explanation and prediction.

Let us suppose that an experiment has been performed that confirms the hypothesis that

(h*)   All F are G
        = For any individual x, if x is F then x is G

An experiment will consist, in its simplest form, at an attempt to falsify this hypothesis. If the experiment confirms the hypothesis, then it finds some individual x which is F and the prediction of the hypothesis, that x is also F, if it turns out to be true, confirms the hypothesis. To replicate the experiment is to find another individual, say y, such that y, too, is F. One produces, in other words, in one's laboratory, another F. If it turns out that y is also G, then the experiment has been replicated, and the hypothesis has been (re-)confirmed. If, however, the attempt at replication fails, that is, if one has produced in one's laboratory an individual, say z, such that z is F but z is not G, then one has falsified the original hypothesis that all F are G.

Why might there have been a failure during the attempts are replication?

It might be the case that instead of F's producing G's, as the hypothesis suggests, the G in the original laboratory is produced by some factor that is unique to the laboratory, say C. C might be some systematic problem with the experimental setup, e.g., cheating or falsifying experiments in an effort to get tenure or win a Nobel Prize. Or it might be a matter of some unknown factor that the experimenters were simply in no position to recognize or take into account.

If there is such a factor then we have a regularity not like (h*) but rather like

(h**)   For any individual x, if x is F then (x is G if and only if x is C)

Whatever C is, if (h*) is true then so is (h**): the former logically entails the latter. But the converse does not hold: (h**) does not entail (h*).

If the experiment is repeated in a second laboratory, where the factor C, idiosyncratic to the first lab, does not exist, and it turns out that producing an F in turns leads to its being G, then we will have the results

Table 4.6		
F	G	C
t	t	f

This will confirm (h*) but eliminate (h**) as false. Note that this is so even if we have no clear idea what C might be. Thus, *replication ensures the elimination of the possibility that the effect in which one is interested depends for its occurrence on unknown factors idiosyncratic to one laboratory or one observer.*

(η) The Comparative Method

Scientists regularly employ the methods of elimination in their experiments. But there is much scientific inference based on these methods which is not experimental. Or, to put it slightly differently, scientists often find that Nature performs for them experiments

that they themselves could not perform. This method of inference has been called the "comparative method."

This method was used and articulated by Georges Cuvier, the greatest anatomist of the early part of the century.

Cuvier, in the introductory "Letter to J. C. Martrud" in his *Lectures on Comparative Anatomy*,[86] noted the fact that controlled experiments were difficult in anatomy.

> The machines which are the object of our researches, cannot be demonstrated without being destroyed. We have no means of discovering what would result from the absence of one or several of their parts, and consequently we remain ignorant of the operation of each of these parts, in producing the total effect (p. xiii).

But natural experiments are possible.

> Fortunately Nature herself seems to have prepared for us the means of supplying that want which arises form the impossibility of making certain experiments on living bodies. The different classes of animals exhibit almost all the possible combinations of organs: we find them united, two and two, three and three, and in all proportions; while at the same time it may be said that there is no organ of which some class or some genus is not deprived. A careful examination of the effects which result from these unions and privations is therefore sufficient to enable us to form probable conclusions respecting the nature and use of each organ, or form of organ.

> In the same manner we may proceed to ascertain the use of the different parts of the same organ, and to discover those which are essential, and separate them from those which are accessory. It is sufficient to trace the organ through all the classes which possess it, and to examine what parts constantly exist, and what change is produced in the respective functions of that organ, by the absence of those parts which are wanting in certain classes ... (pp. xiii-iv).

As we have earlier noted, to speak of "functions" in biology is to speak of effects. Cuvier, too, it is evident, recognizes this point. What it means is that the scientist wants to find sequences or regularities of the form

All F are G

where "F" represents the organ and "G" its effects. We may wish to find which organ it is that produces certain effects. In that case *we are looking for sufficient conditions*. What we want to do is examine a wide variety of cases, examining those cases where the effect G is absent, and noting which organ or organs are simultaneously absent. *This is, of course, the method of difference.* We might also wish to find what the specific effects are for a certain organ. In this case *we are looking for necessary conditions*. What we have to do in this case is look at those cases where the organ F is present, and note those effects alone that are at the same time present when it is present. *This is the method of agreement.*

In short, *the comparative method used by Cuvier is but another example of the use by science of Mill's Methods of experimental science.* This is thus not different in principle from the ordinary methods of science. It differs not in its *logic* but only in the circumstances in which the logic can be applied. In particular, it differs in using natural experiments rather than deliberately manufactured interventions.

Mill himself commented on this method.[87] Whatever we are investigating, we must follow the rule that "the foundation of all the rest," namely, "the Baconian rule of *varying the circumstances*" (III, vii, 2). In what Mill calls "experiment" the circumstances are varied through the deliberate intervention of the experimenter; in what he calls "observation" we limit ourselves to the "variations in the circumstances [that] nature spontaneously offers" (III, vii, 3). The latter has its limitations; it is often an advantage to produce precisely the combination we want as we investigate phenomena. It is not always possible, however, and so we must at times rely upon observation rather than experiment. Logically speaking, however, there is no difference between the two methods.

For the purpose of varying the circumstances, we may have recourse ... either to observation or to experiment; we may either *find* an instance in nature suited to our purposes, or, by an artificial arrangement of circumstances, *make* one. The value of the instance depends on what it is in itself, not on the mode in which it is obtained: its employment for the purposes of induction depends on the same principles in the one case and in the other, as the uses of money are the same whether it is inherited or acquired. There is, in short, no difference in kind, no real logical distinction, between the two processes of investigation (III, vii, 2).

The differences are only practical, not logical.

Nonetheless, there were problems. These have to do with the Principle of Limited Variety (P2). Whether Mill's Methods are to be applied experimentally or observationally, whether we are to use the method of experiment or the comparative method, it is still a necessary condition for these methods to permit us to discover causes that the Principles of Limited Variety and of Determinism hold. The latter is no great problem; we have been reasonably successful in discovering causes. The problem is the Principle of Limited Variety.

In the case of biological functions, the range of possible causes of the functions is limited to the organs and organic structures of the body. In particular cases, the range of possible alternatives can be further limited to more specific regions of the body or more specific forms. Similarly, if one is searching for the functions of a given organ, one is again limited to effects that take place in the body. Again, there can be more narrow limitations to specific areas or specific forms of effects. In the case of the use of the comparative method in biology, then, the relevant Principle of Limited Variety is clear.

It is not so unproblematic in other cases, however. There are, for example, real difficulties when it comes to psychology. This is a topic, however, that we need not pursue.

(θ) The Inverse Method of Agreement

In the (direct) method of agreement we look for a property that is present when the conditioned property is present. It is a method for finding necessary conditions. In the **inverse method of agreement** we look instead for a property that is *absent* when the conditioned property is *absent*. That is, there is an agreement but it is an agreement in respect of absence. Since we know that *a sufficient condition for a conditioned property cannot be present when the conditioned property is absent*, it is clear that the inverse method of agreement aims at the discovery of *sufficient conditions*.

Again consider the four hypotheses

for all x, x is F_1 if and only if x is G
for all x, x is F_2 if and only if x is G
for all x, x is F_3 if and only if x is G
for all x, x is F_4 if and only if x is G

The following data eliminate all but one of these:

Table 4.7

	G	F_1	F_2	F_3	F_4
a	f	t	f	f	f
b	f	f	t	f	f
c	f	t	f	t	f

Event a shows that F_1 cannot be a sufficient condition for G it therefore eliminates the first alternative hypothesis. Event b shows that F_2 cannot be a sufficient condition for G, eliminating the second hypothesis. Event c eliminates F_3 as a sufficient condition and therefore eliminates the third hypothesis. The three events agree only in F_4 being absent whenever the conditioned property G is absent. F_4, or, what is the same, the fourth hypothesis alone is uneliminated. It follows that F_4 is the sufficient condition for G.

(ι) The Double Method of Agreement

Just as one can use the joint method of agreement and difference to discover conditions which are necessary and sufficient, so one can combine the (direct) method of agreement and the inverse method of agreement to a similar end. The (direct) method of agreement yields necessary conditions. The inverse method of agreement yields sufficient conditions. Used in combination, the methods yield conditions which are necessary and sufficient.

(κ) Negative Properties

If "F" represents a positive property then

not-F

represents a *negative property*.

Thus far we have considered only positive properties. But we often have to take account of negative properties. They are in fact often among the necessary conditions in which we are most interested. Not letting one's grade fall too low is a necessary condition for passing a philosophy course; not walking into the path of a car is a necessary condition for remaining whole of limb; not keeping one's hands antiseptic is a necessary condition for infection. We are interested in negative necessary conditions because if we know them then we know what to avoid if we are to attain our goals.

Since the method of agreement is the method for discovering necessary conditions, it can be used to discover negative necessary conditions. The following set of data illustrate this use of the method of agreement.

Table 4.8

	G	F_1	F_2	F_3	F_4	not-F_1	not-F_2	not-F_3	not-F_4
a	t	f	t	f	t	t	f	t	f
b	t	t	f	f	f	f	t	t	t

A conditioning property which is absent when the conditioned property is present is eliminated as the necessary condition. Event a eliminates F_1, F_3, not-F_2 and not-F_4 as the necessary condition. Event b eliminates F_2, F_3, F_4, and not-F_1. The only property that is uneliminated is not-F_3. It follows that this is the necessary condition for G, or, what is the same, that the following hypothesis is true

for all x, if x is G then x is not-F_3

Now, consider once again the inverse method of agreement. The following hypothesis

(+)   for all x, if x is F_4 then x is G

was justified as true by the data. But this hypothesis is logically equivalent to

(++)   for all x, if x is not G then x is not-F_4

This means that we can consider the inverse method of agreement to be a case of the (direct) method of agreement applied to negative properties. The following elaboration of the original Table 4.7 for the data shows how this is so.

	G	F_1	F_2	F_3	F_4	not-F_1	not-F_2	not-F_3	not-F_4	not-G
a	f	t	f	f	f	f	t	t	t	t
b	f	f	t	f	f	t	f	t	t	t
c	f	t	f	t	f	t	t	f	t	t

Table 4.9

The original set of data for the positive properties alone are logically equivalent to the data in the table as enlarged to include negative properties. The data for the positive properties led to the acceptance as true by the inverse method of agreement the hypothesis (+) that F_4 is the sufficient condition for G. In the enlarged table of data, we see that not-F_4 alone is present when not-G is present. It is therefore, by the (direct) method of agreement, the necessary condition for not-G. That is, we can accept (++) as true. But therefore we can accept its converse (+) as true.

(λ) Conjunctive Sufficient Conditions

Besides positive and negative properties, we can have *conjunctive properties*. Thus, for example, if we have the positive properties F and G, we can also have the conjunctive property

F & G

A conjunctive property is present in an individual or event if and only if both conjuncts are present in that individual or event. If either conjunct is absent then so is the conjunctive property.

Conjunctive properties can often be sufficient conditions where neither conjunct alone is sufficient. One may exercise a lot but if one eats only peanut butter sandwiches, one will not be healthy. At the same time, one may eat very healthy food, but if one avoids all exercise one will not be healthy. Neither exercise alone, nor a good diet alone is a sufficient condition for health. It can be argued that together they are sufficient.

The following table shows how a set of data may eliminate all but a conjunctive property.

Table 4.10

	G	F_1	F_2	F_3	F_4	$(F_2 \& F_4)$
a	f	t	f	t	t	f
b	f	t	t	f	f	f

Apply the inverse method of agreement. The conditioning property that is absent whenever the conditioned property G is absent is the sufficient condition for the conditioned property. With the above data, all conditioning properties save the conjunctive property (F_2 & F_4) are eliminated. We may conclude, then, that the hypothesis

for all x, if (x is F_2 & x is F_4) then x is G

is true.

Conjunctive properties can also be eliminated by the mechanisms. Thus, if we had had the above data, but also the conjunctive property (F_1 & F_4) as a possible conditioning property, then we would have had the following table:

Table 4.11

	G	F_1	F_2	F_3	F_4	$(F_1 \& F_4)$	$(F_2 \& F_4)$
a	f	t	f	t	t	t	f
b	f	t	t	f	f	f	f

Event a eliminates (F_1 & F_4) as a sufficient condition for G, because the conjunctive property is present when the conditioned property is absent. The conjunctive property is present in event a because both its conjuncts are present.

(µ) Disjunctive Necessary Conditions

Besides negative and conjunctive properties, one can also have *disjunctive properties*. If F and G are two simple positive properties, then

F or G

is a disjunctive property that is present in an individual if F is present in the individual or if G is present in the individual or both. Conversely, the disjunctive property is absent from and individual if and only if both disjuncts are absent.

There are often disjunctive necessary conditions in which we are interested. Thus, one may seek permanent regular employment in Canada only if one is either a Canadian citizen or a landed immigrant. But this disjunctive condition is not sufficient. For, one might be either and yet ineligible for permanent regular employment because one is underage.

The following table shows how a set of data may eliminate all but a disjunctive property.

	G	F_1	F_2	F_3	F_4	(F_3 or F_4)
a	t	t	f	f	t	t
b	t	f	t	t	f	t

Table 4.12

Since the only conditioning property that is uneliminated is (F_3 or F_4), it follows by the (direct) method of agreement that it is the necessary condition for G. Or, in other words, it follows that the hypothesis

for all x, if x is G then (x is F_3 or x is F_4)

is true.

Disjunctive properties can also be eliminated by the mechanisms. Thus, if we had had the above data, but also the disjunctive property (F_1 or F_4) as a possible conditioning property, then we would have had the following table:

	G	F_1	F_2	F_3	F_4	(F_1 or F_4)	(F_3 or F_4)
a	t	t	f	f	t	t	t
b	t	f	t	t	f	f	t

Table 4.13

Event b eliminates (F_1 or F_4) as a necessary condition for G, because the disjunctive property is absent where the conditioned property is present. The disjunctive property is absent in event b because both of its disjuncts are absent.

(v) Very Complex Necessary and Sufficient Conditions[88]

When we begin to allow conjunctive and disjunctive properties the neatness of the simple forms of the eliminative methods that we have examined begins to disappear. There are to be sure methods of elimination that remain. But they become more complex.

If we have a set of n possible simple conditioning properties, then we have 2^n conjunctive properties each conjunct of which is either one of the simple conditioning properties or the negation of one. Each of these is a possible conditioning property for the conditioned property in which we are interested. So is every disjunction of such conjunctions.

These conjunctive properties are mutually exclusive. Hence, an individual that exemplifies one will exclude all others.

Now let us judiciously choose our events so that each event exemplifies one of these conjunctive properties, and so that in the set of events each such property occurs once. (Since the properties are mutually exclusive, each such property will occur exactly once.) The conditioned property will occur in some of these events and not others.

If such a conjunctive property is present when the conditioned property G is present, it will be absent when G is absent. It is therefore by the inverse method of agreement a sufficient condition for G. If G is present in several events, then there will be several of the conjunctive properties which are sufficient for G. But a disjunction of sufficient conditions is a sufficient condition. Hence, the disjunction of the set of conjunctive properties that are sufficient for G is a sufficient condition for G.

If only one such conjunctive property is sufficient for G, then it will also be necessary. For, G will be absent from all other individuals, and this conjunctive property alone will be present when G is present. Hence, by the method of agreement, it will be a necessary condition.

If several such conjunctive properties are sufficient for G, then none of them will be present whenever G is present. So none of them will be necessary. But their disjunction (and no other disjunction) will be present whenever G is present. Hence, their disjunction will be a necessary condition.

Since the disjunction is also sufficient, it will be both necessary and sufficient.

An example will help to make this clear. Suppose we have the conditioned property G and three possible conditioning properties, together with the following data.

	G	F_1	F_2	F_3
		Table 4.14		
1	f	t	t	t
2	t	t	t	f
3	t	t	f	t
4	f	t	f	f
5	f	f	t	t
6	t	f	t	f
7	f	f	f	t
8	f	f	f	f

These data are such that *all possible combinations of the conditioning properties* are represented among the data. With this rather special set of data, we can use the methods of elimination to discover the necessary and sufficient condition for the conditioned property G.

Thus, event 2 establishes that $(F_1 \& F_2 \& \text{not-}F_3)$ is a sufficient condition. For it is absent whenever G is absent. Hence by the inverse method of agreement, it is a sufficient condition.

Event 3 similarly establishes that $(F_1 \& \text{not-}F_2 \& F_3)$ is a sufficient condition.

Event 6 similarly establishes that $(\text{not-}F_2 \& F_2 \& \text{not-}F_3)$ is a sufficient condition.

Since a disjunction of sufficient conditions is a sufficient condition, it follows that

(%) $(F_1 \& F_2 \& \text{not-}F_3)$ or $(F_1 \& \text{not-}F_2 \& F_3)$ or $(\text{not-}F_2 \& F_2 \& \text{not-}F_3)$

is a sufficient condition.

Any other of the possible conjunctive property is present when G is absent, so none of them could be a sufficient condition. Moreover, any other disjunction of such conjunctive properties will be present when G is absent. Hence, there is no disjunctive property besides (%) which is sufficient for G.

But this disjunctive property is present whenever G is present and is otherwise absent. It therefore follows by the method of agreement that this disjunctive property is also a necessary condition.

Moreover, not one of the conjunctive properties is present whenever G is present; if such a property is present in one event when G is present, then, given that these properties are mutually exclusive, it is absent in any other event in which G is present. So not one of the conjunctive properties is a necessary condition for G. In particular, then, none of the other sufficient conditions is a necessary condition for G.

Since this complex property (%) is both a sufficient condition and a necessary condition, it follows that it is both necessary and sufficient. Since it is the only condition that is both necessary and sufficient, it follows that it is *the* necessary and sufficient condition.

That is: we can conclude that the uneliminated condition is both is necessary and sufficient **provided that we make the assumption — *the matter-of-fact assumption* — that there is some conditioning property that is both necessary and sufficient for the conditioned property**.

In other words, even for complex eliminative inferences of this sort, if we are to draw a conclusion from the eliminative data that the uneliminated relationship is a true generalization, then it is still necessary to introduce lawful matter-of-fact assumptions that do the logical task that the Principles of Determinism and of Limited Variety do in the simpler cases.

Consider the following sort of case. We have certain disease, and various treatments F_1, F_2, and F_3. We want to find which treatments are relevant to the disease being cured; being cured of the disease is the conditioned property G. The following table indicates the results. Some patients who were cured G were given treatment F_1 but not the other treatments; this is case 1. Some who were cured were given treatments F_1 and F_3 but not F_2; this is case 2. Case 3 consists of those persons who were not cured after receiving treatment F_2 and no other treatment. Case 4 is those who were not cured in spite of being given treatment F_3 and no other treatment. Case 5 represents those people who were given no treatment and were not cured.

Table 4.15

	G	F_1	F_2	F_3
1	t	t	f	f
2	t	t	f	t
3	f	f	t	f
4	f	f	f	t
5	f	f	f	f
6	t	t	f	f
7	f	t	t	t
8	f	f	t	t

A property that is present when G is present cannot be a necessary condition for G. Hence, since F_1 alone is present when G is present we can infer, by the (direct) method of agreement that F_1 is a necessary condition for G. In addition, a sufficient condition cannot be present when G is absent. Since F_1 alone is present whenever G is present, it alone, by the inverse method of agreement, can be a sufficient condition. Hence, from the data represented by the events 1 through 5, F_1 is a treatment which is both necessary and sufficient for G, the patient being cured.

Let us suppose, however, that research continues into the possible treatments and we secure the data represented by events 6 through 8. In event 7, F_1 is present when G is absent: these patients do not recover under this treatment. But that means that F_1 could not be a sufficient condition.

But background scientific theory that is available to the physicians testing the treatments may suggest that if different treatments are used in combination, they interact to cancel each other out. F_1 seems effective except in case of event 7. It is when F_1 is given along with F_2 but not F_3 that it seems to do the job of curing (event 1), while when it is given with F_3 but not F_2 it also seems to work. Event 7 thus suggests that when treatment F_1 is given together with F_2 and F_3, the latter interact to as it were cancel out its effect. This suggests that it is not so much F_1 that is sufficient for G but F_1 when in the absence of $(F_2 \& F_3)$, or, in other words, that it is not F_1 that is sufficient for G but rather

(^) $F_1 \& \text{not-}(F_2 \& F_3)$

Inspection of the above table will show that this property alone is present when G is present and absent when G is absent. Since these data fulfill the special condition that *all possible combinations of the conditioning properties are represented*. It is therefore the complex conjunctive property (^) that is necessary and sufficient for G.

(ξ) Causes, Conditions, and INUS Conditions

We ordinarily would say that the striking of the match caused the match to light. Now, a sufficient condition is one the presence of which brings about the effect. Thus, the striking of the match seems to be a sufficient condition. More generally, since causes bring about effects, it would seem that causes are sufficient condtions and sufficient conditions are causes.

However, if we reflect upon the situation in which the match lights upon being struck, it is clear that, in terms of necessary and sufficient conditions, it is misleading to say that the striking of the match is sufficient for the lighting of the match. For, it is clear that *by itself* that is *not* a sufficient condition for the match lighting. *In the context*, striking the match is indeed sufficient to bring it about that the match lights. But the context includes more than this, more than the simple striking of the match. Thus, for example, if the match is to light there must be oxygen present, and the match must be dry. In the absence of either of those conditions, the match will not light when struck. What is sufficient for the match to light is the match being struck in the presence of oxygen and when the match is dry. The striking of the match is in fact by itself only an insufficient part of a more complex condition. It is only the more complex conjunctive condition that is sufficient. But, if the parts of that conjunctive condition other than the striking of the match are present, oxygen and dryness for example, the match will still not light. Those conditions alone will not bring it about that the match lights. In this sense, the striking of the match is necessary if the conjunctive condition is to be a sufficient condition.

Thus, the striking of the match is an insufficient necessary part of a sufficient condition.

At the same time, however, that complex sufficient condition is not a necessary condition for the match lighting. There are other, conditions which are sufficient. For example, one might light the match by using an magnifying glass to focus the rays of the sun on the head of the match. If one were to do this, one would say that so focussing the rays was the cause of the match lighting. But again, by itself that was not a sufficient condition. Oxygen has also to be present and the match must be dry. So the focussing of the rays is an insufficient part of a more complex conjunctive sufficient condition. Again, however, the focussing of the rays is a necessary part of the sufficient condition; in its absence the parts of the conjunctive condition alone will not produce the effect. Since this conjunctive sufficient condition can bring about the light of the match in the absence of a complex condition involving striking, it follows that the latter is not a necessary condition for the lighting of the match. And for the same reasons, neither is the complex condition involving the focussing of the sun's rays a necessary condition: even if the latter is absent the match might still light, e.g., by being struck.

There are other complex conditions that are also unnecessary but sufficient for the lighting of the match. The match might, for example, be thrown into a bonfire. And so on.

We have, then, two complex conditions for the lighting of the match either of which is sufficient but neither of which is necessary.

What is true of the striking of the match is true in general of what we call *causes* in our ordinary everyday discourse about things. As we explained previously, in our ordinary discourse, *a **cause** is an **I**nsufficient **N**ecessary part of an **U**nnecessary **S**ufficient condition*. Or, to put it briefly, a cause in our ordinary way of speaking is an **INUS** condition.

There are many INUS conditions that are present when an effect is brought about. All the necessary parts of the sufficient that brings about the effect must be present, and all of them are INUS conditions. It is only one of these INUS conditions, however, that we pick out as the cause. What is the criterion that we use?

It seems to be this: *as the cause of an event we pick out that INUS condition that is, in the expected course of events, otherwise absent*. The cause is the INUS condition that, as it were, *makes a difference* in the ordinary or expected course of events.

Thus, consider again our previous example. In the ordinary course of events, oxygen may be expected to be present when there are matches about, and given what we aim to do with matches we generally keep them dry, so the match being dry is also what we regularly expect. It is also true that as events pass matches rest unstruck. Being struck is not one of the *standing conditions* in the situation. However, when the situation changes from one in which the match is not struck into one in which the match is struck, then the situation also changes from one in which the match is not alight to one in which it is. The striking of the match which transformed the situation from one in which many INUS conditions were present but not one of the necessary parts to a situation in which all the necessary parts of a sufficient conditions were present. And when all those necessary parts of the sufficient condition came to be present, the match lit. In other words, it was the striking of the match which *made the difference* in the context between the match not being lit and the match being lit.

The context will make a difference as to which INUS condition we pick out as the cause of an event. Suppose we are working in a laboratory and we have a match in a glass container from which all air has been evacuated. In this context, oxygen is absent. Let us now suppose further that in this context the match is normally subjected to fairly high temperatures. Where oxygen is present such temperatures will cause the match to light — that is what is achieved when the sun's ray are focussed by a glass on the head of the match — but in the context we are considering no oxygen is present and so the match does not light. Suddenly, however, it does ignite. What is the cause? We would cite as the cause, as that which made the difference, the presence of oxygen: it has somehow leaked into the container.

In everyday circumstances, the presence of oxygen is a standing condition, and we would not pick it out from among the set of INUS conditions as the cause of a match lighting. In the different context of the laboratory, however, where it is the absence of oxygen which is the standing condition and the along with the presence of a high temperature, it is the sudden presence of oxygen which we cite as the cause of the match lighting.

Suppose, however, that the equipment in the laboratory were normally monitored by an assistant, and the apparatus adjusted when necessary to ensure the absence of oxygen. Suppose, further, that this assistant nods off momentarily, and oxygen does enter the apparatus. At that point the match will light. In these circumstances, one would normally not cite so much the presence of oxygen as the cause of the match lighting as the fact that the assistant was not attending to the task in which, in the circumstances, one might expect him or her to be diligent. The cause, in short, was the *negligence* of the assistant.

Often we believe that a cause in the sense of an INUS condition is present when we do not know what it is. A group of seven visit a restaurant and two come down with food poisoning. What was the cause? What made the difference? Clearly, the method of difference will enable one to isolate the cause, that is, the condition in the context and was sufficient to bring about the illness. But, of course, that sufficient condition will in general be an INUS condition. Though in the context it is sufficient, in fact it is only a necessary part of a more complex condition which is the real sufficient condition.

This sort of thing applies to more complicated cases also.

Consider the case of astronomical explanations of planetary orbits. During the 19th century astronomers were able to successfully predict in great detail the motions of the planets. These successful predictions provided immense and convincing confirmation for Newton's theory.

For the planet Uranus, however, the predictions turned out not to be successful. There are perturbations in the orbit — that is, deviations from exact elliptical form — that cannot be accounted for on the basis of Newton's theory and the assumption that there are eight objects in the solar system, the seven planets Mercury through Uranus and the sun. These anomalous perturbations presented science with a problem. The pattern here is simply the one that we have just noted. The inference patterns are clear. Most planetary orbits have the property of being as predicted by Newton's theory. One planetary orbit — that of Uranus — lacks this property. Since there is a deviation from the normal, we assume that there is something that makes the difference. This condition that makes the difference is what we call the cause. The problem is to find the cause.

In this case the background theory provided in detail an hypothesis about this cause. Astronomers made calculations based on the *composition law* of classical mechanics which enabled them to formulate an hypothesis about this cause. This hypothesis was that of the existence of an eighth planet of a certain mass and in a certain position. When telescopes were trained on the appropriate spot, this planet was observed; astronomers had discovered the cause of the anomalous perturbations of the orbit of the planet Uranus, it was the presence of the planet that came to be named Neptune.

The relevant theory will in general not be so specific in its prediction of the existence of the condition that makes the difference. Thus, consider my watch. Usually when it is wound it works. But sometimes it does not. I take it to the watchmaker and he is able to locate the factor, a speck of dust. In this case, there is nothing as determinate as the prediction of the existence of the planet that came to be known as Neptune. But the watchmaker is not devoid of knowledge. The watch is a mechanism (I am old fashioned in my watches) consisting of gears, levers, a spring, etc. These operate according to the laws of mechanics. The problem is to discover whatever it is that is blocking the transmission of power from the spring to the gears that turn the hands of the watch. There will in the context provided by this theoretical background be a number of alternative places where the watchmaker might look for the blockage, that is, a number of alternative hypotheses. His task is to eliminate them in turn until he finds what it is that is the cause of my watch not working when it is wound. This he does until he locates the speck of dust that is blocking the mechanism. In this case the discovery is effected not by a telescope but by the watchmaker's glass.

The general pattern is clear, however. We have a certain property P, say. And this is sometimes followed by Q and sometimes not. We assume on the basis of some more or less definite background theory that there is some condition, let us say R, such that when P is present, then if R is present then Q is present, while if R is absent then Q is absent. In other words, we have the two regularities

>   Whenever P, then if R then Q

and

>   Whenever P, then if not-R then not-Q

or, what is logically equivalent,

>   Whenever P & R then Q
>   Whenever P & not-R then not-Q

If we think of Q and not-Q as two possible states of the system, then what we aim at when we aim to discover the factor R that "makes the difference" are the complex conjunctive conditions P & R on the one hand and P & not-R on the other. When we have discovered R we shall have discovered the sufficient conditions for these two states of the system.

In general a state of a system will be a more complex condition. Moreover, the number of possible states will in general be greater than two. But we would still like to discover for each of those states the various, undoubtedly also complex, sufficient conditions. Indeed, for each state of the system what we would like, *ideally*, to have is a set of laws that specify the conditions that are both necessary and sufficient.

What we have just described, of course, is *process knowledge*, the ideal of scientific explanation of individual facts. A *process law* for a system, in being able to specify all future and past states of the system given the present state, and in being able to specify what will have to be or would have had to have been for other states to obtain, provides conditions that are necessary and sufficient for those states actual and possible.

It would seem, however, that the very complexity of a process law, that sheer number of necessary and sufficient conditions, that it specifies, would make it impossible in practice to use the mechanisms of elimination. To be sure, they would in principle be available. But the number of times that would be required for them to be applied, for the relevant experiments to be performed, would make it impossible to use them to locate all the members of the set of necessary and sufficient conditions that are included in the statement of the process law.

There thus seems to be a divorce as it were been the logic of experiment on the one hand and the logic of process laws on the other. The former is a method appropriate to simple systems but does not apply to systems described by logically more complex laws, and by process laws in particular.

It turns out, however, that the methods of elimination can also be used in the more complex cases.

Consider that of Newton.

Newton has discovered a generic abstract theory that applies to a wide variety of mechanical systems, and to the solar system in particular. This is the theory constituted by the Law of Inertia and the other basic axioms of classical mechanics. In terms of the logic of elimination, the relevant sort of system P is the solar system. We are given a condition Q for which we have to discover the necessary and sufficient conditions. This event Q is the series of positions of the planet Mars. The theory determines a range of hypotheses as to the force that will account for this event Q, that is, the positions of the planet. The alternatives are R = inverse square, R' = inverse cube, R" = ..., and so on. We thus have

>   Whenever P, then R if and only if Q
>   Whenever P, then R' if and only if Q
>   Whenever P, then R" if and only if Q

In fact the range of alternatives R, R', R", ... is infinite. They are, however, mutually exclusive. If, therefore, the data are such as to confirm one of these alternatives, then the remaining, infinite though they be in number, will be eliminated.

It is precisely this sort of situation which obtains. The relevant data consist in Kepler's Laws of Planetary Motion, the fact that the positions Q all lie on a curve which is an ellipse with one focus at the sun. In this case, the background theory provides one with some powerful tools. Specifically, it enables one to deduce that, given the data, one and only one R, that is, one and only one sort of force is necessary and sufficient for Q. This is the inverse square force. Since the presence of a force of this sort excludes the presence of the other possibilities, it follows that these latter are all eliminated automatically.

The force law, however, is the *process law* for the system. Or rather, mathematically speaking, it is differential equation of the second order, and when it is integrated, we obtain the process law for the system. Since this law is a process law, it yields the necessary and sufficient conditions for any state of the system.

In these inferences the methods of elimination were not applied directly to obtain for each state of the system the necessary and sufficient conditions. That, as we have said, would be impossible: there are infinitely many such possible states. However, we, or rather Newton, were able to take the set of possible process laws as determined by the theory of classical mechanics to be a set of possible conditioning properties relative to the motions to be explained as the conditioned property. It then turned out that the data were ready to hand to eliminate all alternatives but one. This uneliminated condition, that is, the uneliminated force, was thereby established as necessary and sufficient for the conditioned property, that is, the positions of the planet.

The mechanisms of elimination thus remain important in the logic of science even in cases where we have to deal simultaneously as it were with large sets of necessary and sufficient conditions.

(iii) Revolutionary Science[89]

T. Kuhn has presented, in his essay on *The Structure of Scientific Revolutions*, a plausible picture of scientific research. This process he divides into two stages. On the one hand there is *normal science*, on the other hand there is *revolutionary science*. While these phases of the research process can be distinguished, one should not think of them as wholly distinct: they do merge into one another, normal science growing out of revolutions and revolutionary science growing out of normal scientific research.

Much has been made out of these distinctions. They have been used in various ways, in particular by those who wish to defend the claims of pseudoscience. We have already looked at some of Kuhn's points, but given the ways in which they have been cited in debates over the criteria for pseudoscience, it will pay to look at these things once again, in greater detail. We will discover that Kuhn has presented a clear set of methodological proposals, a set of norms for the conduct of scientific research that can in fact be shown to be rationally justified. We will also discover that appeals to Kuhn to try to justify the acceptance of various pseudoscientific hypotheses are spurious.

a. Normal Science, with some reference to Continental Drift and to Astrology

Kuhn's portrayal of this process is rich indeed; with the expository skill of an historian he interweaves themes from logic, epistemology, the psychology of perception and of language, the sociology of science, and, of course, the history of science. What is important for our purposes are the points that describe the logic and methodology of science.

As we have seen, what Kuhn argues, in effect, is that the logic of normal scientific research is that of the eliminative methods. As he presents normal scientific research, it goes roughly like this. It is guided by a theory or, as Kuhn calls it, a paradigm. This theory applies to a variety of specific areas (pp. 17-18). We can represent this theory or paradigm by a schema like this:

(b)   For any species f of genus **F**, there is a species g of genus **G** such that, for any x, x is f only if x is g

This theory has in the past been applied to different areas, and has successfully guided research in these areas; it has led to the discovery of previously unknown laws in these areas. In other words, we have a number of causal laws something like this:

(a)   For any x, x is F_1 only if x is G_1
        For any x, x is F_2 only if x is G_2

where the F_i are of a common genus **F**:

$$F_1 \text{ is } \mathbf{F}, F_2 \text{ is } \mathbf{F}$$

and the G_i are of a common genus $\mathbf{G}$:

$$G_1 \text{ is } \mathbf{G}, G_2 \text{ is } \mathbf{G}$$

The law (b) is a generalization of the laws (a). The past successes of the theory or paradigm in research, that is, in leading to the discovery of the laws (a) testify to its utility as a tool for the researcher (p. 23).

Normal research processes are *guided by theories or paradigms*. Thus, let us suppose that a researcher lets herself be guided by the paradigm (b), and approaches a specific sort of system F_n that has not yet been investigated. Since

$$F_n \text{ is } \mathbf{F}$$

the theory (b) applies to this new area, and we can assert of such systems the law that

(c) There is a species g of genus **G** such that, for any x, x is F_n only if it is g

This asserts, as Kuhn says, that *there is* a law for systems F_n without, however, asserting specifically what this law is. It is the task of the researcher to find this law. Now, if she does *not* find such a law, does not isolate such a **G**, then *that* does not *falsify* (c), nor, therefore, the paradigm (b). This is due to the mixed-quantificational structure of (c) and (b): by virtue of the particular or existential quantifier appearing in (c) and (b), those laws cannot be falsified by observational data that fail to testify to the existence of what is asserted to be there.[90] The research task, or puzzle, is to find the law that the theory asserts to be there. Failure to find the law does not falsify the theory or paradigm; it testifies more to the lack of skill on the part of the researcher (p. 35ff, p. 147).

The research process goes something like this. The law (c) poses a puzzle to the researcher. The latter is moved by her cognitive interest, the passion of curiosity, if you wish, and aims to discover precise laws, laws which make determinate predictions (p. 42). But (c) yields only *determinable* predications. Let us again say that, if **H** is a genus such that $H_1, H_2, \ldots,$ are species under it, then the predicate 'x is **H**-ed' — cf. 'x is coloured' — is defined as short for

there is an h such that x is h and h is **H**

Now consider a particular F_n, say a. From

$$a \text{ is } F_n$$

we can use (c) to deduce *and therefore predict*

$$a \text{ is } \mathbf{G}\text{-ed}$$

But the scientist also hopes to discover a law which yields a more specific or *determinate* prediction. Moreover, (c) asserts that *there is* a law which *can* yield such predictions. So, the task is to find that law.

(c) delimits a *range* of hypotheses that are *worthy* of the scientist's consideration. These are, let us say,

(d)
For any x, x is F_n only if x is G_n^1
For any x, x is F_n only if x is G_n^2
For any x, x is F_n only if x is G_n^3

where each of the G_n^i is **G**. In contrast, if K is not **G**, then

(e) For any x, x is F_n only if x is K

is *not* among the hypotheses that the scientist must consider; (c) deems it to be unworthy of her attention. The task is to find data that confirm one and eliminate the others of the set (d), that is, to discover (so far as one can, given the tentative nature of all scientific judgements) which of (d) is not only plausible — as (e) is not — but also true. Moreover, the scientist proceeds in the knowledge that, if she is skilful enough, she will succeed: the paradigm (b) assures her of that (p. 36)!

The theory or paradigm that is guiding research, when applied to a new, as yet unexamined, specific area, asserts, on the one hand, that *there is* a law of a certain sort there to be discovered, while not, on the other hand, asserting *specifically* what that law is (p. 25ff); and it asserts that this law is within a certain range of alternatives. This theory or paradigm plays the role of a Principle of Determinism and Principle of Limited Variety. When the theory asserts that there is a law without asserting specifically what that law is, it asserting a gappy law. The cognitive aim of the researcher is to fill in this gap, to replace the gappy knowledge with knowledge of the specific law that is asserted to be there, or, in other words, to discover knowledge that more closely approximates the ideal of process knowledge. The cognitive aim creates the problem, that is, the *research problem*. The theory or paradigm tells the researcher where the solution to the problem may be found.

It is worth noting how the discovery and confirmation of the specific law confirms, or, better, further confirms, the paradigm, and strengthens its claim to be useful as a research tool.

If we assume that the hypotheses (d) are *contrary* to each other, then data that confirm one will thereby eliminate the others.[91] Suppose our researcher examines the particular system z, where

(x)   a is F_n

and discovers that

(y)   a is G_n^1

This is what is entailed, and therefore *predicted by*

(d_1)   For any x, x is F_n only if x is G_n^1

The latter is therefore *confirmed*, and the other hypotheses are *eliminated* as false. The researcher has now discovered the specific law that she desired to know, had a cognitive interest in discovering, and which the paradigm asserted was there to be discovered. But we must also notice that these same data *also confirm* both (c), the law predicted by the paradigm (b), and the paradigm (b) itself. For, since G_n^1 is **G**, if (y) is true, then so is

(y')   a is **G**-ed

which is what (c) predicts of a. Moreover, if (x) is true, then so is

(x')   a is **F**-ed

and given this, the paradigm (c) entails, and therefore *predicts* (y'). Moreover, of course, when the laws (a) are used to predict successfully, *those* data, too, count for the same reasons as data confirming the paradigm (b)

which is generalized from the laws (a). Thus, the data testifying to the truth of the specific laws (a) and (d_1) *also* testify to the truth of the paradigm. Thus, while the paradigm — unlike the specific laws (d) — cannot be falsified by particular data, it nonetheless *can be confirmed* by such data.

Another way of putting the same point is equally illuminating. Since (d_1) has (c) as its existential generalization, the former entails the latter. So the data that confirm (d) also testify to the truth of (c). But (c) is a prediction of the theory (b). Hence, the latter is also confirmed. In predicting (c) which goes beyond the laws (a) that led to its proposal, the theory (b) is *theoretically progressive*; when (c) is confirmed by the discovery of (d), (b) is shown to be *empirically progressive*.[92] For Kuhn, such empirical progressiveness is the test for a theory being acceptable for purposes of explanation and prediction, that is, acceptable as true — with, however, the proviso that since these are laws (note the universal quantifier) no set of observational data can *conclusively* testify to their truth.

Kuhn of course accepts this last point about the tentative nature of all scientific judgements: he is careful to allow that however well tested a theory or paradigm might be, it could still subsequently be overthrown and replaced by another — something which has in fact happened often in the history of science. But Kuhn also speaks of these theories or paradigms as functioning in a way that is something akin to "dogma."[93] Since dogma is the stuff of fanaticism rather than reason, the choice of terminology is perhaps not the best: it suggests that scientists, in holding to their theories in the face of certain sorts of contrary evidence, are simply being irrational. That, though, was far from Kuhn's intention. He wanted both to hold that scientists do in fact continue to assent to their theories, and continue to use them to guide research, in the face of apparently contrary evidence, but also to hold that this was a perfectly reasonable or rational thing to do. We have seen that if we construe theories or paradigms as having an abstract generic logical form, then these theories do have the logical resources to do just what Kuhn claims they do.

The evidential support which can accrue to theories of abstract generic form is what justifies treating such a theory in a way somewhat akin to "dogma." This building of evidential support derives from two features of scientific theories.

One of these is the way in which they are tied together evidentially. Thus, the data that support (b) *prior* to the discovery of (d_1) also tend to support (d_1). For those data justify using (b) to predict, and therefore support the prediction (c). But the data eliminate all possible hypotheses save (d_1). Hence, (c) could hold only if (d_1) were true. Hence, when one takes the paradigm (b) and the data that eliminate all hypotheses by (d_1), then (b) predicts (d_1). Thus, the data that justify using (b) to predict also support (d_1). In short, the data supporting the acceptance of the laws (a) also support the acceptance of (d_1). Conversely, of course, once (d_1) is discovered, then the data that confirm it also tend to support the acceptance of the laws (a). In this way, paradigms set up interlocking patterns of inductive support. These are the very same patterns that we noted with regard to the structure of Darwin's theory of evolution by natural selection and also with regard to Newton's theory of mechanics and more specifically of the solar system.

It is important to note that the forms of research as described by Kuhn are essentially those of the eliminative methods of experimental science. We earlier sketched how the inductive methods of science, and those of eliminative induction in particular, could be justified: conformity to these norms is, so far as we can tell, an efficient means to satisfying the cognitive interests of science in true matter of fact generalities. It would thus seem that by construing Kuhnian paradigms as having the logical form (b), as abstract generic theories, we can account for much of what Kuhn says about the nature of the process of research in normal science. Certainly, theories that have in fact guided research, such as classical mechanics and classical thermodynamics, clearly have this logical structure.

In particular, we see that normal research is very much a *problem-solving activity*. Others besides Kuhn have noted this feature of normal scientific practice.[94] Kuhn's achievement is to notice that in normal research these problems have the nature of *puzzles*. They are in fact analogous to, for example, cross-word puzzles. On the one hand, the researcher, like the cross-word afficianado, knows that there is a solution. In the case of science, this knowledge is provided by the background generic theory or paradigm. On the other hand, the researcher also knows, like the solver of cross-word puzzles, that what is there is a test of his or her ingenuity:

if he or she fails, that does not show that there is no solution — after all, it is *known* that *there is* a solution — the paradigm tells one that — ; what a failure to find a solution is not that the theory is false but that the researcher has not searched hard enough — the mixed quantificational structure of the generic theory or paradigm guarantees that is will not (easily) be falsified.

In noting the puzzle solving nature of normal science, it is important to recognize the relevant cognitive interest. This interest is that of filling in gaps in our knowledge. From the background generic theory or paradigm we can derive gappy laws that describe an area of interest. These gappy laws give a generic description rather than a specific. But the cognitive ideal of scientific explanation of individual facts is that of process laws, where one has regularities that describe in full detail the specific workings of the system; the scientist, interested in a particular area, has a cognitive interest in moving as close to this ideal as possible. With this motive, this cognitive interest, the scientist, guided by the paradigm, undertakes research to find those specific laws that the paradigm asserts are there to be discovered.[95] A puzzle is defined by the generic theory on the one hand and the cognitive interest on the other.

One can find in the world, many patterns — regularities, if you wish — that one cannot really claim are lawlike. One of the means by which a "mere" regularity or, as it has also been described, a mere correlation, can be transformed into a lawlike hypothesis is by embedding it in a theory of paradigm which connects it to other laws and theories. An example will make clear how this goes.

In the 1920's, the German geophysicist and meteorologist Alfred Lothar Wegener looked systematically at the geological formations on the east coast of Brazil and west coast of Africa. The shapes of the two land masses look as if they might have fit together at one time, and Wegener discovered a series of remarkable parallels in the geological formations. If "C" represents being a certain sort of geological formation on the Brazilian east coast and "E" represents being the same sort of geological formation except for being in a corresponding location on the west coast of Africa, then what Wegener was able to establish was, roughly, that

Whenever C, and only then, E

that is,

C if and only if E

Wegener proposed that this provided evidence that the two continents, Africa and South America, had once been joined but had since drifted apart.[96]

This latter proposal was rejected by almost all geologists. There was no doubting the fact of the correlations that Wegener had established. But they could doubt the further claim that there was a causal connection that related the geological formations on the two continents. For, there was no background theory that established that such a causal connection obtained. To the contrary, the accepted geological theory explicitly denied that there could be any such causal connection: it denied that the continents could move. The evidence was clear that the continental masses were of lighter material than the underlying masses; and the theory allowed that these masses could move up and down on the heavier material. They would sink into the underlying material, for example, if they were covered with massive ice caps. But they were more or less solidly embedded in the underlying material. The generally accepted account was that of the Earth cooling and shrinking; as the Earth cooled, and therefore shrunk, the solid outer crust contracted, and these contractions resulted in folds that produced mountains. No mechanisms were known that would be able to move the large continental masses relative one to another. Australia, the smallest continent, weighs some 500 million million million kilograms, and there seemed to be no sources for the tremendous forces that would be required to move such weights through the oceans floors. The standard geological theory thus more or less explicitly ruled out there being any causal connection that would transform the correlation Wegener had drawn attention from being a coincidence to being the effects of a causal processes relating C and E. As has been pointed out,

> ... the main reason for disbelief [in continental drift], or at least skepticism, arose because there was no known mechanism at that time which was capable of separating the continents in the manner that Wegener, or his supporters and predecessors, had suggested.[97]

Wegener's proposal remained in effect a curiosity, more or less to be scoffed at by geologists, until the '50's. At that point a number of geologists and geophysicists, led by J. Tuzo Wilson, using additional evidence unavailable to their predecessors, set about reconstructing geological theory. In particular, they were able not only to propose mechanisms which could account for continental drift, but also to provide strong evidence that these mechanisms were indeed at work.[98] Once this revised geological/geophysical theory became available, it became clear that the correlation that Wegener had discovered was indeed the effect of certain underlying causal processes.

Thus, in the absence of a background theory that could relate the regularity Wegener discovered to other parts of geology, geologists quite reasonably rejected the inference that Wegener had proposed, the inference from a correlation or "mere" regularity to causation. The correlation was simply a coincidence. But once the background theory had been developed, once the regularity had been fitted into a research-guiding paradigm, geologists concluded that the correlation was more than a *mere* correlation: it is a correlation deriving from causation.

This illustrates the important point that we cannot infer from a correlation among variables that there is a causal relation among those variables. But it further makes clear that something that once had the status of a mere correlation without causation can become, through the development of a background theory, a correlation that is explained in terms of an underlying causal process.

Now compare astrology.

As we know, astrology claims that such things as stellar and planetary positions influence human events on Earth, determining when it is propitious to take various sorts of action, and also determining whether or not persons have various character traits such as generosity or aggression. But, as we have already noted, there is little to this. We have previously noted the opinion of G. Dean and his colleague A. Mather, at once astrologers but also striving to be scientific, who, after reviewing over 700 astrology books and 300 scientific works on astrology, put the essential point this way:

> Astrology today is based on concepts of unknown origin but effectively deified as "tradition." Their

application involves numerous systems, most of them disagreeing on fundamental issues, and all of them supported by anecdotal evidence of the most unreliable kind. In effect, astrology presents a dazzling and technically sound superstructure supported by unproven beliefs; it starts with fantasy and then proceeds entirely logically. Speculation is rife, as are a profusion of new factors (each more dramatically "valid" than the last) to be conveniently considered where they reinforce the case and ignored otherwise.[99]

As they indicate, it is clear that astrology not only has no evidence to support it but that there is evidence that shows that astrology as traditionally understood is false.

To be sure, as we have also noted, this has not prevented people from accepting it. Fantasy, after all, does often have an appeal. And there are moreover the illegitimate inferences to its truth from the fact that it "works." Typical are the comments of Linda Goodman,

> Alone among the sciences, astrology has spanned the centuries and made the journey intact. We shouldn't be surprised that it remains with us, unchanged by time — because astrology is truth — and truth is eternal.[100]

This, however, is not the language of science, where conclusions as to what are the truth are always tentative. It is, rather, the language of *authority*, the authority of tradition. Dean and Mather concluded, in effect, that it is the method of authority that is behind the beliefs and practices of astrology, not the method of science: as we saw them express it, "Astrology today is based on concepts of unknown origin but effectively deified as 'tradition'." Moreover, as we saw them go on, "astrology presents a dazzling and technically sound superstructure supported by unproven beliefs; it starts with fantasy and then proceeds entirely logically."[101] So there is not only authority but also the *method of fantasy*. But there is still more; another, equally non-scientific method also plays a role. Dean and Mather comment on this state of affairs in this way:

The current chaos in astrology is largely the result of a chronic infatuation with symbolism at the expense of

reason. This is because the majority of astrologers reject a scientific approach in favour of symbolism (based on dubious tradition), intuition, and holistic understanding.[102]

That is, astrologers rely on the *method of resemblance*[103] rather than the *inductive method* characteristic of scientific inquiry. These non-scientific methods of authority and resemblance are supplemented by various devices that can be used in *ad hoc* fashion to save various hypotheses from falsification. Dean and Mather put this last point in this way, as we saw: there "...are a profusion of new factors (each more dramatically 'valid' than the last) to be conveniently considered where they reinforce the case and ignored otherwise." All this is put in the context of "entirely logical" developments of the basic non-scientific premises. These developments include good doses of mathematical calculations, so that the scientifically arbitrary assumptions are cloaked in a veneer of rationality. It is non-science disguising itself as science. It is, in other words, pseudoscience.

We have made these points previously. But there is another important connection which remains to be made, one that ties the criticisms of astrology to the Kuhnian theses about the logical structure of science and the logic of scientific research.

The important feature of theories or paradigms that have been successful in guiding research is that very success, the fact that they *are* empirically progressive. The more that they lead to the discovery of laws in different areas, the harder it becomes to conceive those laws obtaining if the theory were not true: the probability of these many laws being true if the theory were false becomes smaller and smaller. Paradigms which turn out to be empirically progressive thus acquire for practical purposes a degree of probability that amounts to certainty. This shows that it is indeed rational for scientists to treat paradigms as sufficiently certain that they can ignore apparently contrary evidence: the degree of certainty that attaches to the theory outweighs the usually much lower probability of the apparently contrary evidence. The latter only weakly tends to establish that the theory is unworthy of acceptance, and can reasonably be ignored.

We can contrast the sort of "dogmatism" that one finds with regard to paradigms that guide research in normal science with the sort of dogmatism that is characteristic of astrology. In astrology one finds, for example, the views of Linda Goodman, already cited. Here the theory, or, rather "theory," is accepted on the basis of the method of authority. That is why astrologers can continue to hold on to their theoretical framework even after they have acquired empirical evidence that that framework is false: what justifies accepting the theory is the authority and the fact that this authority comes into conflict with empirical evidence is irrelevant. This is in contrast to the inductive method of science where it is precisely empirical evidence that determines the acceptability of theories: if the data come into conflict with what one has hitherto accepted, then one must reject that older theory, and modify it so as to fit the new data. A paradigm continues to be accepted, "dogmatically" if you wish, not because of some authority but rather because the empirical data justify that acceptance: the paradigm guiding normal research has been successful in leading to the discovery of new laws, and the data that justify the acceptance of these laws also justify the acceptance of the theory or paradigm. The new data not only *confirm* the theory or paradigm but *re-confirm* it, and this *repeated re-confirmation* justifies the acceptance of the theory by the scientific community. In particular, it justifies relying on the theory even in circumstances where new research does not immediately lead to its re-confirmation. Failure to find a law that the theory asserts to be there does not falsity the theory nor require its rejection; normally all that it means is that the researcher has not been diligent enough.

There are other features of astrological theories that have been proposed that prevents their falsification. Thus, consider the theory, or, better, "theory" that has been proposed by P. Roberts.[104] This theory, hardly scientific, presupposes that there is an entity which is the counterpart of but pre-exists the person and plays a role in that person's birth. This entity is such that it is on the one hand stimulated in its actions by the planets when they are in certain positions, and, on the other hand, helps determine the future character traits of the person and therefore in turn helps ensure pre-eminence in certain fields. More specifically, this entity is such that

1)   it possesses certain qualities which will ultimately manifest themselves in the child and the adult;
2)   these qualities that it possesses are related to an organ with a flower-like structure with groups of petals, the prominence of each group being correlated with the strength of the associated quality;
3)   it [the entity] responds to planetary stimulation at the moment of birth when there is a joining to the (physical) person;
4)   its special petal groups have their prominence enhanced by the stimulation in ways which accord with the planet involved; and
5)   this stimulation may enhance an existing state of the entity or may introduce a new balance by stimulating groups not already prominent.

This entity is unfortunately non-physical and non-material; it is, in other words, purely metaphysical, quite beyond the world of ordinary experience and ordinary scientific testing. The theory cannot be falsified because it refers, or, better, purports to refer to entities that have no connections to the observable, connections, that is, that enable one to test for their presence or to recognize when changes are taking place in them. The literature of astrology is full of "theories" similar to that of Roberts in depending for its "explanatory" power on appeal to non-empirical entities. One find mysterious forces, energies, and so on, in these "theories," but we are never given ways in which one can test for their presence or detect changes in them. In physics when an appeal is made to, say, some form of energy, then one is provided with a set of rules that enable one to detect empirically the presence of that form of energy and more generally to measure the amount of that form of energy that is present. But in astrology when appeal is made to special entities, or forms of energy, or new forces, we are provided with no rules that enable us to empirically detect and measure them. These "theories" are therefore incapable of either confirmation or falsification by empirical data. The entities to which these theories purport to refer are like the mythical River Ocean which Herodotus rejected as explanatory, and a theory of the sort that Roberts proposes is to be rejected on the same grounds: the theory is simply non-scientific, or, if it purports to be scientific, then it is simply pseudoscience. If there is a method behind the acceptance of the theory by its proponents, then it is the *method of fantasy* rather than the inductive method of science.

Without a theory or paradigm to guide it, astrology presents a very different historical picture from that of normal science, astronomy for example. In the ancient world, astronomy and astrology were parallel; the practitioners of the one were, in general, practitioners of the other. But after Copernicus, through the work of Kepler, Galileo, and especially Newton, a paradigm developed to guide research into the motions of the heavens. Newton was able with his theory to, as he put it, "demonstrate the frame of the system of the world." His theory provided this "frame" or paradigm to guide research. The astronomical exploration of the solar system became a puzzle-solving activity. At this point, astronomy diverged dramatically from astrology. The latter, with no theory to guide it, ceased being, if it ever had been, a puzzle solving activity.

To put the same point another way, if one compares the history of astronomy and the history of astrology, we find that there is to be found in the former a research tradition, a settled practice of puzzle-solving activities. In astrology, in contrast, there is no tradition of puzzle-solving activities.

Not only are there no data that can justify the assertion of various astrological hypotheses, but there is no theory that can guide research. As Kuhn has made clear, the absence of such a theory is evidenced by the fact that there is no research tradition in astrology — in fact, this is, according to Kuhn, one of the features that make it safe to reckon astrology a pseudoscience.[105]

There are times, however, when the practices of normal research break down. These are periods of what Kuhn has called "revolutionary science." These are also periods in which methods of research are adopted which are unlike those of normal science, which is to say that the norms to which scientific practice conforms in these periods are very different from the theory guided research that is presupposed by the experimental methods. Kuhn has suggested that science has discovered a set of techniques to use in such periods. There are two issues. The first is this: What, precisely, are these norms that govern research during periods of revolutionary science? The second is this: Can these norms be given a pragmatic justification similar to that which we have given for the rules of the experimental method during periods of normal science?

We can neatly approach these issues by comparing the Kuhnian approach to scientific inference with one that is at once similar and also different from that which has been proposed by Kuhn. This is the set of methodological rules that has been proposed by Nelson Goodman.

b. Avoiding the Gruesome

Nelson Goodman, in his *Fact, Fiction, and Forecast*,[106] has pointed out (p. 74) that if we define

x is grue

to be short for

(x is green & x is examined before t)
or (x is blue & is examined after t)

then the two generalizations

(i)   All emeralds are green

(ii)   All emeralds are grue

on the one hand make contrary predictions about emeralds examined after t (assuming, of course, that nothing is both blue and green all over at the same time), while, on the other hand, are both supported by the data, collected before t, that all observed emeralds are green (p. 77). This is a simple and apparently trivial example. Nonetheless, it illustrates in dramatic fashion the fact that, on the basis of an observed sample, any number of contrary extrapolations to the total population can be made.

One can give other examples. Thus, suppose we are looking at samples of some gas, say, not realistically but for the sake of the argument, nitrogen. We consider constant volumes of this gas. We make observations of pressure P and temperature T, as recorded in Graph 4.1.

We are here also assuming a convenient set of units of measurement. In any case, for these observations we have the functional relationship

(1)   For any sample of nitrogen gas, $P = k\,T$

where

Graph 4.1

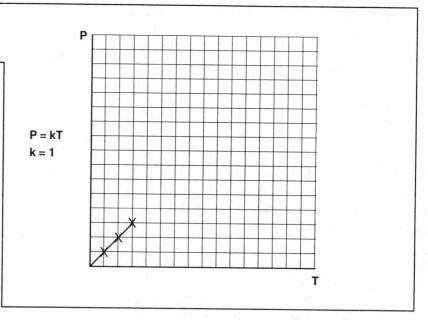

$$P = kT$$
$$k = 1$$

$$k = 1$$

We now make a further series of observations, as recorded in Graph 4.2.

What we seem to have is that formula (1) applies to these observations also, as in Graph 4.3.

Normally we would now inductively extend the functional relationship as in Graph 4.4.

However, there is nothing in the observed sample that *guarantees* that this is the *true* relationship in the population. After all, there are unobserved samples in both at higher temperatures, and also in the intervals between the cases actually observed. For all that is required by the observed data as recorded in Graph Three, we could have either the relationship in Graph 4.5 or that in Graph 4.6.

Once again, in an example slightly more realistic than Goodman's example of possible grue emeralds, we have on the one hand inductive inferences to generalizations that make contrary predictions about samples of nitrogen examined after some time t, which generalizations are, on the other hand, both supported by the data, collected before t.

Or, to give another example, consider an astronomer plotting the path of a planet, let us say Jupiter. After making a series of observations, the astronomer plots the orbit as in Figure 4.1.

But of course, he or she has not observed every point on this path. There is in fact nothing in the observed data that *guarantees* that the planetary orbit is not sinuous in the parts between the observed positions, as in Figure 4.2, or that the orbit is not one that moves off in a straight line after a period of time as an ellipse, as in Figure 4.3.

Once again, in another example also slightly more realistic than Goodman's example of possible grue emeralds, we have on the one hand inductive inferences

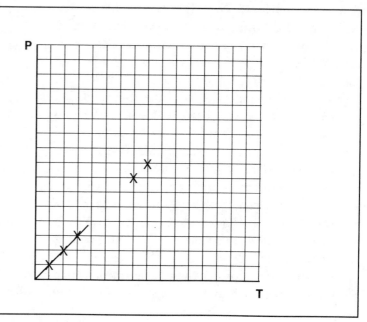

Graph 4.2

Graph 4.3

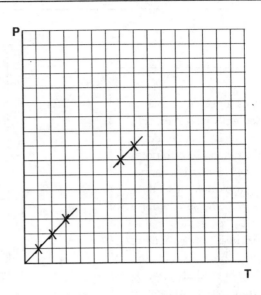

to generalizations that make contrary predictions about orbital positions of Jupiter examined after some time t, which generalizations are, on the other hand, both supported by the data, collected before t.

The point that these examples make clear is that, since there is no necessity in the world we observe in ordinary experience, nothing in an observed sample can guarantee anything in the total population: scientific judgements are always, as Judge Overton has emphasized, tentative. It follows that the available data cannot decide between the contrary extrapolations from the observed emeralds to (i) and (ii), or the contrary extrapolations to curves represented in Graphs 4.4, 4.5 and 4.6, or the contrary extrapolations to the orbits represented in Figures 4.1, 4.2 and 4.3. Nonetheless, it is equally clear that we take some of the possible extrapolations, e.g., (i), or the curve in Graph 4.4, or the orbit in Figure 4.1, to be rendered plausible by the data and others, e.g., (ii), or the curves in Graphs 4.5 and 4.6, or the orbits in Figures 4.2 and 4.3, not so rendered. The first sort may be said to be *lawlike* hypotheses, the latter not lawlike. The problem, then, that Goodman proposes to take up is that of distinguishing lawlike from non-lawlike or gruesome hypotheses (*Fact, Fiction and Forecast*, p. 80). The example of grue is somewhat fanciful, but as the other examples make clear, the problem is in fact a real one.

Goodman argues plausibly that neither syntactical criteria, nor the criterion of containing only purely qualitative predicates, suffice to distinguish lawlike hypotheses form gruesome hypotheses (*Fact, Fiction and Forecast*, p. 83). Some have proposed that there are differences between properties such as green and those such as grue that make it more reasonable to suppose that the former occur in laws than do the

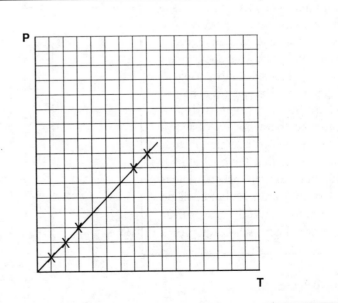

Graph 4.4

Graph 4.5

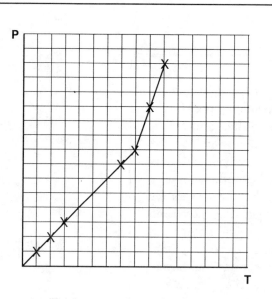

latter.[107] Grue is supposed to involve a temporal element while green does not. No doubt this is true. No doubt also that we can identify a thing as green by looking at it now but cannot in the same simple way identify a thing as grue. But why should these facts *by themselves* establish that green is while grue is not projectible? Why should these facts *by themselves* establish that green is more likely than grue to appear in statements of law? Let us say, following Bergmann, that a predicate is "significant" just in case that it appears in statements of law.[108] Then the claim against Goodman is that green is more likely to be significant than is grue. Now, this is no doubt true: we do so think. In fact, when we normally say of a thing that "it *is* green," that predication includes a projective element; we are giving it to be understood that the thing will continue to be green.[109] But this is to say that we accept as a law the regularity that things which are green, normally and other things being equal, continue to be green. However, to recognize this as a fact is only to say that we do think that green is more likely to be significant than is grue; it is not yet to say what it is about *green as such* that makes it more likely to be significant. The point is of course that in the world of ordinary experience there are no objective necessities, which is, among other things, to say that there is nothing about *green* as such that guarantees or helps to guarantee that it will be connected lawfully to other properties. Those who argue for a

special inductive status for red and green as opposed to grue and bleen simply fail to understand, what Goodman clearly grasps, that the argument of Locke and Hume against objective necessary connections precludes neat properties such as red and green from having any special inductive status.

It is evident that the same point could be made with regard to our examples of nitrogen and Jupiter. We do think of smooth curves as more inductively plausible than

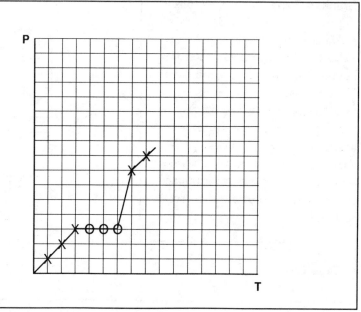

Graph 4.6

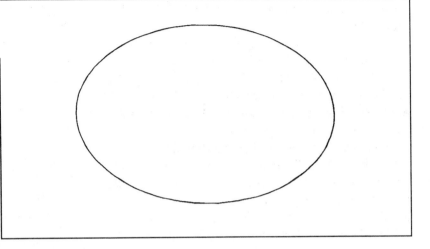

Figure 4.1

sharply twisting curves as in Graphs 4.5 or 4.6, or curves that verve off in odd ways, as in Figures 4.2 and 4.3. The sane sort of functional relationship is assumed more likely to be significant, appear in statements of law, than the other kinds. The point is once again of course that in the world of ordinary experience there are no objective necessities. It follows, again, that among other things, to say that there is nothing about smooth curves as such that guarantees or helps to guarantee that these relationships apply as well to unobserved cases as to observed cases. Those who argue for a special inductive status for red and green as opposed to grue and bleen, or for smooth as opposed to irregular relationships, simply fail to understand, what Goodman clearly grasps, that the argument of Locke and Hume against objective necessary connections precludes neat properties such as red and green or the neat relations of smooth curves from having any special inductive status.

The same point should be made about those who find "natural kinds" to be a fundamental feature of the world.[110] Natural kinds are those properties that naturally occur in lawful relations to others; they are properties that somehow guarantee their own significance. Real definitions are hovering nearby; the Aristotelian roots of this idea are clear. But what was argued in the early modern period when scientists were learning to search for explanations in terms of matter-of-fact regularities, the properties of things that we know by sense experience lack any such logical or ontological feature that would guarantee that they are hooked lawfully to other properties. It is the fact that a predicate occurs in statements of law that make the property a natural kind, rather than its being a natural kind that guarantees its appearance in lawful relations. It is significance that

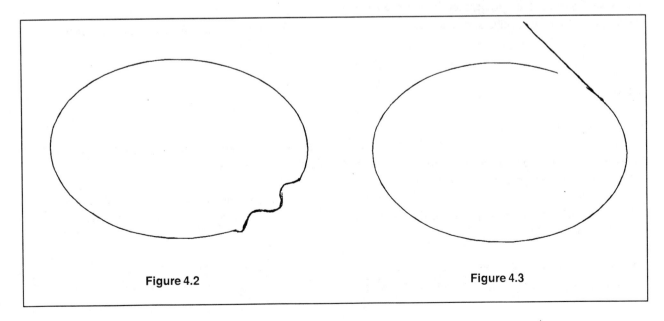

Figure 4.2

Figure 4.3

creates natural kinds, not natural kinds that creates significance.

How, then, do we distinguish green from grue? or what we might call inductively nice curves from those that are inductively implausible? What criterion do we use to distinguish green as picking out hypotheses that are lawlike from grue which does not occur in lawlike hypotheses? why is the hypothesis that the orbit of Jupiter is elliptical lawlike while the hypothesis that it is sinuous not lawlike? why is the hypothesis that the relationship between pressure and temperature in nitrogen samples linear lawlike while the hypothesis that the relationship involves sharp bends not lawlike? These questions are in effect asking what makes for a good hypothesis? How do we select from among all possible hypotheses those that are genuinely worthy of test?

We have already attended to various aspects of this issue. We have in particular seen how Kuhn provides one answer to this question. This answer is that the range of hypotheses that are worthy of test is determined by an background abstract generic theory or paradigm. This is a theory of the form

(b)   For any species f of genus **F**, there is a species g of genus **G** such that, for any x, x is f only if x is g

As we have throughout insisted, a good hypothesis must fit with the requirements of our antecedently accepted background theory. Thus, to go over the point again, we are looking at a new specific sort of system

$$F_n$$

where

$$F_n \text{ is } \mathbf{F}$$

Given this, the theory (b) applies to this new area, and we can assert of such systems the law that

(c)   There is a species g of genus **G** such that, for any x, x is F_n only if it is g

This asserts that there is a law for systems F_n without, however, asserting specifically what this law is. (c) delimits

a *range* of hypotheses that are *worthy* of the scientist's consideration. These are, let us say,

(d)
$$\text{For any x, x is } F_n \text{ only if x is } G_n^1$$
$$\text{For any x, x is } F_n \text{ only if x is } G_n^2$$
$$\text{For any x, x is } F_n \text{ only if x is } G_n^3$$

where each of the G_n^i is **G**. In contrast, if K is not **G**, then

(e)   For any x, x is F_n only if x is K

is *not* among the hypotheses that the scientist must consider; (c) deems it to be unworthy of her attention. We cannot say that the hypotheses (d) as yet are to be accepted as lawlike; none has as yet as part of the research process been confirmed. Only further research will provide data that eliminate all but one of these hypotheses, to justify the assertion of the remaining hypothesis as lawlike. Still, precisely because (e) does not fall within the range delimited by the generic theory or paradigm (b), we are right at the start of the research process able to eliminate it as a candidate for lawlike status: it is in effect gruesome.

Similar remarks hold for the other examples that we provided. The first of these was the relation between pressure P and temperature T of samples of nitrogen at constant volume. Suppose that we have antecedently examined another gas, say carbon dioxide, and have obtained the results recorded in Graph 4.7.

Here we have

(2)   For any sample of carbon dioxide gas, $P = k\,T$

where

$$k = 2$$

We have previously generalized from this result to the generic hypothesis

(3)   For any sample of any sort of gas, there is a constant n such that for any sample of that gas $P = n\,T$

(3) now delimits a range of alternatives when we come to examine samples of the gas nitrogen. Given that

Graph 4.7

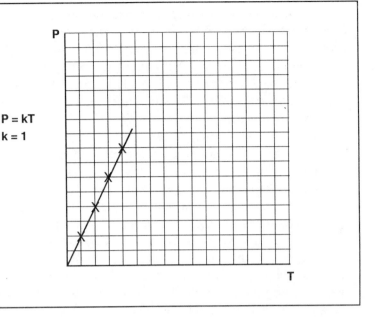

P = kT
k = 1

nitrogen is a gas, we can deduce from (3) that

(4)   There is a constant n such that for any sample of nitrogen $P = n\,T$

Given that the paradigm (3) implies this about samples of nitrogen, it follows that we can rule our hypotheses of the forms represented by Graphs 4.4 and 4.5; curves of these sorts are excluded by the paradigm as not lawlike. (4) asserts that *there is* an equation that connects pressure P and temperature T, and that this is equation has the *generic form* of being linear.

Or, consider again, the case of Jupiter. As we have seen when we discussed criteria for the selection of a good hypothesis, N. R. Hanson has suggested[111] that we need to distinguish

(R1)   reasons for accepting a particular minutely specified hypothesis H,

from

(R2)   reasons for suggesting that, whatever specific claim the successful H will make, it will, nonetheless, be a hypothesis of one *kind* rather than another.

The example which he uses is that of Kepler. He takes for granted that Kepler has established that the orbit of Mars is a certain specific ellipse. This is a regularity to the effect that

(g1)   Any position of the orbit of Mars lies on a certain specific elliptical curve

Kepler now generalizes to the generic hypothesis that

(g2)   For any planet, there is some curve of elliptical form such that any position of the orbit of that planet lies on that curve

We now of course assume that

Jupiter is a planet

and deduce from (g2) that

(g3)   There is some curve of elliptical form such that an position of the orbit of Jupiter lies on that curve.

Thus, as Hanson says, "what makes it reasonable to anticipate that H will be of a certain type is *analogical* in character."[112] The generic theory (g2) asserts that there is a shared logical form or analogy that holds among the laws that describe the orbits of the various planets. The law (g3) that follows from (g2) ascribes this *generic form* to the specific law that describes the motion of Jupiter.

(g3) is a generic law. Kepler's cognitive interest is in a specific law; he wishes to obtain the specific curve that describes the orbit of Jupiter. (g3) tells him where to look. The gappy generic law (g3) about Jupiter's orbit excludes from among the hypotheses that it is reasonable to test those which ascribe to Jupiter's orbit the curves in Figures 4.2 and 4.3, orbits which are sinuous when unobserved or orbits on which the planet after a time moves off on a straight line. More strongly, since (g3) asserts that there is such a curve, it tells Kepler that the search will be successful. This is normal science as a puzzle-solving activity.

Goodman, in his *Fact, Fiction and Forecast*, proposes what seems at first glance to be a rather different criterion for distinguishing lawlike from non-lawlike hypotheses. He in effect turns to Gustav Bergmann's notion of *significance*.[113] A predicate is significant when we are prepared to law-assert a statement of regularity in which it appears. But we are prepared to law-assert a regularity only on the basis of past experience, only if it has been put the test of predictive success and passed that test. Using this notion Goodman argues that there is a pragmatic criterion available to distinguish green and grue, or, what is the same, lawlike hypotheses (those which contain the predicate "green") and non-lawlike hypotheses (those which contain the predicate "grue").

Goodman proposes that there is a certain pragmatic criterion that can do the job. He defines a number of concepts in order to state his criterion. One predicate is said by Goodman to be *better entrenched* than another just in case that it has in fact, in the past, been *successfully projected* more often than the other (*Fact, Fiction and Forecast*, pp. 85-6, p. 94), where to be projected is to occur in an hypothesis that is used to predict (p. 89). The pragmatic criterion that Goodman proposes is that a projection is to be ruled out just in case that it conflicts with the projection of a much better entrenched predicate (p. 96). One hypothesis overrides another just in case the two are contrary and the former contains a better entrenched predicate and is contrary to no further hypothesis with still better entrenched predicates. Thus,

(i)   All emeralds are green
(ii)   All emeralds are grue

are contrary. It is in fact true that 'green' has been projected more often than 'grue.' There is no similar hypothesis which is contrary to (i) but with a better entrenched predicate in place of 'green.' It follows that (i) overrides (ii). This becomes the criterion for an

hypothesis being *lawlike*, and worthy of use in explanations and predictions: Goodman's proposal is that an hypothesis is *lawlike* if all contrary hypotheses are overridden; that it is *gruesome* if overridden; and that it is non-projectible if it is contrary to another and neither is overridden (p. 101).

In order to extend this idea, Goodman goes on to suggest that entrenchment may be earned — in actual projections — but also inherited. One predicate is the parent of another if the extension of the latter is included in that of the former, i.e., just in case that the former is a genus relative to the latter as a species (p. 106). Two predicates that have equal earned entrenchment can be compared with respect to inherited entrenchment by comparing the earned entrenchment of their parents (p. 107). Total entrenchment is a matter of earned plus inherited entrenchment. In particular, if we have two predicates occurring for the first time in an hypothesis, their earned entrenchment will be negligible and equal, but one might have more inherited entrenchment; in that case the hypothesis containing it will be reckoned lawlike, and its competitor gruesome. This allows for judgments of relative projectibility in cases of hypotheses containing novel predicates.

Finally, Goodman notes the role that *overhypotheses* can play. One hypothesis is an overhypothesis of a second just in case that the antecedent and consequent of the former are parents of the antecedent and consequent of the latter (p. 110), that is, just in case that the predictions of the former are generic where the predictions of the latter are specific but within the same genus. If an over- or generic hypothesis is itself lawlike, then it will not only tend to confer that status on more specific hypotheses that it is over, but, in addition, data that tend to confirm it tend also to confirm the more specific hypotheses (pp. 111-12). However, the amount of support an overhypothesis lends to a more specific hypothesis in general varies inversely with its generality: the more sweeping is the generalization, the broader it is, the less effect it has. A more general overhypothesis lends more support than a more specific hypothesis only if it is supported by data other than those supporting the less general hypothesis (p. 116). Thus, in reckoning the support that an overhypothesis gives to more specific hypotheses falling under it, one must take into account not only the data but also its variety.[114]

The role of overhypotheses can be easily illustrated. Consider what has been called "Bode's Law."

This "law" states a relation among the distances of the planets from the sun.[115] More specifically, it purports to state a numerical relationship among the mean distances of the planets from the sun that will enable astronomers to predict the positions of other, hitherto unknown, planets. It was first stated in the 18th century by the German astronomer Johann Elert Bode. A series of numbers is constructed in the following way. Consider a sequence of 4's; add 3 to the second 4, 6 to the third 4, etc., doubling each time; divide each sum by 10. This sequence of numbers represents the mean distance of the planets.

Bode proposed his law in 1772. At that time, neither the three outermost of the major planets nor any of the minor planets or asteroids were known. Uranus was discovered in 1801, roughly in conformance with Bode's rule. Bode himself pointed out that there was a gap in the planets at 2.8, and after the discovery of Uranus a search was undertaken, resulting in the discovery of the minor Planet Ceres in 1801, at roughly the proper distance.

But the rule breaks down for Neptune and Pluto. Moreover, it is arguable that it also breaks down for Mercury, since 1.5 (= half of 3) should be added to the first 4 if we are to keep to the later pattern of the progression.

What Bode discovered was a regularity holding between a certain relational property of distances from the sun, E. This relational property is given by Bode's formula. What Bode proposed is that this relational property of distances from the sun is correlated with C, the presence of a planet. This is a regularity to the effect that

Whenever C then E

or

All C are E

This pattern enabled Bode to infer where hitherto unobserved planets should exist. (It suggests, for example, that there ought to be a planet in the asteroid belt. There is none, but one might speculate, as many have done, that there once was a planet there that somehow disintegrated.) The exact pattern in fact breaks down, making the connection more uncertain than Bode reckoned; indeed, to put it boldly, it falsifies the "law." However, even if it had not broken down so clearly, it would still have been a "mere" regularity or *correlation*. And, being merely a correlation, the claim that there was a certain causal pattern present cannot be inferred. There is nothing that distinguished Bode's law as lawlike, that is, like Goodman's (i), or gruesome, that is, like Goodman's (ii). If "E" represents, as we have suggested, a relational property of distances from the sun, then what Bode would of course like is that this property E be *projectible*. But is it? Is it like *green* in (i) or is it like *grue* in (ii)?

What one would like to have is a background theory that would provide an explanation in terms of the mechanisms of planetary formation. One can imagine what such a theory would be like. Rooted in physics and chemistry, it would describe the processes through which the gaseous envelope of a star condensed into solid objects like Mars and semi-solid objects like Jupiter in the regular pattern discovered by Bode. Because it would describe a variety of processes, it would be a generic theory. Such a theory, if available, would explain why planets are

	Mercu	Venus	Earth	Mars	Aster	Jupit	Saturn	Uranus	Nept	Pluto
	4	4	4	4	4	4	4	4	4	4
add	0	3	6	12	24	48	96	192	384	768
Bode's law	0.4	0.7	1.0	1.6	2.8	5.2	10.0	19.6	38.8	77.2
True distance	0.39	0.72	1.0	1.52	2.77	5.2	9.54	19.19	30.07	39.46

Table 4.14

located at just the intervals predicted by Bode's Law. Because such a theory would be generic, it would constitute an overhypothesis for Bode's Law. Because generic, it would have made predictions in other areas; its predicates would therefore be parents of predicates in Bode's Law. Because the predicates in the theory would be entrenched, the predicates of the specific hypothesis such as Bode's Law covered by the theory would inherit this entrenchment. This would mean in turn that the predicates of Bode's Law would be projectible, and the Law would indeed be a Law, rather than a mere correlation. That is, if such a theory were available, it would render Bode's Law lawlike rather than gruesome — *if* such a theory were available ... Unfortunately, such a theory is not now, nor ever has been available — such a theory simply does not exist. There is thus no reason to suppose that the correlation is the result of anything other than coincidence. As one authority has put it,

> The "law" is probably only a coincidence, no scientific explanation being evident.[116]

The point here is one that we have previously emphasized, that correlation by itself does not establish causation.

Goodman proposes a criterion for predicates being projectible, or, what amounts to the same, for predicates being, in Bergmann's sense, significant. This is the criterion of entrenchment. In the first instance, this is simply the property of being actually used successfully in predictions. But he also makes the point that entrenchment can be inherited from the predicates of a theory, where the latter predicates are generic. Goodman is therefore arguing, quite reasonably, that embedding a regularity or correlation in a background generic theory can transform the former from being a "mere" correlation into being a genuine lawlike hypothesis, one worthy of use in explanation and prediction.

What is interesting about these proposals of Goodman is that, in spite of their apparent difference from those of Kuhn, they in fact generate an account of research that is not essentially different from the account of normal scientific practice given by Kuhn. Goodman's criterion for a good hypothesis selects as good those hypotheses that are picked out by Kuhn's criterion. The

crucial point is that overhypotheses in Goodman's sense are nothing other than Kuhn's background abstract generic theories or paradigms.

To see this, we should begin by noting that predicates such as 'x is **H**-ed' are, in Goodman's sense, parents of the specific predicates H_1, H_2, ..., that fall under the genus **H**.

We next note that the paradigm (b)

> For any species f of genus **F**, there is a species g of genus **G** such that, for any x, x is f only if x is g

entails the law

(β)   For any x, x is **F**-ed only if x is **G**-ed

This is but another way of saying that, from the determinable initial condition (x') that a is **F**-ed, the paradigm (b) yields the determinable prediction (y') that a is **G**-ed. What is interesting is that (β) is, in Goodman's sense, an *overhypothesis* for the laws (a)

> For any x, x is F_1 only if x is G_1
> For any x, x is F_2 only if x is G_2

from which we supposed (b) to be generalized, and for the hypotheses (d)

> For any x, x is F_n only if x is G_n^{1}
> For any x, x is F_n only if x is G_n^{2}
> For any x, x is F_n only if x is G_n^{3}

in the new area to which the scientist was applying (b), as part of the puzzle-solving activity of normal science.

Now, according to Kuhn, hypotheses other than those of the set (d) are not worthy of the researcher's attention. It is easy to see, however, that in Goodman's terms, these other hypotheses are gruesome. For, when the laws (a) are used to successfully predict, then these data also, as we saw, test the theory (b), or, what amounts to the same, the theory (β). However, for Kuhn, normal research is always *paradigm guided*. Hence, to project a law like one of (a) is *ipso facto* to project the theory. Thus, in normal research, when the laws (a) yield successful projections, then the paradigm (b), or, equally, the law (β) *ipso facto* also yields a successful projection. As a consequence, in normal science, when the laws (a) are successfully projected, not only do the specific predicates they contain earn entrenchment, but so do the generic predicates of the paradigm (b), and the determinable predicates of (β). But the predicates of (β) are parents of those occurring in the hypotheses of the set (d). The latter therefore have inherited entrenchment. It is by virtue of this inherited entrenchment that hypotheses other than those of (d) are overridden and thereby judged to be gruesome.

None of the hypotheses (d) is yet lawlike, however. For, they are contraries but none overrides the others. They are, therefore, non-projectible. To decide which one is lawlike one can do no other than actively undertake research, to locate further data to effect a "crucial experiment."[117] This is what our researcher, discussed above, set out to obtain, and, we supposed, she succeeded. The data eliminated all but (d_1) as false. At that point, (d_1) ceased to be non-projectible, and became lawlike.

Moreover, not only is (d_1) thus lawlike, and therefore confirmed by its instances, but it also derives support from the data that confirm the hypotheses (a). For these latter data also support the hypothesis (β) which is an overhypothesis for (d_1). As Goodman says, if (β) were supported merely by the data that support (d_1), then it would yield little further support of (d_1). In Lakatos' terms, to appeal to it would be *ad hoc*, since it would be doing nothing more than explain (predict) the very facts to which it was supposed to give additional support.[118] However, the background generic theory or paradigm (b), or, equally, the law (β), has shown itself to be *empirically progressive*: that is what makes it acceptable for explanation and prediction, and for guiding research, according to Kuhn. And to say that is to say that it has a *variety* of data supporting it. This means, in Goodman's terminology, that the overhypothesis satisfies the condition necessary for it to give additional support to the specific hypotheses falling under it.

We can conclude, I think, that the picture of normal research generated by Goodman's account of science is little different from that given by Kuhn.

c. Revolutionary vs. Conservative Science

What is characteristic of normal science is that it results in the discovery of causes, or, what is the same, laws. However, it is sometimes the case that research does *not* lead to the discovery of causes. For one who is moved by cognitive interest in matter-of-fact truth, this must soon turn out to be an unsatisfying strategy. It will be unsatisfying simply because it does not lead to the discovery of the relevant causes. One moved by the love of truth will therefore at some, not well-defined,[119] point become dissatisfied with this strategy of research and will turn to another that he or she expects, or at least hopes, will yield results that will turn out to be more satisfactory to our scientific cognitive interests, to the passion of curiosity, than are the results generated by the current paradigm or generic theory.

Kuhn has offered an account of what happens in science when a method formerly yielding results satisfactory to the scientist's desire to discover precise laws has ceased to be so. What happens then is that scientists begin to employ a different strategy, one that Kuhn argues persuasively argues is characteristic of *revolutionary periods in science*.

As we have seen, for Kuhn normal science is puzzle-solving. A paradigm is therefore a tool: it poses puzzles and guides one to their solutions. Unfortunately, there are times when it ceases to do this. Sometimes, when scientists repeatedly attack a puzzle, they fail to find solutions, even though the paradigm says that solutions do exist, that there *are there* to be found. To be sure, such failures do not falsify the paradigm. Nonetheless, they do call into question the utility of the paradigm as a research-guiding tool. If such failure is repeated, what was merely another puzzle becomes an *anomaly* (*The Structure of Scientific Revolutions*, sec. VI). With the appearance of an anomaly, a *crisis* in normal scientific practice begins to appear (sec. VII). When this happens, some, at least, among the scientific community begin to cast around for a new tool (p. 76): when the efficiency of a paradigm as a guide in normal research declines beyond a certain point, then scientists begin to look for alternative theories to guide them (p. 83, p. 158, p. 169). Now, by hypothesis, such a search for a new theory will not itself be paradigm-guided. It will therefore not be normal science. Rather, Kuhn suggests, it can reasonably be called "revolutionary science" (secs. IX, XII).

The problem is that when the paradigm is given up, then one has no *theory* to guide one's search. There is therefore no theory to determine the group of hypotheses that it is reasonable to put to the test. There is, in other words, no Principle of Limited Variety that will place reasonable limits upon the area in which to search, nor, indeed, any Principle of Determinism to tell one that there is a solution there to be found. This creates a real problem however for the research process.

This is the problem: Given the infinite range of possibilities — Goodman's simple examples make clear that such an infinite range always exists — , no search can proceed unless *some* guide is available, some limitation on the Variety one must deal with in deciding which hypotheses to put to the test. In part, of course, one must be guided by the idea that the new paradigm must account for the successes of the old one (*The Structure of Scientific Revolutions*, p. 169). This importance of this criterion we have repeatedly emphasized. It was on this basis that we rejected the theories of Velikovsky and of homeopathic medicine, for example.

Nonetheless, even when we apply this criterion, it is still true, as Goodman's examples of emeralds being green and being grue makes clear, an infinite range of contrary possibilities confronts the researcher, each hypothesis in the range compatible with the past data that support the successful theories one has up to now accepted.

Kuhn argues that in their practice the community of scientists have discovered a solution to this problem. He argues that in such situations, while no individual scientist can confront the totality of possibilities, the *community* as such can. It does this by letting scientists in periods of revolutionary science be guided by metaphysical and/or aesthetic considerations that vary subjectively from scientist to scientist. As Kuhn puts it, these considerations are

> ...the arguments, rarely made entirely explicit, that appeal to the individual's sense of the appropriate aesthetic — the new theory is said to be "neater," "more suitable," or "simpler" than the old.

He goes on,

> The early versions of most new paradigms are crude. By the time their full aesthetic appeal can be developed, most of the community has been persuaded by other means. Nevertheless, the importance of aesthetic considerations can sometimes be decisive. Though they often attract only a few scientists to a new theory, it is upon those few that its ultimate triumph may depend. If they had not taken it up for highly individual reasons, the new candidate for paradigm might never have been sufficiently developed to attract the allegiance of the scientific community as a whole (*The Structure of Scientific Revolutions*, pp. 155-6).

Each scientist will confront some group of hypotheses, some Limited Variety. But because the criteria that Limit the Variety vary from scientist to scientist, because the criteria of selection are subjectively variable, the community of scientists as a whole can confront the very broad range of alternatives that is opened up once one decides to search for a new paradigm.

It is important to note that this introduction of subjectively variable criteria of theory choice at the stage of research that occurs when one is searching for a new paradigm is *not* to introduce irrationalism into science.[120] Rather the introduction of such things as aesthetic value is but a *means* to an end[121] — where the end is, of course, the discovery of a new paradigm (*The Structure of Scientific Revolutions*, p.169). In fact, for all that we know, given the Humean limitations on induction, the fact that all scientific judgments are fallible and that all scientific hypotheses must, as Judge Overton has emphasized, be accepted only tentatively, it might turn out that the older paradigm will be able to solve the puzzle after all. To allow for this possibility, the community of scientists permits some researchers to tolerate a lower rate of efficiency than do others; the former will continue to use the old paradigm while the latter search for a new (p. 159). As we have seen, a paradigm or background generic theory will in fact be well-confirmed; this is a consequence of the fact that it has been successful in guiding research in the practice of normal science. This creates in most scientists, even when confronted with an anomaly during a period of crisis, a sense that the old paradigm will ultimately succeed. There will be a resistance to the newer theory.

> The source of resistance is the assurance that the older paradigm will ultimately solve all its problems, that nature can be shoved into the box the paradigm provides. Inevitably, at times of revolution, that assurance seems stubborn and pigheaded as indeed it sometimes becomes. But it is also something more. That same assurance is what makes normal or puzzle-solving science possible (pp. 151-2).

Thus, there is no *precise* or *well-defined* point at which a paradigm is to be given up; that, too, varies subjectively from individual to individual. But that, too, is part of the *rational research strategy* of revolutionary science.[122] In periods of crisis that occur when one is in the process of searching for a new paradigm the methodological rule is:

> **let theories proliferate**[123]**; let scientists be guided in their search by their subjective values**[124]**; and let them accept for purposes of explanation and prediction those theories that satisfy those subjective values**[125]**.**

By following this rule, the community of scientists *as a whole* is able to deal with the vast range of possibilities it must investigate. Thus, this rule is adopted because having its members conform to it in periods of crisis is a *means* for *efficiently achieving* the *end* shared by all members of the community, namely, that of discovering a new paradigm capable of successfully guiding research in the normal way. Thus, the resort to subjectively variable criteria of choice is a *rational response* to crisis in the scientific community.

Now, for our purposes the important point is that *the subjective values provide one with reasons in periods of revolutionary science to take up a theory never previously tested*[126]: to repeat Kuhn's point,

> **If they [those who initially adopt an untested candidate for a new paradigm]**

> **had not quickly taken it up for highly individual reasons, the new candidate for paradigm might never have been sufficiently developed to attract the allegiance of the scientific community as a whole** (*The Structure of Scientific Revolutions*, p. 156).

An example will help.

Kuhn's point can be illustrated very clearly if we look at the considerations that led Einstein to develop the special theory of relativity.[127]

At the end of the 19th century, physics consisted of two theories, classical Newtonian mechanics for mechanical phenomena and Maxwell's theory for electromagnetic, including optical, phenomena. Behind both were a set of common kinematical assumptions about moving bodies. This combined theory had proven itself to be both theoretically and empirically progressive. But as with all comprehensive generic theories, there were unsolved problems or anomalies, where the fit was loose between the observed facts and the sorts of laws predicted by the theory. In particular, there were a number of problems between what Maxwell's theory predicted in the area of optics and the observed facts. Many of these were solved by H. A. Lorentz, when he developed a particulate theory of electrical phenomena, the theory of electrons.[128] There remained other unresolved anomalies, however. These included the results of the Michelson-Morley experiment.[129] This experiment measured the speed of light with a moving apparatus and found that the speed of light was constant whether measured in the direction of motion of the apparatus or measured perpendicularly to it, contrary to the requirement of the theory that these differ. Fitzgerald was able to show that one could account for this fact if one assumed that objects became shorter in the direction of motion in accordance with a certain rule; the relevant rule was developed by Lorentz. These rules are now known as the Lorentz transformation; and the postulated contraction is known as the Lorentz-Fitzgerald contraction.[130] But this hardly helped. On the one hand, there was no independent empirical support for this theory of changing lengths; and, on the other hand, there was nothing in the theory itself that could justify this additional assumption. The most that Lorentz could say for it was that if it were accepted then the facts it was designed to explain could be accounted for, and that "From the theoretical side...there would be no objection to the hypothesis."[131] It was clearly *ad hoc*. Repeated attempts were made to detect the Lorentz-Fitzgerald contraction; they failed repeatedly.[132] There was one phenomenon for which the Lorentz' rules could provide an account. This was the variation of the mass of an electron with its velocity. Lorentz' formula for the variation of mass with velocity was confirmed in a careful series of experiments.[133] Still, the contraction itself remained undetected. Moreover, the contraction hypothesis and the Lorentz rules remained unintegrated into the theory, sitting there as a conjoined set of extra assumptions. In spite of the repeated failures to confirm the contraction hypothesis, and in spite of the lack of theoretical integration of that hypothesis, Lorentz, and most other physicists, continued to law-assert the accepted theory. In particular, the kinematical assumptions were never questioned.

The important point to be noted is that none of the revisions to the theory thought to challenge the kinematical assumptions upon which it was based. All proposed modifications designed to accommodate the theory to apparently anomalous phenomena were made at a much lower level in the theory structure. It was Einstein's genius to develop an alternative that succeeded where the older theory failed. But he could do this only because he was prepared to challenge the kinematical assumptions of the older theory, and develop a new, contrary theory that embodied a new kinematics. These were assumptions that had previously hardly been investigated, let alone challenged — though as Einstein's work showed, they could in fact be challenged on empirical grounds. The new theory was revolutionary, and in order to achieve this success a revolutionary strategy in research was needed.

The young Einstein was not so happy with a theory that was proving itself increasingly less able to solve problems, the anomalies that confronted it. For Einstein the likelihood that the theory was false had increased sufficiently that it called the theory itself into question. In his view, it was time to search for a new theory. Bothered by the increasing number of anomalies, and the inability of the accepted theory to solve them, he proposed an alternative. This was a new generic theory that was radically contrary to the accepted theory at many fundamental points, though it fit with what was predicted by the accepted theory in certain limiting cases. The new theory was that of Special Relativity.[134]

There were two basic assumptions to this theory (p. 2). *First* was the assumption that the result of the Michelson-Morley experiment was basic, that the velocity of light was the same in whatever frame of reference it was measured. *Second* was the assumption that Einstein referred to as the "Principle of Relativity", that basic laws had to have the same form in all frames of reference. The latter provided a generic constraint on the forms that laws could assume. It asserts that *for any* general sort of system, *there are* laws that hold in the system and these laws have a certain generic form. The theory thus has the abstract mixed-quantificational, generic form that we have found Hanson and Kuhn to ascribe to theories. Note that it is an empirical claim that might well have turned out to be false; indeed, if the accepted theory were true, then in fact this generic law would be false.

These assumptions had dramatic implications. First, they eliminated the need to postulate any form of an electromagnetic ether to provide a stationary frame of reference, something that had been taken for granted by electromagnetic theory throughout its history. As Einstein put it, one of the results of his theory is that "the introduction of a 'Lichtäther' will be proved to be superfluous..." (p. 2). Secondly, they entailed a radical revision in the kinematical assumptions behind the physics of moving bodies. Specifically, it implied the relativity of length and time (§ 2). We must distinguish the length of a moving rod as measured by means of a rod by an observer moving with the rod, and the length of a moving rod as measured by an observer in a stationary system. In traditional theory, these two measure the same thing:

In the generally recognised kinematics, we silently assume that the lengths defined by these two operations are equal... (p. 7).

In Einstein's theory the two concepts will yield different results. A similar point holds with regard to time. Where the old theory allowed that the concepts of locally simultaneous and non-locally simultaneous would always yield the same results, on Einstein's theory they differ.

We ... see that we can attach no absolute significance to the concept of synchronism; but two events which are synchronous when viewed from one system, will not be synchronous when viewed from a system moving relatively to this system (p. 8).

This is a radical departure from, and is contrary to, what was implied by the accepted theory.

In fact, the assumption of the equivalence of the local and non-local definitions of length and simultaneity is a gratuitous assumption within the classical framework. Even within that framework, the operational concepts are different. The assumption that they are equivalent has no empirical basis. It derives more from the Newtonian metaphysics that space is as it were the sensorium of God than from any careful logical analysis of the concepts themselves and how they should be related within the logical structure of the empirical theory. The assumption of the constancy in all frames of reference of the speed of light forced Einstein to reformulate the operational definitions for non-local simultaneity (p. 1), and thereby the concept of "non-local length" (p. 2). These new operational definitions proved crucial for the theory. If once again, following Bergmann, we refer to a concept that appears in a statement of law as "significant", then, when the new theory proved itself as empirically progressive, these concepts proved to be extremely significant. As so often, though an operational definition is not a theory, a new operational definition may well lead to the formulation of a good scientific theory. In this case, however, the operational definitions served the further role of eliminating metaphysical assumptions that protected the empirical assumptions of the older theory from being questioned. The new operational definitions enabled scientists to doubt propositions that had been in principle open to challenge but where they could not be challenged because they were protected by non-empirical, or, if you wish, metaphysical dross.

From the Principle of Relativity, the generic constraint on the form of laws in systems, together with the observed fact of the constancy of the speed of light in all frames of reference, Einstein is able to deduce the Lorentz transformations that relate velocities and times "from a stationary system to a system which moves relatively to this with uniform velocity" (p. 8). The deduction that

Einstein makes is analogous in form to Newton's deduction of the law of gravity. On the one hand Newton assumed his basic laws of motion; this theory provided generic constraints on the forms that laws could exemplify. On the other hand Newton took for granted Kepler's laws. From these two things, theory plus data, Newton was able to deduce that the force that moved the planets varied inversely as the square of the distance and directly as the product of the masses.[135] Once the law of gravity had been deduced from the phenomena, it could then be used to explain those phenomena: Kepler's laws follow directly upon the assumption of the generic theory and the law of gravity. Einstein's is a similar deduction of an explanatory law from generic constraints, on the one the hand, namely, the Principle of Relativity, and, on the other hand, observed phenomena, namely, the constancy of the speed of light in all frames of reference. Then once the deduction has been performed, the law that had been derived can be used to explain the phenomena. Specifically, the Lorentz transformations immediately imply, and therefore explain, the constancy of the speed of light in all frames of reference.[136]

The Lorentz transformations also imply that objects *as it were* contract in the direction of their motion. The contraction, however, is not real but only a consequence of the relativistic assumptions of the new theory. What is a real contraction under the old theory is due instead to a change in the local units of length and time, and, since we have merely two different descriptions of the same reality, there is no real contraction. No wonder, then, that it could not be detected by experiments. The theoretical assumptions of the new theory thus provide a non-*ad hoc* account of the Fitzgerald "contraction."

Einstein's theory could thus account for phenomena that the accepted theory could account for only if it made *ad hoc* assumptions. Furthermore, it could account for the behaviour of the electron which the accepted theory could deal with only through *ad hoc* assumptions (p. 10). Einstein's theory thus immediately proved itself, if not empirically progressive, accounting for facts that the hitherto accepted theory could not account for, then at least able to account for empirical facts that the accepted theory could not account for save by means of an *ad hoc* assumption.[137]

But what prompted the initial acceptance of the new theory?[138] What we must recognize is that a *new candidate for paradigm* is a theory that has never been projected. Initially, then, it cannot be taken to be empirically progressive; in this sense it has not acquired any confirmation from predictive success. *Nor is there any background theory to provide guidance as to what form the laws should take.* There is no background theory to confer any prior probability on the new theory. Indeed, much to the contrary. The old theory has not been eliminated, and is contrary to the new theory. Its posterior probability may be decreasing as it fails to solve the anomalies that confront it, but it has been highly confirmed in the past and its probability is far from zero. In fact, were other things equal, that probability would no doubt be sufficient to over-ride the probability of the new theory even after the initial confirmations of the latter, e.g., in this case by solving the problems of the electron that confronted Lorentz. What, then, makes the new theory acceptable prior to the needed confirmations?

The answer that Kuhn gives is that it is chosen on the basis of subjectively variable values. This answer is confirmed in the case of Einstein. In the first place, the new theory is more coherent; as one commentator has noted,

> The greatest triumph of this new theory consists ... in the fact that a large number of results, which had formerly required all kinds of special hypotheses for their explanation, are now deduced very simply as inevitable consequences of one single general principle.[139]

In the second place, there is an important *empirical symmetry* which it reflects in its theory but which the older theory treats as a *theoretically asymmetry*. Einstein considers the relation of a magnet and a conductor. We may either move the magnet with respect to the conductor or move the conductor with respect to the magnet. Whenever there is a relative motion then a current is generated in the conductor. The traditional explanations, based on Maxwell's equations, are different, however. In the first case, the magnet moves while the conductor is fixed relative to the ether. Due to the variation of the magnetic field with time, the older theory asserts that an electric field arises in the whole of space and in particular an electron in the conductor will experience a force and a current will flow in the

conductor. In the second case, the magnet is kept fixed relative to the ether and the conductor is moved. Since the magnet is fixed, the magnetic field is static and no electric field is produced. However, an electromotive force with no energy in itself is generated in the conductor and this in turn causes an electric current of exactly the same force and magnitude as in the first case. Einstein, in the initial sentence of his paper "On the Electrodynamics of Moving Bodies," directs attention to this asymmetry, and clearly gives it to be understood that this is a flaw in the traditional theory.

> It is well known that if we attempt to apply Maxwell's electrodynamics, as conceived at the present time, to moving bodies, we are led to asymmetry which does not agree with observed phenomena (p. 1).

Einstein is going to remove this flaw with his new theory — and does: after he proves his point, he is careful to tell us that "It is further clear that the asymmetry mentioned in the introduction which occurs when we treat of the current excited by the relative motion of a magnet and a conductor disappears" (p. 22).

The crucial point here is that Einstein invokes the *aesthetic criterion* of symmetry as a way of justifying the acceptance of his theory. This is precisely the sort of policy for theory-choice that scientists adopt during periods of revolutionary science, according to the Kuhnian methodology. It is also true that during periods of revolution, different scientists adopt different, subjectively variable criteria for theory choice. Thus, on the basis of certain non-empirical metaphysical considerations, Whitehead rejected Einstein's account of the relativity of simultaneity.[140] In the end, however, it was Einstein's theory and not Whitehead's that proved to be empirically progressive. And in the end, the subjectively variable criteria of the revolutionary period led to the selection of a theory that proved to be empirically progressive. As Einstein's theory did lead to the discovery of new laws, so it eventually formed the basis of a new period of normal science in which the subjectively variable criteria of theory choice no longer had the role to play that they played during the revolutionary period.

Einstein and the discovery of the special theory of relativity thus illustrate very nicely the claims that Kuhn makes about rational procedures in the methods of scientific discovery.

There is another important point to be made about the replacement of an older paradigm by a newer.

Kuhn points out that, contrary to Popper, a paradigm is not easily to be falsified. As Kuhn puts it,

> ...anomalous experiences may not be identifiable with falsifying ones (*The Structure of Scientific Revolutions*, p. 146).

"Indeed, [Kuhn adds,] I doubt that the latter exist." The last remark is no doubt something of an exaggeration. specific hypotheses are open to direct refutation by confrontation with empirical observational data. But the point about non-falsifiability does apply to paradigms. These do not make specific predictions but instead merely state that *there is* a specific law of a certain generic sort. That is, paradigms that guide research have a mixed-quantificational structure; they are, we have argued, of the form (b)

> For any species f of genus **F**, there is a species g of genus **G** such that, for any x, x is f only if x is g

And so, "...falsification, while it undoubtedly occurs, does not happen with, or simply because of, the emergence of an anomaly or falsifying instance" (p. 147).

Since a paradigm of the form (b) cannot be rejected, as can a specific hypothesis, by a single falsifying observational counterexample, there must be other grounds that lead to the rejection of the older paradigm. But it is now clear what these ground are.

The old paradigm (b) is no longer, for a part of the research community anyway, functioning as an adequate guide in research: puzzles that it poses are no longer being solved. For the dissatisfied portion of the research community, there is a search, guided by subjectively variable constraints, for a new paradigm. This paradigm will have the same general mixed-quantificational form as (b). It will make the same predictions about specific laws that the old paradigm does in areas where the latter is successful. But it will also be such that it yields new predictions with regard to the specific laws that hold in the areas that have turned out to be anomalous and that have generated the crisis. The new paradigm will thus apply to the same specific sorts of system as the old: in terms of our schematic example (b), the genus **F** in the new paradigm will be the same as in the old paradigm (b). But it will not predict simply **G**. Rather, it will predict something else. say **G'**, where this genus has many of the same species as the old genus **G**, but also different species for the areas of anomaly. If we recall the Einstein example, both theories apply to motions that are both far less than and close to the speed of light: both apply to systems of the generic sort **F**. The old theory — Newton's — provides generic constraints **G** of the same for both those motions that are close to the speed of light and those that are far less than the speed of light. The new theory — Einstein's — imposes a set of generic constraints **G'** which are the same as those of **G** for motions at speeds much below the speed of light but a set of constraints that are different from those of **G'** for motions close to the speed of light. We therefore have a new theory

> (b') For any species f of genus **F**, there is a species g of genus **G'** such that, for any x, x is f only if x is g

The point, of course, is that the constraints imposed by **G'** on motions near the speed of light are not those imposed by **G**: the two are inconsistent.

It follows that the two theories (b) and (b') are **contrary** to each other: they cannot both be true. To be sure, for most areas they yield the same predictions. But in the anomalous areas they yield contrary predictions. If, therefore, a scientist confirms (b') then he or she will be led to accept that theory and *therefore* to reject the old theory (b). As Kuhn puts it,

> ...paradigm-testing occurs only after persistent failure to solve a noteworthy puzzle has given rise to crisis. And even then it occurs only after the sense of crisis has evoked an alternative candidate for paradigm. In the sciences the [paradigm-]testing situation never consists, as [normal scientific] puzzle-solving does, simply in the comparison of a single paradigm with nature. Instead, testing occurs as part of a competition between two rival paradigms for the allegiance of the scientific community (*The Structure of Scientific Revolutions*, p. 145).

Thus, **an older theory or paradigm is never rejected because of a single observational counterexample; rather, it is rejected only when a *new and contrary* paradigm has been confirmed by observational data and thereby rendered acceptable for purposes of observation and prediction and for guiding research into new areas**.

Let us now return to the comparison of Kuhn and Goodman.

As Kuhn has argued, and as the Einstein example illustrates, in periods of revolutionary science, when the old paradigm has ceased to be effective as a guide in research, the scientific community turns to subjectively variable values to provide a Limited Variety for the individual researchers to put to the test. These subjectively variable values are needed because the new theory has not previously been projected, so there are no empirical tests that it has survived. On this point we can contrast Kuhnian methodology with that of Goodman.

A second point relevant to the comparison of Kuhnian and Goodmanian methodologies is that the new theory or paradigm will involve concepts that did not appear in the old theory. Kuhn in this context refers to the notion of "mass" in classical and relativistic mechanics (*The Structure of Scientific Revolutions*, p. 101). In mechanics this plays a very basic role in most theoretical considerations. In the older theory of Newton, the mass of an object was constant. The result that the mass of the electron varied with velocity was anomalous, though, as we noted, Lorentz was able to advance *ad hoc* adjustments to the old theory that could make it fit these results. Einstein showed that in the new theory mass was variable, and that this variability flowed from the basic assumptions of the theory (*Principle of Relativity*, pp. 32-33). For us, the important consequence is that, since the new theory involved a differently defined concept of mass than did the older theory, then, given that the new theory had not previously been projected, neither had this new concept. The same holds for the newly re-defined "non-locally simultaneous" and "non-local length". Thus, the basic concepts in the new candidate for paradigm will be subtly different in meaning and extension from those in the old (*The Structure of Scientific Revolutions*, p. 49).

Now, from the perspective of Goodman, *the concepts of the old theory will be entrenched where those of the new will not be*. Moreover, since the period is revolutionary, there is no overhypothesis, no background abstract generic theory, to give support to the new theory, and since the competitors are equally generic, there are no other yet more generic parent predicates from which the concepts in the new candidate could inherit entrenchment. The concepts of the new theory that Einstein was preparing to use had predicates that had no entrenchment. They were neither entrenched by virtue of having been projected previously, nor had they acquired entrenchment by way of heredity.

However, an hypothesis is, according to Goodman, lawlike rather than gruesome, only if its predicates are entrenched. So Goodman's rules would exclude Einstein's theory from ever being accepted as worthy of projection, worthy of being put to the test. Thus, *Goodman's rules, if applied to periods of revolutionary science, would forbid the search for radically new paradigms*. If scientists were committed to Goodman's methodology, Einstein would never have proposed his theory to the world; he would have, like Lorentz and most of the rest of his colleagues, restricted himself to variations that fit within the framework of the older paradigm — no matter how much the likelihood of its falsity increased with its increasing tendency to lead skilled attempts to find new laws to end in failure.

Goodman's account of science is, in short, *inevitably conservative*, restricting us to theories using the concepts of an old paradigm or concepts which have these as parents. This is satisfactory for periods of normal science, which *is* in this sense conservative (*The Structure of Scientific Revolutions*, pp. 151-2). But when the old theory ceases to do its job of guiding research then one needs a non-conservative or *revolutionary strategy* that permits one to investigate, test, project, i.e., *treat as lawlike*, hypotheses that are, in terms of the concepts they use, radical departures from the old theory.

We can therefore conclude that, since Goodman's rules do not yield an efficient strategy for carrying out research when a paradigm becomes an inefficient guide, those rules do not yield an adequate general account of how to decide which hypotheses are lawlike, worthy of test, or projection. Kuhn, in contrast, with his general account of science, and his separation of normal and revolutionary research strategies, does provide an adequate account of how to decide which hypotheses are lawlike. Goodman's rules do have, however, limited validity: they amount, in effect, to Kuhn's rules for normal research.

Kuhn's account of revolutionary science has the ring of truth to it; it *seems* reasonable. Yet, is it truly so? Can some defence of it be given?

Scientific research practices of course vary. They vary between disciplines: physics is very different from geology. They also vary between normal and revolutionary science, as we have just seen. Nonetheless, there is a basic set of cognitive interests that covers all of scientific practice, including both normal and revolutionary. These cognitive interests are what moves the scientist to undertake research. What we have attempted to do is to see that these various norms that we find in actual scientific practice can be justified relative to our **cognitive interests**. We are rational or reasonable if we adopt means that are at least tolerably efficient in enabling us to achieve our ends. In the case of science the relevant ends are determined by the *cognitive interests that define science*. A methodological norm in science will therefore be such that it is reasonable to adopt just in case that, if scientists conform their practice to that norm, then they will, so far as one can tell, be able to discover laws that satisfy their cognitive interests. In other words,

a norm of scientific practice is justified as rational provided that conformity to that norm leads to results that satisfy our cognitive interests as scientists.

First among these cognitive interests is *matter-of-fact truth*. This distinguishes empirical science from, among other things, Aristotelian science or from the empirically vacuous "theories" of astrology or from the "theory" that explained the rising of the Nile by suggesting that this river was connected to the great encircling river Ocean. The latter sort of enterprise, unlike that of empirical science, aims to discover things about the world that cannot be known by, nor confirmed by, appeal to ordinary sense experience.

Some have objected to the suggestion of this as a reasonable cognitive goal of science. Thus, Laudan has suggested that "we evidently have no way of ascertaining

which theories are more truthlike or more nearly certain than they formerly were."[141] If by this is meant that we do not, godlike, have some overview of the whole of the universe past, present and future, here, there and everywhere, and that we are somehow able omnisciently to compare what we believe with the whole of this truth to judge how truthlike it really is, then Laudan is of course right. Since we have no such perspective on the universe, it would be unreasonable to propose that as a cognitive goal: *a cognitive goal is reasonable only if we have a tolerable hope of attaining it*. But Laudan also indicates that there is a sense in which one can aim at truth without proposing to aim at Truth as some "transcendent immanent goal."[142] He proposes that we

...view science as aiming at well-tested theories, theories which predict novel facts, theories which 'save the phenomena,' or theories which have practical applications.[143]

It is evident that it is just this which we, following such methodologists of science as Herodotus, Hippocrates, Peirce and Kuhn, have been using when we are proposing that the cognitive goal of science is **matter-of-fact truth**.

But this is not the only thing that ought to be said about the cognitive interests of science: we must refer to other norms or standards if we are to fully appreciate the ongoing process of science.

Kuhn puts the point in this way: After noting how different disciplines have different norms of practice, often shaped by the instruments that are available, and which change over time, he points out, as we have already seen,[144] that there are cognitive ideals that move the scientist to undertake research.

Finally, at a still higher level, there is another set of commitments without which no [person] is a scientist. The scientist must, for example, be concerned to understand the world and to extend the precision and scope with which it has been ordered. That commitment must, in turn, lead him [or her] to scrutinize, either through himself or through colleagues, some aspect of nature in great empirical detail. And, if that scrutiny displays pockets of apparent disorder, then these must challenge him [or her] to a new refinement of his observational techniques or to a further articulation of his theories (*The Structure of Scientific Revolutions*, p. 42).

What the scientist is concerned to discover is *order* in the world that can be scrutinized or observed in the usual sorts of way; the scientist, that is, is concerned to discover matter-of-fact regularities. The regularities that he or she hopes to discover must be as precise as the world permits. Gaps in the order are to be remedied by order, and this order will be cognitively more satisfying the more precise that it is. In other words, what Kuhn is indicating is that what moves the scientist is the cognitive ideal of gapless knowledge, that is, the cognitive ideal of *process knowledge*. Guided by this ideal, the gaps present a problem for the scientist; this problem is to be solved by empirical research.

What we have to justify is the norms or methodological rules that guide this research. What we have suggested is that the relevant norms are those that lead with reasonable efficiency to the filling in of the gaps in our knowledge of matter-of-fact regularities.

We have examined various methodological rules that people have followed in determining which hypothesis to accept as true for purposes of explanation and prediction. Among these were the *a priori* method, the method of tenacity, the method of authority, and so on. We also looked at the rule of simplistic induction. This is the rule that an hypothesis becomes acceptable simply because "it works". We looked at these alternative rules for inductive inferences, and examined them as to whether conforming our inferences to these patterns would have any reasonable chance of satisfying the cognitive interest that motivates science, that is, our passion of curiosity or love of truth. We argued that each of these methods is very inefficient relative to the cognitive ends which define science: conforming our research practices to these norms will lead us to accept for purposes of explanation and prediction hypotheses which are **not, as a matter of fact**, true. Finally, we looked at the experimental method, the method of eliminative induction. Of all the various rules, only those of science, the rules of the experimental method, satisfied the test of providing us with knowledge that satisfies, so far as we can tell, our cognitive interests.

But one can raise the issue whether we have eliminated *all* rules besides those of science. There is no *guarantee* that all alternatives to the rules of science have been investigated. But this hardly counts against our attempt. After all, those are the limits of human reason. And we may say at least this much, that if we have not *conclusively* established that the rules of science alone can be vindicated, then we have, in Madden's phrase,[145] succeeded in *attenuating* suggestions that there are other, equally acceptable or better, rules.

In fact, however, as we now recognize, Kuhn has proposed an alternative that goes beyond the rules of eliminative induction. To be sure, Kuhn *does* recommend the strategy of being guided by a paradigm, and using the rules of eliminative induction during certain periods in science, those periods that Kuhn has called "normal science." But for periods of revolutionary science, he proposes a very different strategy. It is in a strategy the rules for which can reasonably be described as *counter-inductive*.

Kuhn has in effect suggested that, in periods when an old and formerly successful paradigm has started to fail to successfully guide research, we ought, rationally, to follow a counter-inductive strategy.

As we have seen, Kuhn argues that scientific practice is this: if repeated efforts to solve a problem posed by a paradigm, that is, to find a law it asserts to be there to be discovered, fail, then that, while not falsifying the paradigm, begins to call into question whether there is, after all, a law of the sort that the paradigm claims to exist.[146] Such repeated efforts by skilled scientists begin to make it reasonable to assume that, as John Stuart Mill once put it, "a real [instance] could scarcely have escaped notice."[147] At that point, one begins to search for alternative guides for research, and we enter a period of what Kuhn calls "revolutionary science"[148] where research is no longer guided by a *tested generic hypothesis*,[149] i.e., a paradigm.

Normal research is guided by a *confirmed overhypothesis*, that is, a confirmed abstract generic theory. *All* research requires that it be guided by some hypothesis or other. But if, as Kuhn suggests, in periods of revolutionary science it is no longer guided by the confirmed overhypothesis, i.e., the old paradigm, then what sort of hypothesis should guide research? One thing is clear: if it is not the old and confirmed overhypothesis, then it must be some contrary hypothesis. The old rule proposed, very crudely, the procedure

> If all observed A's have been B's, then infer that all A's are B's

The new rule for revolutionary science proposes, very crudely, the procedure

> If all observed A's are B's, then infer that remaining A's are not B's

Kuhn's policy, in other words, is *counter-inductive*. There are still many possible alternatives for guiding research, that is, many possible counter-inductive policies. Which of these ought one to choose? Kuhn in fact has an answer for this, again as we have seen: different scientists should adopt different theories contrary to the older paradigm on the basis of subjectively variable metaphysical and aesthetic considerations.

When a theory or paradigm T continues to be empirically progressive, it in effect becomes sufficiently confirmed as to be morally certain. But when it ceases to be empirically progressive, when it is ceases to be an effective guide to research, then the repeated failures to find data confirming a law of the sort that T predicts to be there has to be included in the evidence E. Failure to find such data is precisely what one would expect if T is false, or, equivalently, if not-T is true. Thus, rather than finding what we would expect if T were true, we find what we would expect to find if T were false. Thus, when T ceases to be empirically progressive, it ceases to be morally certain; and as it increasingly fails to guide successful research, its probability correspondingly decreases.

How rapidly confidence in T decreases will vary from person to person. Here is one source of subjective variability. Moreover, it will vary from person to person at what point the degree of certainty or probability must be at before one ceases to treat the paradigm T as worthy of acceptance for purposes of guiding research. Here is a second source of subjective variability. Because of these subjectively variable factors, different researchers will become dissatisfied with the paradigm at different points: some will continue to hold on to it as a guide for research longer than others.

The (generic) theory that the scientist accepts provides, as we have seen, prior probabilities for the (specific level) laws that are to be put to experimental test. What happens when a paradigm ceases to guide research is that scientists start in certain areas to look for laws of forms different from the forms predicted by the old theory or paradigm; which is to say that they start testing hypotheses of forms different from the forms predicted by the old paradigm. The testing will yield confirmations only if there are some grounds for antecedently accepting the hypotheses. We will have such grounds provided that we have some background theory that places the new hypothesis within the range delimited by the Principle of Limited Variety. Kuhn's insight is that different scientists will pick out different background theories to determine which more specific hypotheses they will test. In effect, what happens when scientists adopt different alternative theories during periods of revolutionary science is that each picks out a new range of Limited Variety according to his or her own subjective criteria. These subjectively variable re-assignments of prior probabilities lead different scientists to test different hypotheses. As Kuhn indicates, through this process a very wide variety of alternatives is considered, in contrast to the narrow range of alternatives that needs to be considered during periods of normal science.

The issue that is confronting us is this: *is this practice one that is reasonable relative to the end determined by our cognitive interest in matter of fact truth?* We can say, in the first place, that the Kuhnian method of research has in fact been tried and has been found to be successful in leading to the discovery of matter-of-fact causal regularities. In that sense, it has stood the test of experience. In the second place, if we look at other methods that have been proposed, then we can see that these methods fail. Feyerabend has proposed the rule that "anything goes".[150] This is parallel to the sort of thing that Kuhn defends in the case of revolutionary science, but it would be disastrous once a new paradigm has become fixed and normal science once again becomes proper. Conversely, the gruesome method of Goodman, as we have seen, amounts to the method that Kuhn defends for normal science but does not allow for the need to change paradigms, that is, for revolutionary science. Neither of these proposed methods allows for the flexibility that the search after causes has been seen to require. As for methods such as the method of authority, these continue to have the defects that we have previously noted. It would seem, then, that Kuhn's method stands the test of experience in the way that Goodman's method or Feyerabend's does not. Kuhn's method therefore serves the end determined by the cognitive interests of sciences, where these other proposals do not. Hence, *it is Kuhn's method that the rational person who shares the cognitive goals of science ought — tentatively at least — to adopt.*

"Anything goes": This methodological proposal of Feyerabend has been adopted by those who would defend pseudoscience. Thus, in defending his suggestion that the planet Earth has in the past been visited by extra-terrestrial beings in UFOs, Erich von Däniken has indicated that

> ...*nothing* is incredible any longer. The word "impossible" should have become literally impossible for the modern scientist. Anyone who does not consider this today with be crushed by the reality tomorrow.[151]

Here is another instance of the same appeal. This time it is from C. D. B. Bryan, who argues that one can accept as reasonable many reports from people who claim to have been abducted by extraterrestrial aliens who travel in UFOs. The objection is the standard one, that such claims are to be rejected because they would require the extraterrestrial space vehicles to travel at speeds that violate known laws of nature. For example, it is plausible to maintain that extraterrestrial space vehicles could travel from other planets to the Earth only if one also maintained that they made the voyage in reasonably short periods of time. But they could do that only if they travelled at velocities exceeding the speed of light. This, however, would require that they violate the law of physics, discovered by Einstein, that the speed of light is an upper limit on any velocity that such a vehicle might obtain. Bryan writes that

> ...just because UFOs and their occupants defy our laws of physics does not mean that there are not further laws of physics we have not yet discovered or do not as yet comprehend....[152]

Similar arguments can be used to justify such other theories as those of Velikovsky and homeopathic medicine. Velikovsky's views are incompatible with the basic laws of physics that describe motions in the solar system, while the claims of homeopathic medicine are incompatible with the basic laws of chemistry. One might also mention in this context the "theories" of creation science which are, of course, incompatible with the laws of physics, geology and biology, in particular Darwin's account of the origin of species in terms of natural selection.

These appeals to justify as scientifically plausible "theories" which are incompatible with basic laws of science are refer in effect to Feyerabend's principle that "anything goes" or to some variation on that theme. Thus, the rule is often expressed as the principle that "anything is possible." This appeal is also often coupled with an appeal to the liberal principle that one ought to keep an "open mind" and be prepared to admit that one could be wrong. If you do not accept that anything is possible or anything goes, then you are being dogmatic and you are failing to recognize the basic point about science, that it is fallible and that its hypotheses must always, as Judge Overton insisted, be recognized as tentative.

Now, there is a point to what is here being claimed. It is often enough true that hypotheses that were once thought to be unacceptable have come in time to be found acceptable. We have already noticed the hypothesis of continental drift which, when first proposed by Wegener, was not acceptable, but became so when it could be fit into a paradigm or theory that had been successful in guiding research.

Indeed, there are some what seem to be simple observational facts that have been rejected as incredible. One that is particularly relevant to von Däniken's thesis is the perhaps notorious example of meteorites. For many years scientists dismissed the idea that there could be meteorites, that is, stones that fell from the heavens to the earth. Among those who so argued was the great chemist Lavoisier: stones could not fall from the heavens because there were no stones up there to fall. Thomas Jefferson was another scientific thinker who rejected the very idea of meteorites. And so, after reading a report by two professors from Harvard in which it was claimed that meteorites had been observed, Jefferson remarked that "I could more easily believe that two Yankee professors would lie than that stones would fall down from heaven."[153]

Those who follow Feyerabend and von Däniken and Bryan suggest that Lavoisier and Jefferson were blinded as it were by their theories. The paradigm which they accepted provided no place in the picture of the universe which it proposed for the hypothesis that there could be stones in the heavens that sometimes fell to the Earth. These thinkers treated the paradigm dogmatically — a feature of normal science which they argue is regrettable rather than, as Kuhn argues, normal and rational. It is regrettable because it prevents people from accepting new hypotheses that run counter to the old paradigm. They point to the fact that Kuhn himself notes that scientists have in the past given up old paradigms and replaced them by new; this is what revolutionary science is about. But this means that what science has thought in the past to be physically impossible has turned out to be after all not only physically possible but actual. And if it has been done before, that we have discovered that what seemed to be physically impossible is really real, then why should we not expect it to be done again. Shakespeare can be cited in this context: "There are more things in heaven and earth, Horatio, than are dreamt of in your philosophy." In any case, the idea is that science will progress more rapidly if we are prepared to embrace a wide variety of new hypotheses all of which run counter to our current paradigm: we ought not to let the constraints of current normal science block us from embracing new hypotheses. "Anything goes" — including what seems to be the wildest pseudoscience.

Thus, by using Feyerabend's maxim, it is possible to transform apparently wild hypotheses into hypotheses which are scientifically acceptable. Thus, appeal to this maxim can be used, as von Däniken and Bryan use it, to justify accepting hypotheses such as that of our being visited in the past by extraterrestrial astronauts. Pseudoscience is after all science. Or, rather, Feyerabend's maxim is used to transform what are in fact bad and unsupported hypotheses into legitimate science. It is used, in other words, to give to bad science the appearance of science. It becomes an effective tool in the defence of pseudoscience.

"Anything goes" means simply that, if an hypothesis has not been rejected, if there are no observational data that imply its falsehood, then it is acceptable for purposes of explanation and prediction. Upon this rule, then, *acceptability is not a matter of evidence in favour but the fact that there is no evidence against.*

Put this way, it is clear that the maxim in effect declares that the *appeal to ignorance* is not, contrary to tradition, a fallacy.

Here are some typical examples of this form of argument, and, as the examples make clear, there seems to be reason to think that it is indeed a fallacy.

* No one has shown that Smith is lying. Therefore, it is reasonable to think that he must be telling the truth.
* No one has shown that the President is telling the truth. Therefore, it is reasonable to think that he must be lying.
* No one has shown that there are no unicorns. Therefore, it is reasonable to think that unicorns exist.
* No one has shown that Bigfoot does not exist. Therefore, it is reasonable to think that he does exist.
* No one has shown that there are no ghosts. Therefore, it is reasonable to think that ghosts must exist.
* No one has shown that ESP is impossible. Therefore, it is reasonable to think that it exists.
* No one has shown that visits in the past by extraterrestrials in UFOs are impossible. Therefore, it is reasonable to think that this happened.
* No one has shown that alien abductions are impossible. Therefore, it is reasonable to think that they have occurred.

It is generally assumed that a claim is acceptable as true and worthy of use in explanation and prediction only on the basis of evidence in its favour, not on the basis of absence of evidence. An absence of evidence shows our ignorance, it does not provide a reason for accepting an hypothesis as true.

This kind of appeal to ignorance relieves the defender of pseudoscience of the burden of proof. Since "anything goes," what they propose ought to be accepted. What is needed is not reasons in its favour — that it is possible is reason enough to accept it for purposes of explanation and prediction. The burden is on those who would reject it: they must supply reasons why it *ought not* to be accepted.

But the appeal, by transferring in this way the burden of proof to the critic of pseudoscience, makes it well nigh impossible to provide the proof of unacceptability.

There can be two ways to reject the claim that an hypothesis is acceptable. One is to argue that there is no evidence in its favour. This point is often plausible; there is, for example, little plausible evidence that the Earth has been visited by extraterrestrials in UFOs. But the defender of pseudoscience can always argue that the evidence simply has not yet been uncovered. Such a person claims that "there is evidence." The critic's objection that none has been provided does not falsify this claim. So the defender of the pseudoscientific hypothesis can argue that the critic has not provided grounds that the hypothesis ought to be rejected.

The other appeal of the critic is to antecedently accepted background theories. Thus, a theory such as that of Einstein can be used as grounds to reject the proposed pseudoscientific hypothesis as unworthy for use in explanation and prediction: the hypothesis is inconsistent with already accepted knowledge claims and must therefore be rejected. But the defender of the pseudoscientific hypothesis rejects this appeal on the grounds that "anything goes."

The defender of pseudoscience thus effectively disarms the critic. The burden of proof is shifted from the former to the latter, and any grounds to which the latter might appeal are rendered irrelevant. It is evident that *the defender of pseudoscience appeals to Feyerabend's maxim that "anything goes" in a way that renders his or her position irrefutable.*

It is not only defenders of pseudoscience who commit the fallacy of the appeal to ignorance. Those who are sceptical of various pseudoscientific hypotheses also often fall into this fallacy. the have been known to argue, for example, that since there is no evidence that ESP, or UFOs, or homeopathic cures, or ..., exist, therefore such things do not exist. But this is the same fallacy. Even if no one has yet come up with evidence that proves the existence of ESP or of UFOs as vehicles manned by extraterrestrials, or what not, we cannot conclude that none will every be found: for all we know, it could turn up tomorrow. The most that the sceptic can argue on the basis of the absence of evidence is that the hypothesis in question *ought not to be accepted* for purposes of explanation and prediction.

But this, surely, is all that the person sceptical of the pseudoscience needs.

What, though, of the appeal by the defender of pseudoscience to Feyerabend's maxim that "anything goes"? After all, has not Kuhn shown that in periods of revolutionary science this is indeed a legitimate maxim?

It is, of course, not quite true that Kuhn recommends just this maxim. To be sure, this is what he proposes for the full scientific community. But what Kuhn recommends for individual scientists is to rely upon their own subjective aesthetic values to pick out a range of Limited Variety to determine the specific hypothesis which they will propose be put to the test. But this seems to imply that, as Lakatos has put it, "there is no explicit demarcation between science and pseudoscience, no distinction between scientific progress and intellectual decay."[154] Or to put it another way, it seems to replace the inductive method of science with the *method of fantasy*: accept as true for purposes of explanation and prediction those hypotheses that appeal to one's imagination. And that is but another way of making it legitimate to accept as reasonable, conforming to the norms of science, the wildest of pseudoscientific hypotheses, Velikovsky, UFOs, ESP, homeopathic remedies, and so on.

In reply to this, and in defence of Kuhn's methodology and of the distinction between science and pseudoscience, **three** things must be said.

First. The suggestion is that we not dismiss some specific hypothesis, such as the hypothesis of alien abductions, or the hypothesis of the efficacy of homeopathic dilute herbal remedies, on the basis of the principle that such an hypothesis conflicts with some well established scientific theory. The argument is that because the well confirmed theory *might possibly* be false, we cannot dismiss alternatives to it. As good thinkers we must keep an open mind, and for such a person "anything goes."

The reasoning here aims to cast doubt on the received scientific theory in order to enable one to accept the pseudoscientific hypothesis that is contrary to it.

However, if we can reason in this way, then we can also reason in the opposite way. *If the mere possibility that an hypothesis is correct is a reason not to reject it, then, by a parallel inference, the mere possibility of error is a reason to doubt.*

But science is fallible; it must always, as Judge Overton emphasized, be taken as tentative. We may conclude, therefore, that *since every scientific hypothesis that we might adopt is possibly erroneous we have a reason to doubt it and therefore a reason to reject it as worthy of use in explanation and prediction.* Thus, *the principle that* **anything goes** *implies equally that* **nothing goes**. In other words, if we take Feyerabend's principle seriously, or, what amounts to the same, if we allow an appeal to ignorance to justify accepting an hypothesis as reasonable, then we are led directly into a radical scepticism.

Such a scepticism is the stuff of epistemology seminars. It is hardly relevant to the task of deciding which hypotheses are acceptable as scientific and which are to be rejected as pseudoscience. For such a radical scepticism is in direct conflict with what everyone, the defender of pseudoscience included, is prepared to accept as true. No one doubts that one is sitting at a desk in front of a computer simply because it is *possible* that they are dreaming it.

We may reject as unreasonable an unlimited application of Feyerabend's principle that "anything goes."

Second. The suggestion is that one who rejects Feyerabend's principle is someone who is not open-minded, someone who is not prepared to recognize that science is indeed tentative and may be wrong.

But to be open-minded is **not** to be someone who accepts just anything. The real mark of an open-minded person is not the willingness to entertain just anything that might strike one's fancy, but rather the willingness to allow one's beliefs to be determined by the evidence that one has available. Open-mindedness is the willingness to defer to the objective evidence rather than to his or her own inclinations or prejudices.

What truly marks an open-minded person is the willingness to follow where the evidence leads.

To put it neatly, *one must be open-minded but not so open-minded that one's brains fall out.*[155]

Third. The third thing to be said against the general application of Feyerabend's principle that "anything goes" amounts to a defence of Kuhn's more modest application of that principle in the methodology of science that he has defended, that is, its use during periods of what he calls revolutionary science.

What Kuhn argues is that the rule of "anything goes" and the appeal to subjectively variable aesthetic values is reasonable *only in a very restricted research context*. In particular, *such an appeal is legitimate only when normal science is in crisis, that is, only when the background theory or paradigm has ceased to be an effective guide in the discovery of hypotheses for use in explanation and prediction.*

In normal science, research is guided by the background generic theory or paradigm. This provides for the researcher the Principles of Determinism and Limited Variety that define the puzzle-solving activity of his or her normal scientific practice. In this context, where the paradigm is in fact successfully guiding research, where it does lead to the discovery of new specific laws, it is simply not reasonable to allow that "anything goes": **in situations where the practice is that of normal, paradigm-guided, research, what goes is what is allowed by the paradigm**. It is the paradigm that determines which hypotheses are reasonable for purposes of scientific inquiry.

Kuhn does allow that sometimes "anything goes," but the contexts in which this is reasonable are rare. They occur only when normal scientific practice breaks down, where the background theory ceases to be a reliable guide to what laws are there to be discovered.

In situations, then, where the paradigm has not broken down, the rational procedure is to reject as unacceptable, as gruesome, any hypothesis that conflicts with the paradigm.

It follows from this methodological norm that one cannot appeal to the rule that "anything goes" or the rule that "anything is possible" to justify the acceptance of the various pseudoscientific hypotheses that we have noted.

Consider several of them. The appeal has been made in the case of Velikovsky's hypotheses. But these conflict with the Newtonian paradigm. Entertaining Velikovsky's claims as possible alternatives to Newton would be reasonable if the context were on in which the norm of "anything goes" is relevant. But the context is not of that sort. Newton's paradigm is still well-confirmed for the solar system, and in that area it still successfully guides research. For example, in plotting the various Apollo missions to the moon, its was precisely Newtonian theory that was successfully used, both in those missions that were successful and in that mission, Apollo 14, which almost ended in disaster. Because the old paradigm has not broken down, we need not entertain any hypotheses that are in conflict with it. We therefore need not entertain Velikovsky's hypotheses.

It is of course true that as a paradigm, Newton's theory has broken down when it comes to the exploration of systems in which objects move with velocities close to the speed of light. In this area there was once a crisis, and this crisis led to the replacement of the old paradigm with Einstein's new paradigm, the theory of special relativity. But for a system such as the solar system, in which speeds of objects are much below the speed of light, the predictions of the old Newtonian paradigm coincide with those of the new Einsteinian paradigm. This, of course, is exactly what we ought to expect. In any case where a new paradigm replaces an older paradigm, the predictions of the former in the areas where the old paradigm was successful must coincide with those of the old paradigm. This means that Einstein's theory must yield the same predictions as Newton's theory for the motions of objects in the solar system. It follows that the fact that Newton's theory broke down in the areas where objects move at speeds close to that of light does not mean that the paradigm had started to break down in the area of the mechanics of the solar system. Since the Newtonian paradigm was and is still successful in guiding research in this area, it is unreasonable for anyone to entertain Velikovsky's contrary hypothesis.

Now consider the possibility that Earth has been visited by extraterrestrial intelligences in UFOs and the possibility that some of these aliens have abducted human beings and delved into what must be to them the exotica of the human reproductive system. For these hypotheses to be plausible, the extraterrestrials would wish to return to the home planets in a reasonable period of time. They could not do this unless their space ships, the UFOs, could travel at speeds greater than that of light. But Einstein's highly successful paradigm reckons that this is simply not possible: it is a basic law in Einstein's theory that no object can travel at a speed greater than that of light. It follows that one ought to reject as unreasonable any hypothesis that UFOs are the vehicles of extraterrestrial intelligences and that there have been alien abductions.

Or consider the claim that the very dilute herbal remedies of the homeopathic physicians have curative powers. The laws of chemistry state that a substance can interact with other substances only if it is present in certain non-diminishing amounts. The homeopathic remedies are so dilute that the substances are for all practical purposes absent. The homeopathic hypotheses that these remedies can interact with the body and cure various diseases is therefore contrary to the highly successful paradigm in chemistry. It follows that it would be unreasonable to consider it possible that these "remedies" have curative powers.

We see, then, that in all these cases the hypotheses that are proposed are all gruesome and pseudoscientific. Kuhn's methodology shows clearly that one cannot transform these hypotheses from fantasy to genuinely scientific by an appeal to Feyerabend's rule that "anything goes" or to the rule that one ought to be open-minded and entertain hypotheses that run counter to current practice. Such an appeal only disguises the silly as somehow scientific. But what results is a "theory" which is simply not science but rather pseudoscience.

One further remark should be made.

Kuhn has spoken of, to refer to the title of his book, *The Copernican Revolution*. This characterization of what happened with Copernicus and in those who succeeded him is certainly not out of place. But the revolution is not an instance of "revolutionary science" in the sense in which we have just been speaking of it. There are serious differences between the Copernican revolution and intervals of revolutionary science of the sort that we find in Einstein's revolution.

The method Einstein used — Kuhn's method of revolutionary science — is one aspect of the inductive method of science. On this method, hypotheses are tested against observed reality; if the data show them to be false, then they are rejected, and if the data support them, then they are accepted, tentatively at least. In normal science, there is one method of hypothesis selection, namely, guidance by the background paradigm or abstract generic theory. In revolutionary science, there is another method of hypothesis selection, namely, resting on subjectively variable aesthetic or other criteria. But in both cases, the test of the hypotheses selected is their fit to objective observational data. There was more than this going on in the case of the Copernican revolution, however.

The scientific method had until then not really been perfected in astronomy and physics. Even Copernicus proceeded by the *a priori* and Aristotelian methods, rather than the inductive method of science. He was fully prepared to defend *a priori*, using Aristotelian arguments, the Circularity Principle. His objection to Ptolemy was not that the latter had violated the inductive method or that he, Copernicus, had formulated a new hypothesis that made a better inductive fit with the observed facts. Rather, his objection was that Ptolemy's system, with the equant point device, had violated the Circularity Principle, a Principle which Copernicus, like Ptolemy, accepted on *a priori* Aristotelian grounds. What was needed to achieve genuine science was not merely the rejection of Ptolemy, or even the rejection of the Circularity Principle. What was needed to achieve genuine science was the inductive method of science, which insists on the priority of observed facts as the test of truth and falsehood and as the criteria for the rejection or acceptance of hypotheses.

Peirce captured the point nicely in a comment on Kepler's work.

> Kepler undertook to draw a curve through the places of Mars, and his greatest service to science was in impressing on men's minds that this was the thing to be done if they wished to improve astronomy; that they were not to content themselves with inquiring whether one system of epicycles was better than another, but that they were to sit down to the figures and find out what the curve, in truth, was.[156]

It was the inductive method that had to be invented, not just the formulation of a revolutionary new hypothesis.

Peirce also captures nicely how early science did not have much of a feel for means by which to select an hypothesis for testing.

> He [Kepler] accomplished this [finding out what the curve, in truth, was] by his incomparable energy and courage, blundering along in the most inconceivable way (to us), from one irrational hypothesis to another, until, after trying twenty-two of these, he fell, by the mere exhaustion of his invention, upon the orbit that a mind well furnished with the weapons of modern logic would have tried almost at the outset.[157]

The Copernican revolution is more than just another period of revolutionary science; it is, rather, that period in which science came to be invented, or, better, that period in which people came to use the inductive method of science in doing physics and astronomy instead of the *a priori* and Aristotelian methods that had previously been used.

(vi) Theories of Instruments, Auxiliary Hypotheses Once Again, and Observing Unobservables[158]

There are scientific theories that can be tested only through the use of instruments. Thus, we use, for example, instruments such as microscopes, telescopes, Wilson cloud chambers, and so on, to test theories. This use of instruments in science has been pointed out often by philosophers of science, who then correctly draw the conclusion that what is tested is not so much a single theory T but rather a conjunction

T & A

where A is a set of auxiliary hypotheses describing the instrument being used. Some have argued that the usual empiricist accounts of scientific theories, as F. Suppe has put it, simply "ignore the role of auxiliary hypotheses in applying theories to phenomena...".[159] This is to assert that empiricism as traditionally understood has no adequate epistemology of the instrument. It can be seen, however, that this understanding is unreasonable. There is no reason why the account of science that we have been developing cannot include an account of how inferences based on instruments determine the assignment of probabilities to theories.

a. Minute Parts

In the early modern period it had become a commonplace of philosophy that the world had many parts that one could not observe. The empiricists made this point repeatedly. Thus, the English philosopher John Locke[160] at one point tells us that

> Blood to the naked eye appears all red; but by a good microscope, wherein its lesser parts appear, shews only some few globules of red, swimming in a pellucid liquor: And how these red globules would appear, if glasses could be found that could yet magnify them a thousand or ten thousand times more, is uncertain (*Essay*, Bk. II, ch. 23, ¶ 12).

There are many similar examples in the *Essay*. Again, we recall that Locke's successor George Berkeley,[161] in arguing for the great wisdom of the creator, tells us of the small, invisible parts of the world, "the clock-work of nature, great part whereof is so wonderfully fine and subtile, as scarce to be discerned by the best microscope" (*Principles*, ¶ 60).When parts of nature cease to function perfectly, we can always find some cause that is interfering with the mechanism, just as we can do with a watch.

> … whenever there is any fault in the going of a watch, there is some corresponding disorder to be found in the movements, which being mended by a skilful hand, all is right again (*Principles*, ¶ 60).

Berkeley's successor David Hume[162] takes up the same example.

> A peasant can give no better reason for the stopping of any clock or watch than to say, that commonly it does not go right: But an artizan easily perceives, that the same force in the spring or pendulum has always the same

influence on the wheels; but fails of its usual effect, perhaps by reason of a grain of dust, which puts a stop to the whole movement (*Treatise*, p. 132).

Hume is in fact using this example to make a point about how to conduct research into causes. Research, he argues, is guided by the principle of causation which asserts that every event has a cause. We have one kind of event, namely, a watch that has been wound up, which is followed sometimes by one sort of event — the watch working — and sometimes by a different sort of event — the watch not working. The vulgar attribute this difference to chance, but philosophers — scientists — do not allow for that possibility. To the contrary, they infer, upon the basis of the causal principle, that there is present some causal factor that they have hitherto not noticed.

The logic is clear. We have a certain sort of event P. Being P is followed sometimes by Q and sometimes by the absence of Q, that is, in a word, by not-Q. The causal principle asserts that there is a causal factor the absence of which explains P being followed by Q and the presence of which explains the P being followed by not-Q. Call this factor T. What the causal principle asserts is that there is a regularity to the effect that

(1)   For any individual x, if x is P then, x is not-T if and only if x is Q

This is logically equivalent to the conjunction of

(2)   For any individual x, if x is both P and not-T, then x is Q

and

(3)   For any individual x, if x is both P and Q then x is not-T

where the latter is in turn logically equivalent to

(4)   For any individual x, if x is both P and T then x is not-Q

If we have a particular watch w that is wound up

$$w \text{ is } P$$

and if it is working

$$w \text{ is } Q$$

then we can use (4) and therefore (1) to conclude that the interfering factor T is absent

$$w \text{ is not-}T$$

Having thus discovered the absence of T in w in this case, we can use it to explain, deductive-nomologically, why the watch is working:

> For any individual x, if x is P then, x is not-T if and only if x is Q
> w is P and w is not-T

Hence, w is Q

Conversely, if w is not working

$$w \text{ is not-}Q$$

then we can use (2) and therefore (1) to conclude that the speck of dust is present

$$w \text{ is } T$$

and we can use the presence of this factor to explain, deductively from a lawlike hypothesis, why the watch is not working:

> For any individual x, if x is P then, x is not-T if and only if x is Q
> w is P and w is T

Hence, w is not-Q [163]

But what is this unknown factor T? In fact, it is misleading to treat it as if it were on all fours with the factors P and Q. To the contrary, so long as we rely upon the causal principle to justify our asserting its existence, its logical status will be quite different. What the causal principles enables one to infer is that *there is* a *unique* factor such

that it is present if and only if the wound watch works. That is, what the causal principle enables us to infer is that

(5) There is exactly one species f such that f is **F** and such that for any individual x, if x is P then, x is not-f if and only if x is Q

where **F** are appropriate generic constraints on the sort of causal factors that are relevant. Hume infers this from the causal principle, which is the fourth of his "rules by which to judge of causes and effects."[164]

b. The Rules of the Experimental Method

We have already discussed in detail the rules of the experimental method. Historically speaking, these rules developed in the 17th and 18th centuries, and were first presented in a fashion that is clear and unequivocal in the section of Hume's *Treatise of Human Nature* entitled "Rules by which to judge of causes and effects." These are the rules that he needs to justify the inferences that he suggests that the artizan makes in the case of the watch that sometimes works and sometimes does not.

Let us once again recall our schematic example of research into diseases. This simplified example, with which we are by now familiar, will help to bring out the logic of Hume's position. Suppose that we have a series of diseases D_1, D_2, D_3, of genus **D**, and that these are caused, respectively, by the presence of germs of species G_1, G_2, G_3, all within the genus **G**. We thus have the laws

(a)
> For any individual x, x is G_1 if and only if x is D_1
> For any individual x, x is G_2 if and only if x is D_2
> For any individual x, x is G_3x if and only if x is D_3

Together with the data that eliminate other possible causes, these entail respectively

> There is exactly one species g such that g is **G** and such that for any individual x, x is g if and only if x is D_1

(b)   There is exactly one species g such that g is **G** and such that for any individual x, x is g if and only if x is D_2

There is exactly one species g such that g is **G** and such that for any individual x, x is g if and only if x is D_3

Given that the D_i are all **D**, we can now generalize again to a *law about laws* — as John Stuart Mill put it, it states that "it is a law that there is a law for everything"[165] — to the effect that

(c)   For any species f, if f is **D** then there is exactly one species g such that g is **G** and such that for any individual x, x is g if and only if x is f

We now run across a new species of disease of the same sort, e.g., D_4. Since

$$D_4 \text{ is } \mathbf{D}$$

we can deduce from (c) the *existential hypothesis*

(d)   There is exactly one species g such that g is **G** and such that for any individual x, x is g if and only if x is D_4

This asserts that *there is* a species of germs that causes disease D_4.

We have observational data that justify the law-assertion of each of (a). These law-assertions justify the law-assertion of the generalities (b). Since we have in the past succeeded in finding data that justify asserting these regularities as laws, we may reasonably infer that in the future we will find data that confirm *analogous laws* in *analogous cases*; that is, our successes in confirming the laws (b) justify our law-asserting a generalization about such generalizations, namely, (c). This law-assertion in turn justifies our law-asserting the existential hypothesis (d). The inference proceeds from what we have found in *parallel cases* to what we may reasonably expect in the present case. The result is a justified assertion of an existential hypothesis which asserts that a certain sort of cause *is there to be discovered*, or, what is the same, that a certain sort or form of law obtains, without, however, asserting specifically what that cause or law it.

The hypothesis thus presents a *research problem*; we know that there is a law there to be discovered, and the task is to discover it.[166]

In this example, the principle (c) functions as the Principle of Determinism and of Limited Variety. This asserts that for any causal factor of the relevant sort then *there is* another factor which explains its presence. It is, as we said, a *law about laws*. It is achieves this status because it *abstracts* from the specific laws a certain *generic form*.

Return now to our, or rather Hume's, watch example. In the simple case concerning diseases, the law (d) amounts to the principle (5) of the watch example. Hume lays out the relevant inference to (5) in his final rule,[167] which states a further eliminative mechanism subsidiary to the methods of agreement and difference:

> 8. The eighth and last rule I shall take notice of is, that an object, which exists for any time in its full perfection without any effect, is not the sole cause of that effect, but requires to be assisted by some other principle, which may forward its influence and operation. For as like effects necessarily follow from like causes, and in a contiguous time and place, their separation for a moment shews that these causes are not complete ones.

This rule states that when there is an object of sort P which is sometimes Q and sometimes not Q, then P is "not the sole cause" but "requires to be assisted by some other principles," that is, in such a case *there is* some other factor which is present when P is Q and absent when P is not Q. The 8th rule, in other words, states in effect (5). (5), it is clear, is an existence claim, one justified by the causal principle.

The same principle, Hume is clear, justifies not only the claim that there is a causal factor that explains why a P is sometimes Q and sometimes not, but further that this factor is unique, that there is but one of them. This means that we can introduce a *definite description* to refer to this factor:

(6)   *the* species f such that f is **F** and such that for any individual x, if x is P then, x is not-f if and only if x is Q

We can use this definite description to refer to the as yet unknown factor because we know that this definite description is in fact successful: the latter is precisely what (5) asserts. Let us now abbreviate the definite description (6) by

(6') $T^\wedge$

where the superscript makes clear that this is not an ordinary predicate but rather shorthand for a definite description. This means that what the causal principle enables us to infer is not so much (1) as

(1')   For any individual x, if x is P then, x is not-$T^\wedge$ if and only if x is Q

and that what we use to explain is not so much represented by "w is T" as by

w is $T^\wedge$

If we expand the definite description in (1') in the way that Russell has taught us to do, then it turns out, we are not surprised, that (1') and (5) amount to the same: they are logically equivalent. What (1') = (5) presents us with, of course, is a *research problem*: we must go through the *range of possible conditioning factors* described by the generic constraint **F** and eliminate all those that do not satisfy (5) until we have isolated that one factor, say T, which does satisfy (5). Then will shall have *identified* the relevant causal factor which (5) asserts to be there and which can explain why P's are sometimes Q and sometimes not-Q. This identificatory claim will be

T = $T^\wedge$

or, what is the same,

T = the species f such that f is **F** and such that for any individual x, if x is P then, x is not-f if and only if x is Q

This would be a simple application of the Joint Method of Agreement and Different, aimed at discovering among the possible conditioning factors **F** the necessary and sufficient condition for something which is P also being

Q. It is these research methods of experimental science, Bacon's methods of eliminative induction, that Hume is describing in his "rules by which to judge of causes and effects." Once we have identified $T^\wedge$ to be T, then we can replace (1') = (5) by (1) in our inferences, and, in particular, in our explanations.

c. Observing Minute Parts

There is a further problem in the case of the watch, however, to which we must now turn. We have already made some of these points in our discussion above of Mendelian genetics, where we looked at the logical structure of the theory that introduced genes as unobservable entities. But it will pay to look at the issues from a slightly different angle.

The definite description (6) is based upon what Hume calls the "relation" of causation. Any definite description based upon a relation Hume refers to as a "relative idea."[168] A "relative idea" is an idea that we form of a thing based upon the relation which that thing bears to something which is known. Since the description is only relative and by means of a generic reference, what we represented by "**F**", it follows, as Hume says, that the entity to which the relative idea or definite description refers may be *specifically different* from the other entities mentioned. It is, in Locke's phrase, "something I know not what." Or at least, it is so *until it is identified*. Once it is identified, then we know *specifically* what the factor is; it is no longer something "I know not what."

In the case of the watch, however, it is no simple matter to identify the relevant causal factor. This factor is, Hume suggests, a speck of dust. This is something normally too small to see. That is why the vulgar overlook it, and attribute to chance rather than to a genuine causal factor the fact that a wound watch sometimes works and sometimes does not. In effect, what we have is not

x is T

but

There is a unique individual y such that y is S and x stands in relation R to y

where "S" represents something being a *small speck of dust* and "x stands in relation R to y" represents something y being *inside* something x. The *very idea* of an S, that is, of a *small speck of dust*, implies, given what we know of the human ability to discriminate things visually, that an S is something that is *too small to see*.

But that is also what creates the research problem: if it is too small too see, how, then, do we identify it? Nor is it just a research problem: it is also a problem in the empiricist logic of science. For, if indeed it is too small to see, should we not say, then, that we do not see it, and therefore conclude that the causal principle, at least as Hume understands it, is false?[169] Or, if we don't do that, then are we not saying that whenever a theory is apparently falsified we can always save it from falsification by inventing things that are too small or too something to be noticed? As Lakatos pointed out, such "adhocery" is hardly the hallmark of progressive scientific research.[170]

But of course the early moderns did have an answer to this question. We have seen Locke and Berkeley refer to it. It is the use of lens to magnify, *to make visible things that are too small to see*. As Robert Hooke[171] put it, central among the problems facing science are the "infirmities of the senses," in particular, those arise from "the disproportion of the object to the organ."[172] These defects may be remedied "by the addition of ... artificial instruments and methods...".[173] The remedies, more specifically,

> ...can only proceed from the real, the mechanical, the experimental philosophy, which has this advantage over the philosophy of discourse and disputation, that whereas that chiefly aims at the subtilty of its deductions and conclusions, without much regard to the first ground-work, which ought to be well laid on the sense and memory; so this intends the right ordering of them all, and the making them serviceable to each other.[174]

This philosophy requires first "that there should be a scrupulous choice, and a strict examination, of the reality, constancy, and certainty of the particulars that we admit...".[175] Beyond this,

> The next care to be taken, in respect of the senses, is a supplying of their infirmities with

instruments, and, as it were, the adding of artificial organs to the natural; this in one of them has been of late years accomplisht with prodigious benefit to all sorts of useful knowledge, by the invention of optical glasses. By the means of telescopes, there is nothing so far distant but may be represented to our view; and by the help of microscopes, there is nothing so small, as to escape our inquiry; hence there is a new visible world discovered to the understanding.[176]

Hume's "artizan," who could explain where the vulgar could not why the wound watch sometimes worked and sometimes did not, was able to spot the speck of dust that interfered with the working by means of a jeweller's glass, a magnifying lens.

The laws governing the formation of images by means of lenses are well known. They are given by the theory of geometrical optics.[177] The principle of the magnifying lens can be seen in Figure 4.4.

A ray diagram to show the principle of the simple microscope. Two points O'- O' placed at the nearest distance of distinct vision produce a small retinal image I' - I' and are not resolved. Placing the points O - O at the focal distance f of the convex lens L, effectively increases the visual angle, producing an erect virtual and magnified image at infinity. The retinal image I - I is now larger and the points resolved as separate.

The inferences recorded in the diagram relate the size of the object viewed to the size of the apparent image. Specifically, of course, with the magnifying lens, the size of the object viewed is smaller than the size of the apparent image. The inferences themselves all depend upon Snell's Law.[179] As Huygens records,[180] this was discovered by Willebrod Snellius through a long series of experimental measurements. He gave the law in the following form. In Figure 4.5, let AB represent the surface of a sample of water, covered by air, and let D be an object at the bottom. If the eye is placed at F then it sees D in the direction FCG at the point G of the vertical line DA. Then the ratio CD/DG is constant. In the case of air and water, the ratio is 4/3. Descartes, who undoubtedly knew Snell's work, as Huygens remarks,[181] presents the law in a slightly different form, one more familiar to us.

Figure 4.4[178]

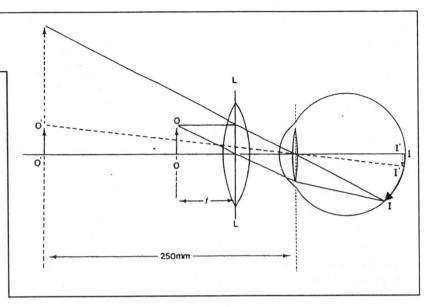

In the plane that contains the incident and the refracted rays, Descartes, Figure 4.6, describes a circle with its centre at B, the point on the surface where the two rays meet. Descartes then records the law of refraction as stating that the ratio BC/BE remains constant, or, as we would now express it, as stating that the ratio

$$\sin \alpha / \sin \beta$$

is constant.[182] This constant, which depends upon the two transparent media involved, is determined empirically.[183]

We must note carefully the logical form of this law. If we take the angles α and β as described, then the law states that

(S)   For all f, g, if f and g are transparent media, then there is a unique number η such that, for any samples of f and g that are contiguous one to the other, if light passes from the one to the other, then
$$\sin \alpha / \sin \beta = \eta$$

This law is an abstractive generic law: it makes an assertion about all kinds or species f falling within a certain genus, namely, that of transparent substances. It is thus a law about laws, stating that for all pairs of transparent media, there is a law of a certain form that describes the path taken by a ray of light as it passes from the one to the other. From this generic law (S) can be inferred more specific laws about particular transparent media. Thus, for example, we can infer a law about air and water:

(SA)   There is a unique number η such that, for any samples of air and water that are contiguous one to the other, if light passes from the one to the other then
$$\sin \alpha / \sin \beta = \eta$$

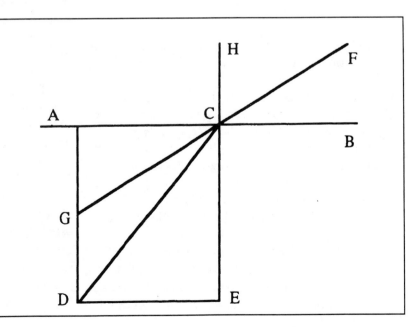

Figure 4.5

Figure 4.6

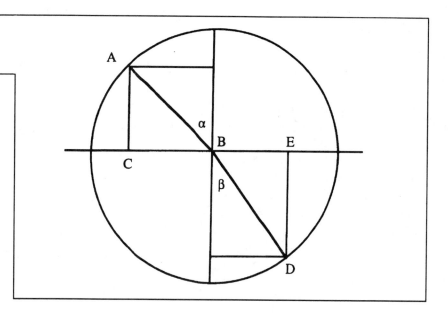

This law, like (5), makes an *existence claim*. Moreover, it makes a uniqueness claim. This permits the introduction of a definite description

> *the* index of refraction for air-water boundaries

It also presents us with a research problem, namely, that of identifying this index of refraction. This is an experimental task which Snell carried out at least for the air-water case:

> (SA*)   4/3 = the index of refraction for air-water boundaries

(SA*) is of course itself a law, which, when the definite description is unpacked, states that

> For any samples of air and water that are contiguous one to the other, if light passes from the one to the other then
> $\sin \alpha / \sin \beta = 4/3$

In effect, Snell discovered (SA*), and then generalized to all transparent media. That is, he inferred (SA) from (SA*), and then generalized to (S). This pattern of inference we are familiar with: it is that of our simple germ example, and that which Hume clearly laid out in his "rules by which to judge of causes and effects."

If one knows the shape of the lens, so one can determine the perpendicular to the surface at any point, and if one knows the index of refraction for the two media involved, then one can produce the construction of Figure 1 for the lens. From this one can deduce the law

> (LL*)   For any lens, if it is of this medium in this other medium and if it is of this shape, then whenever there is a real object of such and such a small size at such an such a place in

relation to the lens then and only then there is an apparent image of the object of so and so a size

where, of course, "so and so a size" is many sizes bigger than the "such and such a size" of the real object. The law (LL*) establishes that there is a one-one correlation between, on the one hand, the size of the object at a standard location relative to the lens and, on the other hand, the size of the apparent image. The law permits the inference of the size of the one from the size of the other, that is, the size of the real object from the size of the image and conversely. It should be noted that (LL*) itself is already generic, stating a whole body of specific laws, one in fact for each specific size of real object.

(LL*) is of course a very specific law, specific relative to the media involved and specific relative to the shape of the lens. It can be generalized. The result is a more generic form. In fact, this generic form is what is summarized in Figure 4.4 by means of the construction. One can find the relevant sort of generic construction laid out by Halley in a paper in the *Transactions of the Royal Society* of 1693,[184] where he showed how the use of algebraic signs enabled one to include many special cases under the general rule, that is, to include many specific laws under the abstractive generic law. Let us use (L) to refer to the generic law that constructions of Halley's sort and those of Figure 4.4 justify.

(LL*) is, as we said, already generic, describing a whole body of specific laws, one for each specific size of real object. Let us suppose that we have a small visible object of size 1, and let (LL*1) be the specific instance of (LL*) that applies. Let (LL*1) imply that an object of size 1 produces an image of size i^1. If we look at the object of size 1 and then look at the image and discern it to be of size i^1. We have then confirmed the specific law (LL*1). We can discover other confirming instances for the laws for objects of size 2, objects of size 3, etc., where we assume that each of the objects of sizes 1, 2, 3, etc., are all visible. From these confirmed specific laws (LL*1), (LL*2), (LL*3), etc. we can generalize to the generic law (LL*).

We can similarly confirm other laws that relate as (LL*) does to specific lens shapes and specific media. From these we generalize to the abstractive generic law (L) describing, as indicated in Figure 1, the general relations between small objects and their magnified images. We can then use this law, as Halley indicated, to design other shapes of lens, and, if we wish, for other media. If these lenses work as predicted, then that confirms those further specific laws, which in turn further confirm the generic law (L).

At the same time (L) receives confirmation from Snell's Law (S) from which ti is deduced. For Snell's law is itself a confirmed law. At the same time, since Snell's Law predicts the law (L), the confirmation of (L) yields further confirmation of Snell's Law.

What we have here is a pattern of confirmation that moves up from specific laws to generic laws and then conversely from the generic laws to the specific. The links are effected by the relations of specific and generic laws. This is a pattern that is familiar in science, as we have seen.

Now let us consider an object of size vsm — very small — so small in fact that it is not visible to the naked eye. The generic law (LL*) implies a specific law about objects of this size. This is the specific law (LL*vsm). This specific law implies that the object of size vsm will produce an image of size i^{vsm}. Or, rather, the prediction is not so much from the object of size vsm to the image but conversely, from the image of the very small object to the existence of that object. For, after all, the very small object is, we are assuming, too small to see.

What happens, of course, is that we observe the image of size i^{vsm} and use (LL*vsm) to infer the existence

of a very small object of size vsm. We can then explain the presence of the image by appeal to the very small object whose existence we have thus inferred. We are familiar with this pattern. Here the issue is the law (LL*vsm) which we use to infer the existence of the very small object, even when we cannot see that object. What right do we have to assume that this law obtains?

But now we know the answer to this question. The law (LL*vsm) is justified because it acquires a probability from the generic law (LL*) of which it is a specific instance. That generic law in turn acquires a certain probability from the fact that various of its specific instances have in fact been confirmed directly by observation, and further by the fact that it can be inferred from the still more generic law (L), which has a variety of specific instances that are confirmed, and which, moreover, can itself be deduced from the even more generic law that we call "Snell's" after its discoverer. We have, then, very strong evidence that the law (LL*vsm) obtains, and this gives us strong grounds for supposing that the very small object is there even though we cannot see it.

But further, *we can explain why we cannot see it.* As we have seen, Hooke spoke of objects being "disproportionate" to the senses. For, by the time of Hooke, *scientists had provided an explanation* of the working of the eye which could establish which objects are and which objects are not disproportionate to the eye. Descartes very skilfully lays this out in his *Optics.*

The argument of the *Optics* is carefully structured. The "First Discourse" provides a general account of light. In the "Second Discourse", he presents the law of refraction. [This he has clearly taken from Snell, as Huygens indicates, but he places it in the context of his own metaphysical speculations concerning material mechanisms. While the latter are essential to his overall programme, it turns out that they are in fact not needed for the rest of the *Optics*. For the latter, all that we need is the empirical or experimental fact that Snell's Law holds.] Descartes is going to end up with the design of instruments such as telescopes and microscopes as artificial aids to improve and extend our vision. But before turning to the issues of improving the senses by means of lenses, he examines vision itself, to determine through the use of natural science, the sort of improvements that might be needed. Specifically, he carefully and empirically examines the eye, and shows how the law of refraction can help us explain, *scientifically*, the mechanisms of vision. In the "Third Discourse," Descartes presents the physiology of the eye, and, after a general discussion of the senses in the "Fourth Discourse," he goes on in the "Fifth Discourse" to examine "the images that form on the back of the eye." He shows how, with careful dissection to remove part of the rear of the eye and the use of clear parchment, we can view the image projected onto the retina. In the "Sixth Discourse," he shows these are created by the physiological lens acting in conformity with the law of refraction. He shows how nature, within the limits of the lens of the eye, has introduced mechanisms that minimize error.[185] But there are still limits.

> ...she [Nature] has not so completely provided ... that something cannot be found to add to it: for granted that she has not given each of us the same means of so curving the surface of our eyes that we may see very close objects distinctly, e.g., at a distance of an inch or half an inch, yet she has failed more seriously with some, to whom she has given eyes of such a shape that they can use them only to look at things that are far away.

Thus, "in order for us to be able to remedy these deficiencies through art,"[186] we have to design lenses of glass that will improve the vision of the far- and of the near-sighted. This is discussed in the "Seventh Discourse," "Of the Means of Perfecting Vision," which deals first with simple magnifying lenses and then goes one to discuss instruments with multiple lenses such as telescopes. These latter are useful not simply "for the improvement of seeing *better*," but also "for the convenience of seeing *more*."[187] Finally, Descartes turns in the·"Eighth Discourse" to the use of the law of refraction, that is, of the relevant scientific theory to best design those lenses that we need in order to improve our vision in order to see better and more.

Thus, Descartes shows how some objects are, while other objects are not, "proportionate" to the senses. Those which are not can be made accessible using certain instruments designed in conformity with the laws of light, the law of refraction in particular. We can therefore explain why we cannot discern visually with the naked eye the objects seen in the microscope, but also explain why the images that we view in the microscope are in fact images of objects that are very small. Or, in the case of telescopes, too far away to be seen.

d. Confirmation Relations for Theories about Unobservables

The confirmation and falsification of theories by means of observations made by means of instruments can be understood in terms of the inference patterns that we have been examining.

Suppose that we have some theory, say T_1, which predicts the existence of some data E. T_1 has a prior probability derived from a generic theory. Suppose, further, that for some reason we cannot observe the fact E; it is, perhaps, too small to see. Thus, we cannot use this as evidence E to try to confirm T_1.

However, we may further suppose that we also have instruments available that can used to detect whether E obtains. Thus, if the predicted data E are too small to see, then the instrument we need would be a microscope. By means of this instrument we can detect E. That is, we observe the instrument and certain facts about it, and infer, that is, predict, that E obtains. Thus, if the instrument is a microscope, then we observe an image and infer from this image that there is a small object that is causing that image. In order to infer the data E from what we observe about the instrument, we need a second theory, let us call it T_2. This is the theory of the instrument. This theory too has a prior probability deriving from a confirmed generic theory. We observe the instrument, and use T_2 to predict E. If the instrument is the microscope, then what is predicted is the existence of the very small object of the relevant sort. Again, however, we do not observe E — it is too small to see — and so also with regard to this theory we cannot use the prediction as evidence E to try to establish a posterior probability greater than the prior probability of the theory.

If, however, we consider the conjunction of the theory T_1 and the theory of the instrument T_2,

$$T_1 \text{ and } T_2$$

then *this* theory does entail certain evidence, call it E', namely, the reading of the instrument. T_1, together with certain initial conditions, does predict E. From E in turn, by means of the theory of the instrument T_2, we predict E'. But, we are supposing, where E is not observable, E' *is* observable. So, while we cannot confirm the theory if we work with E alone, it turns out that with E', which we *can* observe, we are in a position to confirm our theory. We can obtain predictions that *can* raise the probabilities of the theories. From T_1 we can deduce that E. From this in turn we can predict by means of T_2 something that we can observe, namely E'. This latter can function as evidence for the conjunction. The observation of E by means of the instrument reading E' and as predicted by the conjunction of the theories will ensure that the probability of the conjunction of the two theories will be increased. Since the theories are independent of each other, this increase in the probability of their conjunction will ensure that the probability of the theory T_1, which is the object of our interest, is increased — though it ensures as well that the probability of the theory of the instrument is also increased.

Further, the confirmation of these specific theories will increase the confirmation of the generic theories of which they are instances.

We thus see that observations based upon instruments can be used to confirm, and to increase the probability of, the theories that are so tested.

Let us now see how such falsifying instance affects the probability of a theory.

We have observational data O as initial conditions, and these together with T_1 entail the existence of evidence E_M that cannot be observed, because, let us say, it is too small too see:

$$(O \ \& \ T_1) ==> E_M$$

where " ==> " represents the relation of logical implication or entailment. Further, if the magnifying instrument is turned on E_M, then this and the theory of the instrument T_M entails that one will observe through the lenses an image constituting evidence E.

$$(E_M \ \& \ T_M) ==> E$$

Thus,

$$(O \ \& \ T_1 \ \& \ T_M) ==> E$$

However, instead of observing image E, we observe instead image E', where this implies the presence of an object E_M' different from, and excluding E_M:

$$(T_M \ \& \ E') ==> E_M'$$

where

$$E_M' ==> not\text{-}E_M$$

In that case we can infer not-E_M, and therefore that

$$not\text{-}(O \ \& \ T_1)$$

However, we did observe O. Since this has been accepted, it follows that we must reject the theory being tested:

$$not\text{-}T_1.$$

The observations using the magnifying instrument thus in one way lead directly to the rejection of the theory we are interested in testing. That is, they so lead, *provided that* we have grounds for maintaining T_M. For we could equally infer, if we accept T_1, that we should also accept E_M, and therefore conclude that not-E_M'. In that case we would conclude that

$$not\text{-}(T_M \ \& \ E')$$

But we have observed E', and we must therefore conclude that

$$not\text{-}T_M$$

In other words, from the fact that the prediction E turns out to be false, we may conclude that either T_1 or T_M is false, but we have no grounds for concluding which of the two is false.

In our case, what we have is a conjunctive theory

$$T_1 \ \& \ T_M$$

where T_1 is the theory one is interested in testing and T_M is the theory of the instrument, the microscope. The conjunction predicts the existence of an image of a very small speck of dust; this is the evidence E. If E does not obtain, if we do not see the speck in the magnifying glass, then the probability of the conjunction of theories goes to zero: the conjunction which entails that E exists turns out to be false. That means that at least one of the conjunctions, that is one of T_1 or T_M is false. One at least of the two hypothesis must be considered to be eliminated.

Which one precisely is to be eliminated will be determined by other considerations. Thus, one may have a greater prior probability than the other, and this will affect how the inferences go. If our instrument is well tested, and the theory about it works is very secure, then we would take the theory T_1 to be falsified. But if our theory T_1 were very secure and we were instead trying to establish the reliability of some new instrument that we were developing, then we would keep this theory and reject T_M.

An example from Hooke will make illustrate some of the sorts of inference that are involved.

In the inferences at which we shall look, Hooke is concerned to establish the general point against Cartesian *a priori* reasoning that experiment is necessary to establish reasonable principles of explanation. Descartes had formulated an hypothesis to account for why sparks fly when a piece of flint strikes steel.[188] This involved the striking of the flint causing the pores in the steel to become smaller and force out small particles of earthy matter, what Descartes called the "third sort of matter,"[189] which were moved about rapidly by other particles of a still more minute nature. The latter were what Descartes called the "first sort of matter," and his account of fire is that it consists of particles of earthy matter being moved about rapidly by the first sort of matter.[190] [His account of air, that is, of the atmosphere, is that it too consists of earthy matter, save that this is now moved about by other matter of an intermediate size, what he calls the "second sort of matter."[191]] Descartes' "explanation" of the sparks caused by flint striking steel therefore requires these sparks to consist of matter different in kind from the matter of either the flint or the steel. Hooke made a series of experiments, all involving the microscope, and drew the conclusion that Descartes' theory was wrong. Hooke's general conclusion is that

... we see by this instance, how much Experiments may conduce to the regulating of *Philosophical notions*. For if the most Acute *Des Cartes* had applied himself experimentally to have examined what substance it was that caused the shining of the falling Sparks struck from a Flint and a Steel, he would certainly have a little altered his *Hypothesis*, and we should have found that his Ingenious Principles would have admitted a very plausible Explication of this *Phænomenon*; whereas by not examining so far as he might, he has set down an Explication which Experiment do's contradict.[192]

More specifically, Hooke argues, "we need not trouble our selves to find out what kind of Pores they are, both in the Flint and Steel, that contain the Atoms of fire, nor how those *Atoms* come to hindered from running all out, when a dore or passage in their Pores is made by the concussion...".[193] Hooke provides a series of inferences designed to refute the Cartesian hypothesis, on the one hand, and to confirm an alternative, on the other.

When flint and steel are struck, sparks are given off. The latter become small particles of matter that can be collected. Hooke collected these and examined them through his microscope. He observed that these are metallic and have a variety of special shapes. For convenience let us refer to these as CSM particles (characteristically shaped metallic particles). That the particles are CSM is something that cannot be discerned by the unaided eye; it can be observed only via the images of the microscope.

Now, Hooke knows the generic law something like this:

(h1)   For all sizes f of pieces of iron or steel, there are factors g among which is heat but excluding light such that, if y is a sample of steel of size f and a factor g is applied to y, then y's composition is changed.

This law provides a range of factors for affecting the composition of samples of steel, and has many specific instances, and it asserts that these factors are sufficient to change the composition of samples of steel. It thus functions as a Principle of Determinism and of Limited Variety for sufficient conditions for changing the composition of samples of steel. Where the sizes f can be examined by the unaided senses, these specific laws have been confirmed, e.g., by iron workers. The confirmation of these specific laws tends, in the fashion that we have seen, to confirm this generic law. This confirmation of the generic law (h1) in turn provides confirmation of other specific instances, *including instances that apply to particles too small for their composition to be determined by the unaided senses.*

Hooke proceeds to take small particles of steel and pass them through a candle flame.[194] These particles emerge from the flame changed to CSM particles, as he confirms using his microscope. The particles that do not go through the flame do not change to CSM particles; those that go through the flame do change. The flame provides heat, which, by (h1), is the only relevant factor that is present. Hence, by the method of difference one can conclude from the experiment that the heat of the flame is what has produced the special CSM composition of the particles.[195]

In the context of this experiment there is only one relevant factor present, namely, the heat. We can therefore use a *definite description* to refer to the characteristic composition of the particles that result from a process of this sort. Specifically, we can refer to the particles as having "the composition caused by the application of heat," or, more briefly, we can refer to them as "burned particles."

The point of the experiment with the candle is, of course, to draw an inference with respect to the particles given off when flint and iron or steel are struck together. Both processes yield particles of the CSM sort. But *the same effects derive from the same causes.* More specifically, we have the law (h1) which delineates the possible relevant factors for explaining the presence of CSM particles. But, Hooke argues, such heat is the only factor that is present. It is present due to the force of the blow of the flint and iron or steel being struck together. First, we know that striking iron or steel can in fact make it hot.

.... either hammering, or filing, or otherwise violently rubbing of Steel, will presently make it so hot as to be able to burn one's fingers.

And we know, moreover, that a very strong blow occurs when the samples of the two substances are struck together.

...the whole force of the stroke is exerted upon that small part where the Flint and Steel first touch: For the Bodies being each of·them so very hard, the puls cannot be far communicated, that is, the parts of each can yield but very little, and therefore the violence of the concussion will be *exerted* on that piece of Steel which is cut off by the Flint.[196]

We can therefore reasonably assume from this physical knowledge that heat is present when the blows are struck. It is the only relevant factor that is present, and therefore constitutes the difference that accounts for the production of CSM particles when flint and steel are struck together. Again we can use the method of difference to conclude that the CSM particles produced by the flint and steel being struck is the heat produced by the blow.

Hooke is therefore able to conclude that he has confirmed the following hypothesis:

(T[H])   ...the spark appearing so bright in the falling, is nothing else but a small piece of Steel or Flint but most commonly of the Steel, which by the violence of the stroke is at the same time sever'd and heatt red-hot, and that sometimes to such a degree, as to make it melt together into a small Globule of Steel; and sometimes also is that so very intense, as further to melt it and vitrifie it; but many times the heat is so gentle, as to be able to make the sliver only red hot, which notwithstanding falling upon the tinder ... it easily sets it on fire.[197]

This hypothesis is confirmed; the observational data support it. Moreover, it contradicts the Cartesian hypothesis, since Hooke's confirmed hypothesis asserts that the sparks are pieces of steel while Descartes holds that they are particles of some other substance. The Cartesian hypothesis must therefore to be rejected.

We have three theories, the Cartesian hypothesis T^D, Hooke's hypothesis T^H, and the theory of the magnifying instrument T^M. The hypothesis T^D leads one to conclude that the particles given off when flint and steel are struck are not steel. If this were correct, then the image that one observes in the magnifying instrument, according to the theory T^M of the instrument, would not be an image of a particle of steel. But in fact we do observe the image of a particle of steel. This evidence E therefore requires us to reject the conjunction (T^D & T^M). Now, T^D is a mere hypothesis, with no support from any sort of observational data. The theory T^M of the magnifying instrument, in contrast, has considerable observational support. The prior probability of T^M is thus very much greater than the prior probability of T^D. Hooke therefore opts to reject the Cartesian hypothesis T^D rather than his theory T^M of the instrument.

However, in addition, he observes in the magnifying instrument exactly what his own theory T^H would lead one to expect. Thus, the conjunction (T^H & T^M) leads to evidence E that confirms the conjunction, raising the posterior probability of both T^H and T^M higher than their prior probabilities.

This confirmation of the hypothesis T^H further justifies the rejection of the Cartesian hypothesis T^D. For, T^H contradicts T^D concerning the composition of the particles. Since T^H is to be accepted, it follows that we must reject its contradictory T^D. [198]

I. Lakatos has used the notion of auxiliary hypotheses about instruments to provide a different criticism of empiricism. He has argued that this calls into question both the empiricist account of science and also the alternative falsificationist view of Popper. If a theory T entails observational data E, and E is not observed, then, instead of concluding that not-E, we may instead always add an auxiliary hypothesis to explain the non-observation of E. [199]

Now, first, if this line of argument is accepted, then it shows that no theory can be falsified. It follows that we must reject Popper's philosophy of science, which argues that what demarcates science from non-science is that the propositions of the former are falsifiable in principle. But the example, if successful, also shows, second, that any theory can be made to fit the observed facts, if we but add sufficient auxiliary hypotheses about the powers of our instruments. It follows, therefore, if Lakatos is

correct, that every theory can always be made to fit the facts, and must therefore count as confirmed. And from this it would follow in turn that we must also reject the empiricist account of science.

We have seen, however, that the scientists of the early modern period were prepared to provide an answer to Lakatos. According to Lakatos, when we have a theory T that entails evidence E and E is not observed, then we can always add an auxiliary hypothesis A to explain why E is not observed, thus saving T from falsification. Note how A works, however. A is an existential hypothesis, asserting the existence of some entity that interferes with our ability to discern the facts E. A is *not* inferred by means of some hypothesis from *what we do observe*, that is, from *observed initial conditions*. Rather, A is simply inserted in order to save the theory T. It remains an hypothesis, not a prediction. There is, therefore, *no explanation for why* E *is not observed* — only a guess. Moreover, A does not provide any independent evidence that E obtains; it is only inserted in order to provide an hypothesis why E is not observed. The only basis for regarding E being there is the prediction of the theory T. *The auxiliary hypothesis A provides no independent reason for predicting that* E *obtains*. Now note the difference between this and the case of the watch. In the latter, the Hume's theory leads to the prediction when the watch is not working that there is a speck of dust present. We do not see it, but infer instead that it is too small to see. However, the artizan is using his jeweller's magnifying glass. On the basis of the theory of the lens, and *also on the basis of observed initial conditions* provided by the seen image, we are able to infer again, independently, that there is an object present that is too small to see. Finally, a confirmed theory of vision also testifies that the object whose existence is revealed by the magnifying glass is indeed one that we can reasonably expect to be one that is too small to be discerned visually. Thus, **first**, the theory predicts the existence of certain entities; but the theory of the instrument provides independent evidence for the existence of those entities. For Lakatos, in contrast, there is no inference to E apart from the theory T. Moreover, **second**, there is an explanation, based on initial conditions and confirmed laws, why the entities cannot be discerned by means of the (unaided) senses. For Lakatos, in contrast, there is no explanation, based on initial conditions and confirmed laws, why the entities predicted cannot be discerned by means of the unaided senses.

It may be that any theory can be "saved" from falsification by the sort of ad hoc moves that Lakatos indicates. Certainly, the elaboration of hypotheses in the way that he indicates could hardly be counted as contributing to the progress of science. As he correctly sees, any methodology that is proposed, if it is to be adequate, must eliminate such moves. What we are arguing is that the early moderns knew what they were about so far as concerns the theory of the instrument. Where Lakatos permits theories to be "saved" by the introduction of mere hypotheses, the early moderns insisted that *a theory could be counted as saved only on the basis of explanations and predictions of the deductive-nomological sort based on observed initial conditions.* Here, as elsewhere, "hypotheses non fingo." It is clear that all that it takes is imagination if one is to construct hypotheses capable of saving a theory. Lakatosian adhocery is always possible. But imagination alone cannot construct theories which are, on the one hand, explanatory and which, on the other hand, raise the probability of the theory that is being saved. By placing these restrictions on an adequate epistemology of the instrument, the early moderns were able to present a methodology that allowed for the introduction of entities that cannot be discerned by the senses, but allowed also for conditions under which the theory is to be counted as falsified.[200]

What we have seen, then, is that the empiricist has a good reply to this objection.[201] We may therefore conclude that, contrary to what Lakatos suggests, an empiricist account of the auxiliary hypotheses that are involved in the theory and epistemology of the instrument can easily allow for the falsification of theories.

It follows, moreover, that we have shown the unreasonable nature of a method like that suggested by Lakatos for saving theories from falsification, which proceeds upon the basis of mere hypotheses and does not insist upon confirmation.

e. **Extending Our Instruments**

The use of instruments, then, can enable us to overcome the objection to one's hypothesizing, in order to explain observed data, objects or entities that are too small to see, or otherwise too something to enable them to be discerned by our unaided senses.

Descartes carefully builds up from single lenses to instruments with compound lenses. This, to be sure, is partly expository, part of his general — and wise — strategy for both research and pedagogy, to begin with simple cases and proceed to more complex ones. But there is another aspect to this, having to do with confirmation.

For the simple lens, there is a generic law that applies to all objects which it is used to observe. There are specific instances of this law, themselves laws, as we have seen, that apply to objects of various sizes. For larger objects, these specific instances can be confirmed visually. These confirmations tend to confirm the generic law. This generic law then renders probable the specific instances that hold for the objects that are too small to be discerned visually. Microscopes involve systems of lenses. These render observable objects that are too small to be discerned by means of a single lens. But they also render larger those objects which can so be discerned by a single lens. The laws of the single lens confirm the existence of the small objects. These specific laws from the single lens thus yield confirmation of some specific laws for the system of the microscope. These in turn confirm the generic law for the system of lenses. This generic law is also supported by the still more generic theory of refraction. With the generic law for the system of lenses, we can infer specific laws for objects too small to be discerned by a single lens. These specific laws receive support in the way we have gone over from the generic theory. We thus infer, when we observe images through the microscope, that there are objects of an extremely small size that cause those images.

Some instruments take us some way into the realm of the indiscernible. Other, stronger, instruments take us even further. There is an area of overlap, and in this area the confirmed laws of the theory of the weaker instrument provide confirmation of the specific laws for the same area for the stronger instrument. These confirmed specific laws for the stronger instrument confirm its generic law. This in turn support the specific laws that enable us to infer the existence of objects of the appropriate sorts in the area that is beyond the strength of the weaker instrument. Descartes' strategy of exposition is also a strategy whereby patterns of confirmation lead us legitimately deeper and deeper into realms inaccessible to the unaided senses.

We can now of course go beyond the optical microscope, the limits of which can be predicted on the basis of the wave theory of light. But there is an overlap with the realm that can be explored by the electron microscope, which can, however, take us to entities too small to be discerned by the optical microscope. The logic of the situation nevertheless remains essentially the same.

There are other ways to detect still smaller entities. Protons, for example, can be detected by means of photographic plates, or in the Wilson Cloud Chamber. We introduce such entities for much the same sort of reason that Hume introduced the specks of dust too small to see, namely, in order to fill in gaps in the causal, or at least lawful, picture of the world.[202] Our instruments do the job of aiding the senses in discerning these entities, much as the jeweller's glass could help the artizan. The theories of the instruments are much more involved than the theory of geometrical optics. But to repeat, the logic of the situation nevertheless remains essentially the same.

One thing is perhaps worth noting. The speck of dust could not be seen, but it is nonetheless something that *could* be seen, if our eyes were a bit more acute, which is to say, if the lenses were a bit stronger in a way that could be predicted by means of the theory of geometrical optics. We are as it were just too far sighted. The entities, such as electrons and protons, not to say quarks, that are now part of the theory of matter are different sorts of thing altogether. For, entities such as these are entities that simply *could not* be seen.

What the eye responds to are colours. These colours may be chromatic — red, yellow, etc. — or non-chromatic — white, the grays and black — but in any case colours. Now, an object has a colour only if it absorbs and radiates certain masses of photons. These masses of photons strike the retina and create there the image that Descartes described. The point is that an object such as a photon or an electron or the even smaller quark is, *according to the theory of matter*, just *not the sort of entity that could absorb and emit a mass of photons*. It follows immediately that it is not the sort of entity the image of which could be created on the retina of the eye. It is therefore *not the sort of entity that could be seen*.[203]

The instruments that justify the claims of the modern theory of matter thus take us to a realm of entities that is not only beyond the reach of our senses but to a realm that is totally outside the world of the senses. They not only take us to that world, they provide us with the tools to *confirm its existence*.

Such a world might seem strange. It might, too, seem strange that we live in such a world. It might seem even stranger to say that we *know*, or, at least, reasonably believe on the basis of good evidence, that we live in such a world. But, given the epistemology of our instruments, it is not at all strange that we know that such a world exists. We understand how those instruments work, the theory for that understanding of the instruments is available, and those instruments and that theory provide the basis for our coming to know, and to understand, the world in which we live but which is totally outside any world that we could know by our senses unaided by the artifice of instruments.

f.   Parapsychology

It is often claimed that some persons have the capacity to know things that seem unknowable by ordinary means. This capacity was labelled "ESP" for "extrasensory perception" by J. B. Rhine. It is also claimed about these persons that they can foretell future events. This capacity has been called "clairvoyance." Some also claim to move physical objects other than their own bodies by simply willing it. This capacity has been called "PK" or "psychokinesis." Other similar capacities are attributed to many people. Dr. J. B. Rhine, who spent many years defending the claims that these sorts of power do exist, thought that it was vitally important that this defence be made. In his view, the ordinary scientific view of humankind amounted to materialism, and the widespread acceptance of this view would have disastrous social consequences.

The most far-reaching and revolting consequence lies in what would happen to volitional or mental freedom. Under a mechanistic determinism the cherished voluntarism of the individual would be nothing but idle fancy. Without the exercise of some freedom from physical law, the concepts of character responsibility, moral judgment, and democracy would not survive critical analysis. The concept of a spiritual order, either in the individual or beyond him, would have no logical place whatever. In fact, little of the entire value system under which human society has developed would survive the establishment of a thoroughgoing philosophy of physicalism.[204]

To baldly assert this, as Rhine here does, is to ignore a large body of literature that argues to the contrary. But in any case, the present point is that he thinks that the evil consequences of physicalism could in his opinion be counteracted by recognizing that people do have these special sorts of psychic powers. The concerns that Rhine here exhibits and the claims that he makes are similar to those of the creation scientists, who also argue that it is a virtue of their position that it can counteract materialism and humanism.

Rhine's beliefs that many have a variety of psychic powers were formerly shared by only a relatively small number of people. They were publicized either by journals of the committed or by supermarket tabloids such as the *National Enquirer*. These beliefs have now become more respectable, however, and find their boosters in much more respectable circles in regular newspapers, books and TV programmes. Many have become convinced that there is in fact *scientific evidence* that some people have these extra-ordinary capacities.

In spite of this belief, it remains true that after almost a century of attempts to obtain scientific evidence supporting the claims of parapsychologists, the investigators have never been able to convince any group of independent experts. The increasingly general impression that there is scientific evidence is belied by the facts. The general impression is due less to any evidence than it is to heavy propaganda by financially interested parties, including those who profit by having supposed "psychics" demonstrate their self-claimed extrasensory, clairvoyant, or psychokinetic capacities on TV shows.

Parapsychologists have endeavoured over the years to provide statistical tests aimed at showing that certain individuals do have "psychic powers."

These tests go something like this. A person who is supposed to have such psychic powers of extrasensory perception is placed in, let us say, a special room. Call this person "Jones." Another person, "Smith," is located elsewhere. These locations are so arranged that there can be no communication between them. Smith has a set of cards. There are various symbols on these cards, such as a triangle or a circle or a cross, and the same number of each kind of card. These cards are shuffled by Smith, to put them into a random order. Smith draws a card, and presses a buzzer that tells Jones to identify the card drawn. The experiment is repeated many times. The claim is made with regard to some subjects, some Joneses, that they can identify the card drawn in a way that is incompatible with chance.

If C is the event of drawing the card and calling it, and E is the event of calling the card correctly, then the claim is that in the population of C's,

$$Pr(E) > \frac{1}{2}$$

The issue is whether the subject Jones is guessing or not. If Jones is guessing, then we can expect that he will call the cards correctly roughly half the time. More specifically, if Jones is guessing, then the hypothesis

$$Pr(E) = \frac{1}{2}$$

correctly describes the situation. If Jones is not guessing then

$$Pr(E) \neq \frac{1}{2}$$

Those who argue that there is scientific evidence that supports the claim that certain persons have ESP powers advance statistical tests to defend such claims. Their tests are, of course, statistically sophisticated , and have become more sophisticated over time. The results have over the years been disputed: it remains unlikely that the defenders of parapsychology have succeeded in demonstrating that some persons sometimes achieve results in calling cards that would not appear on the hypothesis that it all was mere chance. At best, it is uncertain. But if there are problems with the statistics, there are even deeper problems with the claims of the defenders of the parapsychology and the claims that there are people with powers of ESP. So let us waive the concerns with regard to the statistics. For the sake of the argument, and the critical evaluation of the worth of parapsychology as a science, *let us assume that the statistical tests yield results that would normally justify the rejection of the hypothesis that the results happened as. a matter of chance.* Let us assume, in other words, that the results are such as would be consistent with the hypothesis that Jones was not guessing and that he did indeed have powers of ESP. **The important point is that even if one accepts that the results have this sort of probative force, there are still sufficient problems with the results given us by parapsychologists to justify the claim that what they are doing is a pseudoscience.**

The sort of experiment just described deals with mass events. There is on the one hand the sequence of card draws. these are the events C. The long sequence is a mass event. Call it M. M consists of a long sequence of events C. For purposes of statistical evaluation, it is essential that the sequence of draws be random. On the other hand there is the sequence of events E in each of which the experimental subject, our Jones, correctly describing the card drawn. This too is a long sequence, a mass event. Call it R. R consists of a long sequence of events E. A trial consists of an event C followed by an event E. For purposes of establishing that there is ESP, extrasensory perception, Jones must be so located that he or she cannot physically view or otherwise have by ordinary means information about the card drawn on the trial prior to he or she giving the report. We are assuming that the event R is one in which the proportion of correct descriptions is greater than would result from chance. We are in other words taking for granted that we have established the regularity with regard to Jones that

(s)   Whenever M then R

The claim of parapsychologists is that establishing (s) amounts to establishing that Jones has ESP powers, that we have in Jones a person who has powers of extrasensory *perception*. This way of putting it carries the suggestion that there is some sort of process involving Jones and the cards that can explain why C is followed by E in a way that is contrary to the assumption that Jones is merely guessing, that the call is simply the result of chance. The use of the term *perception* suggests that the process is somehow parallel to the processes by means

of which we gather information about the world in which we live by means of ordinary sensory perceptual processes, e.g., seeing, hearing, touching, and so on. The two sorts of perceptual process, the sensory and the extrasensory, will involve the transmission of information from the environment to the subject. In the case of Jones, the information will enable Jones to described the card drawn on a trial. The process is not infallible; Jones sometimes gets the call wrong. But neither are ordinary sensory perceptual processes infallible; sometimes, as in the case of various kinds of illusion, we do misperceive things. Because the process of extrasensory perception is not infallible, the percentage of E's in a long sequence of C's is less than 1. But — this is what we are assuming — the percentage is greater than ½, greater, that is, than would result if Jones were merely guessing.

Now, if there is this sort of connection between C and E, then we immediately have an explanation of the regularity (s): (s) obtains as a consequence of the obtaining of certain underlying causal processes, processes of *extrasensory perception*.

At the same time, however, it must be emphasized that by itself the regularity (s) does not establish that such a causal connection exits. By itself (s) is a "mere" regularity, a correlation, and, as it is important to remember, *correlation does not establish causation.*

What one can infer from the fact, if it be a fact, that (s) obtains is that this would be a matter of course if there are a causal process that relates C and E. But by itself (s) does not establish that there is such a process. In other words, from the fact, if it be a fact, that (s) obtains one *cannot* infer that there exists a process of extrasensory *perception*, a process in which information is somehow transmitted by extrasensory means from the card drawn to the subject attempting to describe that card.

What is important for evaluating the claim that parapsychology is indeed a science rather than a pseudoscience, is therefore the claim that there is some sort of causal process of extrasensory perception. It is here that we come up against the central criticism of parapsychology as a science: *there are no known theories with any reasonable plausibility that suggest there might be some sort of causal connection between the causes and effects that are being investigated in parapsychological experiments.* Even if we grant what we are here assuming, that there are some statistically significant results, then these facts are, as has one sympathetic discussant has put it, "facts in search of a theory."[205] Another such discussant, H. H. Price, has made the same point. As he has expressed it,

> The theoretical side of psychical research has lagged far behind the evidential side. And that, I believe, is one of the main reasons why the evidence itself is still ignored by so many ... Such an explanation is needed for its own sake; and it is also needed to get the evidence attended to and considered.[206]

More strongly, given what we do know about our abilities to know facts about the things around us, it is clear that the known processes of sense knowledge — seeing, hearing, touching, feeling, smelling — all involve the transmission by physical processes of information from external objects to the brain where the information is de-coded. If we try to generalize from these causal facts to cover the case of knowing which card was drawn, there seems to be no way in which information could be conveyed by a physical process from the card to the brain of the subject, Jones, whose task it is to call it.

Milton Rothman has put the relevant point this way:

> Transmission of information through space requires transfer of energy from one place to another. Telepathy requires transmission of an energy-carrying signal directly from one mind to another. All descriptions of ESP imply violations of conservation of energy [the principle that mass energy can be neither created or destroyed in physical processes] in one way or another, as well as violations of all the principles of information theory and even of the principle of causality [the principle that an effect cannot precede its cause]. Strict application of physical principles requires us to say that ESP is impossible.[207]

Rothman concludes that it is therefore unreasonable to entertain any hypothesis to the effect that some person has powers of ESP, or, for that matter, any other suggested form of non-natural "psychic powers."

Thus, generalizing from what we do know quite well about all the different sorts of ordinary perception to the very different case of Jones identifying cards under conditions in which there is no sensory input, we have to conclude that there should be a similar causal process, some sort of physical conveying of information, from the drawn card to the brain of the subject. But there is, so far as we can tell, no such process. Our generalization from the specific cases of our ordinary senses to this next specific case requires certain constraints of a generic nature on what could be taking place. These generic constraints are not fulfilled: there is no physical process conveying the required information — that at least seems to be precluded from the way the experiment is set up, to prevent any communication, or other physical conveying of information, between the person drawing the card and the person calling the card. We must therefore conclude that it is reasonable to hold that there are no causal relations connecting C and E.

There inference here is similar to the inferences that lead one to find that the theories of, say, Velikovsky or of homeopathic physicians, are not acceptable as scientific theories: *they conflict with clearly justified background scientific theories*. It seems that we can reasonably add parapsychology to the list of theories that are on this basis rejected as pseudoscientific.

This problem, the lack of a background theory, has been recognized by the defenders of parapsychology. Dr. Rhine himself proposed that we adopt a theory to the effect that some sort of *non-physical energies* are at work. He refers things such as ESP, telepathy, precognition, clairvoyance, and similar phenomena as "psi phenomena." Concerning these he writes:

> Might not the same logic, that has produced the concepts of the various energies involved in physical theory profitability be followed to the point of suggesting that psi energy be hypothesized? ... It is no great jump from the broad concept of energy as it now prevails in physical theory over to the notion of a special state of energy that is not interceptible by any of the sense organs....It may tentatively be proposed, then, that back of the phenomena of psi must exist an energy that interoperates with and interconverts to those energetic states already familiar to physics.[208]

The suggestion is that we can begin to understand why Rhine obtains the results that he does (we are assuming that he does obtain these results) if we introduce as part of our theory the rather strange entity which he calls "psi energy." Is this so strange? After all, in order to understand the workings of the observable world science introduces not only the minute specks of dust to which Hume referred but things such as electrical energy which have no counterparts in the visible world, and beyond that really weird things such as electrons and even quarks. If physicists can do it, why not parapsychologists?

In order to evaluate Rhine's suggestion it is necessary to compare his proposed "theory" concerning "psi energy" with the sorts of theory just examined in which physics introduced entities which lie beyond the capacity of our senses to detect.

In the theories that we have examined the entities that lie beyond the limits of perception are clearly connected to entities that we *can* perceive. These connections are established by the background theories that are available. In the example we looked at in detail, the relevant theory was that of geometrical optics, but the point is applicable to all similar theories in physics, chemistry, etc. Rhine suggests that there are similar connections in the parapsychology. The concept of energy is, as he correctly points out, a generic concept. There is electrical energy, heat energy, kinetic energy, and so on, various species of energy. He suggests that we introduce another species of this generic concept in order to account for ESP and other psi phenomena. Now, in the theories of physics that introduce inperceptible entities, it is necessary to refer to them and their properties by means of definite descriptions that rely upon generic concepts to achieve their reference. Generic concepts are thus essential if we are to refer to entities and to the properties of such entities when these lie beyond the limits of ordinary perception. Rhine has so far captured the logic of the situation.

But there is a further point. In the examples from physics, the generic concepts are used as part of a system of abstract generic laws. These generic laws have as instances specific laws *that have been confirmed experimentally by observational data*. Thus, in the example we developed, the relevant theory is that geometrical optics. This theory has specific laws which describe phenomena observable in ordinary sense experience. These laws have been confirmed in considerable detail. The theory of geometrical optics is a generalization from these results. The observational data that confirm the specific laws also provide, in the way with which we are now familiar, confirmation of the generic theory. The confirmed generic theory is now applied to unexamined cases, indeed to cases that are unexaminable. The conclusions which we deduce about the unexminable cases it turns out fit into other aspects of the theory than those on the basis of which they were inferred. This leads to predictions of otherwise unexpected events. These predictions turn out to be successful. We have thus further confirmed the theory through which the unobservable entities were introduced. More importantly, we have also thereby confirmed the hypothesis, based on those theories, that those unobservable entities exist.

What about Rhine's proposal? What theoretical connections does he provide to link his new specific concept of energy to similar concepts in physics?

Consider the case of Hume's minute speck of dust that caused his watch to not work correctly. This speck had a certain property, namely, a shape. This shape was of course specific. But since the speck was not given in sense experience, neither was this specific shape. Yet we do know that this specific shape falls under the genus *shape*. Moreover, it is related geometrically to shapes that we can observe. The shape of the speck might, for example, have been that of a cube. As such it would have square sides, the angles at the corners would be right angles, and so on. Further, it would be similar to the shape exemplified by cubes that we can discern by means of our senses — in fact it would be exactly like the sensible cube only smaller. The relevant point for our purposes is that the non-sensible property of the non-sensible speck is a species falling under a genus. This genus is known to us by sense since in fact we are aware in sense of various specific properties which fall under this genus. Thus, although the properties of the minute speck are not given to us in sense, these properties are members of a genus other members of which *are* given in sense. Since we know by sense what the genus is, and since we know by sense how members of the genus are related to each other, we can form concepts, definite descriptions, by means of which we can refer to the properties of the minute particle.

Not only does science have available clear concepts that refer to the properties of the unobservable speck, but in addition our theory of matter contains laws that describe the behaviour of minute particles. The relevant laws are generic. We have confirmed specific instances of these generic laws in the case of observable entities. These observational data then provide empirical support tot he specific laws that apply to the minute particles. These laws that apply to minute particles enable us to explain how they have causal effects including casual effects that can be detected at the level of sense experience, e.g., the stopping of one's watch.

Without going into the details of the case, similar things can be said about energy. There are various species of energy, mechanical, electrical, heat, and so on. What these have in common is that any form of energy can perform work; this is the generic feature that they have in common. This means that each form of energy can be converted into mechanical energy. When science introduces a new concept of energy, there are clearly specified regularities that describe how this conversion can be effected. There are, in other words, connections that relate various species under the genus one to another. There are, further, generic laws which apply to all forms of energy, the most important of which is the Law of Conservation of Energy. And there are laws which relate energy in its various forms to observable effects.

Turn now to the "psi energy" that Rhine introduces. We have two sorts of event, C and E. The latter is sometimes consequent upon the former, rather as Hume's watch sometimes stops subsequently to being wound. In order to explain why Hume's watch sometimes did not work, appeal was made to the law (1)

> For any individual x, if x is P then, x is not-T if and only if x is Q

Rhine envisages a similar appeal, suggesting in effect that there is a law something like

(R)   For any individual x, if x is C then x is ψ if and only if x is E

Given this, we can infer the presence of ψ ("psi") in an individual j who is in situation C just in case that j is E.

In the case of Hume's watch, the law (1) is not confirmed directly; the speck is too small to see. But there is a background generic theory that provides empirical support for the law (1). This confirmed background generic theory enables the scientist to reliably use (1) for purposes of explanation and prediction.

What about Rhine's proposed law (R)? It can be used for purposes of explanation and prediction only if it there are data which confirm it. These data con either be direct or indirect, by means of a background theory. *Ex hypothesi* the confirmation cannot be direct: as Rhine himself states, his hypothetical psi energy is not "interceptible." That means the confirmation of Rhine's proposed (R) must be indirect, by means of a background generic theory. Is there such a theory? Alas, we are given none.

To be sure, Rhine informs us that the relevant cause mentioned in (R) is a form of *energy*, and we all know what energy is, do we not?

To this the answer is that we do indeed know what energy is — it is in fact a well-defined concept in physics and chemistry. But for Rhine to appeal to this concept of energy as relevant to establishing that (R) is capable of confirmation by means of some background theory, he must establish that his new form of energy, psi energy, is a species that falls under the generic concept that all will agree is a legitimate and, in Bergmann's terms, a significant concept in science.

Rhine's appeal to the generic concept of energy gives an aura of scientific legitimacy to his proposal (R). But the legitimacy is illusory.

If we look at the other forms of energy that are described in well confirmed scientific theories, then, as we just said, it turns out that are all inter-related one to another. Specifically, they are interconnected through their capacity to be converted into mechanical energy and thereby into work. How is "psi energy" so related? Count Rumford was able to demonstrate that heat was a form of energy by showing how it and mechanical energy could be converted into one another. Rhine is no Count Rumford. Rhine provides no techniques by means of which psi energy and mechanical energy can be converted into each other. Nor does he show how his new form of energy can be harnessed to perform mechanical work. "Psi energy" may well in some sense of the term 'energy' be a form of energy, but it clearly is not a member of the same genus *energy* that is used in physics and chemistry. Rhine suggests that his form of energy and those of science are all members of the same genus. But this is spurious; he has provided no such connection whatsoever. This means that the concept ψ (psi) that appears in (R) has no scientific import; it has not be specified in a way that makes it one of the species falling under the generic scientific concept of energy. By hypothesis, (R) cannot be confirmed directly; it now turns out that neither can it be confirmed indirectly, by fitting it into a background generic theory about energy.

Rhine's problem was that of concluding that the correlation which (we are supposing) he observed was more than a *mere* correlation. The way to do this, as we know, is to fit the observed regularity into a background generic theory which has antecedently has been confirmed in other specific instances. Rhine's (R) is his attempt to do this. We now see that the proposal is in fact vacuous. We can conclude that he has failed in his claim that there is a causal process which can, even tentatively, explain the apparent result that some people, Jones among them, can identify randomly selected cards at a rate of success greater than chance. The result remains a mere correlation.

We may safely conclude that physics has every right to introduce weird entities such as quarks into the theory of why the world behaves as it does. There is evidence that justifies their introduction. In contrast, Rhine has no evidence that justifies, even in the most tentative way, the introduction of his extra-physical "psi energy." His proposal has no basis in fact.

Rhine's suggestion concerning forces or energies that are somehow non-physical are not better than similar theories in astrology and in alternative medicine. Recall the proposal of Mr. Ted Saito of the Toronto Shiatsu Centre, as reported by Pat Moffat in Toronto's national newspaper, that we can explain the presence of disease in people through an inbalance in their energy field.[209] "When a person's *chi* [energy] is out of balance, [Saito's] theory states, disease may develop." She reports Mr. Saito as claiming that

> I can sense [the energy], I can draw the meridians [the passages through which the energy is supposed to flow in the body] exactly. But other people can't detect it, so this is a problem.

Here we have another hypothesis that sounds empirical because the word 'energy' appears in it. But like that of Rhine, we are not given any means by which to detect the energy, nor, therefore, any means for undertaking research that could put the hypothesis to the test.

Mr. Saito and Professor Rhine might equally well have appealed to bumbies in their proposals. We would have been given something equally testable, and equally reasonable.

Rhine's theory thus turns out to be straight forwardly non-scientific. But this non-scientific theory is so disguised by language that it has the appearance of science. It is, in other words, pseudoscience.

We have assumed for the sake of the argument that Rhine actually has obtained success rates in card identification trials which are greater than would be expected by chance, that is, greater than would be expected if the subject mere merely guessing.

From these results Rhine draws his (illegitimate) conclusions that there are "psi energies" at work.

However, if we want to understand in a deeper way, that is, scientifically, the results that we are assuming that he has obtained, we need not postulate those weird sorts of non-empirical non-scientific entities.

To the contrary, *there are in fact a series of considerations deriving from background knowledge that make it more than reasonable on good scientific grounds to expect non-chance results in parapsychological experiments.* This could be so if there were some theory that could establish a connection between C, the event of the drawing of the card, and E, the event of the identification of the card by the subject, that would make it reasonable to assume that there is something other than chance at work. If we reflect upon the situation, it is in fact easy to conceive a theory that would raise the prior probability of $H_1(s)$ to a high degree. That is the theory that *there has been cheating*, or, to put it a bit more charitably, that *there are flaws in the experimental set-up.*

Those who defend ESP want to establish that their subjects like Jones can achieve a non-chance result with regard to calling the cards, and can do so even in the absence of the physical conveyance of information as in ordinary sensory processes. That is why it is called *extrasensory* perception. They wish to establish that their subjects have some non-ordinary power.

This means that they must rule out cheating. This is why tests in parapsychology must be much more rigid than those in ordinary research. A psychologist who runs an experiment using white rats, or a medical researcher testing a new drug, does not need such rigid controls against cheating: the subjects either, like the rats, do not cheat, or like the patients, at least have no motive to cheat. But there is often a strong motive to cheat in the case of parapsychology — indeed, increasingly so, since there are often significant monetary rewards for those who can establish that they are "psychics" and for those who can provide "scientific evidence" that these people have unusual powers.

In fact, the whole area of ESP and parapsychological research has been bedevilled by cheating. Perhaps the most notorious was that reported in 1974:

> Dr. Walter Levy, head of the [Institute for Parapsychological Research at Durham, North Carolina] and ... heir apparent to Dr. Joseph B. Rhine, the noted writer [and parapsychologist] ... has been fired in a scandal involving irregularities in statistics.... Dr. Rhine [said] he has returned to the Institute as acting director because ... 'of the shake-up we've had. The situation is not one that is needed to be known [by the public].' Dr. Levy was fired in June, but his dismissal has been kept secret since then.[210]

Critics have found other flaws. Zusne and Jones have explained that

> Chance was clearly not producing Rhine's results. It was opportunities to establish the identity of the cards by sensory means. They were so numerous and so readily available that much of Rhine's work during the 1930's may be safely ignored.[211]

Among other difficulties they cite the following:

> The instructions that accompany the ESP cards, which were made available to the public in 1937, indicate that an 18 x 24 piece of plywood would be sufficient for screening purposes. It is decidedly not. A small screen still allows the percipient to see the faces of the cards if the agent wears glasses, and even if the agent does not, because the card faces are also reflected from the agent's corneas. Changes in facial expression give away clues that are not concealed by small screens. Larger screens still allow the percipient to hear the agent's voice. If the agent also serves as the recorder, which was routine in Rhine's experiments, voice inflections are as useful a source of information as are facial expressions.[212]

They note that

> When the distance between the percipient and the cards was increased, scores dropped.[213]

Others have come to the same conclusion. A recent detailed examination of the available literature has concluded that "the best scientific evidence does not justify the conclusion that ESP — that is, gathering information about objects or thoughts without the intervention of known sensory mechanisms — exists." "Nor," it was also concluded, "does scientific evidence offer support for the existence of psychokinesis — that is, the influence of thoughts upon objects without the intervention of known physical processes."[214]

In most cases, the prior probability of cheating or some other fault in the set-up is sufficiently high to make that a plausible explanation of the results in those cases where the subject in fact scores at a rate higher than chance. For most of the tests where success has been claimed, there is, unfortunately, almost no way in which we can go back and examine them to know exactly what controls were in force to prevent cheating and to know if they were tight enough to actually succeed in preventing cheating. Such an evaluation, if it were to be done well, would require immense amounts of time and effort. Raw data would have to be scrutinized. All the participants would have to intensively interviewed. The experimental set-up would have to carefully examined, and put to further tests. And so on. Who has the time? who has the funding? Really, would it be worth it?

In fact, would it be worth either the time or the funding? On the one hand, the prior probability of cheating has a high prior probability. This prior probability is made higher by the fact that cheating has regularly been uncovered when the efforts at independent verification have been made.

There is another reason to suspect that cheating or neglected faults in the set-up can account for results that are improbable upon the hypothesis of chance. This is the fact that it has usually been impossible for others to replicate the results of the experiments run by the parapsychologists. It turns out that high results are generated only when the experiments are done by those committed to the belief that ESP does really occur — people like Dr. J. B. Rhine.

We have seen how he dealt with the case of Lady Wonder, the clever horse. He tested the horse, with lax controls, and concluded that the horse had telepathic powers. When he returned to give further, and more stringent tests, the horse did not perform nearly as well. Instead of suspecting that his first set of data was problematic, he proposed that the horse had lost its telepathic talents. Data that give negative results, results that call into question ESP or telepathy or whatever psychic power is at issue, are not put together with positive results, to give a total picture. Rather, *they are explained away*, argued to be irrelevant. When another experimenter tries to replicate the experiments of the parapsychologists, and the results are negative, a reason is found for not taking those results seriously.

It is clear that *under those conditions in which negative evidence is always dismissed as irrelevant it is not difficult to accumulate evidence that is overwhelmingly positive.*

We have already suggested how this might go. Recall the parapsychologist named Danube. He runs an experiment on Jones and finds that Jones can in a long run of draws call the cards correctly at a rate greater than that which would be expected on the hypothesis of chance. A sceptical psychologist Dr. Cam now comes along and runs a similar set of tests on Jones, only with more rigid controls. Jones' success rate is now about what would be expected on the hypothesis of chance, on the hypothesis that he was guessing. Dr. Danube and his colleagues explain the failure by citing the fact that when Jones gets bored he does not perform well, and the rigid controls, alas, bore him. Or maybe Jones gets distracted by the visitors. Or maybe he is ill. Or maybe the experimenter, Dr. Cam, is in a negative state of mind,

and Jones, via his telepathic capacities is aware of that state, e.g., of scepticism, and is disturbed by it. If the scores are high these factors are, of course, irrelevant. But if the scores start to drop, the search is for the factor that is the cause of the drop.

Previously what was being tested was a statistical hypothesis about a mass event. The mass event consists in Jones calling repeatedly the sort of card that has been drawn. This mass event has a certain property, namely, that the run of successes is more than would be probable under the assumption of chance. The hypothesis being tested, in other words, is this:

> Whenever Jones is subjected to a long sequence of tests, then his success rate is greater than that expected by chance.

Let us abbreviate this by

> Whenever Jones is F then Jones is G

But it turns out to be false, or, at least, when Dr. Cam comes along, he obtains a result that, within the limits of the statistical test, leads us to reject this hypothesis. He finds that he cannot accept that the mass event has the property G. Dr. Danube retorts, however, that we have not taken into account all the relevant variables. We also have to include the variable of, say, boredom, call it "B":

(%)   Whenever Jones is F then Jones is G if and only if Jones is not B

And so, if Jones is F and it turns out that he is also not G — doesn't get enough right — we can explain that negative result *ex post facto* by appeal to B. However, *is there any evidence that this new hypothesis is true?* After all, *we can rely upon this to provide* ex post facto *explanations that are acceptable only if it itself has a reasonable level of probability.*

A related technique used by parapsychologists to obtain positive data is this. Suppose Jones has a long run of positive results. He then has a string of negative results. The former are saved as positive, indicating the presence of ESP (or whatever form of psychic powers one is testing for). The negative results that came later, however, are dismissed as the result of Jones becoming bored, or perhaps tired, or something like that. The gambler would have expected a long run of good luck to be followed by a run of bad luck; he or she would not be surprised at that in any case. The gambler knows that in the long run the good and the bad balance out. That is the nature of objective chance. But this technique of the parapsychologists fails to allow this balancing to occur. Unless all scores are taken into account, it is not hard to recognize that one will soon obtain a good set of non-chance results.

The parapsychologist can make his or her case only if he or she can provide independent evidence for the presence of the special factors which are introduced to explain away negative results, factors such as that of the boredom that was used to explain away Dr. Cam's results. Such evidence need not, of course, be direct. It could be evidence in the background that has provided confirmation for a regularity like (%) to which appeal can be made in order to introduce *ex post facto* the factor of boredom or whatever other factor to which appeal is made in order to explain away the negative results.

There are parallels to this in the arguments of the critics of parapsychology.

Critics argue that there is considerable cheating or at least unintentional fraud when the success rates are high. In fact, they can quite correctly point out that *there is considerable evidence* to support this view. For, after all, much cheating and unintentional fraud have been discovered in the past. This gives a strong prior probability to the hypothesis that

(%%) Whenever Jones is F then Jones is G if and only if there is something fraudulent in the experimental set-up Jones is in

Since there is independent evidence that (%%) is true, since it does have a quite high prior probability, then it is possible to explain the positive result *ex post facto* by citing fraud. To be sure, we may not have detected exactly what the fraud is in the present case; we only have reason, on the basis of the observed effect, to think it very likely that *there is* fraud, without being able to say precisely what that fraud is. Failure to find it does not falsify our claim — note the existential quantifier! — ; it only means that we have not looked hard enough.

The point is, that such fraud or at least careless experimental set-ups have been discovered often enough in the past to make it reasonable to suppose that there is such fraud in this case too, even though we have not yet discovered precisely what it is.

Parapsychologists often suggest that critics dismiss all positive results as being consequences of cheating or fraud. They make the charge that *under those conditions any positive evidence is dismissed as irrelevant and it is not difficult to accumulate evidence that is overwhelmingly negative.* Of course, there is no good data according to our critics, the parapsychologists charge; our critics merely dismiss all positive evidence as irrelevant. This, of course, is parallel to the charge of the critics that the parapsychologists set up the game in such a way that they can always dismiss apparently negative results as irrelevant, due perhaps to boredom, to continue with our example. The parapsychologists present this *tu quoque* not only as an attempt to refute their critics but also to embarrass them, depicting them as dogmatic. The parallel charge by the critics of parapsychologists is that the latter are gullible. But in a context in which were are talking about science, we have to talk about being open-minded. After all, did not Judge Overton emphasize what many philosophers of science have said, that judgements in science are always *tentative*, and subject to revision in the light of further evidence? Parapsychologists may be gullible, but at least that represents a sort of open-mindedness. It is the parapsychologists that were exhibiting the cognitive virtues of science, not their critics. That at least is the argument.

We should not too quickly accept that line, however. After all, the critics do have a good deal of evidence to support (%%) and therefore to support their claim that positive results should be rejected on grounds that they are likely due to fraud.

The parapsychologists have an effective *tu quoque* only if they provide an similar sort of strong support for their premise (%).

The important here is that the critics do have solid evidence that positive results are the consequence of fraud or otherwise faulty experimental set-ups. This is the evidence that supports (%%). The issue is, does (%) have any similar previously supporting data and would give it some reasonable prior probability? If there is no independent evidence for (%), for the role of the factor B, no background theory with some prior probability, then we are *not* entitled to introduce this hypothesis as a way to explain away negative results.

What is the criterion for boredom B? If it is merely Jones' say-so, then the possibility of cheating is clear. Especially if Jones has some interest in achieving a better than chance result. All that he has to do, when faced with a result no better than that expected by chance, in order to preserve his reputation as someone with psychic powers, is to claim that he was bored. A criterion of boredom independent of Jones' say-so is needed if we are to eliminate the hypothesis of cheating.

Even worse are attempts to explain away negative results with hypotheses such as

> Whenever Jones is F then Jones is G if and only if Jones is not rigidly tested

or

> Whenever Jones is F then Jones is G if and only if Jones is not subject to the bad vibes of a sceptical tester

The problem is that neither of these hypotheses remains scientific: they are, as we have noted previously, in fact both tautological. Without any criterion of a rigid test other than its being one that prevents the result C, then of course the first hypothesis is true. Except that it is no longer a scientific hypothesis: every test that is sufficiently unrigid is going to confirm the hypothesis of necessity. Similarly, the second hypothesis is tautological, since every negative result, every result that purports to falsify the hypothesis, will be one where bad vibes are present. Again, every test will confirm the hypothesis: negative results will simply not apply. The hypothesis has become sufficiently elastic as to be tautological. But we should not be surprised that a tautological hypothesis is confirmed by the data and that it is not falsifiable. At the same time, neither does such an hypothesis explain anything, nor, especially, yield any astounding predictions or any ascriptions of astounding special capacities.

In principle, if the notions of "rigid test" or of "bad sceptical vibes" were independently defined then even these hypotheses could be subjected to a variety of tests.

And these tests would enable one in principle to begin to put the real question to the test, of whether Jones has powers that, even in special circumstances — "no bad vibes" — , would permit him to call the cards at a rate better than that which can be expected if he were only guessing.

Such independent definitions of these notions are never given by the parapsychologists. Nor, really, should that surprise us. It is clear that they are introduced more to explain away adverse criticism than to try to advance scientific research. It is evident that parapsychologists are moved more by the desire that their hypotheses regarding psychic powers are true than they are moved by the desire to subject them to the tests of the inductive method.

There is another, related technique that is often used to generate apparent support for strong non-chance statistical outcomes. The subject Jones does not always guess the card that is turned over, called the "target." Rather, he sometimes calls the card that is turned up next after the target. Parapsychologists refer to this as "forward displacement." Jones might also call the card that was turned up immediately previous to the target. This is "backward displacement." If these displacement effects are allowed to go back or forward even a few cards, then it is clear, on statistical grounds alone, that the probability of getting a high score is increased. Or conversely, if a chance score is obtained, it can be explained away by appeal to the phenomenon of displacement.

But again, we need independent tests for whether displacement is or is not occurring. Without that, the appeal to the statistics alone is worthless.

Parapsychologists also sometimes argue that when a miss has occurred, the subject fixed upon another target that is nearest "in some respect" to the target intended by the experimenter. This is a generalization of the notion of displacement. But, of course, we have to know antecedently what the limits are on "some respect". We need some generic description at least, so that we can begin to sort out which respects are relevant and which are not, and to control for those which are. If we do not have that information, the test is worthless. One can always find respects in which things are similar, and so one could always find after the fact some way in which the actual target and the intended target were "near" to each other. Anyone who has played the lottery or gambled with dice will have noticed that there are a good many results that other than the one desired but frustratingly "near" it. Again, it becomes tautological that Jones will get a high score by hitting the intended target or some other target that is "nearest" to it "in some respect."

A group of British physicists sympathetic to claims that psychic ("psi") phenomena exist observed, and were impressed by, the psychokinetic powers apparently exhibited by Uri Geller, who had regularly exhibited his ability to make spoons bend apparently without touching them. There has been considerable scepticism about whether Geller was a systematic cheat. Certainly, those who claimed to observe him exercising his psychic powers generally used very imperfect controls. Even if cheating and fraud did not occur, the controls did not exclude them, and could have been evaded by any competent magician (and Geller was in fact a trained magician!). The tests were therefore tainted, just as a dirty test tube can taint a chemical experiment.[215] But the British physicists were optimistic: "It should be possible," they stated,

> to design experimental arrangements which are beyond any reasonable possibility of trickery, and which magicians will generally acknowledge to be so.

Unfortunately,

> In the first stages of our work we did in fact present Mr. Geller with several such arrangements, but these proved aesthetically unappealing to him.[216]

The physicists suggest that obtaining success in tests for psychic powers depends upon the relations among the participants, both the subject and the experimenters. Everyone must be in a relaxed state, and all must sincerely want the subject to succeed in displaying his or her psychic powers. Above all, "the experimental arrangement [should be] aesthetically or imaginatively appealing to the person with apparent psychokinetic powers."[217] While it is no doubt reasonable to try to make the alleged psychic feel relaxed, this last condition puts that person in control of the experiment. If such a person were

determined to cheat, he or she would be in a position to veto any arrangement in which he or she was unable to cheat. The result is that any experiment conducted under these conditions is suspect. It suffers from the dirty test tube fault.

The physicists add further conditions that make things even worse. The alleged psychic should be treated as a co-worker in the experiment, to create a feeling of trust which is supposed to make the appearance of the psychic powers more likely. All involved must believe that the paranormal or psychic powers are there, even if not always manifest. Any hint of suspicion or scepticism could stifle the exercise of the powers. No particular outcome should be anticipated and looked for, since that would interfere with the relaxed state of mind necessary for the exercise of the psychic powers. Even concentrated attention on aspects of the experiment might interfere and so should be avoided; participants, both the subject and the experimenters, should talk and think about matters totally irrelevant to the experiment.

But it is clear that these conditions simply play further into the hands of the alleged psychic, if he or she is determined to cheat and fool the experimenters. In order to make the alleged psychic feel comfortable, conditions are established that make scientific observation impossible: if the observers are not to attend to what is going on in the experiment, then the possibility of deception is large indeed. Again, the alleged psychic is in control.

Moreover, if the possibility of deception is not controlled for, neither is the possibility of self-deception. If the participants are to be believers, and no hint of scepticism is to be allowed, then the likelihood that the experimenters will find what they are searching for is undoubtedly high. They are, moreover, encouraged to be emotionally involved with the rest of the team in this voyage of discovery aimed at finding that in which they all devotedly believe. In other words, a high emotional involvement is encouraged rather than a disinterested search for truth. Finally, since no one is to attend to anything in particular, and to talk and think about things other than the experiment, there will nothing particularly controlled about the experiment; to the contrary, there will be a sort of general chaos surrounding the test.

It is in this context that there arises, for both the alleged psychic and the observers, the possibility of self-deception. When observations emerge from a context where emotions are at once high and very personal, involving deep commitments, they are more than likely to be the result of bias and pre-established beliefs. One could not find a set-up more faulty than that of the sort described by these physicists.

Under these conditions, one would have every reason to suppose that a regularity such as (%%) would hold of the alleged psychic. Certainly, under those conditions one could never safely conclude that positive results in favour of extrasensory perceptual of psychokinetic powers had been obtained.

However, let us assume for the sake of argument that Dr. Danube has established an experimental set-up which does control for these very many factors, and that it has provided strong statistical evidence that his subject, Jones, can identify the cards at a rate that is greater than chance. What follows?

First, it will be necessary to conclude that Jones, at least to a certain extent, has a power to identify cards. In this sense, he will have a "psychic power." Note, however, that to say that is not to *explain* why Jones can do what he does: it is merely to redescribe the situation in different terms. One cannot explain why opium puts us to sleep by citing its "dormitive powers," and one cannot explain why the unmarried man is unmarried by citing the fact that he is a "bachelor": similarly, one cannot explain why a person can identify cards without sensory input by claiming that the person has "psychic powers."

A scientific or causal explanation of the facts would require some scientific theory that would provide a connection between the drawing of the card and its identification.

But second, the mere fact that there is statistical relationship does not imply that there is any causal relationship. We have seen this already in our discussion of Bode's "law" and of Wegener's hypothesis of continental drift.

In the case of parapsychology, the statistical result is that the hypothesis that the events occur by chance is false. That is to say, for example, if the subject is asked to guess Heads or Tails on the toss of a coin, then it is false that the percentage of correct calls is $\frac{1}{2}$. They occur to the contrary with some other probability, say p^*. If this is greater than $\frac{1}{2}$ the conclusion is that there is positive ESP; if it is less than $\frac{1}{2}$ then the conclusion is that there is negative ESP. But p^* is still a probability, and to ascribe this probability to an event in a mass event is a matter of objective chance, that is, objective *chance*. To say that the result is due to chance just in case that the probability is $\frac{1}{2}$ is misleading in the extreme. For the result that it occurs with some other probability p^* is still a statement of probability, and without any background theory to say otherwise, it is still a "matter of chance," a mere correlation. A probability other than $\frac{1}{2}$ is still a probability.

A non-chance result, that is, one other than $\frac{1}{2}$, does not imply that there must be a causal explanation that renders the correct result non-chance in the sense of causal. There is an ambiguity here: on the one hand, 'non-chance' means a result other than $\frac{1}{2}$; on the other hand, 'non-chance' means causal. Non-chance in the first sense does not imply non-chance in the second sense. Hence, obtaining a non-chance statistical result does not by itself imply that there is any sort of causal connection, whether ordinary or extra-ordinary. In this sense, it may simply be chance that we obtain a non-chance result. That the subject has powers of ESP in the absence of a theory means nothing more than the fact that there are non-chance results. There is nothing to establish that there is some process going on that is parallel to ordinary processes of perceiving, save that it proceeds by non-sensory means, spiritually perhaps. The subject may be getting good results, but all that follows is that he or she is getting good results. Nothing about this by itself implies that these results are a matter of anything other than events which are the result of the accidents of chance.

Endnotes

1   Cf. C. S. Peirce, "The Fixation of Belief," in Charles S. Peirce, *Collected Papers*, 6 vols. (Cambridge, Mass.: Harvard University Press, 1931-35), vol. IV, pp. 223-47.

2   Cited in W. J. Overton, "The Decision of the Court," in A. Montagu, ed., *Science and Creationism*, p. 393n7.

3   Henry W. Morris, *Studies in the Bible and Science* (Grand Rapids: Baker Book House, 1966), p. 114.

4   H. W. Clark, *Fossils, Flood and Fire* (Escondido, Calif.: Outdoor Picture, 1968), pp. 18-19.

5   Quoted in Theodosius Dobzhansky, "Nothing in Biology Makes Sense Except in the Light of Evolution.," in J. P. Zetterberg, ed., *Evolution vs. Creationism* (Phoenix, AZ: Oryx Press, 1983), p. 18.

6   Erich von Däniken, *Chariots of the Gods?*, p. 31.

7   *Ibid.*, p. 25. Italics added.

8   *Ibid.*, p. 52.

9   On the relation of this method for belief formation and pseudoscience, see Paul Thagard, "Resemblance, Correlation, and Pseudoscience," in Marsha Hanen, Margaret Osler, and Robert Weyant, eds., *Science, Pseudoscience and Society* (Waterloo, ON.: Wilfrid Laurier University Press, 1980), pp. 17-27.

10 John Stuart Mill, *System of Logic*, Eighth Edition (London: Longmans, 1970), p. 501.

11 See Lawrence Jerome, *Astrology Disproved* (Buffalo: Prometheus Books, 1977), p. 70.

12 See R. Rosenthal, "On the Social Psychology of Psychological Experiment: The Experimenter's Hypothesis as Unintended Determinant of Experimental Results," *American Scientist*, 51 (1963), pp. 268-83.

13 "Les Gourous dans l'Ombre du Pouvoir," *L'Actualité*, Février, 1997, p. 14.

    The article goes on to mention the Astrologers Fund investment firm, which is apparently quite successful. But is the success due to astrology?

14 John Paulos, "Coincidences," in his *Beyond Numeracy* (New York: Knopf, 1991), p. 38.

15 *Note: We shall have more to say about the method of science below, when we come to discuss what Kuhn has called "revolutionary science."*

16 Judge William J. Overton, "The Decision of the Court," in A. Montagu, ed., *Science and Creationism*, p. 380.

17 William Benjamin Carpenter, *Principles of Mental Physiology* (New York: D. Appleton, 1874), pp. 292-93.

    See also his *Mesmerism, Spiritualism, &c., Historically and Scientifically Considered* (New York: D. Appleton, 1877).

18 Michael Faraday, "Professor Faraday on Table Moving," *The Athenæum*, no. 1340, July 3, 1853, pp. 801-803.

19 Cf. I Hacking, "Some Reasons for Not Taking Parapsychology Very Seriously," *Dialogue*, 32 (1993), pp. 587-94.

20 Oskar Pfungst, *Clever Hans*, trans. C. L. Rahn, with intro. by R. Rosenthal (New York: Rinehart & Winston, 1965), p. 254.

21 Judge William J. Overton, "The Decision of the Court," p. 380.

22 "Lady, A Clairvoyant Mare, Solves Problems, Gives Advice and Writes a Headline for 'Life'," *Life*, 33 (1952), Dec. 22, pp. 20-21.

23 J. B. Rhine and Louisa E. Rhine, "An Investigation of a 'Mind-Reading' Horse," *Journal of Abnormal and Social Psychology*, 23 (1928-29), pp. 449-466.

24 J. B. Rhine and L. E. Rhine, "Second Report on Lady, the 'Mind-Reading' Horse," *Journal of Abnormal and Social Psychology*, 24 (1929-30), pp. 287-292. The quoted statement is on p. 289.

25 See Martin Gardner, *Fads and Fallacies in the Name of Science* (Dover: New York, 1957), p. 352.

26 *Ibid.*, p. 292.

27 Judge William J. Overton, "The Decision of the Court," p. 380.

28 Pat Moffat, "A Lesson in Eastern Medicine," *The Globe and Mail*, Tuesday, Sept. 22, 1998.

29 See, for example, Shen Ziyin and Chen Zelin, *The Basis of Traditional Chinese Medicine*, (Boston: Shambhala, 1996), p. 244.

30 Augustus De Morgan, *A Budget of Paradoxes*, 2 vols. (London: Longmans, Green & Co., 1872), vol. I, p. 87.

31 Augustus De Morgan, *A Budget of Paradoxes*, vol. I, p. 86.

32 Morris Cohen and Ernest Nagel, *An Introduction to Logic and Scientific Method* (New York: Harcourt, Brace and World, 1934), p. 221f.

33 N. R. Hanson, "Is There a Logic of Scientific Discovery?", in H. Feigl and G. Maxwell, eds., *Current Issues in the Philosophy of Science* (New York: Holt, Rinehart and Winston, 1961).

34 See T. Kuhn, *The Structure of Scientific Revolutions*, second edition.

35 John Stuart Mill, *System of Logic*, eighth edition, Bk. III, ch. v, sec. 1.

36 What (j) asserts is that the **G**-ish cause of D_4 is the cause of D_4. This statement is not tautological contrary

to appearances, since it implies that there is a **G** which is the cause of D_4, and that statement, i.e., (d), is an empirical matter of fact generalization.

This is parallel to the point that the statement that

(*)   the F is F

is not tautological, but rather matter of fact since it asserts that there is exactly one F, a clearly empirical statement. If we put (*) into Russell's notation in the symbols of formal logic

(**) F(ιx)(Fx)

and expand this according to the Russellian rule that

$$F(\iota x)(Fx) =_{Df} (\exists x)[Fx \,\&\, (y)(Fy \to x = y) \,\&\, Fx]$$

then we obtain

$$(\exists x)[Fx \,\&\, (y)(Fy \to x = y) \,\&\, Fx]$$

which is clearly logically equivalent to

(***) $(\exists x)[Fx \,\&\, (y)(Fy \to x = y)]$

Thus, (*) = (**) is logically equivalent to (***). But (***), which asserts that there is exactly one F, is clearly an empirical truth. Therefore so is (*) which is logically equivalent to it.

37 For those who like to see these things done, here is a formalization of the proof.

(f) [**D**$f \to$ (E!g) [**G**g & (x) ($gx \leftrightarrow fx$)]]	
($\exists f$)(**D**f & fx)	/ $\therefore$ ($\exists f$)(**G**f & fx)
**Df & fa	EI Ass
Df	Simp
D$f \to$ (E!g) [**G**g & (x) ($gx \leftrightarrow fx$)]	UG
(E!g) [**G**g & (x) ($gx \leftrightarrow fx$)]	MP
* Gg & (x) ($gx \leftrightarrow fx$)	EI Ass
(x) ($gx \leftrightarrow fx$)	Com, Simp
$ga \leftrightarrow fa$	UI
fa	Com, Simp
ga	Equiv, MP
Gg	Simp
Gg & ga	Conj
($\exists f$)(**G**f & fx)	EG
* ($\exists f$)(**G**f & fx)	EI
** ($\exists f$)(**G**f & fx)	EI
QED	

38 K. Popper, "Autobiography: ' sec. 37 Darwinism as Metaphysics," in P. A. Schilpp, ed., *The Philosophy of Karl Popper*, 2 vols. (La Salle, Ill.: Open Court, 1974), vol. 1, p. 137.

39 *Ibid.*, p. 138.

40 *Ibid.*, p. 137.

41 *Ibid.*, pp. 135-6.

42 *Ibid.*, p. 136.

43 M. Ruse, "Karl Popper's Philosophy of Biology," *Philosophy of Science,* 44 (1977), pp. 638-61, at 645.

44 T. Kuhn, *The Structure of Scientific Revolutions*, secs. VI, VII.

45 *System of Logic*, p. 574.

46 *The Structure of Scientific Revolutions*, Secs. VIII and IX.

47 *Ibid.*, p. 156.

48 *An Introduction to Logic and Scientific Method*, p. 222.

49 William Harvey, *The Circulation of the Blood and Other Writings*, trans K. Franklin (London: J. Dent, Everyman's Library, 1963), p. 176.

50 K. Popper, *The Logic of Scientific Discovery* (New York: Basic Books, 1959), p. 317.

51 "$=_{Df}$" means "is defined as short for". Because the left hand side is short for the right hand side, the one can replace the other in all ordinary contexts. (The exception is contexts which report beliefs. These exceptions we need not here explore.)

52 N. M. de S. Cameron, *Evolution and the Authority of the Bible* (Exeter: Paternoster Press, 1983), p. 16.

53 K. Popper, "Autobiography: ' sec. 37 Darwinism as Metaphysics," in P. A. Schilpp, ed., *The Philosophy of Karl Popper*, vol. 1, p. 137.

54 See Joyce Arthur, "Creationism: Bad Science or Immoral Pseudoscience?" *Skeptic*, 4 (1996), pp. 88-93.

55 *Ibid.*, p. 89.

56 Duane Gish, *The Amazing Story of Creation from Science and the Bible* (El Cajon, Calif.: Institute for Creation Research, 1990), p. 43.

57 Karl Popper, "Replies to My Critics," in P. A. Schilpp, ed., *The Philosophy of Karl Popper*, vol. 2, p. 985.

58 K. Popper, *The Poverty of Historicism*, Second Edition (London: Routledge and Kegan Paul, 1960).

For some detailed criticisms of the characteristic Popperian theses of *The Poverty of Historicism*, see L. Addis, *The Logic of Society* (Minneapolis, Minn.: University of Minnesota Press, 1975).

59 Cf. J. W. N. Watkins, "Between Analytic and Empirical," *Philosophy*, 32 (1957), pp. 112-131; and "Influential and Confirmable Metaphysics," *Mind*, n.s. 67 (1958), pp. 344-65.

60 K. Popper, "Autobiography: ' sec. 37 Darwinism as Metaphysics," in P. A. Schilpp, ed., *The Philosophy of Karl Popper*, vol. 1, p. 137.

61 But not by all positivists. Neurath never took up those issues. Neither did some younger members of the Vienna Circle such as Gustav Bergmann.

62 I. Lakatos, "Falsification and the Methodology of Scientific Research Programmes," in *Criticism and the*

Growth of Knowledge, ed. I. Lakatos and A. Musgrave, pp. 100-1.

63 This is the de Morgan rule that "p & q" is logically equivalent to "it is not the case that either not p or not q".

64 The term is Lakatos'.

65 This terminology is also from Lakatos.

66 J. Nicod, *Foundations of Geometry and Induction* (London: Kegan Paul, Trench, Trubner, 1930), pp. 219-220.

67 C. G. Hempel, "Studies in the Logic of Confirmation, I and II," *Mind*, n.s. 45 (1954), pp. 1-26, pp. 97-121.

68 K. Popper, "Science: Conjectures and Refutations," in his *Conjectures and Refutations* (New York: Harper and Row, 1963).

For some other aspects of Popper's views, see D. Rothbart, "Demarcating Genuine Science from Pseudoscience," in P. Grim, ed., *Philosophy of Science and the Occult* (Albany, N. Y.: State University of New York Press, 1983).

69 Cf. the discussion in J. Agassi, "Corroboration versus Induction," *British Journal for the Philosophy of Science*, 9 (1958-59), pp. 311-17.

70 Cf. Hempel, "Studies in the Logic of Confirmation."

71 We will let the symbol ' ==> ' represent the logical relation of entailment.

72 For an extended discussion of this topics of this subsection, see G. H. von Wright, *A Treatise on Induction and Probability* (London: Routledge and Kegan Paul, 1951).

See also the important discussion in J. L. Mackie, "Mill's Methods of Induction," in P. Edwards, ed., *The Encyclopedia of Philosophy*, 8 vols. (New York: Macmillan, 1967), vol. 5, pp. 324-332.

73 John Stuart Mill, *System of Logic*, Book III, Ch. viii, sec. 2.

74 And British sailors, and latterly all Englishmen, came to be referred to as "limeys".

75 Richard Solomon, "An Extension of Control Group Design," *Psychological Bulletin*, 46 (1949), pp. 137-150; the "Brief History" is on pp. 137-40.

See also E. G. Boring, "The Nature and History of Experimental Control," *American Journal of Psychology*, 67 (1954), p. 573-589.

76 E. L. Thorndyke and R. S. Woodworth, "The Influence of Improvement in One Mental Function upon the Efficiency of Other Functions," *Psychological Review*, 8 (1903), pp. 247-261, pp. 384-395, pp. 553-564.

77 Solomon, pp. 137-8.

78 John Stuart Mill, *System of Logic*, Book. III, Ch. viii, sec. 1.

79 Cohen and Nagel, *An Introduction to Logic and Scientific Method*, p. 257.

80 Francis Bacon, *Novum Organon*, ed. with an intro. by F. H. Anderson (New York: Liberal Arts Press, 1960), Bk. II, aphorism xv. Von Wright, *The Logical Problem of Induction*, second edition (Oxford: Blackwell, 1965), p. 206, justly refers to the usual Popperian claims as "exaggerations."

81 David Hume, *Treatise of Human Nature* ed. L. A. Selby-Bigge (Oxford: Oxford University Press, 1888), p. 153.

82 Pat Moffat, "A Lesson in Eastern Medicine," *The Globe and Mail*, Tuesday, September 22, 1998.

83 John Stuart Mill, *System of Logic*, Book III, Ch. viii, sec. 4.

84 Cf. I. Copi, *Introduction to Logic*, Seventh Edition (New York: Macmillan, 1986), p. 442.

85 Semmelweis met with very strong opposition from the other physicians in his hospital and he was forced to leave Vienna. He returned to Hungary and became professor of obstetrics at the University of Pest. The bitter and often violent controversy over his theories continued, however, and he eventually fell into insanity. He died in an asylum in 1865, as a result of the infection of a dissection wound that had occurred prior to his confinement.

The Hungarians erected a monument to Semmelweis in Budapest in 1906, honouring him as a pioneer in surgical antisepsis.

86 G. Cuvier, "Letter to J. C. Martrud," in his *Lectures on Comparative Anatomy*, trans. W. Ross (London: T. N. Longman and O. Rees, 1802).

87 John Stuart Mill, *System of Logic*, Bk. III, ch. vii.

88 For the material developed in this part, see J. L. Mackie, "Mill's Methods of Induction," in P. Edwards, ed. *Encyclopedia of Philosophy*, vol. 5, p. 328.

89 Material in this section is adapted from F. Wilson, *Hume's Defence of Causal Inference*, (Toronto: University of Toronto Press, 1997), Ch. 3, sec. 9.

90 Lakatos, in his "Methodology of Scientific Research Programmes," insists that non-falsifiability is due to the alleged fact that the observer can always, by fiat, reject an observation as non-veridical (p. 107f), while insisting that all theories are of the naive falsificationist *logical form* "All A are B" (p. 110). Kuhn rejects this account of the logical form of theories, insisting that greater logical complexity is what prevents falsification. See *The Structure of Scientific Revolutions*, p. 147, and "Logic of Discovery or Psychology of Research," pp. 14-15:

91 It is only slightly more complex if we do not assume that hypotheses are contrary.

92 Lakatos, "The Methodology of Scientific Research Programmes," p. 118; *The Structure of Scientific Revolutions*, p. 154.

93 See T. Kuhn, "The Function of Dogma in Scientific Research," in A. C. Crombie, ed., *Scientific Change* (London: Heinemann, 1963), pp. 347-69.

94 Cf. L. Laudan, "A Problem-Solving Approach to Scientific Progress," in his *Beyond Positivism and Relativism: Theory, Method and Evidence* (Boulder, CO.: Westview Press, 1996), pp. 77-87.

95 Laudan, *loc. cit.*, fails to bring out clearly the cognitive interest that moves scientists in normal research, that is, the explanatory ideal of process knowledge. Nor does he bring our in any clear fashion, as Kuhn in contrast does, the way in which an abstract generic theory or paradigm guides research in normal practice.

96 See A. Hallam, "Alfred Wegener and the Hypothesis of Continental Drift," in *Continents Adrift and Continents Aground*, Readings from *Scientific American*, introduction by J. Tuzo Wilson (W. H. Freeman: San Francisco, 1976).

97 Don and Maureen Tarling, *Continental Drift*, Revised Edition (Anchor/Doubleday: Garden City, New York, 1975), p. 7.

98 See J. Tuzo Wilson, "Continental Drift," in *Continents Adrift and Continents Aground*, Readings from *Scientific American*.

99 G. Dean and A. Mather, *Recent Advances in Natal Astrology: A Critical Review 1900-1976* (Para Research: Rockport, Mass., 1977), p. 1.

100 Linda Goodman, *Linda Goodman's Sun Signs* (Bantam Books: New York, 1971), p. 475.

101 Geoffrey Dean and Arthur Mather, *Recent Advances in Natal Astrology: A Critical Review 1890-1976*, p. 1.

102 Dean and Mather, p. 2.

103 As we noted, this is made clear in Paul Thagard, "Resemblance, Correlation, and Pseudoscience," and "Why Astrology is a Pseudoscience."

104 P. Roberts, *The Message of Astrology*, (Wellingborough: The Aquarian Press, 1990).

105 T. S. Kuhn, "Logic of Discovery or Psychology of Research?" in I. Lakatos and A. Musgrave, eds., *Criticism and the Growth of Knowledge*, p. 9.

106 Nelson Goodman, *Fact, Fiction, and Forecast*, Third Edition (Indianapolis, Indiana: Bobbs-Merrill, 1973).

107 See S. Barker and P. Achinstein, "On the New Riddle of Induction," in B. Brody, ed., *Readings in the Philosophy of Science* (Englewood Cliffs, N.J.: Prentice-Hall, 1970).

108 Cf. G. Bergmann, *Philosophy of Science* (Madison, Wisc.: University of Wisconsin Press, 1956), Ch. 1.

109 Cf. Bernard Grunstra, "The Plausibility of the Entrenchment Concept," in S. Luckenbach, ed., *Probabilities, Problems, and Paradoxes*, (Encino, Calif.: Dickenson, 1972), p. 285.

110 Cf. J. O. Nelson, "Induction: A Non-Sceptical Humean Solution," *Philosophy*, 67 (1992), pp. 307-327. For discussion, see W. Waxman, *Hume's Theory of Consciousness* (Cambridge: Cambridge University Press, 1994), p. 311n1.

111 N. R. Hanson, "Is There a Logic of Scientific Discovery?" in H. Feigl and G. Maxwell, eds., *Current Issues in the Philosophy of Science* (New York: Holt, Rinehart and Winston, 1961), pp. 620-625.

112 *Ibid.*, p. 624.

113 See G. Bergmann, *Philosophy of Science*, Ch. I.

114 Cf. E. Nagel, "Carnap's Theory of Induction," in P. A. Schilpp, ed., *The Philosophy of Rudolf Carnap* (LaSalle, Ill.: Open Court, 1968); and F. Wilson, "Mill on the Operation of Discovering and Proving General Propositions," *Mill Newsletter*, 17 (1982), pp. 1-14.

115 See Hugh S. Rice, Article "Bode's Law," *Collier's Encyclopedia* (New York: Collier Books, 1950), vol. III.

116 Hugh S. Rice, *ibid.*

117 Kuhn, *The Structure of Scientific Revolutions*, p. 117.

118 Lakatos, "The Methodology of Scientific Research Programmes," pp. 177-80.

119 It is important, as we shall see, that it *not* be well defined.

120 Cf. Lakatos, "The Methodology of Scientific Research Programmes," pp. 177-80.

121 T. Kuhn, "Comment on the Relations of Science and Art," in his *The Essential Tension* (Chicago: University of Chicago Press, 1977), pp. 340-52, at p. 342.

122 L. Laudan, *Progress and Its Problems* (Berkeley: University of California Press, 1977), charges that Kuhn's position is inadequate because he does not give, once for all, a definite degree of inefficiency at which one and all should start searching for a new paradigm. We see that such a demand is unreasonable. Laudan's criticism is plain mistaken.

123 *The Structure of Scientific Revolutions*, p. 83.

124 *Ibid.*, pp. 155-6.

125 *Ibid.*, p. 156, p. 158.

126 Neither Lakatos, "The Methodology of Scientific Research Programmes," nor Laudan, *Progress and Its Problems*, ever suggest criteria for when it would be rational to take up and use, put to the test, a theory

never hitherto worked on. This is a serious deficiency in their positions. Kuhn in contrast, we see, has no such gap in his account of scientific practice.

127 For some of this history briefly stated, see P. C. Mahalanobis, "Historical Introduction," in A. Einstein, *The Principle of Relativity*, trans. into English by M. N. Saha and S. N. Bose, with an historical introduction by P. C. Mahalanobis (Calcutta: University of Calcutta Press, 1920).

128 See *ibid.*, p. ix-xi.

129 See *ibid.*, pp. iv-vi.

130 Cf. H. A. Lorentz, "Michelson's Interference Experiment," in H. A. Lorentz, A. Einstein, *et al.*, *The Principle of Relativity*, trans. W. Perrett and G. B. Jeffery (London: Methuen, 1923).

131 *Ibid.*, p. 6.

132 See P. C. Mahalanobis, "Historical Introduction," p. xv.

133 *Ibid.*, pp. xv-xvi.

134 Einstein, "On the Electrodynamics of Moving Bodies," in A. Einstein, *The Principle of Relativity*. Also translated in H. A. Lorentz, A. Einstein *et al.*, *The Principle of Relativity*. Page references are to the former.

135 Cf. F. Wilson, "The Rationalist Response to Aristotle in Descartes and Arnauld," in E. Kremer, ed., *The Great Arnauld* (Toronto: University of Toronto Press, 1994).

136 Cf. J. Dorling, "Einstein's Methodology of Discovery was Newtonian Deduction from the Phenomena," in J. Leplin, ed., *The Creation of Ideas in Physics* (Dordrecht, The Netherlands: Kluwer, 1995).

137 E. Zahar, "Why Did Einstein's Programme Supersede Lorentz's? (I) and (II)," *British Journal for the Philosophy of Science*, 24 (1973), pp. 95-123 and pp. 223-262, somehow misses this point.

138 Zahar, *ibid.*, never attempts to answer this question, thus exemplifying a major failure of the Lakatosian account of the "methodology of research programmes" which Zahar takes up. In contrast, by answering this question, Kuhn's methodology exhibits its superiority.

Zahar's suggestion that Kuhn accounts for the triumph of Einstein's theory through the dying out of the protagonists of the old theory (p. 238) is simply a caricature that shows an incapacity to read Kuhn carefully.

139 P. C. Mahalanobis, "Historical Introduction," p. xvi.

140 A. N. Whitehead, *The Principle of Relativity* (Cambridge: Cambridge University Press, 1922), p. 46.

141 L. Laudan, "A Problem-Solving Approach to Scientific Progress," p. 78.

142 *Ibid.*

143 *Ibid.*

144 We have discussed this point right at the beginning, in the Introduction.

145 E. H. Madden, "The Riddle of Induction," *Journal of Philosophy*, 55 (1958), pp. 705-18.

146 T. Kuhn, *The Structure of Scientific Revolutions*, secs. VI, VII.

147 *System of Logic*, BK. III, Ch. xxi, sec. 4.

148 *The Structure of Scientific Revolutions*, secs. VIII, IX.

149 *Ibid.*, p. 156.

150 P. Feyerabend, *Against Method* (London: New Left Books, 1975).

151 Erich von Däniken, *Chariots of the Gods*, p. 30.

152 C. D. B. Bryan, *Close Encounters of the Fourth Kind: Alien Abduction, UFOs, and the Conference at M.I.T.* (New York: Knopf, 1995), p. 422.

153 Saul-Paul Sirag, "The Skeptics," in *Future Science*, ed., J. White and S. Krippner (Garden City, NJ: Doublday, 1977), p. 535.

154 I. Lakatos, "Science and Pseudoscience," in his *The Methodology of Scientific Research Programmes*, ed. J. Worral and G. Currie (Cambridge: Cambridge University Press, 1978), p. 1-7, at p. 4.

155 Whoever made this quip deserves credit for it. Unfortunately, it is a maxim that I seem always simply to have known, and I cannot provide the reference it deserves. My apologies to whomever.

156 C. S. Peirce, "This Fixation of Belief," Charles S. Peirce, *Collected Papers*, 6 vols. (Cambridge, Mass.: Harvard University Press, 1931-35), vol. IV, pp. 223-47, at p. 234.

157 *Ibid.*

158 Material in this section is adapted from F. Wilson, "Empiricism and the Epistemology of the Instrument," *The Monist*, 78 (1995), pp. 207-229.

159 F. Suppe, "The Search for Scientific Understanding of Scientific Theories," in F. Suppe, ed., *The Structure of Scientific Theories*, Second Edition (Urbana, Ill.: University of Illinois Press, 1977), p. 106.

160 John Locke, *Essay concerning Human Understanding*, ed. by P. H. Nidditch (Oxford: Oxford University Press, 1975).

161 George Berkeley, *Principles of Human Knowledge*, in vol. II of *The Works of George Berkeley*, edited by A. A. Luce and T. E. Jessop (Edinburgh: Nelson, 1950ff).

162 David Hume, *Treatise of Human Nature*, ed. L. A. Selby-Bigge (Oxford: Oxford University Press, 1888).

163 Concerning these sorts of inference, see F. Wilson, *Explanation, Causation and Deduction*, sec. 2.5 and *passim*; and *Empiricism and Darwin's Science*, p. 28ff, p. 101ff, pp. 167-72.

164 D. Hume, *Treatise of Human Nature*, Bk. I, Part 3, sec. 15.

165 John Stuart Mill, *System of Logic*, Bk. III, ch. v, sec. 1.

166 We have already noted how this amounts to the thesis defended by T. Kuhn, who also argues that abstract generic theories, or, as he calls them, paradigms function as guides to research. See his *The Structure of Scientific Revolutions*, Section IV, which describes paradigms as providing puzzles to researchers, guaranteeing that *there is* a solution to the problem which it is the task of the researcher to discover.

See also F. Wilson, *Empiricism and Darwin's Science*, Ch. 2.

167 The 6th rule is, as we indicated, the method of difference; the 7th is the method of residues.

168 On this point, see D. Flage, "Locke's Relative Ideas", *Theoria*, 142-59 and "Hume's Relative Ideas", *Hume Studies*, 7 (1981), 55-73.

See also F. Wilson, "Was Hume a Subjectivist?" *Philosophy Research Archives*, 14 (1989), pp. 247-82, where this notion is shown to be important in securing a reading of Hume as a critical realist rather than a mere sceptic.

169 This is the conclusion of Lewis White Beck, "A Prussian Hume and a Scottish Kant," in D. F. Norton *et al.*, eds., *McGill Hume Studies* (San Diego: Austin Hill Press, 1979), pp. 63-78. Or, rather, he concludes that since in fact we do not let the principle be falsified by such examples, therefore we do not affirm the principle on empirical but instead on *a priori* grounds. He concludes that Hume in fact implicitly accepts the Kantian position that the causal principle is *a priori*. For a discussion of Beck's position, see F. Wilson, "Is There a Prussian Hume?" *Hume Studies*, 8 (1982), pp. 1-18

170 I. Lakatos, "Falsification and the Methodology of Scientific Research Programmes."

171 Robert Hooke, *Micrographia* (London: Martyn and Allestry, for the Royal Society, 1665; reprinted with a new Preface by R. T. Gunther, New York: Dover, 1961).

For a discussion of the significance of this work, see A. R. Hall, *Hooke's "Micrographia," 1665-1965* (London: The Athlone Press, 1966).

172 *Micrographia*, (Hooke's) Preface, p. a2; this "Preface" is unpaginated.

173 *Ibid.* p. a1.

174 *Ibid*, p. a3.

175 *Ibid*.

176 *Ibid.*, pp. a3-a4.

177 A good discussion of this theory can be found in E. Mach, *The Principles of Physical Optics*, trans. J. S. Anderson and A. F. A. Young (New York: Dover, 1953). Mach's treatment has the virtue, absent in most recent treatments, of developing the theory independently of the theory of wave optics.

The logical structure of the theory is briefly discussed in F. Wilson, *Explanation, Causation and Deduction*, p. 102ff, p. 275ff, p. 279ff.

178 From Savile Bradbury, *An Introduction to the Optical Microscope*, revised edition (Oxford: Oxford University Press, 1989), p. 6.

179 For a discussion of some of the background, see A. C. Crombie, "The Mechanistic Hypothesis and the Scientific Study of Vision: Some Optical Ideas as a Background to the Invention of the Microscope," in E. S. Bradbury and G. L'E. Turner, eds., *Historical Aspects of Microscopy* (Cambridge: Heffer, 1967), pp. 3-112.

180 Christiaan Huygens, *La Dioptique*, in vol. 13, fasc. 1, of the *Œuvres Complètes de Christiaan Huygens* (La Haye: Martinus Nijhoff, 1916), p. 6.

181 *Ibid*.

182 R. Descartes, *Optics*, in his *Discourse on Method, Optics, Geometry and Meteorology*, trans. P. J. Olscamp (Indianapolis: Bobbs-Merrill, 1965), Second Discourse, p. 75.

183 Huygens, p. 8.

184 E. Halley, "An Instance of the Excellence of Modern Algebra, in the Resolution of the Problem of finding the Foci of Optick Glasses Universally," *Philosophical Transactions of the Royal Society*, Nov. 1693, pp. 960-969.

185 Descartes, *Optics*, p. 115.

186 *Ibid.*, p. 116.

187 *Ibid.*, p. 125.

188 See R. Descartes, *Principles of Philosophy*, trans. V. R. Miller and R. P. Miller (Dordrecht, Holland: D. Reidel, 1983), Part IV, Prop. 84.

189 For the three sorts of matter, see Descartes, *Principles*, Part III, Prop. 52.

190 For the account of fire, see *Principles*, Part IV, Prop. 80-83.

191 For the difference between air and fire, see *Principles*, Part IV, Prop. 80.

192 *Micrographia*, p. 46.

193 *Ibid*.

194 *Micrographia*, p. 46.

195 Applications of this sort of the method of difference are important since they in effect permit causal inferences from a single experiment, as Hume was aware but as

others equally are not aware. See F. Wilson, "Hume and Ducasse on Causal Inferences from a Single Experiment," *Philosophical Studies*, 35 (1979), pp. 305-309; and *Explanation, Causation and Deduction*, pp. 118-126.

196 *Micrographia*, p. 45.

197 *Ibid.*

198 Kuhn has emphasized the important point that theories are often rejected only when an alternative has become confirmed. See Kuhn, *Structure of Scientific Revolutions*, Sec. XII; and also F. Wilson, *Empiricism and Darwin's Science*, p. 52ff.

199 I. Lakatos, "Falsification and the Methodology of Scientific Research Programmes," in *Criticism and the Growth of Knowledge*, ed. I. Lakatos and A. Musgrave, pp. 100-1.

200 I have made the same point in a rather different context in "Is Hume a Sceptic with regard to Reason?" *Philosophy Research Archives*, 10 (1984), pp. 275-320.

201 We will let Popper and his epigones fend for themselves.

202 Cf. K. Lehrer, Review of W. Sellars' *Science, Perception and Reality*, *Journal of Philosophy*, 63 (1966), pp. 266-77.

203 Of course, we often *say* that we "see" these literally invisible entities when we discern them by means of an instrument. But contrary to what some have suggested, no epistemological weight ought to be placed upon this. Cf. F. Wilson, "Discussion of Achinstein's *Concepts of Science*," *Philosophy of Science*, 38 (1971), pp. 442-452.

204 J. B. Rhine, "The Science of Nonphysical Nature," in J. Ludwig, ed., *Philosophy and Parapsychology* (Buffalo: Prometheus Books, 1978), p. 126.

205 C. W. K. Mundle, "Strange Facts in Search of a Theory," in J. M. O. Wheatley and H. L. Edge, eds., *Philosophical Dimensions of Parapsychology* (Springfield, Ill.: Charles C. Thomas, 1976), pp. 76-97.

206 *Enquiry*, July 1949, p. 20.

207 Milton A. Rothman, *A Physicist's Guide to Skepticism* (Buffalo, NY: Prometheus Books, 1988), p. 193.

208 J. B. Rhine, "The Science of Nonphysical Nature," in *Philosophy and Parapsychology*, ed., J. Ludwig (Buffalo, NY: Prometheus Books, 1978), p. 126.

209 Pat Moffat, "A Lesson in Eastern Medicine," *The Globe and Mail*, Tuesday, September 22, 1998.

210 *Baltimore Evening Sun*, July 24, 1974, reported by the paper's science editor, Jon Franklin. For greater detail, see J. B. Rhine, "Second Report on a Case of Experimenter Fraud," *Journal of Parapsychology*, 39 (1975), pp. 306-325.

211 Leonard Zusne and Warren Jones, *Anomalistic Psychology* (Hillsdale, NJ: Lawrence Erlbaum Associates, 1982), p. 374.

212 *Ibid.* p. 375.

213 *Ibid.*

214 D. Druckman and J. A. Stews, eds., *Enhancing Human Performance* (Washington, D. C.: National Academy Press, 1988), p. 207.

215 Cf. James Randi, *The Magic of Uri Geller* (New York: Balantyne Books, 1959).

216 J. Hasted, D. Bohm, E. W. Bastin, and B. O'Regan, "Experiments on Psychokinetic Phenomena," in C. Panati, ed., *The Geller Papers*, (Boston: Houghton Mifflin, 1976), p. 194.

217 *Ibid.*

DATA

The cognitive interest of science is in matter-of-fact regularities, and the method of science, the inductive method, aims to provide, so far as it is attainable, knowledge of such regularities. The method is a set of norms or standards. If we conform our inferences to these norms, then we will be led to conclusions which satisfy our cognitive interest. These inferences have as their starting point statements of individual fact reporting observations. These are the data to which our inductive inferences are regularly adjusted.

Scientists more or less take for granted the observations that are made in the laboratory. There is in fact little need to question these, though they are not as unproblematic as they appear to be when they turn up in papers published in scientific journals. They are not so much given to the scientist as extracted with difficulty. And given this difficulty, they are subject to negotiation within the group of researchers in the laboratory.

Even so, these data are more or less unproblematic compared to the observational claims that are used to support the claims of pseudoscience. Simply consider the observational claims that are used to support the claims that certain individuals have psychic powers or the observations that are used to defend the claim that UFOs are vehicles piloted by presumably extraterrestrial intelligences.

If we are to separate science from pseudoscience, it is necessary to look in greater detail at the nature of good scientific data.

One point can be made emphatically right at the start.

Consider the prediction by the self-proclaimed psychic Tamara Rand of the attempted assassination of President Reagan. This attempt on his life was made on March 30, 1981. Four days later on April 2, it was announced to the world NBC's "Today" show, on ABC's "Good Morning America," and on CNN that the Los Angeles self-proclaimed psychic Tamara Rand *had predicted the attempt three weeks earlier.* The TV shows all aired a videotape which they announced had been made weeks earlier, on January 6. On this tape Rand predicted the assassination in great detail, indicating for example that Reagan would be shot by a sandy haired young man with the initials "J. H.", that the President would be wounded in the chest, that there would be a "hail of bullets," and that the attempt would be made in the last week of March or the first week of April.

Many people were impressed by this report. Many were confirmed in their belief that some people do indeed have psychic powers of precognition. Others, perhaps initially sceptical of such claims, were convinced that their doubts were misplaced, that such psychic powers are indeed real.

However, there is more to the story.

Here is the follow-up as told by Kendrick Frazier and James Randi:

On the morning of April 2, 1981, you will remember — four days after the assassination attempt on President Reagan in Washington — the NBC-TV "Today" show and ABC-TV's "Good Morning America" joined the Cable News Network in broadcasting a tape that was claimed to have been made on January 6, 1981, by Tamara Rand, a noted Los Angeles "psychic" who makes a handsome living by advising movie stars whether or not to sign contracts. On that videotape, viewers saw and heard Rand predict that Reagan would be shot in the chest by a sandy-haired young man with a wealthy family. The assailant, said Rand, would have the initials "J. H.." first name possible Jack, last name something like "Humley." She foresaw a "hail of bullets" as well. All this was to take place during the last week of March or the first week of April.

It was an impressive apparent prediction, the kind of thing that can turn a local "psychic" into a world-class celebrity, with the fame and fees that go with that exalted status. Yet to veteran psychic watchers the prediction was too precise.

The wire services had already helped broadcast Rand's miracle prediction world-wide, but Paul Simon, an Associated Press reporter in Los Angeles, contacted CISCOP (Committee for the Scientific Investigation of the Paranormal) chairman Paul Kurtz in Buffalo. Simon was skeptical about Rand's claims. Kurtz asked whether it was certain that the videotape had been prepared before the assassination attempt as claimed and urged Simon to check the time sequence. Simon called Kurtz back several times that day to report on his investigation. He said he had contacted station personnel at WTBS in Atlanta, Georgia, which Rand claimed had broadcast the tape Saturday night, March 28, two days before the predicted event. They denied that such a tape had been aired. He then called KTNV in Las Vegas, where the tape was said to have been prepared, and technicians there told him that they had taped it on Tuesday night, March 31, more than 24 hours after the shooting of Reagan.

Simon said Arthur Lord, bureau chief in Los Angeles for NBC, admitted that he had accepted the word of the producer of the TV tape and of Tamara Rand and that he has been bamboozled. Unfortunately, at the same time, NBC-TV officials in Los Angeles were telling the press that they "stood by every word" of the broadcast. That evening Kurtz sent telegrams on behalf of CSICOP to the "Today" show (NBC) and to "Good Morning America" (ABC) requesting that they retract the "prediction" story since all evidence pointed toward a hoax.

Fortunately, most of the network programs did eventually acknowledge that the affair had been a hoax and they had fallen for it.

The coup de grace arrived when Dick Maurice, host of the faked TV show, who had maintained that the prediction was genuine, finally admitted that it was an outright hoax. In a front page column in the Los Angeles Sun on April 5, Maurice proclaimed:

"I am sorry.

"I have committed a terrible wrong. I have committed the cardinal sin of a columnist. I have perpetrated a hoax on the public and feel very much ashamed.

"My interview with Tamara Rand in which she predicted the assassination attempt on President Ronald Reagan is a lie. Ed Quinn's (vice president and general manager of KTNV-TV) statement about the actual taping taking place on March 31 is the truth."[1]

It is clear that the fact that an event is reported on TV is not by itself a good reason for believing it. *Things are not always what they seem to be* — especially on TV. This goes equally for other favourite TV themes, e.g., UFOs. Presentations on FOX Network and elsewhere of such things as the so-called Rosswell incident or of the notion that aliens built the pyramids have more often than not been designed apparently more to impress the credulous and those who have a capacity to believe than they are to provide a rational evaluation of the evidence.[2]

Unfortunately, there is a common human tendency to believe what one wants to believe. This is the method of fantasy, and is one of the fallacious methods that leads to the acceptance of pseudoscience as truth.

Some people have the need to believe in paranormal phenomena. This need is in fact often sufficiently strong that they simply refuse to accept the confession of someone to having hoaxed them in a supposed demonstration of such powers. Consider this example reported by Gustav Jahoda:

I found myself in the company of six other people after dinner, and conversation veered toward the supernatural. An impromptu seance was proposed, and all of us settle around a large circular table. The idea was that questions would be asked, and the spirits would answer by rapping once for "yes" and twice for "no." The first question was asked, and nothing happened. We sat for several minutes in semi-darkness, with tension rising. Getting rather stiff, I shifted in my chair, accidentally knocking the table, and was staggered to find that this was taken as the expected answer. After a brief struggle with my conscience, the desire to experiment gained the upper hand, I told myself that after a while I would reveal the deception and pass it off as a joke. For another half-hour or so I knocked the table quite blatantly with the tip of my shoes, without arousing the slightest suspicion. I was just about to summon my courage to come clean, when one of the persons present asked the spirit to materialize. Another long tense silence followed, then one person whispered, "He's there, in the corner — a little grey man." It was said with such conviction that I almost expected to see something when I looked. There was in fact nothing except a faint shadow cast by a curtain moving in a slight breeze. Two others claimed to see the homunculus quite clearly .... About a year after the seance I met one of the participants. Recalling the evening, he said that he had previously been sceptical about the occult, but the experience had convinced him.

On hearing this my guilt feelings were thoroughly aroused, and I decided to make a clean breast of it. Once more I had badly miscalculated — he just could not believe me.[3]

Given that many have this sort of capacity for belief in the occult, it is likely that many more came away from the Tamara Rand announcement believing her to have special powers than were dissuaded from such beliefs by the subsequent revelation of fraud.

It is clear that if our interest is in matter-of-fact truth, then we will not want to base our scientific inferences on the sort of "data" that we have just discussed. It is no doubt that this is the sort of reason the physicist Sir Arthur Eddington had in mind when he once proposed that the results on which physics builds its theories ought to be "pointer readings." These pointer readings were taken to provide a secure set of data on which physicists could begin to makes their inferences. The point was made in a popular way, and some years ago. Today we can speak of electronic sensors, computer print-outs, and computer enhanced images as ways in which we can obtain data. All these would fit Eddington's description of "pointer readings."

Eddington made his remark, one presumes, because pointer readings, like our more sophisticated ways of recording data, have two features that help to generate their certainty.

One is the fact that such results are *reproducible*. In fact, it is evident why this is a feature of observations that is desirable. If the experimental situation can be reproduced, then the results can be checked by other scientists. The possibility of fraud and hoax will thereby be eliminated.

The other feature of "pointer readings" is that *the human observer has been eliminated*. There is in such circumstances no possibility other than an objective result. The tendency recorded by Jahoda, for an observer to see what he or she wants to see and to hear what he or she wants to hear, is absent.

Upon reflection, however, neither of these features could be essential to scientific observation. They are to be sure features that are to be valued; if it is possible to obtain results of these sorts, then so much the better. There are, however, many observational results in science that are good and form a solid basis for scientific inference but which do not meet these standards.

The problem with the first of these standards, reproducibility, is that, if we were to restrict acceptable observations to it, then we would rule our all observation in geology and astronomy, to mention only two cases. Who can reproduce an eclipse or a mirage? Who could reproduce an earthquake? Who could reproduce the death of an archaeopteryx? Reproducibility is nice, but we cannot always use it.

As for the second ground for liking "pointer readings," that we eliminate the human observer, adherence to this standard too would rule out many observations that are not only solid but important. If we were to adhere to it, how could Darwin have collected the geographical, the geological, the zoological and the botanical data that were essential to the inferences that led to his theory of evolution by natural selection? In fact, most data in geology and in the biological sciences, for example, are not collected using instruments. Even in astronomy until Galileo introduced the telescope, the data were obtained without instruments, by naked-eye observations. Even so, the data were sound. Brahe's data obtained without telescopes enabled Kepler to discover his laws of planetary motion. William Harvey did not have a microscope available, yet his data obtained by ordinary observational means enabled him to discover the circulation of the blood. So the use of instruments is a good thing, but here again we cannot always have observations that meet this standard.

What, then, are the conditions for good scientific observation? We have to try to achieve some clarity on this topic.

The means we shall use in this investigation should perhaps be obvious. Science has a certain cognitive aim. This is the discovery of matter-of-fact truth. Those methods of observation will count as good if they provide us with data that we can reliably count on as being true as a matter of fact.

In this context, where we want to throw a bit more light on the nature and role of observational evidence in the process of scientific inquiry, two further introductory points need to be made before we proceed.

First, we are concerned with the search after truth, not simply with debate.

Debate is often one of the means that can bring about greater clarity in the search after truth. But much goes on in debates that is not relevant to the search after truth. More strongly, much goes on in debates that obscures the search after truth. Stephen Jay Gould once made the point this way in commenting on his own debates with creation scientists:

> Debate is an art form. It is about the winning of arguments. It is not about the discovery of truth. There are certain rules and procedures to debate that really have nothing to do with establishing fact — which they are very good at. Some of those rules are: never say anything positive about your own position because it can be attacked, but chip away at what appear to be weaknesses in your opponent's position. They are very good at that. I don't think I could beat creationists at debate. I can tie them. But in courtrooms they are terrible, because in courtrooms you cannot give speeches. In a courtroom you have to answer direct questions about the positive status of your belief. We destroyed them in Arkansas. On the second day of the two-week trial, we had our victory party![4]

One might cite, for example, the appeal that creation scientists often make to the fact — and it is a fact! — that science has in the past fallen into many errors. They cite among other errors the so-called Piltdown Man. In this case a human skull and the jawbone of an ape, appropriately doctored to look ancient, were placed in gravel deposits where they would be found by physical anthropologists digging for evidence of human ancestors. The object was apparently to hoax a rather pompous and self-important researcher. It turned out that not only was the researcher taken in by the hoax but also the whole community of physical anthropologists. It was only many years later that the hoax was uncovered. In the interval, researchers strained to understand this odd "missing link." This is just one of the blunders that have been made in the name of science. So-called "Nebraska Man" or *Hesperopithecus* is another such blunder, also at times cited in debate by creationists as showing how mistaken and untrustworthy scientific conjectures can be.[5] *Hesperopithecus* was a conjectured American hominid, proposed on the basis of a single tooth discovered in Nebraska in 1922; but further research revealed five years later that the conjecture was specious and the tooth most likely that of an extinct peccary, a pig-like animal.

The argument of the creationists cites mistakes such as those of the Piltdown and Nebraska Man, and then goes on to the effect that science clearly cannot be trusted and that we have no reason to suppose that present theories are any better than these discredited older ones.

It is a paradox is worth noting, that creation scientists on the one hand appeal to the authority of science, e.g., in their arguments based on the Second Law of Thermodynamics, while also attacking science as a basically flawed process. But in any case, the fact that science has in the past been deceived and the fact that it likely will be deceived in the future makes a good debating point. But it has little force when it comes to the search after truth. In fact, when understood appropriately it makes exactly the opposite point. The inductive method of science is a self-correcting process. It is one which endeavours to uncover error and to correct it. In an important sense, progress in science consists, first, in making mistakes and then, second, correcting them. There are dishonest mistakes like the Piltdown Man and many more honest mistakes like the Nebraska Man. In due course research in conformity to the inductive method comes to expose them. They are then corrected, the errors replaced by truth, or, if not truth then at least by beliefs that are less inaccurate. The bloopers are simply relegated to the past, to the story how science really has made progress in achieving its aim of matter-of-fact truth.

The essential point about science on the one hand and creation science on the other has been nicely made in this way:

> ...what creationists ridicule as guesswork, and trial and error, and flip-flopping from theory to theory, is the very essence of science, the stuff of science. Error correction is part of the creative element in the advance of science, and when disagreement occurs, it means not that science is in trouble but that errors are being corrected and scientific advances are being made. Creationism comes on the scene arguing that the Bible is inerrant as a source of scientific truth and that "creation science" cannot admit of error because it simply does not exist.

We cannot conceive of two more diametrically opposed methods of explaining the world around us. One uses the correction of error as an inherent part of the process of searching for the truth, or ultimate reality in nature; the other rejects error or cannot admit of its existence. Although it may be human to make mistakes, it is scientific to correct them. That is the nub of the issue between creationism and science.[6]

Creation scientists are not really interested in how error has not only been part and parcel of science but is part of the way in which science progresses in its aim at understanding the way that world works. In public forums what the creationists are interested in is not truth but only in wining the debate, only in making science seem wrong. Citing an error without citing the progress that resulted when the error was detected will have that effect. So creationists, to win the debate, will take things such as the Piltdown Man or Nebraska Man and distort their real significance, the positive contribution they actually made to the progress of thought.

Or, to take another, rather different example, creation scientists will often cite the existence of so-called "living fossils" such as the coelacanth and horseshoe crab in the attempt to establish that there has in fact been no development over time, and that creatures therefore have been created all at once.

The facts they cite really are fact, but to cite them alone is to ignore the fact that besides these creatures there are many species which have shown long histories of development from simpler to more complex. If the coelacanth has not shown development, the sea mammals in contrast have exhibited very rapid and dramatic development. To cite the former and ignore the latter is good at winning a debate but it is fundamentally at odds with any serious attempt to discover the truth of the matter.

Moreover, it should be added that the existence of unchanging forms such as the coelacanth, forms which show little development over time, is not at all at odds with the theory of evolution by natural selection. All that is established by creatures such as these is that they inhabit an unchanging ecological niche. On the theory of natural selection, each sort of creature has in general a form and set of behaviours that enable it to survive and reproduce in the environment in which it exists. If this environment is relatively unchanging, then there will be no forces of natural selection that will cause it to change. To the contrary, in such a context the selective pressures will be to remain the same!

In their selection of evidence, then, creation scientists are too often more interested in so citing data that they can win at debates. This, however, is not our concern. Rather, what we are interested in locating are the criteria for evidence that will aid in the search after matter-of-fact truth.

One ought not to have the impression, however, that creationists are the only ones who prefer the ploys of rhetoric to the search after truth. Consider, for example, the columnist Pat Moffat of Toronto's national newspaper, *The Globe and Mail*. Her piece of September 22, 1998, aimed to persuade her readers of the acceptability of certain theories of medicine based on the non-empirical speculative physiology deriving from ancient China.[7] Specifically, she aimed to convince her readers that it was not unreasonable to entertain as plausible the views of Mr. Tetsuro (Ted) Saito of the Toronto Shiatsu Centre.

She introduces this practitioner this way:

Mr. Saito learned the healing art of finger-pressure massage in his native Japan from this century's masters, Shizuto Masunaga and Tokojiro Namikoshi.

Finger massage no doubt feels nice, and it has no doubt proved relaxing at the very least for those upon whom it is practised. What is at issue, however, is its validity as a medical practice. Does it in itself (and not through, say, the placebo effect) have a curative power? Is it really efficacious? And have the theories upon which it is based passed the rigorous sort of testing to which all medical theories ought to be subjected if they are to be counted as rationally acceptable? However, instead of addressing the issue head on, Moffat provides a pedigree for Mr. Saito. He studies under this century's "masters." Now, to be a master, there must have been something there to be mastered. To refer to these persons as "masters" is already to assume the validity of the theory and the efficacy of the practice. It is to assume from the beginning the very point which is to be established. It is a nice debater's point aimed at getting the audience to accept without question the proposition being advocated. But if one's concern is, as it ought to be, with the truth of

the matter, and with *rational* persuasion, then such ploys are out of place.

Moffat continues:

I figured that if anyone could explain the seemingly mystical ideas that underlie shiatsu, acupuncture and Chinese herbal medicine, it would be this gentle guru.

Now, if the ideas really are mystical then they cannot be explained: the mystical is beyond explanation. But they are only "seemingly" mystical, not really. That is why they can be explained. But maybe it will turn out that Mr. Saito does have a hard job explaining his ideas, and in trying to make them sufficiently clear that one could figure out not only their empirical applications but how to put them to tests that could confirm or falsify them. Which as it turns out he does: It turns out with regard to something *chi* that he suggests is a form of energy that, as he tells Moffat, "I can detect it …. But other people can't detect it, so this is a problem." It possible, however, to explain away this problem by appeal ot the mystical nature of the notions: they can't really be explained after all. The language is a way of ensuring that in the course of the debate one will be able to have it both ways: the theory both can and cannot be subject to test. Neat, but not a line of argument that could be used by one genuinely concerned to uncover what really is the truth of the matter.

Moffat declares Mr. Saito to be a "gentle guru." The alleged gentleness is no doubt relevant to the estimation of his capacities in finger-massage. But since he is introduced as someone who can explain medical theories we might hope for an estimation of his capacities for sound thinking.

We also discover that he is a "guru." Some have suggested that Einstein, too, has acquired the status of a guru, a teacher of wisdom. That status, however, is irrelevant to the evaluation of his theory of relativity as a scientific hypothesis. For the latter we want to know the evidence to which Einstein appealed, not his status as a guru. Similarly with regard to Mr. Saito's theories. Here, too, if we are interested in whether these theories are true or not, what we want is the evidence on which they can be evaluated, not Mr. Saito's status among sages.

Ms. Moffat's point is good in debate. It parallels the remarks often made by creationists to the effect that

Marx was an evolutionist and that evolution is akin to communism. Nice stuff that will get some at least on your side. But not the sort of consideration that is relevant if your concern is that of science, the discovery of matter-of-fact truth.

The **second** introductory point to be made in this context, where we are concerned to shed further light on the nature and role of observational evidence in the process of scientific inquiry, is to firmly remember that in general *the absence of evidence is not evidence. The failure to observe something is not evidence that that thing does not exist.*

This point can be made by reference to the mystery of the so called Bermuda Triangle. It is claimed with regard to this area of the mid-Atlantic Ocean that there are many unaccounted for disappearances of airplanes and ships. This is but one instance of the more widespread tales of ships mysteriously lost at sea. In all these cases, including those of the Bermuda Triangle, the ships or airplanes were well built and presumably well navigated and sailed. There was no reports of bad weather that they might have encountered. Nor were there any other reasons to explain their disappearance. But disappear they did!

In the absence of evidence for a naturalistic explanation of such events, the temptation is to provide more outlandish hypotheses. Formerly one might suppose that some gigantic sea monster swallowed or otherwise destroyed the missing ships. Nowadays one might resort to the hypothesis of alien visitors coming from elsewhere in the galaxy to abduct us.

One should ask, however, the parallel question, how many transport trucks have disappeared? It turns out that, by comparison with ships and airplanes at sea, very few. In fact, it is hard to find any at all. To be sure, every once in a while a truck will really disappear. But then it usually turns out that it re-appears a few days or weeks later in some out of the way parking spot where it has been emptied of its cargo of cigarettes or high fashion dresses. The disappearance is not a matter of mystery; it is just that the truck has been highjacked and its cargo stolen.

The important contrast is between what happens in the middle of the ocean and what happens on land. If a ship or airplane has something happen to it in the middle of the ocean, there is no one about to witness the event

nor therefore even to record where it might have occurred. In contrast, if sometime goes wrong with a large transport truck, then it is all over highway 401. Even if people do not witness the event, the evidence remains and people not only come to see it but have to cope with it, clear up the mess, and nearly always they figure out why the disaster happened.

It is hardly surprising then that the number of mysterious disappearances at sea is greater than the number of mysterious disappearances on land. Those disappearances at sea do occur, but it is safe to assume that they have naturalistic explanations even if we do not know what they are, and safe therefore to assume also that there is no need to invoke aliens or the occult.

It is worth noting that with the development of navigational devices, the mysterious disappearances of ships and airplanes is becoming a much more uncommon occurrence. But the point is clear and needs to be emphasized: the fact that we lack evidence which provides the basis for explaining the disappearance of a ship or airplane at sea does not mean that the event must be explained by extraordinary or supernatural hypotheses. The absence of evidence is not evidence, and, in particular, not evidence of the supernatural.

There are other cases where the absence of evidence cannot be used as evidence. Bigfoot is such a case. The aboriginals in Canada's west refer to this legendary beast as a Sasquatch. Though there have been claimed sightings, no one has every provided solid evidence for the existence of such a creature.[8] Some have claimed that there have been footprints that the creature has left in the snow. These may well be more a matter of fakelore than folklore, but that has not been proven. It is arguable, however, that these marks, if they really are genuine, are those of ordinary animals modified and enlarged by a variety of evaporation phenomena which occur at the higher altitudes at which the Sasquatch is supposed to exist. Be that as it may, the relevant point of logic, once again, is that the fact that there have been no sightings of Bigfoot and nor solid evidence, e.g., a corpse, does *not* imply that Bigfoot does not exist. It might after all be there, high in the Rockies, still waiting to be discovered.

After all, at the beginning of the 19th century, gorillas were supposed to be mythical creatures. The stories about them that had come down from the Greeks were thought to be distorted stories about wild men in the forest. We now know that they really do exist.

What counts against Bigfoot is not the fact that it has not been seen but rather the fact that it is supposedly a large primate living at high altitudes. In the first place, there are no other large primates in the western hemisphere to which it might be biologically related. In the second place, there is no food supply in its supposed area of habitation that would adequately support even a small population of large primates.

It is unlikely, then, that Bigfoot exists. Still, the logical point remains: the absence of evidence for the existence of Bigfoot is not evidence for its absence.[9]

In this connection one might contemplate the case of sea monsters. Everyone has seen reproductions of renaissance and early modern woodcuts which show some monstrous creature attacking a ship. Some of these supposed sea monsters even had names. One was the "kraken," a giant creature with arms like an octopus which was capable of destroying small ships.

No one had ever sighted such a creature; there was an absence of evidence. It was therefore assumed at the beginning of the 19th century by all clear thinking and reasonable people that these creatures were simply mythical. But then three of the beasts washed ashore in Newfoundland in the late 19th century. These were giant squid over 80 feet long. Normally such animals live deep in the ocean, but in this case unusual currents happened to wash them ashore, where they could be measured. The result was that stories of kraken attacking small sailing vessels had to re-evaluated: perhaps the mediaevals who spoke of these monsters had after all seen something that actually existed.

In this case, as in the case of gorillas, the absence of evidence was not a safe basis on which to infer non-existence.

And logically, maybe Bigfoot is like gorillas or the kraken.

The same point of logic can be made with regard to certain of the arguments offered by creation scientists. One of the considerations that they often advance is the absence of fossil data pointing to the existence of the transitional forms that must at one time have inhabited the Earth if evolutionary theory is correct. If the process of evolution is, as Darwin believed, one which everywhere consists of slow and incremental change, then there ought

to have been many forms transitional between lower forms and more complex higher forms. But these transitional forms do not occur in the fossil record. There is therefore missing from the record data that would testify to occurrence of evolution as a long and stately process from the incomplex to the complex. The absence of evidence is taken as evidence that such forms never did exist.

But, once again, this argument, the argument from absence of evidence, is fallacious: it does not establish the non-existence of transitional forms, at most what follows is that we simply have not observed them.

In this context, however, one can make stronger points against the creationists. Given the fact that the conditions for fossilization occurring are relatively rare, the existence of major gaps in the fossil record is not at all surprising. Not only is it not reasonable to infer the non-existence of the transitional forms and the falsity of Darwinian theory from the absence of fossils, that absence to the contrary is precisely what one would expect if our theories of geology are correct. Since those theories are in fact highly confirmed, the absence of transitional forms in the fossil record is precisely what one would expect: far from disconfirming Darwin's theory, it to the contrary confirms it.

In point of fact, however, one does find in the fossil record, contrary to the creationtists, evidence of transitional forms. These are, as one would expect, rare. But they do exist. *Archaeopteryx*, for example, is one such form: it is transitional between reptiles and birds. *Ambulocetus natans* is a transitional form in the evolutionary history of whales (see illustration, Figure 5.1).

There are two living whale lineages, the toothed whales or *Ordontoceti* and the baleen whales or *Mysteceti*. Together these comprise the mammalian order *Cetacea*. *Archaeocetes* are a "scrapbasket" group of extinct Eocene whales. The closest relatives of whales are an extinct group of ungulates, *Mesonychid condylarths*.

The newly discovered *Ambulocetus natans*[11] is intermediate between the Archaeocetes and the Mesonychids. "Recovery of a skeleton [of this extinct species]…documents transitional modes of locomotion, and allows hypotheses concerning swimming in early cetaceans to be tested. The fossil indicates that archaic whales swam by undulating their vertebral column, thus

Figure 5.1
Whale family tree[10]

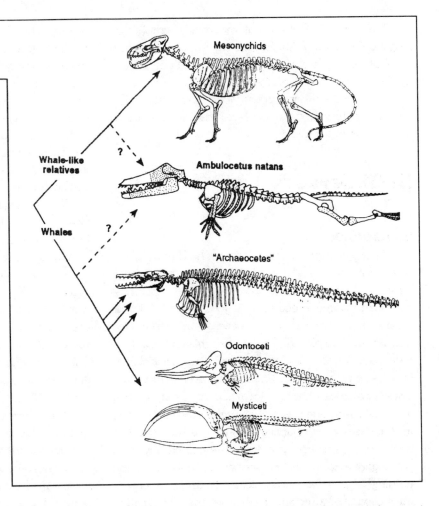

forcing their feet up and down in a way similar to modern otters. Their movements on land probably resembled those of sea lions to some degree, and involved protraction and retraction of the abducted limbs."[12]

Creationists have two debating ploys that they introduce at this point. One is that of simply ignoring the evidence, and denying that the form, such as *archaeopteryx* or *ambulocetus natans*, really is a transitional form. This is hardly a cogent form of argument, however appealing it might be in the context of a debate.

The other ploy is to insist once such a form is given that there are gaps between the transitional form and the two forms on either side of the gap that it is supposed to fill. Thus, when *archaeopteryx* is cited as filling the gap between reptiles and birds, creationists reply that there are gaps that have not been filled between reptiles and *archaeopteryx* on the one hand and between *archaeopteryx* and birds on the other. This is a neat ploy since it has the consequence that the evolutionist can never meet the demand for evidence: every time the evolutionist fills a gap two more gaps are created! But once the absurdity of the creationists' ploy is recognized, we no longer take it seriously.

There is a further point that ought to be made against the argument of the creationists concerning the fossil record. If evolution has been the rather slow and even stately process that Darwin supposed, then the absence of transitional forms in the fossil record might, in spite of its lack of significant logical force, succeed in raising some modest degree of doubt about the adequacy of the theory. The account of evolution which is thereby called into question is, however, not correct. Darwin was wrong,

not about evolution and not about the role of natural selection in the origin of species, but in the belief that the process was everywhere one that was slow and incremental. However, in 1972 Niles Eldredge and Stephen Jay Gould argued that the process was not of this sort after all. They argued persuasively that speciation occurs not in large populations but in small isolated "founder" populations. In these groups changes occur in time spans which are, for geological times, relatively short. In these periods new species emerge, but it occurs with sufficient rapidity and in relatively small numbers that we can normally expect that there be few if any transitional forms preserved in the fossil record. Far from being the negative evidence against there being a process of biological evolution as claimed by the creationists, the lack of fossils is positive evidence for just such a process — though one not of slow and incremental but one of rapid and episodic change.

The point stands, then, that we wanted to emphasize, that the absence of evidence is not the same as evidence.

But we may further conclude that creationists show in their arguments that their capacity to believe far exceeds their capacity to marshal evidence.

Let us turn now to look at what is involved in observation in science.

(i) Observers

a. Laennec[13]

Since the ancient world physicians had known about noises in the chest, and from Hippocrates onward through Galen to the early modern world listening to the chest had been recommended as a diagnostic tool. But until the end of the 18th century, however, its use was limited and not well explored. There were reasons for this. Patients were not the cleanest, even the upper classes but especially the poorer. This was true even in hospitals. These included no tubs for regular bathing. Clean bedding was rare unless one was rich enough to afford it — and in that case one was unlikely to go to a hospital. The wards were generally quite dirty. The trained physician was a gentleman — only a gentleman could afford the training! — and was usually cleaner than his patients in the hospitals. Moreover, throughout the 18th century gentlemen wore rather elaborate wigs. This included the physicians, and no one would be willing to allow their luxurious headgear to become infected by lice by placing it close to a poor patient's chest to listen to the noises. Patients too had problems with the use of this diagnostic tool. The noises could be heard only if the ear of the physician was placed firmly on the chest, and many of the patients, especially women, did not relish him — it was always a "him" — doing this. There were, then, considerable difficulties with regard to physicians listening to the chest in order to diagnose illness.

This was something to be regretted. For, the noises in the chest can be quite revealing. If the heart is normal, there is a regular thump. This is something with which most of us are familiar. But if there are heart problems then beat will be irregular, in both tempo and pitch. If the valves leak a faint murmur can be heard. The sound of the lungs is also revealing. Normal breathing sounds

through the rib cage something like the blowing of a gentle breeze. But if the air tubes are clogged then the sound is one of bubbling and boiling. And if the air sacs are closed by inflamation then a deep breath that forces them open will cause a crackling sound. At the beginning of the 19th century, however, these points were ill-understood. The incapacity to listen carefully to the sounds meant that the significance of their differences had escaped physicians. Disagreement abounded as to what was meant by the various sounds.

This changed in 1819. A young French physician from Brittany, Theophile Laennec, was working in the Necker Hospital in Paris. More dedicated than most of his colleagues, he was distressed when he was unable to diagnose a patient's illness. One patient for which he could not effect a diagnosis was a young woman. She was in great pain and short of breath. But she was overweight and the layers of fat tended to blanket the sounds of the chest. She was moreover of a modest disposition and this made her unwilling to let Laennec get close enough and certainly not close enough for any length of time to hear clearly the noises.

The scientist is moved by a cognitive interest in matter-of-fact truth; his or her aim is that of knowing. If this passion for truth is strong enough, he or she will be moved not only to improve their knowledge but to locate the means for such improvement. Laennec was such a person. He was moved by this passion for truth. And it was more than idle curiosity, though that was no doubt part of it. He also had a pragmatic interest in the truth: he was deeply concerned for his patients, wanting to relieve them, so far as he could, of their pain, their disease.

We know what happened. Laennec, after attending to his patients in the hospital, was accustomed to taking a walk in the Gardens of the Louvre. They were not as well kept or as beautiful as they had once been and as they have now again become. The various upheavals of the French Revolution had wrought their havoc. Debris and timber lay scattered about. But as unfortunate as this might be for someone who enjoys gardens, in this case it had the welcome effect of providing Laennec with an idea that he could use in trying to diagnose the young woman who was his patient.

What he noticed was two young boys playing on a long beam of wood. One of them was at one end with his ear placed tightly on the wood. The other was at the other end lightly tapping the beam. To the delight of the boys the wood transmitted the noise very clearly. To us it was a telephone, however crude. To the boys, it was a toy, and great fun. To Laennec, it provided the idea that led to the solution of his problem.

The problem was that of being able to listen to the sounds of the chest carefully and systematically without coming into physical contact with the chest. The solution consisted in finding a device that would transmit the sounds from the chest to the ear. The sport of the boys suggested the needed device. Returning to the ward, Laennec grabbed a paper note book, a *cahier*, rolled it up into a tight tube, and placed one end on the chest of the overweight girl and the other end at his ear. The experience of the boys was replicated: the sounds came through clearly. As the boys heard the tapping, so Laennec was able to hear clearly the noises of the chest. The stethoscope had been invented.[14]

The physics of sound transmission was well known so the theory of the instrument was clear. Laennec began systematically to improve this new diagnostic tool. The material of his "cornet de papier" proved rather flimsy, so, following the boys, he made a tube of wood to replace it. To do this he had to become a technician: to perfect his new wood stethoscopes Laennec had to acquire the carpenter's art of turning on a lathe. He also experimented with various kinds of wood, discovering that an instrument made of beech or linden was best for transmitting the sound. He also added a hinge so that it could conveniently fold and fit into his pocket. (The two-ear stethoscope was introduced only in the 1850's.)

The new tool revolutionized parts of medical science not because it was complicated or sophisticated new technology but simply because it enabled physicians to overcome some simple difficulties that lay in the way of systematically studying the noises of the chest. Previously, largely because of the difficulties in listening to those noises, their study had been relatively aimless and certainly haphazard. Laennec's new instrument made a systematic and methodical study possible. Laennec undertook that study.

Laennec began a regular study of the sounds in the chest of the hundreds of patients in the Necker Hospital. He listened to patients with all sorts of diseases, and did this day after day, taking very careful notes in each case.

Besides having available his new instrument, Laennec was well suited to undertake this investigation. Laennec was in fact a close student of the varieties of sound. He enjoyed a sound reputation as an amateur flutist; his ear for music was superb. His musical practice translated easily into his medical practice. What he heard were crackles and bubbles, what earlier listeners had heard, but these noises also had pitches and tones. No one before had tried to study these. Laennec did, and his ear served him well. He was able to relate the noises in their nuances to the disease conditions that caused them.

Up to that time, it was "consumption" that was the serious disease of the chest. When it appeared there was little physicians could do to relieve the dis-ease of the patient. It could only be endured and was usually fatal. What Laennec was able to do was relate this malady to several other conditions of the chest that physicians had identified. He showed that these were all of them the same as consumption, only its earlier stages.

This was of major importance in the treatment of tuberculosis. If the disease is discovered early, the chances of successful treatment and recovery are much better than if the illness is discovered in its later stages.

But physicians also acquired a re-conceptualization of the very concept of disease. Prior to Laennec, what physicians were concerned to treat were dis-eases, conditions in which the patient was actually suffering physical pain and discomfort. When Laennec connected consumption with earlier-existing conditions in the chest, he was able to establish that the condition existed prior to its being one of dis-ease, discomfort. He was able to establish that the disease existed before the dis-ease. By creating the idea that a disease, the cause of dis-ease, could exist prior to the appearance of the symptoms, prior to the dis-ease, Laennec radically re-defined that concept. Physicians could now try to locate and to fight the disease prior to there being any dis-ease. With this there were to be far greater likelihoods that the dis-ease could be prevented from arising in the first place.

Laennec made another contribution. This had to do with scrofula, the so-called "King's Evil" that was supposed by many, Samuel Johnson included, to be a disease that could be cured if at all only by the miraculous touch of the King. Hippocrates had many centuries earlier dismissed the notion that there were diseases with non-natural causes when he criticized the ancient view that epilepsy was the "sacred disease," caused not by ordinary causes but by the gods. Laennec made the same point with respect to scrofula: it is methodologically just wrong to suppose that there is anything miraculous about either it or its cure. What Laennec was able to show was that scrofula is in reality a species of tuberculosis, one that occurs in the lymph glands in the neck.[15]

Finally, Laennec taught his students to observe what he had observed. He provided them with stethoscopes, at first rolled-up *cahiers*, and later the more sophisticated versions of the instrument that he created. He taught them how to listen using these tools and, more importantly, he taught them what to listen for. Using his own listening skills he had noticed facts that no one had noticed before, facts that he discovered had significance for the cognitive ends of the scientific and medical community. He then taught those same listening skills to his students so that they too could notice and record such facts.

Prior to Laennec, the crackles and bubbles, the many nuanced noises of the chest, were all facts. But they were not in any seriously way treated as *data*. In this sense they had not made their appearance as scientific *facts*, facts which could be used for purposes of explanation and prediction, and in particular, since this was medicine, as facts which were importantly relevant for the treatment of disease. The facts had to be discovered before they could become facts useful to the scientific community. And before they could become facts, they had to be observed. In a way they were like the giant squid. At one time, this creature, not being observed, did not, for all right-thinking people, exist. Now it does. Prior to Laennec, the bubbles and crackles in the chest did not exist as facts, in the sense of observational data. Now they do. This was Laennec's achievement. And, once more, before these facts could be observed, scientists had to acquire the observational skills that are needed if one is to notice them, observe and record them, to as it were bring them into existence. Laennec had to learn to use his instrument and had carefully to train his ear before the various different bubbles and crackles would be counted among his colleagues as genuine *data*. There was a process of trial and error, a process of learning to observe, before these facts entered the domain of science as *facts*.

This way of putting the point is, of course, not quite accurate. The facts after all were there in the world, really in existence, prior to their being observed. But for the scientific community, for the medical community, they did not exist, they were not real, in the sense that they were not — not yet — data that could be used for explanation and prediction. *Before facts can become data useful to the scientific community for purposes of explanation and prediction they must be observed.*

But further, such facts are not simply waiting there, imposing themselves when noticed on the casual observer. They are, rather, *gathered with difficulty*. Obtaining data is not simply a matter of just looking. *Observation is a skill that is acquired through learning, through training and discipline.*

b. Observation and Inquiry

Science proceeds by putting hypotheses to the test. These tests, as we know, need not be deliberate experiments. They can be experiments that nature performs for us. But in any case, the sort of observation that goes on in science cannot be separated from the process of hypothesis formation and hypothesis testing.

The following example gives a clear sense of this.

John Horner is an American paleontologist who, in a famous dig in the Rockies near Bozeman, Montana, uncovered the first fossil dinosaurs eggs found in North America.[16] Horner notes that palaeontology is not a science in which deliberate experiments can be performed. To be sure, it formulates hypotheses and puts them to the test. But the tests are by Natural, not contrived, experiments. As Horner puts it, "Paleontology is not an experimental science; it's an historical science." He explains:

This means that paleontologists are seldom able to test their hypotheses by laboratory experiments, but they can still test them (Horner and Gorman, p. 168).

The method, of course, is the comparative method first described by Cuvier and which we described above.

As Horner describes it, the process of inquiry that he undertook had two stages. The first was "getting the fossils out of the ground…". The second was "to look at the fossils, study them, make hypotheses based on what we saw and try to disprove them" (p. 168). In this case, the first phase makes clear that data are not simply found, they are, to repeat, *gathered with difficulty*. In this case the first phase consisted of unsheathing the fossilized bones from the surrounding stone. This is work that can only be described, in its initial stages, as backbreaking, involving heavy labour with jack hammers and pickaxes. It proceeds from there in gradual steps to the use of dental tools and small brushes.

What Horner discovered in 1981 was a site that contained approximately thirty million fossilized fragments of bones from dinosaurs of the species *Maiasaur*. The team did not actually dig up thirty million fragments. The bed that was being examined was approximately 1 1/4 by 1/4 miles in extent. The team looked at certain areas of this bed that were exposed, and their conclusion, about thirty million fragments, was an extrapolation from these observations. From this estimate they inferred that the bed represented "the tomb of ten thousand dinosaurs" (p. 128). This raised the question, "What could such a deposit represent?" (p. 129).; what, in other words, was the cause of this effect?

Some hypotheses could be ruled out quickly. One hypothesis was that the dinosaurs had died from the action of predators, perhaps some cousin of the *Tyrannosaurus rex*. It is true that the fossils indicated that many of the bones had been broken in half. But as the example of lions makes clear, predation involves chewing on bones and this leaves marks. These marks would have appeared in the fossils. But there were no such marks. So the observed results could not have derived from work of predators.

Further progress in narrowing the range of alternatives required noticing details that were not at first apparent. Various details turned out to be important. The bones had all been arranged from east to west, the long dimension of the bed in which the fossils were located. Small bones had been separated from bigger bones. There were no bones of baby *Maiasaurs*, only those of *Maiasaurs* between nine and twenty-three feet long.

These data made clear that the cause to be discovered had to satisfy certain conditions. The cause had to make the bones splinter lengthwise. The cause would have to separate the small from the big bones. The cause would have to be one which could explain why no young had been killed.

The inferences that were being made, and which came to be made, were of course all about the past. To which the investigators had no direct access. In this they were in the same position as Herodotus when he tried to draw inferences about the rising of the Nile from what he could infer about what was for him the unobservable interior of Africa. Both had to rely on what they did know about causes acting in the present to infer what the causes could be in places and times to which they had no direct access.

One hypothesis considered by Horner and his colleagues was that the deposit was the result of a mudflow. This was easily ruled out: "it didn't make sense that even the most powerful flow of mud could break bones lengthwise ... nor did it make sense that a herd of living animals buried in mud would end up with their skeletons disarticulated, their bones almost all pointing in one direction..." (p. 129). The data in fact eliminated any idea that there was a single cause. "It seemed that there had to be a twofold event, the dinosaurs dying in one incident and the bones being swept away in another" (p. 129). There was, first the event that killed the dinosaurs and which at the same time killed off predators and scavengers that might otherwise have left marks on the bones. And then there was the event coming some time after that distributed the bones in the lengthwise fashion in which they fossilized remains were found. Both these events had to satisfy the further condition that they could explain why there were no young in the mass that had been killed.

As for the first event, the fact that there was a layer of volcanic ash half a metre thick above the bone bed suggested that volcanic activity was the cause of death. This hypothesis received further support from the fact that volcanoes "were a dime a dozen in the Rockies back in the late Cretaceous" when the event occurred (p. 130). Horner concluded that

> A herd of *Maiasaura* were killed by the gases, smoke and ash of a volcanic eruption. And if a huge eruption killed them all at once, then it might have also killed everything else around...

which, as he adds, "would explain the lack of evidence of scavengers or predators gnawing on the bones" (p. 130).

Then for the second event, he hypothesized, "there was a flood" (p. 131). The breach of a lake could provide the water for this innundation. In any case, a large flood would carry the rotting and partially fossilized bodies downstream. The movement of the water would explain why the big bones were separated from the lighter small bones, and it would explain the uniform orientation.

> Finally the ash, being light, would have risen to the top in this slurry, as it settled, just as the bones sank to the bottom. And over this vast collection of buried, fossilized dinosaur bones would have been left what we now find — a thin but unmistakable layer of volcanic ash (p. 131).

Finally, there is the issue of why there were no young. Here, various hypotheses are possible. Perhaps the smaller bones of the young were so smashed up by the flood that all trace simply disappeared. Perhaps the young stayed in groups, perhaps with their mothers, but in any case apart from the herd. Perhaps the dinosaurs did not produce young every year and this was one of those years. As Horner admits, "nobody knows for sure" (p. 133).

It is clear how observation, then hypothesis formation, then attempts at explanation, go hand in hand, even in non-experimental sciences.

The story goes on, however. Horner and his colleagues did discover *Maiasaura* young and were able to establish in their excavations that these young were hatched in colonies. This suggested behaviour akin to that of birds such a penguins which live in large colonies. And this in turn suggested that *Maiasaura* were, like birds, but in contrast to reptiles, warm-blooded. In order to investigate this hypothesis Horner went to Paris to work with an expert in histology, Armand de Ricqlès who had been among the first to suggest that dinosaurs were warm blooded (p. 172). Others had already shown that "dinosaur bone looks more like the bone of mammals and birds that it does like that of reptiles" (p. 173). Now, we know from living things that warm blooded creatures such as birds and mammals grow more rapidly than do cold blooded creatures such as reptiles. Further, rapid growth leaves one sort of trace in the bones, slow growth another. By examining the bone structure of the young dinosaurs, Horner would be able to test the hypothesis

of warm bloodedness. If there were evidence of rapid growth, then the alternative of cold bloodedness could be eliminated (p. 173).

As it turned out, the results of the Paris investigation were to confirm the hypothesis of warm bloodedness (p. 178ff). We need not go into details. The point for our purposes that observational techniques were needed that were very different from those used in the field work.

For this investigation, very thin sections of the fossil bone had to be made and carefully examined under the microscope. But, "looking at a real maiasaur tibia is very different from looking at diagrams or reading textbook descriptions. You need a guide, an interpreter, someone to show you the way — to show you what 'conversational' histology is like. Ricqlès was our guide." Horner and his colleagues had to *learn how to see.*

> We catalogued each slide that we made. Then we examined it under a microscope and recorded a description of the cross section in a notebook. First, of course, we looked to see if we'd gotten the slide thin enough so that light would shine through the cross section and illuminate the internal structure of the bone. Then we had to learn to see what we were looking for. Just as untrained eyes don't notice the difference between bits of fossil bones and bits of rock, or don't see an anticline or a deformation of sedimentary rock beds, untrained eyes don't recognize Haversian canals, lamellar zonal bone and plexiform bone. With Ricqlès' help, we trained our eyes (p. 177).

Once again, to repeat to make the point emphatically, *observation is a learned skill.*

Now, evolutionists are not the only ones who cite the fossil record in support of their position. One finds similar appeals by creation scientists. The latter often cite examples of fossils that are supposed to be in the wrong order for evolution, attempting thereby to prove that the sequence to which evolutionists appeal is in fact a misordering. In particular, they often claim that human footprints have been found in deposits alongside those of dinosaurs or other creatures who evolutionists argue existed only millions of years before humans appeared on the Earth. It is claimed that these results provide observational data that support the thesis of the creationists that the story of the Noachian flood accurately describes the history of the Earth.

The example most often cited in creation science literature is that of apparent human and dinosaur footprints side by side in a Cretaceous limestone deposit near the Paluxy River close to Glen Rose, Texas.[17]

During Cretaceous times, a withdrawing sea left behind a swampy area in which dinosaurs left footprints in the mud. The sun baked these prints, which were later covered by more sedimentary mud, and then fossilized. The area today is traversed by the Paluxy River, often fast-flowing enough to move downstream four ton blocks of limestone, but also often enough dried up during drought conditions. When the river dries up, dinosaur tracks become exposed to view. Geologists agree that most of these are genuine.

Alongside some of these there are also depressions that have somewhat the appearance of human footprints. It is these that are cited in creationist literature. The best of these prints in fact have turned out to be probable or actual hoaxes, carved as money-making tourist attractions during the Great Depression. The Seventh-Day Adventist geologist Berney Neufeld has stated with respect to two collections of claimed human fossil footprints from the Paluxy basin:

> Local old-timers in the Paluxy River area tell that the tracks were both excavated and carved as a source of income during the depression years. Both these collections [that of the private collector Clifford Burdick and that of Columbia Union College, Takoma Park, Maryland] may well be carvings of that period.[18]

Creationists admit that some of the impressions were fakes. This does not, however, deter them. John D. Morris writes that

> These counterfeit tracks do not, of course, disprove the genuine tracks. In fact, it could only have been the existence of genuine tracks that made the manufacture of counterfeit tracks profitable.[19]

The claim that profits could be made from fakes only if there were genuine tracks present in the stone is rather dubious, but the logical point is quite correct: the fact that some impressions are provably or likely fakes does not imply that there are no genuine fossilized human footprints.

But what one does see is not convincing. There are a variety of marks and impressions to which creationists appeal. Several things are to be said. In the first place, many of the examples seem to be dinosaur footprints that have eroded. Dinosaurs are three-toed and most of the weight was one the middle toe. That means that the imprints of the side toes were a little more shallow and eroded away to leave a single elongated impression that could be construed as being of non-dinosaur origins. Even so, in these impressions one can still notice traces of the side toes.[20]

In the second place, when a heavy creature leaves a print in swampy ground, mud squishes up the outsides. Fossilized tracks show traces of this mud. A number of the impression which creationists cite as human tracks do not exhibit such squish marks and are therefore more reasonably construed as nothing more than the result of erosion occurring in the undulating rock.

In the third place, a number of marks to which creationists appeal have been given their characteristic shapes by applying oil. This is supposed to bring out the details of the impression, making it more clear to the human eye, but is more reasonably construed as creating the foot shape out of sets of impressions and marks that in fact exhibit no such shape in themselves.

Of these marks the geologist Berney Neufeld writes that

> In my opinion, these footprints are not tracks at all but represent random erosion marks in the surface of the limestone plate. The surrounding surface is covered with erosional marks of almost every imaginable shape. Individuals have reported visualizing the tracks of practically any mammal species on this surface .... It is only with a great deal of imagination that a bipedal trackway can be seen at all.[21]

There is, then, nothing in these observations that could reasonably count as supporting the claims of the creationists.

What is even more important, however, is the way in which evolution scientists on the one hand and creation scientists on the other respond to their data.

A scientist is moved by a cognitive interest in matter-of-fact truth, in finding the causes and effects that explain the data that he or she has collected. We have seen how Horner systematically elaborated hypotheses about the fossils that he uncovered and then devised ways in which these hypotheses could be put to the test.

The creationist, confronted by the data that he has supposedly located near the Paluxy River, has things that need to be explained, namely, as he would have it, the co-existence of men and dinosaurs and their walking together in swamps forming adjoining trackways. The flood hypothesis poses a major constraint on the set of hypotheses that he can entertain. This constraint means that the creationist has to hold, first, that the great Flood deposited some two thousand metres of sediment in the Paluxy valley; he must hold, second, that dinosaurs and men then walked side by side in this deposit; and he must hold, third, that there almost immediately occurred another huge tidal wave which both covered and preserved them. The details of the causal processes and mechanisms that effected all this are hard indeed to envision. John D. Morris admits the difficulty:

> The main problem of geologic origin for biblical catastrophists stems from the fact that underlying the Paluxy River basin is nearly eighty-five hundred feet of sedimentary rock. According to the catastrophic model, this must all have been laid down by the flood of Noah's day. The problem is how could man and dinosaurs witness such massive deposition at the beginning stages of the flood and survive long enough to leave their prints so high up in the geological column?[22]

Morris has some vague hypotheses about what might have happened. Since the rocks are limestone, Morris had to admit that they are of marine origin rather than postdiluvian. There is a precambrian rock structure about twenty-five miles from Glen Rose, the Llano Uplift, and Morris speculates that men and dinosaurs survived together on this after the first stage of the flood, only to be destroyed in a later stage when they ventured into the now swampy area below the Uplift. This suggestion has its own problems, however. How, for example, could a tidal wave that laid down deep layers of sediment not also have splashed over the Llano Uplift?

The appearances cited by creationists as supporting their account of history require one concerned about the causes of things to form a variety of hypotheses. Most of these are not only unconfirmed and untested but highly implausible. In fact, given the data that supports the explanations offered by the evolutionists one has little choice but to reject the creationist hypotheses as false.

In this situation of conflicting hypotheses, one would expect persons concerned with matter of fact truth, that is, scientists, to try to elaborate hypotheses that will fill in the details of their clearly *gappy knowledge* and to try to devise tests, either experimental or natural, that will either confirm or falsify these hypotheses. For one with the cognitive interests that move scientists, data contribute the process of inquiry through which our tentative explanations of facts come gradually to be less and less gappy. This, as we have seen, is how the paleontologist Horner moved to undertake further and more detailed inquiry in regard to his discovery in Montana of a massive dinosaur kill. *It is precisely this sort of process of inquiry that creation "scientists" do not undertake.* In this respect creation scientists are much like astrologers. These pseudoscientists too, and in contrast to astronomers, undertake no programme of research. In Kuhn's therms, there is no tradition of normal scientific research, nor any paradigm to guide such research. When problems arise, and often, as in the case of creation science, there are many, there is no attempt made to initiate a process of inquiry that could resolve the problems. This absence of a normal scientific research tradition shows that, in spite of the appropriation by creationists such as Morris of the term 'science' in 'creation science,' these persons are not genuinely *scientists* in the way that Horner is a scientist: *the creationist is not moved, as he or she would be were he or she genuinely a scientist, by a cognitive interest in discovering simple matter-of-fact truth.*

This, of course, is but another way of recognizing that creation science so-called is not science but pseudoscience.

c. Blondlot[23]

René Blondlot (!849-1930) was an eminent French physicist. He was a respected professor at the University of Nancy and a member of the French Academy of Sciences. The turn the century was a time for the discovery of various forms of radiation. Alpha, beta, gamma and X rays had all made their appearance. It was also a time of intense French-German competition; the memory of the Franco-Prussian war was still fresh, and X-rays had famously been discovered in Germany. In 1903 Blondlot announced a rival French discovery of "N-rays," named after his university.

Blondlot announced that his research indicated that N-rays were emitted by certain kinds of source, not only electric-discharge tubes but also ordinary sorts of household gas burner, as well as heated pieces of silver and iron. The sun was another source. Interestingly enough, the sources did not include the Bunsen burner. Various materials, including wood and thin strips of assorted metals such as iron and silver, turned out to be transparent to these new rays. Aluminum prisms were constructed to bend and focus the rays. The presence of N-rays could be detected because they increased the brightness of a spark and when they were directed at objects coated by luminous paint they increased the brightness of those objects. These effects, though small, could be discerned by the human eye. Indeed, it seemed that when N-rays were present in situations of dim light, they helped the human eye to see better.

Other French researchers took up Blondlot's lead and quickly confirmed his results. Research continued and Blondlot and other scientists reported other properties, many of which were quite strange and amazing.

At the same time, however, there were severe problems. While French scientists seemed to have no problem detecting N-rays, the results could not be replicated by scientists elsewhere, including Lord Kelvin and Sir William Crookes in Britain. One of those who was unable, in spite of efforts, to replicate Blondlot's results, was the American physicist R. W. Wood, a professor at Johns Hopkins University. Wood was not only an expert in optics and spectroscopy, he was also a relentless pursuer of fraud. This included exposed spiritualistic mediums. One of the latter claimed that he was in touch with the then dead British theoretical physicist Lord Rayleigh. To expose him, Wood framed some abstruse questions in electromagnetism for the medium to ask the ghost. As might have been expected, there was no response. With this background, Wood decided to visit Blondlot's laboratory.[24]

Wood witnessed various experiments. Several things became apparent. In the first place, in spite of the reports by Blondlot of spots of increased brightness indicating the presence of N-rays, Wood himself could not detect any such increase.

In the second place, when Wood's incapacity to notice the changes in brightness was explained as due to his insensitivity, he moved his hand in the experimental situation in a way that was supposed to interfere with the passage of N-rays and therefore with the supposedly detectable shifts in brightness on a screen. "In no case was a correct answer given, the screen being announced a bright and dark in alternation when my hand was held motionless in the path of the rays, while the fluctuations observed when I moved my hand bore no relation whatever to its movements."[25]

In the second place, the experiments were not foolproof. The set-ups were such as to admit experimental error. Since some of the experiments involved moving photographic plates for set periods of time, the possibility of bias on the part of the experimenter had not been eliminated.

In the third place, Blondlot had contrived a device for detecting by perceived shifts in brightness N-rays that had been refracted by an aluminum prism. This sort of experiment was conducted in a darkened room. During one such experiment, Wood surreptitiously removed the prism, the most important element in the experimental set-up. Wood reported that "The removal of the prism … did not seem to interfere in any way with the location of the maxima and minima in the deviated ray bundle."[26]

Finally, Blondlot and his associates claimed their eyesight was improved by N-rays. The greater acuity was stimulated by a steel file acting as an N-ray source. In the darkened room in which the experiment was performed, Wood, again without the knowledge of the experimenter, substituted a piece of wood for the file. Wood reported that "the substitution of a piece of wood of the same size and shape as the file in no way interfered with the experiment."[27]

What Wood managed to do in each of these latter two cases was create *experimental controls*. These controls established that the instruments used by the French physicists did not perform as Blondlot had claimed.

The failure of Kelvin and Crookes to replicate Blondlot's results showed that something was wrong. It implied that the observed results were due to some idiosyncratic feature of Blondlot's laboratory at Nancy. Wood had succeeded in bringing out the relevant causal factor: the observer was seeing what he expected to see, that is, what he expected to see given his favoured theory, that is, what would confirm his favoured theory.

Blondlot and his colleagues nonetheless continued even after Wood published his report to claim that they could detect various shifts in brightness which they alleged were indicative of the presence of N-rays. The observer, they said, had to acquire the ability to notice these changes, just as the impressionist painter had to train himself to notice subtle shifts in brightness and colour tone in a landscape. Some people might be such that they never could acquire such skills. Perhaps Wood, they implied, was such a one.

The problem, then, according to Blondlot and his disciples was the sensitivity of the observer rather than the validity of the experimental set-up. A certain chauvinistic tone entered the debate. Proponents of N-rays argued that only Latin races possessed the sensitivities (intellectual as well as sensory) necessary to detect manifestations of the rays. It was alleged that the perceptual capacities of Anglo-Saxons were dulled by continual exposure to fog, while those of Teutons were dulled by the constant ingestion of beer.[28] Arguments of this sort, turning on the sensitivity of observers, have once again become familiar; advocates of extrasensory perception have discovered how such appeals can prevent the falsification of their supposed experimental results. In any case, by 1905 the number of papers in scientific journals describing N-rays and their effects had fallen to zero. By then only French scientists were defending N-rays, and with the exception of Blondlot himself they too were soon to see that N-rays had been debunked.

In retrospect it is clear that Blondlot and his colleagues seeing not what was there in the phenomena but what they expected to see. Blondlot's observations followed a clear pattern. In the case of the experiment involving the prism, if Blondlot believed that the prism was in place then he saw changes in brightness. It was similar in other experiments. *What was observed depended not upon the apparatus being correct but upon whether the observer* **believed** *it to be correct.*

Blondlot and his associates were not lying about what they saw. They were not engaged in some sort of hoax. They didn't imagine their experience. It was just that their strong belief in N-rays affected how they saw things. In the experiments they devised, there were no N-rays nor, more importantly, no sifts in brightness, there to be detected. The French physicists simply misperceived the situation, imagining that there were shifts in brightness when there were not. It was their beliefs that produced, unbeknownst to them, this misperception.

There are two points that are to be made.

One. For the trained observer, such misperception is unlikely to occur in ordinary situations. We do not misperceive that the room becomes brighter when we turn the light on. But the events that Blondlot was attempting to describe occurred at the *threshold of perception*. The problem is that *visual judgments of light intensity are notoriously unreliable*. Blondlot and his associates did not carefully enough allow for this fact, and this opened the way for them to *misobserve* what was going on. They themselves were aware of the problem created by the fact that accurate observation is an acquired skill. That is why they could try to explain away some of Wood's concerns as due to an insensitivity on his part. Nonetheless, they in fact did not sufficiently take it into account.

Facts are there in the world, there to be observed. It is by means of observation that they come to be introduced into the process of scientific inquiry, where they are used as the basis of further inquiry. *If the process of observation through which the data are to be introduced into the process of inquiry is itself unreliable, then those results ought not to be admitted into the process of inquiry as determining matters of fact.* In particular, *data acquired by observation conducted at the threshold of perception ought not to be introduced into the process of inquiry as a basis for ongoing research.*

Two. The same point can be made in regard to the apparatus used to effect the observations. The use of tools in observation is a skill that is acquired. Both instruments and those that use them have to be tested, and the observers will require a period of training and discipline. At the beginning of the learning process, it will not be clear exactly what is the significance of what one perceives as the result of the process. If several persons are involved in the process, *there will be period of negotiation as it were before it is reasonable to count what has been observed as a fact to be admitted into the process of inquiry as the basis for further and ongoing research.*

Blondlot had proposed that his observations had provided facts before this sort of process of negotiation had come to its reasonable end.

There is another moral that needs to be drawn, however. Blondlot had been too ready to try to transform his observations into fact. Error was the result. But the method of science is one in which hypotheses, including the hypotheses that are part of the observational process, are put to the test, and either confirmed or rejected. Blondlot, as a good scientist, submitted his results to the process. The process found them wanting: it tested the relevant hypotheses and rejection was the outcome. *Once again we recognize how the scientific method as a self-correcting process leads to the discovery of error and thereby serves the cognitive interest of science in matter-of-fact truth.*

d. Observers as Instruments

The example of Blondlot illustrates that we commonly see what we expect to see or what we believe to be there even when it in fact is not there to be seen.

Past experience also affects how we perceive things. Thus, we often perceive an object as having a certain colour because we know from past experience that the object is supposed to be that colour — and we do this even when it isn't the colour past experience tells us it is supposed to be. In one experiment well known to psychologists, subjects were shown cutouts of trees and donkeys. The subjects perceived these as green and grey, just as they would expect them to be. They so perceived them even though all the cutouts were made from the same green material and lit by a red light that made them appear grey.[29]

Size is another constant in perception. People perceive the size of familiar objects as roughly the same whether they are nearer or closer, even though the image on the retina becomes smaller as the object recedes into the distance. In other words, we perceive the object as constant in size in spite of the changing size of the image, the immediate sensory stimulus, because we have learned that the size or shape of objects is normally constant, and therefore does not change with distance.

The capacity to recognize the correct size of things no matter the distance is one that is learned. There are groups that live in thick jungle. In this environment, where the only objects that can be clearly distinguished are all only a short distance away, the members of these groups have no opportunity to learn size constancy. When such people are taken out in an open plain they cannot distinguish between, on the one hand, large objects which appear small because they are seen at a distance and, on the other hand, small objects seen close up. The anthropologist Colin Turnbull actually performed this experiment, taking members of the Ba Mbuti tribe of pygmies from the jungle where they normally lived onto an open plain. These people identified distant buffalo as nearby insects. At first they refused to believe that the apparently small creatures were buffalo which were in fact twice the size of the buffalo which they were used to seeing. To convince them, Turnbull drove them closer to the buffalo. The animals appeared to become larger and larger as the Ba Mbuti approached them. The Ba Mbuti at first assumed that there were changes of size occurring, and became quite frightened, attributing the changes to witchcraft. "Finally," Turnbull writes of his companion, "when he realized that they were real buffalo he was no longer afraid, but what puzzled him still was why they had been so small, and whether they *really* had been small and had so suddenly grown larger, or whether it had been some kind of trickery."[30]

Under these circumstances it is rather amazing that our perceptual judgments are as good as they are. We have in the first place the sensory input that creates the proximate perceptual stimulus such as the image on the retina of the eye. This is then transformed by the eye-brain connection. Then the result of this process has to be compared through memory with data acquired in the past. The apparatus which does all this is extremely efficient but extremely complicated.

To transmit a television picture by wire a special cable called a coaxial cable is necessary. This sort of cable is equivalent to at least 500 telephone lines that carry voice alone. A television image cannot be sent down an ordinary phone line; there is simply too much information to fit. The nervous system is something like this, transmitting information from the eye to the brain. The "wires" that do this are known as neurons. These are very small, however: it takes tens to hundreds of neurons to do the work of one telephone line. And so, if the brain is to "receive" the information using only neurons, the distal stimulus must cause the image on the eyeball as the proximate stimulus, and the information in this image must somehow be broken up into wee bits, encoded in electrical impulses, transmitted down as many as several hundred thousand neurons, and then reconstituted as a perception of the distal stimulus. This must be done in both eyes so that we view things in stereo. In perception we are conscious of structured things in a structured environment. In becoming conscious of things, we also identify them by comparing what is seen to what we remember and what we have learned about the world. Given the complexity of the process of perceiving, given the fact that memory is not perfect, and given the fact that sometimes we learn what is not really the case, it is hardly surprising that sometimes things go wrong and that we sometimes perceive things that are not there.[31]

And so, where things are vague, we tend to fill in details according to patterns familiar to us. We may not know what we see, but we tend to see what we know. So 100 years ago the American astronomer Percival Lowell saw through his telescope some vague patterns on the surface of Mars. In observing these, he filled in details. The result was the claim that canals had been observed. There are other cases that are clearly of this sort. Recently several persons in Cape Breton Island perceived an image of the Virgin on the outer wall of a Tim Horton's doughnut shop. The observers found in their experience definiteness that was no doubt not there in the things themselves but put there by their own hopes, beliefs, and expectations. Lowell's misperceptions did not create an instant religious shrine as did the misperceptions at Tim Horton's in Cape Breton, but there were other effects. The idea of "men form Mars" had been given a basis in scientific fact; it is an idea that we live with yet, even though the basis in observation has completely been eliminated: Lowell was just wrong, as wrong as Blondlot.

Lowell, like Blondlot, had been making observations at the threshold of perception. In both cases, vagueness was made more definite by expectations. There are other circumstances which equally lead us to misperceive our environment. It is harder than we think to say just that we do not know. And if we are frightened for some reason or other, we are liable to read the situation as being the same as one we remember even though the latter is really dissimilar in various important respects. Or if we are in a situation which for whatever emotional reasons we want passionately to be in a certain way, we are liable not to notice that it lacks features we want to be there. Selective attention can lead us to overlook structures and events that really are there. These sorts of things happen especially when our most cherished beliefs are at stake. If we really want ESP to exist, if we really want there to be people with psychic powers, if we really want there to be N-rays, then we are liable to misperceive situations and discover evidence that confirms those beliefs or we are liable to misperceive other situations and overlook evidence that disconfirms those beliefs. For similar reasons we are liable to find ourselves agreeing with vague but overall favourable descriptions of our personality that are presented by astrologers, palmists or psychics.

The point is that the human observer is like an instrument, one that works quite well generally but sometimes goes wrong. It is necessary to test this instrument and to find the circumstances in which it works well. *Only in those circumstances where perception is reliable ought we to rely upon observation to introduce facts into the process of scientific inquiry.*

And so, we ought not allow observations made at the threshold of perception to count as adding data to the body of scientific knowledge. We ought not allow observations made in conditions of emotional excitement to count as adding data to the body of scientific knowledge. We ought not allow observations made in conditions where we have not yet learned accurately to discriminate details to count as adding data to the body of scientific knowledge.

e. Constructing UFOs
Recall the definition of a UFO: it is

... the stimulus for a report made by one or more individuals of something seen in the sky (or an object thought to be capable of flight but seen when landed on the earth) which the observer could not identify as having an ordinary natural origin, and which seemed to him sufficiently puzzling that he undertook to make a report of it to police, to government officials, to the press, or perhaps to a representative of a private organization devoted to the study of such objects.[32]

There are such private organizations. In fact, there is a whole UFO industry. It involves a large body of semiprofessional persons, such as publishers, TV producers and the like, even a few professors. These persons all earn their income on the basis of the supposition that there are UFOs not simply in the sense of being unidentified phenomena appearing in the sky but in the sense of being vehicles navigated by extraterrestrial intelligences. The income generated by this industry is considerable, more indeed than many large scientific granting agencies. There is here a vested interest in maintaining the notion that UFOs really are piloted by extraterrestrials. There is therefore a vested interest in making sure that there will be continued observations of UFOs that meet the condition imposed on sighting by this industry, that the observations somehow testify to what is sighted being vehicles of alien origin.

The existence of this sort of entertainment-industrial complex ensures that people will know what to look for. Already to say that the visual stimulus one has sighted is an unidentified flying *object* is to identify it as an *object*. Being an *object* is to be something substantial and enduring. These are features which are likely not to be clear from the stimulus itself, particularly if it is viewed late at might and in obscure and lightly inhabited portions of the world. But given the way in which we almost all hook into the UFO industry, it is as *objects* that we identify and conceptualize the stimuli. And not only do we expect to see an object when we have a UFO experience, we also have come to expect these things to be *disc-shaped* or *cigar-shaped*. The visual stimulus is likely hardly so definite, but that is how we experience it, that is how we identify it, finding structure where and of the sort we have come to expect.

Consider first an incident widely reported in the press.[33] In this case a pilot of a private plane near Seattle on the night of July 7, 1968 encountered *nine* UFOs. He was headed in a westerly direction when he confronted head on a squadron of nine objects apparently headed directly towards him on a collision course. To avoid collision, the pilot turned sharply to the right. At the same time the UFOs also turned to their right, in a way that would equally avoid a collision. The action of the pilot was a case of "intelligent control"; the action of the objects seemed similarly to be evidence of the same sort of "intelligent control."

At first the pilot thought that he had encountered a formation of military aircraft. But on radioing the Seattle air control tower, he was surprised to learn that there were no such craft in the area. The night was clear and moon lit, with good visibility, and so the better to view the strange phenomenon, the pilot began to circle the objects. While he did this, he saw the UFOs begin to fire what appeared to be rockets in what seemed to be some sort of attack on Seattle itself.

Then of a sudden the objects, to the pilot's distress, once again started heading directly towards his plane. He quickly turned on his landing lights in an attempt to warn the opposite craft of his presence, to avoid the impending collision. The pilot reported that the UFOs once again exhibited intelligent control by first stopping their flight and then backing up.

Finally, the UFOs one by one began to disappear. When they had all vanished, the pilot set out again for the Seattle airport. But then he noticed that his gyrocompass was 170° in error, that is, almost completely backward. It seemed to the pilot that the UFOs had emitted some sort of mysterious force that could introduce gross errors into his gyrocompass. The badly shaken pilot later said, "I wouldn't believe it myself, but I saw it."[34]

The incident was reported in banner headlines in the Seattle paper the next morning, July 8. But on subsequent days, the story had disappeared from the front pages. Nothing further would seem to have happened.

That is, nothing further would seem to have happened unless you had happened to chance on a story buried on page 12 on July 9. It was here, hidden in back pages, that you would have found the explanation of the attack on Seattle which had so terrified the pilot and several other local citizens. As it turned out, the whole thing was the result of a hoax by several teenage boys.

The UFOs were in fact thin plastic garment bags transformed into hot air balloons by small candles that both provided the hot air that lifted them and the illumination that made them visible. For added excitement the boys had attached railroad flares with long fuses. It was these which, when they eventually went off, were perceived as the "rockets" noticed by the pilot and the other observers.

The "intelligent control" was simply nothing more than the balloons responding to the shifting wind conditions in the atmosphere. In the darkness of the night it would have been difficult for the pilot, who at this point was very frightened, to notice the exact changes of path; what were in fact more or less random shifts acquired the appearance of precise responses and "intelligent control." One is reminded how Lowell transformed vague data into "canals" on Mars.

What of the mysterious force that affected the plane's gyrocompass, creating the $170°$ error? In fact there is no need to appeal to such "forces" to account for that event; a perfectly reasonable naturalistic explanation is ready to hand. The gyroscopes in private airplanes are relatively inexpensive and are not "slaved" to the magnetic north as are the more expensive gyroscopes in larger aircraft. As any pilot experienced with small planes understands, as does any gyroscope designer, if the plane moves in circles for any extended period of time, the instrument will exhibit what is sufficiently well known to have its own name, "gyro turn error." Since the pilot had flown in circles for some time, the better to view the "UFOs", what happened to his gyrocompass is more easily and plausibly understood as "gyro compass error" than as mysterious forces directed at him by extraterrestrials from what were in the event not space craft after all but only fakes. Certainly, that alternative has not been eliminated, and until it is the conclusion that the event was the result of "mysterious forces" is unjustified.

But not every case where UFOs are observed by intelligent and reasonable people is the product of hoaxers.

Here is one such case. On March 3, 1968 a UFO was observed by multiple independent witnesses, all intelligent and educated, all honest and upright citizens, in several midwestern United States.[35]

In Tennessee, three of these observers, including the mayor of Nashville, reported a light in the night sky moving rapidly towards them. They described it as passing overhead at a height of about 1000 feet, and identified it as a huge metallic craft moving with complete silence and with orange-coloured flames shooting out from its tail. It was further described as having many square-shaped windows that were lit from inside the object. One of the witnesses, in a report to the U. S. Air Force, stated that the observed craft was shaped "like a fat cigar ... the size of one of our largest airplane fuselages, or larger." One of the women of the group reported that the three had later discussed what they had seen and speculated that it might have been "a craft from outer space."[36]

The same UFO was spotted about the same time by six people near Shoals, Indiana, about 200 miles north of Nashville. In their report to the U. S. Air Force, the observed object was described as cigar-shaped, as moving at treetop level, as shooting a rocket-like exhaust from its tail, and as having many brightly lit windows.

Again at about the same time, the same UFO was sighted by two people in Ohio. However, in their report they described three luminous objects rather than one. One of these witnesses, a business executive who insisted in his report to the USAF that he was not a "kook," indicated that the UFO or rather three UFOs appeared to turn or manoeuver as a team in a way that suggested "intelligent control."

Another Ohio witness who reported the object was a school-teacher from Columbus with a background in science and several academic degrees. See too saw three objects flying in formation. She had had a pair of binoculars with her and she used these to observe the objects more closely. In her detailed report to the Air Force she described them as shaped like "inverted saucers."

The UFOs were also described as having affected both her dog and herself. Her dog fell to whimpering as if it were "frightened to death," while she herself fell into a deep sleep upon returning to her house. It was as if some strange force emitted by the UFOs had mysteriously affected this mature well-educated woman.[37]

So, was it three objects or one? Was it shaped like a saucer or like a cigar? What, indeed, was it?

In this case, we do know what it really was that all these people, and many others, saw that might in March, 1968. The North American Defence Command (NORAD) keeps track of objects in space, using radar and other means. The object(s) reported on the date in question were in fact flaming fragments from a Russian rocket that had been used to launch a Zond-4 spacecraft on a translunar trajectory. When this large rocket re-entered the atmosphere at high speed the friction heated the fragments into incandescence. NORAD had tracked this object along the same southwest to northeast trajectory that had been noted by the witnesses. It is clear that this is what the several observers actually witnessed.

When the observers in Tennessee and Indiana saw the flaming objects unlike anything in their previous experience, they concluded that there were windows in a giant craft. Since this craft was noiseless it could not be an airplane. It was a UFO which, since it was a UFO, had to have the cigar shape or saucer shape customarily ascribed by the industry to UFOs. What the motions were that were observed by the Ohio business executive is not clear; perhaps they were simply an optical illusion. In any case, the trajectory of the fragments was not the result of any sort of "intelligent control."

The case of the woman's dog is interesting. The woman at one point in her letter reports that the night was "clear and cold," while elsewhere in the same letter she noted that the dog *hated* cold. There is no firm answer to why the dog whimpered, but there is no need to introduce mysterious forces; one can more plausibly speculate that the dog had simply been kept out too long by the women entranced by her UFO sighting and was whimpering to get home to where it was warm. Again, until the latter alternative hypothesis is eliminated, one cannot conclude that the dog's behaviour was caused by mysterious forces.

Nor is there any need to introduce mysterious forces to try to explain the fatigue of the woman; one can more plausibly hypothesize that she was simply exhausted and emotionally drained from the experience of having seen a real live flying saucer. This hypothesis cannot after the fact be directly confirmed. But given that we do have considerable experience of human behaviour and know on this basis that people do often respond to emotionally charged events with fatigue, then this background knowledge provides the hypothesis with a degree of probability that is totally lacking for the effectively non-empirical hypothesis of *mysterious* forces.

But what of all the interesting details — the giant craft, the inverted saucers, the square-shaped windows,

the metallic cigar-shape? As one commentator indicates, they were all in fact simply the result of perceptual construction, non-existent details added as it were by the observers in order to make the vague sensory data more definite with details that they had come to expect should be there.

> These additions and embellishments were purely the creation of the witnesses' minds, not because they were crazy, drunk, or stupid, but because that is the way the human brain works. It can be said that these witnesses did perceive what they said they did. This doesn't mean, however, that what they perceived was the same as what was really there. Note, too, how inaccurate was the estimate of the object's altitude .... [Witnesses] estimated about 1,000 feet while, in fact, the reentering rocket as miles high and scores of miles away. This type of gross inaccuracy frequently occurs when one sees a light in the sky with no background, as in the case at night. Under these circumstances, the many cues the brain uses to judge distance are not present, so no accurate basis for judgment exists.[38]

Once again, recall Lowell and the "canals" of Mars.

Perceptual error of the sort just described, in which the mind adds details that it expects to find, can occur even in broad daylight. Here is a case involving pilots of commercial aircraft who can be expected to have training to give them a certain expertise in observing objects in the sky.

This event occurred on June 5, 1969 on American Airlines flight #112 over St. Louis as the plane headed east on a flight from San Diego to Washington. The plane was cruising at thirty-nine thousand feet in broad daylight when, at about 6 p.m., it encountered a squadron of UFOs in what turned out to be a very frightening experience.[39]

The pilot's seat was occupied by a senior Federal Aviation Administration (FAA) air-traffic controller, who was flying as a cockpit observer, the pilot at this time having gone back to sit for a period in the passenger compartment. The co-pilot suddenly yelled out, "Damn, look at that." The controller looked where the co-pilot indicated and later reported what he saw: "There it was — a flight of four whatever-they-were flying in square formation." There was a larger hydroplane-shaped UFO flying at lead spot in the formation, followed by three smaller objects. They seemed to be propelled by rocket engines that emitted long tails of blue-green flame. The objects in the formation were all headed in a westerly direction *and seemed to be in on a collision course with the jetliner*.

However, moments before the UFOs would apparently collide with the jetliner, they veered away in apparent evasive action that exhibited "intelligent control." It was a close call, however: the FAA controller estimated that the objects had come within 300 feet of the airliner.

The visual sightings were confirmed by others. The co-pilot radioed the control tower in St. Louis to inquire if it had noted any unidentified targets on its airport radar. It radioed back to say that there were two unidentified targets to the west of the airliner.

This radar confirmation was supported by further visual sightings. A few miles west of the American Airlines jetliner there was an eastbound United Airlines plane cruising at thirty-seven thousand feet. They had monitored the conversation between the American Airlines flight and the St. Louis control tower, and shortly thereafter reported that "We see it, too."

Then a few miles to the west of the United Airlines plane there was an Air National Guard jet fighter plane cruising eastbound at forty-one thousand feet. The pilot of this plane had also monitored the communications with the control tower in St. Louis, and a few minutes after the interjection by the United Airlines plane, he broke in to report, "Damn, they almost got me!" He reported that the UFOs seemed to headed directly for his plane until at almost the last moment they exhibited "intelligent control" and abruptly altered their course to move upwards out of the plane's path.

In addition, two pilots at the airport in Cedar Rapids, Iowa, saw the object or objects and reported that they flew over the east-west runway at an altitude of only one thousand feet.

In this case, the observations took place in broad daylight and were made by experienced flight crews. Even so, it is not necessary to infer that the objects really were craft under the "intelligent control" of mysterious beings, perhaps from elsewhere.

In fact, in this case it has been determined with certainty what these experienced flight crews had

observed. As it turned out the objects had actually been photographed by an alert newspaper photographer named Alan Harkrader in Peoria, Illinois. This was 125 miles north of St. Louis, but the photographs make clear the real nature of the objects that had been observed. There were in addition many persons on the ground who were able to report on the same phenomenon that the flight crews had observed.

> The large "UFO" in the lead was really a giant flaming meteor/fireball, coming out of the east and headed west, which was also seen by numerous ground observers in Illinois and Iowa. Flaming fragments periodically would break off and fall into trail, to form a "squadron." From these many ground observations and the photo taken in Peoria, scientists have been able to establish the approximate flight path of the fireball.[40]

What the commercial airline crews and Air National Guard pilot all reported as being in a near-collision close to St. Louis was an object that 125 miles to the north. The Cedar Rapids pilots had equally misjudged the location of the object; they had seen the same fireball, but it was 100 miles to the south. Moreover, the objects had an altitude of above forty-one thousand feet, even though the pilot of the Air National Guard plane had reported that the UFOs had passed directly above him.

There remains the radar sightings made by the St. Louis control tower. There is a reasonable hypothesis about what had been noted. Today airport radar automatically displays the identity and altitude of any aircraft in its area of surveillance. This was not so at the time of the incident in 1969. At that time, aircraft that were flying over but not intending to land were not required to identify themselves to the control tower. It was only planes intending to land that would be identified by small plastic markers on the radar screen. We may suppose, then, that when the air traffic controller in the tower heard the report of the American Airlines liner of the UFOs he hurriedly looked at his screen and noted two unidentified objects to the west of the liner. These would be there on his screen as the images of the United Airlines jetliner and the Air National Guard airplane, images that would be unmarked because these planes were not coming in to land. The moment was one of great excitement for the planes and was undoubtedly equally so for the control tower. One would not be surprised that in that moment of excitement blips on a screen caused by overflying aircraft would, because of their lack of identification, be transformed into images of UFOs. At any rate, here is an alternative hypothesis, and, once again, unless it is eliminated then one cannot safely conclude that the radar screen sightings were of some sort of mysterious flying objects

UFOs are many things. These include hot air and weather balloons. They include meteors and fireballs. They include actual aircraft performing unusual flight-test manoeuvrers. They include swarms of fireflies. They include atmospheric phenomena such as sundogs. They include celestial phenomena seen through haze and clouds. They include optical illusions created by unusual atomospheric conditions. They include outright hoaxes. There are dozens of visual stimuli waiting to be converted into UFOs, especially at night.

No one is immune from the possibility of constructing UFOs from vague perceptual impressions — not pilots, not astronomers, not honest people, not the educated and intelligent, not otherwise reliable witnesses, not the pillars of the community. This is not surprising. If the Ba Mbuti have difficulty identifying the characteristics of buffalo when these are seen at a distance on a plain, then why should be we be any different when it comes to identifying the characteristics of things seen in the sky. As the Ba Mbuti have had no experience that would enable them to identify the characteristics of things seen at a distance on a plain, so we have had no experience that would enable us to identify the characteristics of things seen against nothing more than a background of the sky. Meteors and fireballs often seem much lower than they really are. Observers of things in the night sky simply have no means of estimating their size and distance; there are no cues that the might have learned on which to base such judgments, and this can lead to errors of the grossest magnitude. Neither jungle walls nor celestial backgrounds provide visual clues that enable one accurately to locate and identify the sizes and shapes of objects that are visually stimulating us.

But where the Ba Mbuti have been culturally determined to think of the strange things they see on the plain in terms of witchcraft, for us the determination has been effected by the UFO industry and the way in which reports are treated in the news media. Recall how the Seattle incident was reported on the front page of the paper, the revelation that it was a hoax buried deep on the inside. TV programmes such as "The X-Files" and uncritical "reportage" on things such as the so-called Rosswell incident prepare us for the discovery of UFOs. UFOs are constructions deriving from the imagination of the viewer and nothing more. To be sure, not every UFO sighting has been explained in ordinary terms. But given what we do know, we should expect these events in the unexplained residue to also have ordinary explanations. Past success in discovering these ordinary explanations makes it perfectly reasonable to reject any claim that they can be explained only by invoking that they are alien craft piloted by ETs.

The personal experience from which such claims derive is in fact a poor guide to the reality of what we seem to see. It is appeal to this sort of unreliable personal experience that forms the basis of many of the pseudoscientific claims that one finds as the basis of the UFO industry.

f. Personal Experience

Personal experience of phenomena viewed under poor conditions, e.g., at night and in the sky, or at the threshold of perception, is a poor guide to separating what *really is* from what *merely seems to be*. There are other sorts of personal experience which are equally inappropriate for providing data which are worthy for use in explanation and prediction and in the on-going process of scientific inquiry.

Consider the following example where it is argued that personal experience provides support for the existence of psychokinesis (PK). These experiences occur in what have been called "spoon-bending parties" or "PK parties." These are often organized on a profit-making basis. The fifteen or so persons in attendance pay a fee for the opportunity and bring their own silverware. The leader who accepts the fees guides the attendees through various rituals and encourages those present to believe that if they cooperate with them they can achieve a mental state in which the spoons and forks that they have brought will soften and bend through the non-material agency of their minds.[41]

The 1988 report of the National Academy of Sciences makes clear that although there have been many such parties and many claims of spoon bending by non-material forces; in fact no claim of observed PK has been validated.

> Since 1981, although thousands of participants have apparently bent metal objects successfully, not one scientifically documented case of paranormal metal bending has been presented to the scientific community. Yet participants in the PK parties are convinced that they have both witnessed and personally produced paranormal metal bending. Over and over again we have been told by participants that they knew that metal became paranormally deformed in their presence. This situation gives the distinct impression that proponents of macro-PK have consistently failed to produce scientific evidence, have forsaken the scientific method and undertaken a campaign to convince themselves and others on the basis of clearly nonscientific data based on personal experience and testimony obtained under emotionally charged conditions.[42]

Consider those conditions. J. Houck, who originated the PK parties, makes every attempt to exclude critics on the grounds that scepticism and attempts to conduct objective observations will interfere with, or even prevent, the paranormal phenomena from appearing. It is claimed, moreover, that those phenomena will appear only if the participants are in a state of peak emotional experience. To this end they are encouraged to shout "Bend!" at their silverware and to "disconnect" from their bodies, that is, to deliberately avoid looking at what their hands are doing. If they see that their spoon seems to be bending they are encouraged to jump up and down and scream to direct the attention of others to the event. The objective is to create a state of emotional chaos and pandemonium throughout the party. It is precisely this that makes the situation hardly one in which observation is scientific.

> A PK party obviously is not the ideal situation for obtaining reliable observations. The conditions are just those which psychologists and others have described as creating states of heightened suggestibility and implanting compelling beliefs that may be unrelated to reality. It is beliefs acquired in this fashion that seem to motivate persons who urge us to take macro-PK seriously. Complete absence of any scientific evidence does not discourage the proponents; they have acquired their beliefs under conditions that instill zeal and subjective certainty. Unfortunately it is just those circumstances that foster false beliefs.[43]

In PK parties the heightening of emotion is explicitly cultivated. This is because in fact the role of emotion in the experience of the paranormal can hardly be exaggerated.[44] It figures in such experiences even outside the contrived environment of the PK party. Nor is this the only sort of case in which a paranormal interpretation of events is created by an emotional experience. It is also often true that some event will because of its emotional impact strike us as paranormal. Everyone, even those who reject as unfounded all claims of the paranormal, do from time to time have odd experiences which can be taken to be instances of the paranormal. You are driving along with your partner and think of some event. Just as you are about to remark on it to her, she mentions it first. It seems almost to be a case of mind-reading or telepathy. In fact, if we reflect on these things, we should expect events of this sort to happen. When partners have lived together for a while, they being to share habits of thought. Often enough, then, some event in the environment, a billboard for example, is likely to provide an unconscious cue that will trigger the same thought in both. Vivid dreams occur fairly regularly, and some of these are very emotionally charged. Such a dream in which one experiences Uncle Abner's death can be very disturbing. If, coincidentally, Uncle Abner really does die, it will be hard to resist the inevitable thought that one had had a premonition of the tragedy.

Such experiences are the grounds that many cite for belief in the paranormal. "True believers" in the paranormal almost always report having had an incident sometime in their life which "confirmed" that belief.[45] How do you explain the fact, one is asked, that I dreamed of Uncle Abner on the night that he died? To which the answer is that it can't be explained. If it is a coincidence then for that reason it can't be explained and if it is not a coincidence then we really don't know the causes and for that reason it once again can't be explained. When we have such experiences the best thing to do is to set them to one side and resist the temptation, however strong, to count them as data for our interpretation of the world. Nor should we allow as data any such reports from others, honest though they probably are. It is experience itself that has made abundantly clear that experience of that sort is not to be relied upon!

Parapsychologists themselves have long ago learned that it is unwise to take seriously anecdotal reports and personal experiences. During the early days of parapsychological research, in the heyday of the Society for Psychical Research in the late 19th century in Great Britain, much time and effort was put into the investigation of such reports. But invariably, in those cases where it was possible to check the details of the story against objective evidence, such glaring discrepancies were found that no part of the report could be taken seriously. This was of course true of the many hoaxes that were uncovered, but it was equally true of the reports of individuals whose honesty and intelligence were above reproach.[46] This is why those who are still serious about investigation of the paranormal rely on controlled experiments and statistical tests.

There are often reasons for doubting what we seem to experience. These include poor observational conditions — poor lighting, threshold conditions, vague stimuli, and so on. There are also conditions that can impair us physically — alcohol, fatigue, poor eyesight, and so on. There are also conditions such as those of heightened emotions such as fear or enthusiasm in which we fall into beliefs with such subjective certainty that no amount of objective reflection can dislodge them.

Science attempts to the best of its ability to get around the subjective limitations of personal experience. It tries to ensure that, where possible, there is replication. It attempts to replace subjective judgments with "pointer readings," objective measurements. It insists that, wherever possible, data provided by one person receive corroboration from others. Public scrutiny, not private validation, is required. And so on.

In situations where subjective limitation might be causing distortions in our experience, our personal evidence is necessarily flawed and is to be rejected as a possible source of data worthy of acceptance for purposes of science. In those circumstances we might have experiences that can be enjoyed, experiences that are genuinely inspiring, experiences from which we can learn much. These experiences might even be such that they indicate the need for further investigation of an empirical nature, according to the methods of science. But in the cognitive interests of science the claims that such experiences actually produce credible data ought simply to be set aside. In the absence of further evidence, the best that one can do is to accept that "I don't know."

(ii) Relativism

As Clemenceau is reported to have said, history will not record that it was the Belgians who invaded Germany in 1914.

There is a view that has become popular that runs counter to this. It is the view that truth is a matter of what you believe, that my truth is my truth and that your truth is your truth, and that they are both equally true — even when they contradict. We have all no doubt been halted in an argument with the proponent of some New Age theory who states that "This is *my* truth, and the fact that your have your truth and that it disagrees is irrelevant." It is hardly a new position. The ancient philosopher Protagoras (c. 490BCE - c. 421 BCE) is held to have defended such a subjectivist account of truth when he declared that "man is the measure of all things, of things that are that they are, and of things that are not that they are not."

This is a position against which in the end it is impossible to argue. If you argue the defender will simply reject your claims on the grounds that that is not *my* truth, it may be *yours* but it is not *mine*. New Agers are often inconsistent on this point, trying to convince those who disagree with them that the details of their particular New Age belief system really are credible, that there is evidence, for example, that they have been in communication with the Ice Age warrior Hunk Ra. Given the subjectivism, however, it is easy to understand why the proponent of such a system sometimes lapses from argument into dogmatism: there is nothing else to which he or she can resort.

At the same time, however, the subjectivist, New Age or otherwise, cannot reject someone else's defence of objectivism. The belief that subjectivism is false would be just as true as the belief that it is true.

Indeed, if subjectivism were true, there could never be any disagreement. At least, it would be pointless. In any case, we would all be infallible; error, that is, erroneous belief, would be impossible since the mere fact that we believed something would make it *ipso facto* true. However, the intellectual and moral superiority often claimed and felt by the New Ager is entirely out of place: on their view, no belief is ever superior to any other.

But of course, Clemenceau is correct: there is objective truth. Coming to the meal is followed by our leaving it and not the other way around. It was the cup that fell and broke and not the saucer. Apples nourish and pebbles don't. Fire warms and martinis make you drunk. The pavement will support you, the water in the lake won't. Objectively, it's a pretty sure bet that the sun will rise tomorrow; what New Ager would gamble otherwise? And much as they might dislike it, New Agers are fallible. Like everyone else they make their mistakes, sometimes dialling a wrong number, sometimes betting on the wrong horse. And some, like some "psychics," have been known to go bankrupt.

We need not spend much time, then, with this sort of relativism. As a target either it is easily refuted or it eludes refutation only to be ignored as hopeless. Logically speaking, it is simply not possible to spend much time with it. But there are other more subtle forms of relativism at which we must look.

a. Science and the Sociology of Knowledge

Some argue from the viewpoint of sociology that science does not have a privileged status. Thus, Barry Barnes and David Bloor have argued that

Far from being a threat to the scientific understanding of forms of knowledge, relativism is required by it. Our claim is that relativism is essential to all those disciplines such as anthropology, sociology, the history of institutions and ideas, and even cognitive psychology, which account for the diversity of system, of knowledge,

their distribution and the manner of their change. It is those who oppose relativism, and who grant certain forms of knowledge a privileged status, who pose the real threat to a scientific understanding of knowledge and cognition.[47]

On their usage, "any collectively accepted system of belief" counts as "knowledge." As they point out, "Philosophers usually adopt a different terminological convention confining 'knowledge' to justified true belief."[48] But different collectively accepted systems of belief will have different standards of justification and their argument is that we ought not grant privileged status to one of these systems of norms, and in particular not to that of science. The norms of the scientific method are just one set of norms among others and from the viewpoint of a sociologist ought not to be given privileged status.

In advancing this position, Barnes and Bloor are making a very important and reasonable point. Some philosophers have argued that there are two ways to understand change in the history and practice of science. On the one hand, there are explanations in terms of reason. On the other had, there are explanations in terms of social or other forms of cause. The latter are to be invoked where the former do not seem to be available.

There are cases in the history of science where theories have been accepted even where the evidence does not justify such acceptance.

Thus, for example, during the 19th century in Britain and America, and later by theorists in Austria, economists accepted as a law that wages for working classes would inevitably under competitive pressures come to equilibrium at the subsistence level.[49] This proposition was derived from certain assumptions. These were of two sorts. One set concerned the social institutions. It was assumed, among other things, that the social institutions were those of the time in which governments did not interfere in the workings of a free market, and in particular in the market for labour, and that combinations of labourers, that is, trade unions, did not exist. The other set of assumptions concerned the motivation of those in the market, that they aimed to maximize their profits or their wages as the case may be. The principle that the economists deduced, that wages would tend to subsistence levels, they referred to as the *Iron Law of Wages*. It was argued that attempts to alleviate the poverty of the working classes, to attempt to increase their wages above the subsistence level, were pointless because this was a *law*, and as such *could not be violated*: laws are patterns that are *exceptionless*.

This then became the basis for political action on the part of those who employed labour. The employers acted politically to minimize government efforts to institute regulations that would alter the labour in market, by legislating against child labour, for example, or by legislating hours of work. They also used their political power to prevent the legalization of trade unions. The theoretical basis to which they appealed to justify this political action was the claim that the subsistence level of wages was a *law* that permitted *no exceptions*.

If we think carefully about this, however, we see that the very fact that political action was needed to prevent governments from acting so as to bring it about that the wages of labourers were at a level higher than subsistence shows that the Iron Law of Wages was *not* a law in the scientific sense. The institutional changes against which they agitated would bring about wages that would be above the poverty line. This would produce an exception to the Iron Law of Wages. It would, in other words, produce a counter-example that would falsify the so-called Iron Law. But if it is false, if there really are exceptions, then it is *not* a law.

The logical point is that the Iron Law of Wages is a principle that correctly describes the market in labour only given certain assumptions. We have noted the sorts of assumptions that were made. These did indeed describe the conditions under which labour markets operated in Britain and America in much of the 19th century. But these conditions are by no means universal, and if they do not hold universally then neither does the Iron Law of Wages. And in that case it is not a *law*.

To seek what is as it were a violation of the Iron Law of Wages is to seek to raise the wages of labourers above the poverty level. To do that what is required is that the institutional conditions change to permit this, e.g., by making trade unions legal and giving them the power to negotiate levels of wages.

The 19th century economists argued that the Iron Law of Wages was a law. As such it could have no exceptions. This was then cited as grounds to prevent the institutional changes that would produce "violations." The very fact that there could be such violations established

that it was not in fact a *law* in the sense in which science. It was not, in other words, an *exceptionless matter-of-fact regularity*. The claim that it was a *law* was thus a claim that was *false*. They were not justified in referring to this "law" as a law since it in fact held only under certain conditions, conditions which did not obtain universally. That the conditions did not obtain universally was evident from the fact that they had to undertake political action to prevent the political action that would bring about changed social conditions, e.g., trade unions, that would produce a so-called violation, that is, non-subsistence wages.

The 19th century economists were thus making a claim that a certain principle was a law when in fact that claim was unjustified. Not only was their no evidence to justify that claim, their own actions in the political arena provided ample evidence that the claim was false.

One can see, however, what these economists were up to. They were in fact acting as spokespersons for those who employed labour, the capitalists, who had a vested interest in keeping wages as low as possible. It may not have been a law of nature that wages of labourers had to be at the subsistence level, but it was something that the capitalists *wanted* to be true. The economists provided arguments establishing that it had to be true, and that action to change it, action contrary to the vested interests of the capitalists, was pointless.

The economists, as spokespersons for this class, shared their interest in making the conditions described by the Iron Law so. They, too, wanted it to be true. Bergmann[50] has proposed, not unreasonably, that we call "ideological" those matter-of-fact claims which we think we accept because we have evidence that they are true, where that evidence is objectively inadequate, and where we in reality (mistakenly) accept the claims because we want them to be true. This makes the claim on the part of 19th capitalists and economists that the Iron Law of Wages is a *law* an ideological claim.

Here we have a clear example where science, or, more accurately, what practitioners claim to be science, is used to support the interests of one class over another, in this case the capitalist class over the working class. Ideological claims are often of this sort. In this way science becomes classist in the theories that it advances

Other examples of the same sort of thing come readily to mind. Various anthropological theories in the

19th century aimed to justify claims that the white race is superior to others including of course the black and the colonized. Such claims we now know to be ideological. But at the time they were used to justify institutions such as slavery or colonial expansion in the name of benevolence on the part of the so-called superior races. In this way science became racist in the theories that it advanced.

Or again, male physicians, again in the 19th century, developed physiological theories in which women were supposed to have weaker "constitutions" than men. These theories were then cited as justifications for preventing women from undertaking higher education or entering the professions alongside men. In this way science became sexist in the theories that it advanced.

The ideological nature of these theories has now been exposed. The method of science is one in which it corrects its errors. These corrections may not always occur immediately. But in the longer run, the inductive method has this effect. And so we have come to reject the classist, racist and sexist theories that people accepted in earlier ages.

There are no doubt other theories of a similar classist, racist, sexist or other ideological sort that we accept today. Given the fallibility of human nature, this is to be expected. And given the nature of ideology, we will not recognize the ideological nature of these views; we will — we do — continue to accept them under the impression — the mistaken impression — that they are scientifically justified. Only subsequently, as the method of science continues its self-correcting process, will these mistakes be uncovered and then corrected. At that point we will see that the acceptance was not based on the rational grounds of science but on some other basis.

We must of course carefully distinguish those theories that we accept on the basis of reasonable scientific evidence and those which we accept on other grounds. At the time that we accept those beliefs we cannot always say which is which. That is why we can fall into ideological claims which we mistake for genuine science. That is why others can honestly but mistakenly accept pseudoscience as science. But whether or not we can distinguish the two sorts at the time, they are for all that to be distinguished.

Some philosophers of science have gone on to argue that when it comes to explaining science, there are in fact two sorts of explanation. On the one hand, if a belief can be explained as being the result of the rational evaluation of the evidence available then that fact ought to be accepted as the correct explanation of the belief. On the other hand, if no such rational explanation is possible then we should account for the belief by appeal to social or some other form of cause, e.g., pressures of a classist, racist or sexist sort, that constitute the non-rational basis for accepting the belief. In the jargon, one often speaks in this context of an "internal/external" distinction. The philosophers with whom we are now concerned hold that *in providing explanations in the history and practice of science* one ought to seek an external account only if no internal account can be found.

This methodological principle has explicitly been adopted by Laudan, who refers to it as the "arationality principle":

> ...basically it amounts to the claim that *the sociology of knowledge may step in to explain beliefs* if and only if those beliefs cannot be explained in terms of their rational merits....Essentially, the arationality assumption establishes a division of labor between the historian of ideas and the sociologist of knowledge; saying, in effect, that the historian of ideas, using the machinery available to him [or, one presumes, her], can explain the history of thought insofar as it is rationally well-founded and that the sociologist of knowledge steps in at precisely those points where a rational analysis of the acceptance (or rejection) of an idea fails to square with the actual situation.[51]

Newton-Smith has put the point in a slogan: "Sociology is only for deviants."[52] The sociologist is to step in and offer causal explanations of cognitive decisions in science when and only when *there is some deviation from the norms of rationality*.

On this view, scientific rationality *has a privileged status*. It is this position that Bloor and Barnes are concerned to dispute.

Now, from the point of view of the practice of science, Bloor and Barnes are in fact on sound ground. Science itself is a practice that occurs in the world of ordinary experience. What happens in science are *matters of fact*. Hippocrates argued that epilepsy is something that happens to people and that it is wrong to give it a privileged status among diseases as "the sacred disease." It is to the contrary a disease among disease and the medical practitioner ought to seek out mater-of-fact causes for it. Bloor and Barnes are insisting, and rightly insisting, upon the same point when it comes to science. When one comes as a sociologist committed to the practice of science to apply the norms of science to science itself, then what one ought to search out are matter-of-fact causes. One should not distinguish some parts of science as privileged, and refuse to apply to them the methods of science, restricting the methods of science to the investigation and explanation of the irrational, or, in Newton-Smith's terms, the deviant.

The point upon which Bloor and Barnes are insisting is the Hippocratic point. And they are surely correct. It is wrong to suppose as some do that science is "the sacred enterprise." Jarvie, for one, is hardly adopting the stance of science when he states that

> Perhaps, when we do science, and even more so mathematics, we participate in the divine...[It is] awe at the transcendental miracle of mathematics and science that has moved philosophers since Ancient Greece.[53]

Laudan and Newton-Smith are not so extreme in their statements, but their views are of a piece with those of Jarvie: there is something about science, on their view, that makes it out of this world. Barnes and Bloor are right to object. Science is to the contrary something of this world, it is an enterprise that is very human and nothing more than human.

It is a fact — an *empirical* fact — that some people some of the time conform their cognitive practices to the norms of empirical science, to the norms of the inductive method. There are matter-of-fact regularities that explain why this happens. There are regularities of a psychological sort that connect forms of training and education on the one hand with conformity to the norms of science on the other. If there were no such causal patterns how could education on a systematic basis ever be rational? There are also patterns of a cultural and social sort. Some societies are structured to produce scientific inquirers; others are not. Universities in western

cultures will often produce biologists who aim to develop in conformity to the norms of science the scientific theories of Darwin. Such persons are unlikely to be produced in communities in which fundamentalist Christian religious views are dominant.

Again, for historical reasons which sociologists ought to endeavour to explain, the practice of science came to be institutionalized in western European communities in the 17th and 18th centuries but not in, say, mediaeval Chinese culture, or in 14th century meso-American cultures. This fact has something to do with the fact that already by the end of the middle ages, Western European culture was the most advanced technologically that the world had known. It has something to do with the fact mercantile capitalism had come to be the dominant economic form in Europe at that time, dominating society in a way that such a form had not dominated any other society. It had something to do with the fact that in spite of, or perhaps because of, the wars of religion, western Europe turned away from theological dispute, not only in the process becoming more tolerant of disagreement, but also turning their minds to issues such as science that were less divisive. And in all this they found that science was useful, and came to pursue deliberately something which they had come originally to pursue only accidentally. We cannot here explore these themes. The point is that there are many causes and effects that account for the fact that science as a rational enterprise comes to be institutionalized in western Europe in the 17th and 18th centuries.

Rational behaviour, then, the conformity of persons in their cognitive activities to the norms of science, is a set of events that can be explained scientifically. Sociological explanations of science, explanations in terms of matter-of-fact causes, are possible not only for the deviant but for the rational. As S. Shapin has insisted, empirical science as providing a set of cognitive standards or norms does have a social history: "...practical decisions about precision and variation in factual reports are moral judgements and ... they too have a social history."[54]

We may recall the metaphysical theories of Aristotle. In the ancient world and in western Europe in the period through Copernicus these theories were applied to the explanation of ordinary events. But the explanations were not scientific. They appealed not to matter-of-fact causes but to non-empirical natures or forms or essences. These natures were unanalysable powers the exercise of which by the objects were taken to be explanatory. But not every event which occurred in the object was so explained. There were also events that occurred that were contrary to the nature of the thing changing. These were unnatural events, and were explained by the activity of some external object. The nature determined not only how the object did behave, it also provided the standard about how it ought to behave. When the object behaved as it ought to behave then the explanation was in terms of internal principles. When the object behaved as it ought not to behave then the explanation was in terms of external causes.

Those like Laudan and Newton-Smith who defend the arationality principle and insist upon a privileged status for scientific reason are in effect insisting that among the events that occur in the world there are some to which the patterns of science are not relevant and that we must instead use Aristotelian patterns of explanation. This special area of fact to which the Aristotelian patterns apply is that of science itself. There is on the one hand behaviour that conforms to the norms of science. This is natural, this is how people ought to behave, and it is to be explained internally. In contrast, and on the other hand, there is behaviour that fails to conform to the norms of science. This is unnatural, this is how people ought not to behave, and it is to be explained externally, in terms of non-rational causes.

Barnes and Bloor are quite correct in holding, with Hippocrates — and with more recent philosophers of science including logical positivists such as Gustav Bergmann, whose definition of 'ideology' we have just cited — , that such an appeal to Aristotelian patterns in the understanding of the practice of science is at variance with that practice. They are correct in their claim that those such as Laudan and Newton-Smith who defend the arationality principle are the ones who "pose the real threat to a scientific understanding of knowledge and cognition." From the perspective of science coming to understand science, the norms of science have no privileged status. From this point of view, all bodies of knowledge are to be understood in the same way, on the basis of matter-of-fact regularities that are to be investigated by the inductive method science. As Barnes and Bloor put it,

Because [the sociologist of knowledge who attempts to explain science by means of the methods of science] thinks that there are no context-free or super-cultural norms of rationality he does not see rationally and irrationally held beliefs as making up two distinct and qualitatively different classes of thing. They do not fall into two different natural kinds which make different sorts of appeal to the human mind, or stand in a different relationship to reality, or depend for their credibility on different patterns of social organization. Hence the ... conclusion that they are to be explained in the same way.[55]

Rational behaviour and irrational behaviour are thus both subject to explanation by the sociologist of knowledge. Science on the one hand and ideology and pseudoscience on the other are equally the subject of the anthropologist's study. It is important to note, however, that to speak of sociological and anthropological explanations is not to speak in terms of such things as class or imperialist interests alone. Such social causes are no doubt important. *But so is the cognitive interest in matter-of-fact truth.* Barnes and Bloor argue that

...regardless of whether the sociologist evaluates a belief as true or rational, or as false and irrational, he must search for the causes of its credibility. In all cases he will ask, for instance, if a belief is part of the routine cognitive and technical competences handed down from generation to generation. Is it enjoined by the authorities of the society? Is it transmitted by established institutions of socialization or supported by accepted agencies of social control? Is it bound up with patterns of vested interest? Does it have a role in furthering shared goals, whether political or technical, or both? What are the practical and immediate consequences of particular judgments that are made with respect to the belief? All of these questions can, and should, be answered without regard to the status of the belief as it is judged and evaluated by the sociologist's own standards.[56]

What Barnes and Bloor seem here to neglect is that not all interests are vested interests, not all beliefs are formed on the basis of the method of authority. There is also the cognitive interest in matter-of-fact truth. There is also the inductive method of science as the standard for forming belief. It is likely in fact that these are the standards that explain the practice and beliefs of Barnes and Bloor themselves, as *scientific* sociologists of knowledge. More definitely, these are the standards and motives that move the scientists that Barnes and Bloor aim to study. They do accept the

> ... general conclusion ... that reality is, after all, a common factor in all the vastly different cognitive responses that men produce to it. Being a common factor is not a promising candidate to field as an explanation of that variation.[57]

In this they are correct; it is a simple point in the logic of science. Sometimes my watch works when it is wound, sometimes it does not. What is common cannot explain the difference. To explain the difference one needs a difference. So a shared reality does not explain why people have the beliefs that they do, given that they have different beliefs. What are the relevant differences? Among them are different cognitive interests, and different methods for judging propositions worthy of belief. Some people are concerned to discover matter-of-fact truth, others are not. Some people rely on the method of tenacity or on the method of authority, others rely on the inductive method of science. *The interests and standards of science are among the causes of beliefs.*

In fact, of course, there are often both cognitive interests and also social causes that shape the development of science. Consider the debate about phrenology that occurred in early 19th Edinburgh. This has been examined in some detail by G. Cantor and S. Shapin.[58] Cantor has given a detailed account of the methodological debates between the proponents of phrenology as a science and its critics. He points out how methodologically weak was the defence of phrenology, relying for example on inferences based on simplistic induction,[59] and on elastic hypotheses.[60] Shapin points out that this is not the whole story. He establishes that there was great enthusiasm for

phrenology among the working and middle classes with a variety of popular lectures elaborating and defending it. The University of Edinburgh, in contrast, was highly critical of phrenology.[61] Shapin argues that the defenders of phrenology as a scientific psychology were "outsiders" and that

> British phrenology was a social reformist movement of the greatest significance. Coombe [George Coombe, the leading Edinburgh phrenologist] and his circle vigorously, and to some extent successfully, agitated for penal reform, more enlightened treatment of the insane, the provision of scientific education for the working classes, the education of women, the modification of capital punishment laws and the rethinking of British Colonial Policy.[62]

Shapin agrees with Cantor that there was what he (Shapin) refers to as a "technical debate" in terms of evidence, argument, inference, and so on. However, where Cantor keeps his account of the debate and disagreement to this level, Shapin insists that there is more, that this debate must be placed in a broader social context. To restrict the explanation of the debate and argument to the technical level is to rest content with an explanation that leaves out many relevant factors. It is, in other words, to rest content with a *gappy* explanation. Shapin correctly notes that "to say there was a technical debate is not to say that it can be or was separated from the social conflict, nor that such a technical debate does not reflect social and institutional divisions."[63] If one wants a sociological explanation of the debate concerning phrenology which is less gappy than one that restricts itself to the technical debate, then one must turn to such factors as class and social interests. Nonetheless, it is also true, as Cantor insists, that there *was* a debate and that observational data, scientific inferences, and arguments about the adequacy of these *also* were relevant factors in the on-going process. The scientific inquiry was, as Shapin insists, part of a social process, but it was for all that a *scientific* inquiry in which cognitive factors played a causally significant role. Cantor suggests that "Shapin appears to believe that *only* social factors affect and determine the cognitive realm." This in not entirely fair; Shapin does allow that there was the debate that Cantor discusses and which Shapin calls the "technical debate." But the slightly dismissive connotation of this term gives point to Cantor's charge. Be that as it may, it is also true that Shapin allows that the debate did play a causally important role in the process. But Shapin also insists that a scientific explanation of the process couched in these terms alone is *gappy*, and that these gaps must be filled by adverting to other factors, including the social interests of those who participated in the debate. Still, Cantor is surely correct in holding that "The two belief systems and the historical process of interaction may be influenced by social factors, but social factors do not constitute necessary and sufficient conditions for explaining the debate."[64]

We may agree with Cantor, then, that, to repeat, *the interests and standards of science are among the causes of beliefs*. But we may also agree with Shapin that *social interests are also among the causes of beliefs*. Cantor's apparent position that social interests are not relevant to explaining debates in science is not justified; it is to adopt the arationality principle of Laudan, Newton-Smith and others that the standards of science have a privileged status.

It is clear from Cantor's own discussion that to restrict oneself to cognitive factors is to rest content with a gappy explanation. He makes evident that those who defended phrenology were on weak evidential grounds. They claimed to be relying on the inductive method of science to justify their positions. But their case was in fact weak, as their critics from within the University were able to point out. At the same time, it is clear from Shapin's discussion that those who defended phrenology relied upon its theories to justify their efforts at social reform. Accepting that body of knowledge served their social interests. It is evident that, while those who accepted phrenology aimed to justify that acceptance on evidential grounds, in fact they accepted it because they wanted it to be true: if it were true then they could more easily justify their efforts at social reform. In other words, their belief was in Bergmann's term *ideological*. One can become clear on this ideological dimension of the debate only by going beyond the cognitive issues to which Cantor restricts himself and taking into account the social interests to Shapin directs our attention.

It also served the social interests of the upper classes ensconced at the University of Edinburgh to reject the claims of the phrenologists. Considering only the arguments in the "technical debate," the Edinburgh philosophers had the best of it so far as concerns the standards of science. The knowledge system of the University critics of phrenology was thus objectively acceptable. But it was also one which served their social interests. It is important to insist, however, with Shapin, that the explanation of the knowledge system of the critics of phrenology cannot end simply by citing the observational data and inferences which they used to justify the acceptance of that system. If one goes no further, the explanation will remain gappy, and science demands more. That more, in terms of social interests, Shapin attempts to sketch.

Note that to defend the claim that the knowledge system of the defenders of phrenology is ideological is not to hold that the standards of science have a privileged status, in the sense of Bloor and Barnes. Nor is the claim that the knowledge system of the critics of phrenology was objectively justified relative to the standards of science to hold that the standards of science have a privileged status. For, these standards played a role in the debate in the sense that those who participated in the debate accepted them as the standards in terms of which they aimed to justify their knowledge claims.

The interests and standards of science, then, have no privileged status. Barnes and Bloor are right to reject the claim of apparently Cantor and certainly Laudan, Newton-Smith and others that these standards do have such a status. But at the same time, it is important to recognize, with Cantor, that they are for all that factors that are important in the growth of knowledge.

Problems can arise, however. This sort of position in the sociology of knowledge and the sociology of science, unless the appropriate qualifications are made, will provide comfort to the defenders of pseudoscience.

Consider traditional Chinese medicine. We have seen again and again that as a form of medical theory and medical practice it must be reckoned pseudoscience. At least it must be so reckoned to the extent that it makes claims to be true. Given the norms of science, traditional Chinese medicine falls short. But to this sort of criticism it can be replied that the standards of science *have no privileged place*. Those standards can, to be sure, be used to evaluate western science. But traditional Chinese medicine *has its own standards*. Relative to those standards, its theories can practices can be shown to be justified. *And so it would seem that one must conclude that tt is wrong to reject accepting traditional Chinese medicine on the grounds that it violates the standards of western science.*

The same sort of argument can be deployed in other areas. Groups among the First Nations in Canada believe that they have always existed on the land which they inhabit. Academic anthropologists argue that their ancestors came originally from Asia by way of a land bridge in the area of the Bering Strait when the Ice Age had lowered the level of the sea. If the anthropologists are correct, then the beliefs of the First Nations groups are false. It is possible, however, to try to defend those beliefs by holding that the standards of the anthropologists are those of western science and that the First Nations groups have their own standards for justifying or rejecting beliefs. Some such groups, for example, are very deferential to their elders. They consider it inappropriate to question what the elders tell them. In such a culture, then, what is accepted is what has been passed down by elders from generation to generation. Its method of inquiry is the method of authority. Science, in contrast, adopts the inductive method which accepts that all propositions, conclusions and premises alike, can be accepted only tentatively. But *because western science is not a privileged form of knowledge and because the First Nations groups have their traditions of knowledge, it is wrong to reject their beliefs on the grounds that they violate the standards of western science.* So it is not possible to challenge the claim that ancestors of the First nations groups have always inhabited their present domain on grounds that such a belief conflicts with the archaeological evidence collected and evaluated by the standards of western science or that it conflicts with the contrary hypothesis, also supported by evidence gathered according to the standards of western science, that the ancestors had immigrated eons ago by crossing a land bridge in the area of the Bering Strait.[65]

The arguments that there are no privileged knowledge claims elevate the beliefs and cultural traditions of First Nations groups or of the practitioners of traditional Chinese medicine to the level of science for purposes of explaining reality. Or, what amounts to the same, they reduce science to the level of these systems of traditional practices and beliefs.

We seem once again to have fallen into a relativism that is not just pernicious but downright dangerous.

It is pernicious in the way that any relativism is pernicious. If truth is merely that which accords with the cognitive norms of the group, then it is not possible to criticize that group. Those who favour change could never correctly argue that their claims are correct, those of the group wrong.

But conversely, just because the social reformers form a group, then belief will also be justified, that is, justified by their norms, but in any case as equally justified as those with whom they disagree.

Criticism, critical evaluation, thus become impossible where one holds, as the scientist seems required to hold, that science has no privileged status. One would seem to be in the position that political considerations take precedence over disinterested evaluation of knowledge claims. Without granting science a privileged status it would seem that where standards lead us to have beliefs that conflict, the resolution of the disagreement is not by reason but by other means. Rhetoric rather than reason, politics rather than logic, would appear to be the order of the day.

But of course we all think that there really is more to discourse than this. We all want to believe that the racial theories of the Nazis were not only false, but objectively false, and that this is so even though there was widespread support among the citizens of Germany for what these evil persons did. We are not prepared to accept the relativism that seems to be required by the claim that science has no privileged status.

The point is that there is nothing that makes science intrinsically more valuable than other forms of knowledge. Values are relative to culture, and so that makes science valuable to some cultural groups and not to others. So whether we conform to the norms of science depends upon the culture of which we are a part. Similarly, to accept the results of science as true rather than, say, the deliverances of some authority, depends upon the culture of which we are a part. However, what it is necessary to recognize in this context is that *even though what one accepts as true is culture bound, truth itself is not culture bound.* Clemenceau is right. The relativist him- or herself admits as much. On the one hand, such a person wants to say that the body of knowledge adopted by his or her group is the supreme authority on matters of truth. Criticism by others is not possible because one's own standards are authoritative. But they acquire their authority from the social group. There is no higher authority that gives them privileged status. It follows that for other groups the standards which they have are equally authoritative. Relative standards can be authoritative to a group only because no set of group standards is privileged, all sets of group standards are equally authoritative. And so on the other hand, the relativist also wants to say that the alternative bodies of knowledge adopted by other groups are equally authoritative. But you can't have it both ways. As the philosopher W. V. O. Quine explains:

> Truth, says the cultural relativist, is culture bound. But if it were, then he, within his own culture, ought to see his own culture bound truth as absolute. He cannot proclaim cultural relativism without rising above it, and he cannot rise above it without giving it up.[66]

Facts are there, truth is not relative, and the way to truth, or at least the best way that we know, is the inductive method of science. C. S. Peirce had already made this point. The methods of tenacity and authority are both versions of relativism. What I hold to tenaciously need not be what you hold to tenaciously; there is no transcendent way to adjudicate. What one authority prescribes will differ from what another authority prescribes; there is no transcendent way to adjudicate. Both these methods lead to people or groups holding different and inconsistent views. So does the method *a priori* that prescribes that you accept as true any proposition the contrary of which is to you inconceivable as true. Descartes' *a priori* truth is not the same as Leibniz' *a priori* truth. To resolve such conflicts, Peirce argues, "To satisfy our doubts, … it is necessary that a method be found by which our beliefs may be caused by nothing human, but by some external permanency — by something upon which our thinking has no effects."[67] Such is the method of science.

> Its fundamental hypothesis, restated in familiar language is this: There are real things, whose characters are entirely independent of our opinions about them, those realities affect our senses according to regular laws, and, though our sensations are as different as our relations to the objects, yet, by taking advantage of the laws of perception, we ascertain by reasoning how things really are, and any man, if he has sufficient experience and reason enough about it, will be led to the one true conclusion.[68]

There is, then, matter-of-fact truth — objective matter-of-fact truth. But having said that, it must also be said that neither matter-of-fact truth nor science has a privileged status; it is valuable only to those who do value it. Contrary to Aristotle, there is no metaphysical structure of natures and forms that confers objective and absolute value upon these things. In the world of fact there is nothing to confer absolute value on science, to confer upon it a status of intrinsic superiority to other bodies of knowledge. However, and this is the important point, **if one is interested in matter-of-fact truth then, if one is reasonable, then one ought to adopt the means that is best adapted to achieving that end; and so far as we can tell conforming our cognitive practices to the norms of empirical inductive science is the best means available.**

Science will not determine that one *ought* to do science. But, if you want to know what the facts are, if you have a cognitive interest in matter-of-fact truth, then it is the standards of science that you ought to adopt. This is not to say that you ought to desire to know matter-of-fact truth. But it is to say that, **if** you have that cognitive interest, **then** you **ought** to accept the standards of science as the standards of rationality.

Some cultures seem not to have a concern for matter-of-fact truth. Some in fact seem to have a fear of it. They seem to have a fear of it because in fact it would seem to challenge or call into question certain cherished beliefs. The standards of these groups differ from the standards of those groups that accept the norms of science. So long as there are no shared standards, the resolution of conflicts can only be political: since one side eschews rational discourse, rational discourse will not settle issues.

At the same time, however, it is hard to see how any group can have no interest whatsoever in matter-of-fact truth. Do they really reject any interest in how better to alleviate pain and cure disease? That is hard to believe.

And so long as there is such an interest, *so long as there is a cognitive interest in matter-of-fact truth, then there **is** a set of standards, those of empirical science, which rationally are best for settling disputes.* The norms of science are not intrinsically privileged, but *relative to a shared cognitive interest in matter-of-fact truth* then they do have a privileged status.

Moreover, from the fact that science has no privileged status, it does not follow that groups that reject the standards of science cannot be criticized on the basis of those standards. Of course, when the sociologist of knowledge is examining the beliefs of another group, he or she must simply search for causes rather than dividing the beliefs of the groups into those that are rational and those that are irrational and explain the two kinds differently. But it is *also* possible to critically evaluate those beliefs, and to accept or reject them according to one's own standards, and to urge members of the other group to accept or reject them according to one's own standards, that is, according to the standards of science. The members of the other group are unlikely to accept such criticism. After all, they do not share the same standards. Agreement could be achieved only if their was a shared interest in matter-of-fact truth. But if that is absent, if there really are no shared standards, then agreement would not be possible. It still does not follow, however, that it is unreasonable to evaluate the beliefs of others in terms of the standards of science even in cases where they do not share those standards.

Because thou art virtuous, there shall be no more cakes and ale?

Finally, there is another fallacy that ought in this context to be dismissed.

It is often suggested that to claim the right to criticize others leads to, or even is synonymous with, intolerance and arrogance. It is even more often suggested that this is the consequence of the claim that there is an objective truth that can form the basis of such criticism. Here the thought is that one is claiming that one's beliefs are objectively true and that others who hold opposed beliefs are therefore to be criticized as objectively wrong. Then the inference is that such a one will tend to think that he or she has a monopoly on truth and will be tempted to use force or other political means to oppress and repress those who disagree with them.

It is no doubt true that some have fell into that temptation. This is particularly true of those groups whose members accept beliefs on the basis of the method of authority. Certainly, there are many people who are convinced that they have a firm grasp of the truth, the whole truth, and nothing but the truth, and who despise those who disagree with them.

But this fact, that some people are intolerant, does not affect the more basic issue: from the fact that one accepts the standards of science and accepts that matter of fact truth is objective, does it *necessarily* follow that one is intolerant and arrogant and is likely to go beyond the logic of science to resort to force and other political means to get others to accept one's beliefs?

To this the answer is clearly *no*. It is of course true that one who accepts science might at times be tempted to go beyond those standards and to impose his or her beliefs by force. Persons who do science are as human as any others. And like others they might be tempted to do evil. They might even do it. Sometimes the donor of a research grant might try to impose his or her views on the recipients not by rational means but by threatening to withdraw the grant. There is no monopoly on dogmatism.

But the issue is whether there is a *necessary connection* between using the standards of science for critical evaluation, on the one hand, and intolerance and arrogance, on the other. To which, to repeat, the answer is *No*. As Roger Trigg has put the point,

> There is no reason why someone who believes that basic disagreement *can* admit of solution firstly should arrogantly assume that he himself has a monopoly of truth, and secondly should then make others accept his views by force. The mere fact that a disagreement is capable of solution does not of itself suggest which side is right. When two sides contradict each other, whether in the fields of morality, religion or any other area, each will recognize (if they are objectivists) that at least one side must be mistaken. There need be no contradiction between strongly believing that one is right and yet realizing that one could be wrong. Arrogance is not entailed by any objectivist theory.[69]

Tolerance is a virtue. So is conformity to the standards of science for the evaluation of beliefs, both one's own and others'. There is no reason why one should not have both these virtues. Many do.

b. Science as Social Construction

The sociological study of science has over the last years produced a series of studies that have as their aim not merely the scientific understanding of the social processes of the scientific laboratory, nor even the deflation of the alleged superiority of natural over social sciences, but more deeply the notion held by the natural scientists — but not by the anthropologists doing the research — that they are investigating objective matters of fact, and that the upshot of their work will be an improvement in our knowledge of objective truth.

Thus, to take one example. Latour and Woolgar have proposed that physical scientists "make sense of the laboratory in terms of a tribe of readers and writers who spend two-thirds of their time working with inscription devices."[70] Most readers will likely at first glance suppose that these authors are simply trying to be a little naughty, trying to shock either the natural scientist or the reader with an extreme statement of a position they will subsequently give a more serious formulation. This is not so, however. They really do seem to mean the reader to take quite literally what they say. They continue with the remark that scientists

> ...appear to have developed considerable skills in setting up devices which can pin down elusive figures, traces, or inscriptions in their craftwork, and in the art of persuasion. The latter skill enables them to convince others that what they do is important, that what they say is true, and that their proposals are worth funding. They are so skilful, indeed, that they manage to convince others not that they are being convinced but that they are simply following a consistent line of interpretation of the available evidence.[71]

It is usually thought that publication is the means by which scientists communicate their results to one another and to the world, but that what is central is the results themselves. But Latour and Woolgar seem to be suggesting that it is publication itself which is the end of science, as it is the end of novelist. That they are serious about this is made clear by the subtitle of their book, "The Social Construction of Scientific Facts.": the facts that are recorded in the publications of science are nothing more than social constructions. One might agree

with this, if one were to agree that they are the result of a series of negotiations the aim of which is the production of truth. But according to Latour and Woolgar this is not the aim of science. "What," they ask,

> ... drives scientists to set up inscription devices, write papers, construct objects, and occupy different positions? What makes a scientist migrate from one subject to another, from one laboratory to another, to choose this or that method, this or that piece of data, this or that stylistic form, this or that analogical path?[72]

The answer that they suggest is that scientists aim at peer recognition. Now, it is true that scientists want to achieve the approval of other members of their community. They want to be referred to in footnotes, they want to get promoted and receive tenure, they want prizes, perhaps even the Nobel Prize. This is no doubt true. But Latour and Woolgar suggest that this is the only aim of scientists, that what they include in their literary creations is precisely what they calculate will achieve peer recognition.

> It would be wrong to regard the receipt of reward as the ultimate objective of scientific activity. In fact, the receipt of reward is just one small portion of a large cycle of credibility investment. The essential feature of this cycle is the gain of credibility which enables reinvestment and the further gain of credibility. Consequently, there is no ultimate objective to scientific investment other than the continual redeployment of accumulated resources. It is in this sense that we liken scientists' credibility to a cycle of capital investment.[73]

The analogy is to the activities of those who use their capital to make money which then enables them to re-invest in the process and make more money. Money is invested to make money which is then re-invested to make more money. Similarly, according to Latour and Woolgar, scientists undertake their activities as a form of investment to create credibility which is then re-invested to make gain more credibility. There are to be sure some capitalists who make money simply by investing money in money. So there are some scientists who get more credit than they deserve. But those capitalists who are real entrepreneurs invest in processes that produce something that people find desirable, pins, for instance, to use Adam Smith's example. They gain money, profits, because the produce something that has value for consumers. It would seem to be much the same for scientists. They really do want peer recognition. But what they do to obtain this is create something of value that others want. Scientists get credit because they do what scientists value, namely, good science: they succeed in uncovering previously unknown matter-of-fact truth. Latour and Woolgar are wrong to suggest that scientists are simply a "tribe of readers and writers who spend two-thirds of their time working with inscription devices." What is important is not just the writing, but *what* is written: it has to be good science, science that meets the standards of the inductive method.

The objection to Latour and Woolgar is not that they are attempting a sociological or anthropological analysis of science and the practice of science. It is that they offer *bad* sociology. They propose to explain the activities of scientists in terms of interests. But they mention only one such interest, the interest in peer recognition. They fail to mention, and to introduce into their analyses, the cognitive interest in matter-of-fact truth. In fact, they fail to note the obvious point that *scientists can satisfy their interest in peer recognition only if they satisfy the cognitive interest that the community has in matter-of-fact truth.*

A student aims to obtain recognition from his or her professors. He or she aims to get good grades. But in order to get good grades they have to turn in good laboratory reports. It is best for the student to concentrate on writing a good lab report. In fact, if the student simply aims to get good grades, then he or she is likely to fall short. Students of course sometimes get good grades by cheating on their reports; they concentrate on the style and not on what they do in the lab, pulling the wool over the eyes of their profs. But that sort of fakery does not go on for long: the fakery is uncovered. In the end and overall, good grades are a consequence of performing well in the lab, solving the problems that are posed. Similarly for these scientists once they have matured. They will receive peer credit by solving the problems that they confront in the process of scientific inquiry. And

what 'solving the problem' means is the discovery of objective matter-of-fact truth.

In looking at whether or not the student merits a good grade, the prof will look not simply at the behaviour of the student, the actions that he or she performed, but also at what was in the test tube or what the instruments actually said. The prof would have to refer not just to the actions but to the facts — the objective facts — that those actions were designed to produce. The picture of science that Latour and Woolgar seem to describe is one in which peer recognition depends only on the actions of the scientists. As H. M. Collins, another who defends just this picture, has put it, we think otherwise because "much of our knowledge seems so 'solid' as to require a justification in terms other than those which describe human actions."[74] But this view is mistaken, Collins suggests, and scientific facts are not so much *facts* but *artefacts* "which are built by human actors, to be subsequently erected [in their display cases] by a trick invented and worked by human actors…".[75]

The example that Collins uses to argue his case involves the replication of experiments designed to determine whether gravity waves exist.[76] Gravitational waves are predicted by Einstein's general theory of relativity. The various experimenters were all agreed on this, and agreed further both that the energies required for the generation of gravitational waves that could be detected and that there was some possibility of their being detected through some sort of astronomical catastrophe. Even in this latter case, however, serious problems remain, and it is difficult for scientists to replicate the experiments of others. What Collins discovered in his anthropological study of the scientists involved in these experiments is that they were not strongly moved to make their experiments isomorphous to those of the originator of the research.

> There are many possible ways of explaining this, but a convincing interpretation of their actions is that in the absence of general agreement of what is to count as a 'working experiment' in this field, secondary experiments which do not show the same results as the original experiments may still be seen as 'competent' so there is no special impetus to copy the original experiment. Scientists' actions may then be seen as

negotiations about which set of experiments in the field should be counted as the set of competent experiments. In deciding this issue, they are deciding the character of gravity waves.[77]

We can compare the gravity wave experiments with the experiment performed by a student physicist aiming to replicate Galileo's rolling balls down an inclined plane to measure the acceleration of heavy objects in conditions of free fall. There is agreement on how the experiment is to be competently performed. It is known what the result will be when the experiment is performed in the competent way: scientists know the rate of acceleration of objects in free fall. Students have to learn how to do they experiment, and before they have mastered the performance they often get results that are different from those of Galileo. These results are dismissed as the result of incompetent experiments. These sorts of things cannot be said in the case of gravity waves. Here there is no agreement upon how exactly to perform the experiment, nor agreement on what the outcome ought to be. The outcome of the students' experiment, that is, the determination of the acceleration of objects in free fall, is defined *antecedently* to the performance of the experiments. One therefore has antecedently to these experiments a criterion for deciding whether or not the experiment was performed competently: if it does not give Galileo's result it is incompetent and certainly not a replication. But in the case of gravity waves, the outcome of the experiment, that is, the detection of gravity, is *not* defined *antecedently* to the experiments. To the contrary, exactly which experiments are to be counted as competent is decided only as the result of a process of negotiation. The final conclusion whether gravity waves were measured and thereby detected is the result of a kind of social or political negotiation in which the scientists search for the deal that will best secure them peer recognition. But the way in which the detection proceeds is determined by the nature of gravity waves. Hence, in negotiating which experiments are to be counted as competent and as replications of each other the scientists are negotiating the nature of gravity waves. Gravity waves are not as it were there prior to the set of experiments to provide a criterion for determining whether our test is correct or incorrect. Rather, which experiments are correct, that is, are to be counted as correct is determined through negotiation.

Gravity waves are not there antecedently to the experiment, they are the products of the experiment. Facts are not there independent of the social process of science, they are rather the products of that process. Such at least is the argument.

There is, however, a difference between the case of the students measuring the acceleration of freely falling bodies and the case of gravity waves. It is not that we know in the former case what the objective facts are whereas in the latter case we do not yet know such facts. It is not that we have got hold of the objective facts in the one case and have not yet done so in the other. To be sure, it does seem that way. But, the argument is, that appearance is wrong: in neither case are there objective facts to be got hold of. Rather, in both cases the "fact" is a product of negotiation. It is just that in the one case, that of gravity waves, the process of negotiation that produces what comes to be counted as the fact is open, whereas in the other case, that of the acceleration of objects in free fall, the process by which the fact has been produced has been hidden away and is now itself something not open to observation. As Latour and Woolgar put it,

> *Our argument is not just that facts are socially constructed. We also wish to show that the process of construction involves the use of certain devices whereby all traces of production are difficult to detect.*[78]

The inference that Collins is making is this — that of Latour and Woolgar is parallel.[79]

> Gravity waves exist if and only if certain forms of experimental detection are accepted.
> Which forms of experimental detection of gravity waves are accepted is a matter of social negotiation.
> ――――――――――
> Hence, gravity waves are not objectively there to be discovered but are rather a social construction.

Neither of the premises is acceptable. Nor therefore need we accept the conclusion.

It is clear that Collins on the one hand and Latour and Woolgar on the other intend their analysis to be general. They are not interested in gravity waves simply by themselves; they are interested in these things as illustrative of a more general thesis about the nature of science. One must therefore understand their first premise to imply that

> There is no phenomenon without an experimental form by which to detect it.

The absurdity of this is clear. It implies that objects did not undergo free fall at a certain acceleration until Galileo designed the experiments by means of which it was first measured. The premise makes sense only if it is understood to mean that

> Without an experimental form by which to detect it a phenomenon cannot be said to exist (that is, cannot be reported to exist).

But when the premise is understood this way it leaves it open that the phenomenon exists independently of the tests by which it is detected and independently of the experimenters who design those tests.

Collins and Latour and Woolgar rely upon the second reading to make their first premise plausible. But this reading is in effect inconsistent with the first reading. It is, however, the first reading that they need if their argument is to be successful. It is not the first time, nor likely the last, in which we are given arguments that attempt to make unacceptable conclusions plausible through the use of subtly ambiguous premises.

As for the second premise, to the effect that

> Which forms of experimental detection of gravity waves are accepted is a matter of social negotiation.

this makes the claim that the existence of gravity waves and the form of detection that is the result of the negotiation are things that are so intimately connected that they stand or fall together. Again, Collins and Latour and Woolgar obtain their general claim only if this premise is also true in general, that is, only if the following principle holds:

> Which forms of experimental detection of X are accepted is a matter of social negotiation.

But this general version is clearly false. What we believe about gravity waves may well be largely the result of the forms of detection which are agreed upon in the process of negotiation. However, this is not so in the case of objects undergoing free fall, where most of the things that we accept about such motions are not closely tied to the method of detection and measurement used by the students. The latter example, then, shows that the general claim which these philosophers require for their general conclusion is false.

Our anthropologists of science try to make plausible their claim about the nature of science through the use of carefully chosen examples, such as that of gravity waves, where the phenomenon on the one hand and the method of detection and measurement on the other hand really are closely tied together. Even so, it is easy to see that the falsehood of the claim even when it is understood to make only a restricted claim about the particular instance of gravity waves. We know about gravity waves from Einstein's theory. We know from this objectively confirmed theory some at least of the properties that these waves are supposed to have. We know that there are certain forms of instrumentation that can possibly detect waves of the sort that the theory predicts, and we know that there are other forms which will simply not do. No one or at least no one in her right mind would use the techniques of a dowser. Nor would one use a stethoscope or a geiger counter. And so on. One appeals, to the contrary, to what the theory suggests might be relevant, certain sorts of stellar catastrophe. Given what the theory says *about the objective facts of the matter*, we can *justify* using certain forms of experiment as techniques of detection and not using others. So it is simply not true that which forms of detection of gravity waves are accepted is a matter of social convention: it is also a matter of objective fact which methods of detection are to be counted as good. To be sure, the appeal to the objective facts may be a fallible appeal; it does not for all that cease to be an appeal. Collins, when he goes through his example, in effect simply ignores this appeal. So do Latour and Woolgar when they go through theirs.[80]

But once one recognizes the role of this appeal in the process of justification one also recognizes that facts in science are after all *not* simply a matter of social construction.[81]

If we are to provide an anthropological or sociological understanding of the process of science then, as Collins, and Latour and Woolgar, and Bloor and Barnes all insist, we must understand the process though which scientists negotiate which facts come to be accepted into the body of scientific knowledge. In this process of justification, scientists appeal to the standards of the inductive method of science. As part of this process of appealing to the standards of the inductive method, scientists appeal to objective facts. The fact that the process in which this appeal is made is a social process ought not obscure the fact that *the objective facts of the matter that science aims at understanding are not created by us but are there to be discovered.*

To suppose otherwise is to fall into a relativism about facts that is as pernicious, as dangerous, and as silly as other forms of relativism about facts. Anthropological practices in the sociology of knowledge do not occasion such a fall: we can defend the sociological analysis of science while also defending the objectivity of scientific facts. The anthropology of Collins and of Latour and Woolgar that claims otherwise is simply bad science — bad science disguised as good. It is, in other words, yet another case of pseudoscience.

c. Unmasking Observer Bias

Many 19th century economists accepted the "Iron Law of Wages" as a law. George Coombe and his followers in Edinburgh accepted phrenology as a science. Both groups aimed to justify their claims by appeal to the cognitive standards of empirical science. In neither case was this standard met. Their beliefs were in fact ideological, accepted not because they were rationally justified but because the groups wanted them to be true. In both cases the ideology was classist. The belief of the 19th century economist served the interests of the capitalist class against the interests of the working class. The belief of the phrenologists served the interests of the working class against the interests of those who oppressed them.

In both cases, commitment by science to the inductive method was able to correct these errors. Science *is* a self-correcting process, and if it is pursued systematically, as it was in these cases, then instances where objectively unjustified beliefs are accepted come to be uncovered and are corrected.

B. Barnes has proposed that there is a role here for the sociology of knowledge. The latter aims to provide a scientific explanation of the process of science itself. Barnes' suggestion goes beyond this. It is a proposal that the sociology of knowledge can play a role in the process which it aims to explain. Or, to put it another way, the explanations which it offers of some research processes can help such processes achieve the cognitive goals at which they aim.

Sometimes scientists succeed in conforming their behaviour to the standards of science and sometimes they do not. This is parallel to my watch when wound sometimes working and sometimes not working. One must discover in the latter case the cause that *makes the difference.* So also in the former case. In the cases of both the 19th century economists and the Edinburgh phrenologists the scientists did not succeed in conforming their behaviour to the standards of science. Part of what the sociology of knowledge aims to discover are the causes that make the difference between these cases and others where the attempt to conform to the standards of science is successful. The relevant factor that made the difference in the one case was the class interest of the capitalists; the factor that made the difference in the other case was the class interests of the workers.

[To say that the sociology of science aims to discover the causes of research behaviour that deviates from the cognitive norms of science is *not* to say that the *only* explanatory task for the sociology of science is the explanation of cases where there is deviance from the norms of science. The latter was suggested by Laudan and Newton-Smith. But we have already argued to the contrary, that the sociology of science aims to explain both "normal" and "deviant" scientific behaviour. However, having said this, it is also true that the sociology of science does aim at explaining, among other things, deviant behaviour, behaviour that does not conform to the norms of science. What Barnes is arguing is that this *part* of what the sociology of science does can perform a useful function in ensuring conformity to the norms of the scientific method.]

Scientists are moved by a cognitive interest in matter-of-fact truth. They try to conform their behaviour to the standards of science in order to achieve this aim. Classist, sexist, racist, and other ideological factors sometimes interfere with attempts to achieve the cognitive interests of science. Given one's interest in matter of fact truth, one will try to eliminate the ideological factors that interfere with its pursuit. In order to eliminate such factors, in order to ensure that they do not influence one's observations and inferences, these factors must be uncovered. Barnes' proposal is that the sociology of knowledge, precisely because it does uncover such factors, will enable the scientists to eliminate them and thereby the better to fulfill their cognitive interest in matter of fact truth. In effect, Barnes is proposing the sociology of science as a method for eliminating pseudoscience.

His suggestion is this:

> It remains possible in many instances to identify the operation of concealed interests by a subjective, experimental approach. Where an actor gives a legitimating account of his adhesion to a belief or a set of beliefs we can test that account in the laboratory of our own consciousness. Adopting the cultural orientation of the actor, programming ourselves with his programmes, we can assess what plausibility the beliefs possess for us. In so far as our cognitive proclivities can be taken as the same as those of the actor, our assessment is evidence of the authenticity of his account.[82]

Barnes is thus proposing that the sociologist attempt to use his or her capacity for empathy or sympathetic understanding to as it were enter into the thought processes of the scientist whose work is the subject of the investigation. This attempt at empathetic understanding will generate hypotheses about the factors that are work in the inferences of the scientist. In particular, it will generate hypotheses about those non-rational factors that might make the difference between sound scientific thinking and ideological thinking. These hypotheses will then be open to testing, and if they are confirmed, if the factors they suggest are relevant really are so, then the scientist will be able to eliminate their influence.

It is in this way that the sociologist of knowledge can help the process of science by alerting researchers to possible sources of bias which the researcher can eliminate once they are drawn to his or her attention.

While this suggestion of Barnes is helpful, there is more that can be said. In particular, it is important to recognize that the sociologist of knowledge must bring to bear some sort of theory or framework into which he or she will fit his or her hypotheses. There will have to be some abstract generic theory to render the hypotheses plausible. What if this theory itself is held on what are partially at least ideological grounds? In that case, bias will enter into the very attempt to eliminate bias. If this bias is culture-wide, then it will be difficult to eliminate that bias. To be sure, science as a self-correcting process will uncover these sources of error provided that one's commitment to working within the framework of scientific rationality is sufficiently strong. Still, as Sandra Harding has indicated,

> ...culture-wide assumptions which subsequently are among the most difficult to identify make their way into the research process and shape the claims that result.[83]

These biases enter into the process at that point where one begins to formulate hypotheses to put to the test, in which has been called the "context of discovery." As Harding suggests,

> It is here that problems are identified and designated as scientific ones, concepts selected, and hypotheses formulated.

The same problem arises when it comes to evaluating evidence and the conclusions drawn therefrom. This is the "context of justification" where "peers ... certify research decisions, processes, and outcomes."

> Scientific communities that are designed (intentionally or not) to consist only of like-minded individuals lose exactly that economic, political, and cultural diversity that is necessary to enable those who count as peers to detect the dominant culture's values and interests. The main problem here is not that individuals in the community are

androcentric, Eurocentric, or economically overprivileged (though that certainly doesn't help), but, instead, that the normalizing, routine conceptual practices of power are exactly those that are least likely to be detected by individuals who are trained not to question the social location and priorities of the institutions and conceptual schemes within which their research occurs.[84]

She is surely correct when she states that the norms of scientific practice which aim at securing conformity to the standards of objective scientific rationality are too weak when they "can identify only those values and interests that differ within a homogeneous scientific community, and when ... [there are] no strategies for gaining causal, critical accounts of the dominant cultural standards."[85]

We ought all to be sensitive to the temptation to claim objectivity for our views. The progress of science depends upon such sensitivity. We are unfortunately all too familiar with the refrain that "My views are objective science, yours are pseudoscientific ideology." This may of course be a justified claim, as it was in the case of those at the University of Edinburgh who were critical of phrenology. At other times such a call is unjustified, as it certainly was in the case of the 19th century economists who claimed that they had objective data and arguments to justify their proposal that the "Iron Law of Wages" really was an exceptionless law of nature. Racist theories in anthropology and sexist theories in physiology were also claimed, with equal lack of justification, to have the status of objective science, when they were to the contrary ideology.

At times the appeal to "objective standards" was used to silence opposition. We have argued that the commitment to the thesis that there are objective standards in science does not *of necessity* lead to *arrogant* attempts to suppress opposition. It does not *of necessity* lead to the attempt to suppress rational discourse by political means. However, even though there is no *necessary connection*, it is also true that claims of objectivity, especially when they are themselves ideologically based, often lead to political efforts to prevent the introduction of hypotheses and theories which are opposed to those for which "objective" status is claimed.

What is needed here is of course the recognition of the practitioners of all science that their claims must always be understood to be, in Judge Overton's term, "tentative." This follows from the logic of the matter. The cognitive interest of science is in *matter-of-fact regularities* which it uses for purposes of explanation and prediction. But a matter-of-fact regularity always makes a claim about a population where the data available constitute only a sample. It is this logical fact, as we have emphasized, which is the logical basis justifying Judge Overton's characterization of science as always *tentative*.

Once the tentative nature of scientific claims is recognized, then what is called for is *discipline*. Thomas Haskell has made the point nicely:

> There very possibility of historical scholarship as an enterprise distinct from propaganda requires of its practitioners that vital minimum of ascetic self-discipline that enables a person to do such things as abandon wishful thinking, assimilate bad news, discard pleasing interpretations that cannot pass elementary tests of evidence and logic, and, most important of all, suspend or bracket one's own perceptions long enough to enter sympathetically into alien and possibly repugnant perspectives of rival thinkers. All of these mental acts — especially coming to grips with a rival's perspectives — require *detachment*, an undeniably ascetic capacity to achieve some distance from one's own spontaneous perceptions and convictions, to imagine how the world appears in another's eyes, to experimentally adopt perspectives that do not come naturally — in the last analysis, to develop, as Thomas Nagel would say, a view of the world in which one's own self stands not at the center, but appears merely as one object among many.[86]

The sort of sympathetic understanding that Barnes recommends as useful in uncovering bias and forestalling pseudoscience is possible only with a careful self-discipline of the sort that Haskell describes.

But there are other strategies that can be implemented and made part of the normative structure of science. Susan Harding has made clear that science requires strategies that would permit one

> ... to detect the social assumptions that (a) enter research in the identification and conceptualization of scientific problems and the formation of hypotheses about them (the "context of discovery"), (b) tend to be shared by observers designated as legitimate ones, and thus are significantly collective, not individual, values and interests, and (c) tend to structure the institutions and conceptual schemes of disciplines. These systematic procedures would also be capable of (d) distinguishing between those values and interests that block the production of less partial and distorted accounts of nature and social relations ("less false" ones) and those — such as fairness, honesty, detachment, and, we should add, advancing democracy — that provide resources for it.[87]

Among the strategies that might be adopted is that of entertaining hypotheses that "start thought from marginal lives."

> ...if one wants to detect the values and interests that structure scientific institutions, practices, and conceptual schemes, it is useless to frame one's research question or to pursue them only within the priorities of these institutions, practices, and conceptual schemes. One must start from *outside* them so as to gain a causal, critical view of them. One important way to do so is to start one's thought from marginal lives.[88]

This is to propose that if one is, as Barnes suggests, to use the sociology of knowledge to aid the process of research, then one needs a sociological theory which effectively locates the causes that lead people to deviate from scientific rationality. Barnes does not elaborate such a theory beyond suggesting that it must take account of interests, such as class interests, which are other than the cognitive interests that motivate science. Harding proposes a slightly more definite structure to the theory that the sociologist of knowledge brings to bear as he or she attempts to understand the process of research. This theory will take account of the role of marginalized groups

in society. These groups are sited at determinate, objective locations in the social structure.

> Such locations are not just accidentally outside the center of power and prestige, but necessarily so. It is the material and symbolic existence of such oppositional margins that keep the center in place: the rich can only be rich if there are others who are economically exploited; masculinity can only be an ideal if it is continuously contrasted with a devalued other: femininity.[89]

What needs to be developed in detail and brought to bear on the scientific understanding of the process of research is a theory that takes into account as relevant variables the *social relations* that connect members of these groups.

Such a theory will explain in less gappy fashion than presently available theories the processes of research and discourse in science. A theory of this sort will explain those processes.

But what Harding is proposing is that it can serve a further function beyond that simply of understanding in scientific terms the process of scientific research. The aim is *not* so much to produce theories about marginal groups but something more, a theory that can be usefully employed by both marginal and non-marginal people in generating better scientific research, one free of classist, racist, sexist and other forms of bias. The aim is

> ...to generate scientific problems not from within the debates and puzzles of the research traditions, not from the priorities of funders or dominant policy groups, but from outside these conceptual frameworks, namely, from the lives of marginalized peoples; and to develop this thought through democratic dialogues between knowledge-producing groups.[90]

In order to overcome bias, one needs to consider theories that do not conform to the presently accepted paradigm, the abstract generic theory that one is using to guide one's research. But there are many alternatives theories. Goodman's introduction of "gruesome" hypotheses makes this clear. In order to select a theory to guide one's research one will need some criteria. Harding is proposing the Kuhnian solution of using subjectively variable criteria. Each marginalized group will appeal to its own criteria and will develop the theory that is thus selected. If it does locate variables that are objectively at work, such a theory will, like that of Einstein, eventually come to be recognized by all as useful in the explanation and prediction of the phenomena that one is attempting to understand. Harding is recommending that science, in the hope of detecting and overcoming bias in the process of research, use the techniques of revolutionary science recommended and defended by Kuhn.

We have emphasized, however, that science is rightly justified in rejecting theories, such as that of Velikovsky or that of homeopathic medicine, that conflict with well established theories of science, Newtonian theory in the first of these cases and the law of mass action in chemistry in the other. This is the rational thing to do. Revolutionary strategies of the sort advocated by Kuhn, strategies in which "anything goes," are rationally to be adopted only when the processes of normal research break down.

We ought not to view the proposals of Harding as challenging these methodological points. The sociology of science, the discipline with which she is concerned, is relatively undeveloped. There is no well worked out generic theory here as the Newtonian theory is well established in planetary astronomy or the law of mass action in chemistry. In the absence of a well confirmed abstract generic theory to guide research, in what Kuhn has called "pre-paradigm" science, the sorts of strategies that are used in revolutionary science can also be used here to guide research. This is Harding's proposal. It is entirely reasonable. It is one way for us to detect pseudoscience. Conflict with a well established paradigm is another. Both must be used.

In the end, those who are committed to scientific rationality must be vigilant and disciplined in its pursuit. Nothing else will enable us to fulfill the cognitive interests of science. It is only in this way that we can avoid falling into the traps of pseudoscience. Nor is this a guarantee. We might still fall into irrationality and error: scientists, too, are human and fallible. But it is the best that we can do. One cannot do more. One must not do less.

Endnotes

1   Kendrick Frazier and James Randi, "Predictions After the Fact: Lessons of the Tamara Rand Hoax," *Skeptical Inquirer*, Fall 1981, pp. 4-7.

2   See Richard C. Carrier, "Flash! Fox News Reports that Aliens May Have Built the Pyramids Of Egypt!" *Skeptical Inquirer*, vol. 23, September/October 1999, pp. 46-50.

3   Gustav Jahoda, *The Psychology of Superstition* (Baltimore, MD: Penguin, 1981), pp. 50-51.

4   Stephen Jay Gould, Lecture at California Institute of Technology, 1985, quoted in Michael Shermer, *Why People Believe Weird Things* (New York: W. H. Freeman, 1997), p. 153.

5   See John Wolf and James S. Mellet, "The Role of 'Nebraska Man' in the Creation-Evolution Debate," *Creation/Evolution*, vol. v (1985), pp. 31-43.

6   Wolf and Mellet, "The Role of 'Nebraska Man' in the Creation-Evolution Debate," p. 41.

7   Pat Moffat, "A Lesson in Eastern Medicine," *The Globe and Mail*, Tuesday, September 22, 1998.

8   For an evaluation of a claimed 1967 film record of Bigfoot, see D. J. Daegling and D. O. Schmitt, "Bigfoot's Screen Test," *Skeptical Inquirer*, vol. 23, May/June 1999, pp. 20-25.

9   Compare D. M. Zuefle, "Tracking Bigfoot on the Internet," *Skeptical Inquirer*, vo. 23, May/June 1999, pp. 26-28. "Two years of 'hunting' Bigfoot in cyberspace tells you little about the never-confirmed giant bipedal creature but a lot about those who hunt for it. Some are sincere searchers; for others, the idea is a complex and flexible belief system that serves multiple needs and rules" (p. 26).

10 See Analisa Berta, "What is a Whale?" *Science*, vol. 263, January 14, 1994, pp. 180-181.

11 J. M. G. Thewissen, S. T. Hussain, and M. Arif, "Fossil Evidence for the Origin of Aquatic Locomotion in Archaeocete Whales," *Science*, vol. 263, January 14, 1994, pp. 210-212.

12 *Ibid.*, p. 210.

13 See Jacalyn Duffin, *To See with a Better Eye* (Princeton, NJ: Princeton University Press, 1998).

14 The name 'stethoscope' for this new instrument was coined by Laennec himself; it means "chest examiner."

15 Due to antibiotics, tuberculosis is today relatively uncommon, but in the form of scrofula it is even more uncommon. Both forms have reduced frequency because most of the milk we use is now pasteurized, a process that kills the bacilli that are the cause of the disease and which in earlier times entered the milk from tubercular cows.

16 See John Horner and James Gorman, *Digging Dinosaurs* (New York: Workman Publishing, 1988).

17 See John D. Morris, "The Paluxy River Tracks," *Acts and Facts*. ICR Impact Series, no. 35, May 1976.

18 Berney Neufeld, "Dinosaur Tracks and Giant Men," *Origins*, 2 (1975), no. 2, pp. 74-76; quoted in C. G. Weber, "Paluxy Man — The Creationist Piltdown," *Creation/Evolution*, vol. 2, fall 1981, pp. 16-22, at 18.

19 John D. Morris, "The Paluxy River Tracks."

20 Berney Neufeld, "Dinosaur Tracks and Giant Men."

21 Neufeld, "Dinosaur Tracks and Giant Men," quoted in Weber, p. 19.

   Note that Neufeld is a Seventh-Day Adventist. This is a fundamentalist denomination and its members have an interest in establishing the truth of the Genesis account of creation.

22 John D. Morris, "The Paluxy River Tracks."

23 See I. Klotz, "The N-Ray Affair," *Scientific American*, vol. 242, May 1980, pp. 168-175.

24 See R. W. Wood, "The N-Rays," *Nature*, vol. 70, no. 1822, September 29, 1904, pp. 530-531.

25 *Ibid.*, p. 530.

26 *Ibid.*, p. 530.

27 *Ibid.*

28 Klotz, "The N-Ray Affair," p. 175.

29 Cf. K. Duncker, "The Influence of Past Experience upon Perceptual Properties," *American Journal Psychology*, 52 (1939), pp. 255-265.

30 C. M. Turnbull, "Some Observations Regarding the Experiences and Behavior of the Ba Mbuti Pygmies," *American Journal of Psychology*, 74 (1961), pp. 304-308.

31 Cf. Bruce Murray, "The Limits of Science," in *UFOs and the Limits of Science*, pp. 255-265.

32 E. U. Condon, "Summary of the Study," in E. U. Condon, *Final Report of the Scientific Study of Unidentified Flying Objects* (New York: E. P. Dutton, 1969), p. 9.

33 For this case, see Philip J. Klass, "UFO's: Fact or Fantasy?" *Humanist*, vol. 36, July-Aug. 1976, pp. 9-13.

34 *Ibid.*

35 P. J. Klass, "UFO"s: Fact or Fantasy?"

36 *Ibid.*

37 *Ibid.*

38 Terence Hines, *Science and the Paranormal* (Buffalo, NY: Prometheus Books, 1988), p. 175.

39 Klass, "UFO's: Fact or Fantasy?"

40 *Ibid.*

41 See the report from the Committee on Techniques for the Enhancement of Human Performance, Commission on Behavioral and Social Sciences and Education, National Research Council, *Enhancing Human Performance: Issues, Theories and Techniques* (Washington, D. C.: National Academy Press, 1988), pp. 304-305.

42 *Ibid.*

43 *Ibid.*

44 Cf. James Alcock, *Parapsychology: Science or Magic? A Psychological Perspective* (New York: Pergamon Press, 1981), Ch. 4.

45 F. Ayeroff and R. P. Abelson, "ESP and ESB: Belief in Personal Success at Mental Telepathy," *Journal of Personality and Social Psychology*, 34 (1976), pp. 240-247.

46 Cf. D. H. Rawcliffe, *Occult and Supernatural Phenomena* (New York: Dover, 1959).

47 Barry Barnes and David Bloor, "Relativism, Rationalism and the Sociology of Knowledge," in M. Hollis and S. Lukes, eds., *Rationality and Relativism* (Oxford: Blackwell, 1982), pp. 21-47, at pp. 21-22.

48 *Ibid.*, p. 22, n. 5.

49 This example is adapted from G. Bergmann, "Ideology," *Ethics*, 59 (1951), pp. 123-138.

50 *Ibid.*

51 L. Laudan, *Progress and Its Problems* (Berkeley, Calif.: University of California Press, 1977), p. 202.

52 W. Newton-Smith, *The Rationality of Science* (Oxford: Oxford University Press, 1981), p. 238.

53 I. C. Jarvie, "Laudan's Problematic Progress and the Social Sciences," *Philosophy of the Social Sciences*, 9 (1979), p. 496.

54 S. Shapin, *A Social History of Truth* (Chicago: University of Chicago Press, 1994), pp. 310-311.

   Shapin notes how the practice of science rests importantly on the capacity of scientists to trust the correctness of the observation reports of others participating in the research process. He argues that this institutionalization of trust has it origins in the development of the conventions of civility in Britain in the 17th and 18th centuries. "At one extreme, vouching was not deemed necessary, and any demand for the provision of bona fides would be grounds for resentment. A gentleman's word was his bond .... At the other extreme, testimony might only be put into circulation if it or its source was vouched for by others of known standing, sincerity, or skill" (pp. 305-306). The latter he describes as the "calibration of one dubiously trustworthy source by others assumed to be trustworthy" (p. 21). Shapin notes how shared practices and beliefs are important in creating and justifying as well-placed the required sense of trust. Those involved in the process "seek to discipline trust by *plausibility*, by comparing the claim in question with an overall ordered sense of what the world is like" (p. 21). This fits with the picture of the growth of normal science presented by T. Kuhn in his *The Structure of Scientific Revolutions*. An hypothesis is plausible just in case that it fits with the paradigm, or, in our terms, the abstract generic theory that is accepted by the research community.

It should be noted that Shapin, without actually requiring a relativistic position, does present his book from the perspective of what he refers to as a "liberal" notions of truth (p. 5), a notion that equates truth with consensual belief: "...truth is a matter of collective judgment and ... is stabilized by the collective actions which use it as a standard for judging other claims" (p. 6). He contrasts this with what he refers to as a "restrictive" notion of truth (p. 5), which distinguishes truth and belief, and maintains the truth of a belief is a matter of the mind-independent reality that the belief purports to describe. Nothing much in the discussion turns on the rejection of the restrictive notion of truth, and one who rejects the relativism of the "liberal" notion could give the work the title *A Social History of Certain Forms of Consensual Belief* and be none the worse for it, easily able to accept almost all of Shapin's arguments. Shapin more or less agrees, pointing out (p. 4) that nothing in his book provides an argument for relativism.

However, Shapin does suggest that relativism is important methodologically for the historian. He indicates his view (pp. 4-5) that adopting an anti-relativist position creates a bias in favour of one's own set of beliefs, thereby preventing one for being curious about culturally dependent variations in belief systems. Relativism, in contrast, encourages the historian to leave aside his or her own belief system and to develop a curiosity about culturally different systems. It also

encourages, he further suggests (p. xvi), the disposition to interpret historical actors in their own terms. His idea is that good history requires relativism. The case is hardly convincing. We will be looking in greater detail below at some of these points, particularly concerning the supposed relation between relativism and tolerance, but there are aspects of Shapin's discussion that deserve attention in their own right.

In particular, we should note that Shapin never makes clear why the anti-relativist is somehow prevented from developing an interest in why different groups have culturally different sets of beliefs. Nor does he explain why it is impossible for anti-realists to refrain from using their own knowledge when trying to explain the beliefs of historical actors in their own terms. In these respects his case is indeed unconvincing.

In any case, it is simply not true that an epistemological relativism should require that we be epistemologically tolerant. No more than the fact that one is a moral relativist entails that tolerance is among the set of one's moral principles. Among the principles that a relativist may accept as his or her own, whether epistemic or moral, are principles of intolerance.

If the relativist really does hold that his or her view is the truth about truth then he or she falls into the paradox that renders relativism simply unacceptable. Shapin does nothing to avoid this problem. Moreover, he himself wants to distinguish between what really holds of a community and what the actors in the community think holds of it. If he is to maintain consistently a relativist position that insists on seeing the actors' world in the actors' own terms he cannot make this distinction.

But there is a more specific case to be made against Shapin that relativism is good for history. For, after all, the sort of case that an historian such as Shapin wants to build up concerning the social origins of science is based on testimony, and that requires the historian, unless he or she is completely credulous, to distinguish between that testimony that is truth and that which is false; selective acceptance of some testimony as more reliable than other testimony makes no sense from the relativist position. Conversely, relativism makes for credulity which makes for bad history.

Moreover, all the actors about whom Shapin writes were realists in the sense of being anti-relativists. Shapin the relativist, to write about them consistently in their own terms, must adopt that realism. The paradoxes of relativism constantly re-assert themselves.

The only way out is to reject relativism, and recognize that what Shapin is writing about is a social history of belief, not a social history of truth. At the same time one should also accept what he is trying to achieve with his methodological relativism, and simply refuse when one is writing as an historian of thought to interpret the beliefs of past actors in terms they did not employ.

That, by the way, will not conflict with the fact that sometimes one will want to *explain* what the actors are doing in terms that they did not themselves use. One might want, for example, to explain the outcomes of certain actions in terms of the actors' beliefs being objectively false, where the actors did not themselves recognize that objective falsity. One might believe that boiling water will not hurt one. The result upon putting one's hand into boiling water will be a scald. One must explain why the actor did what he or she did in terms of this or her own belief system which held that boiling water is not harmful. But one must explain the scald also in terms of the objective fact that rendered the belief false. Blondlot recorded the results he did because his beliefs about N-rays were objectively false; but at the same time one must also interpret his actions in his own terms, including his own belief that there are N-rays and that these have certain properties.

A discussion making some other relevant and similar points can be found in Peter Lipton, "The Epistemology of Testimony," *Studies in History and Philosophy of Science*, 29 (1998), pp. 1-31. See also M. Feingold, "When Facts Matter," *Isis* , 87 (1996), pp. 131-139.

55 Barnes and Bloor, "Relativism, Rationalism, and the Sociology of Knowledge," pp. 27-28.
56 Barnes and Bloor, p. 23.
57 Barnes and Bloor, p. 34.
58 G. Cantor, "Phrenology in Early Nineteenth-Century Edinburgh," S. Shapin, "Phrenological Knowledge and the Social Structure of Early Nineteenth-Century Edinburgh," and G. Cantor, "A Critique of Shapin's Social Interpretation of the Edinburgh Phrenology Debate," *Annals of Science*, 33 (1975), pp. 195-218, pp. 219-243, and pp. 345-56, respectively.
59 Cantor, "Phrenology in Early Nineteenth Century Edinburgh," p. 211.
60 *Ibid.*, p. 212.
61 The critics of phrenology were surely correct. Whatever basis it might have had in objective fact had long since disappeared. Nor did the methodological criticisms the phrenologists made of their critics have much substance. See F. Wilson, "Mill and Comte on the Method of Introspection," *Journal for the History of Behavioral Science*, 27 (1991), pp. 107-129, and "Some

62 Shapin, "Phrenological Knowledge and the Social Structure of Early Nineteenth-Century Edinburgh," p. 232.
63 Shapin, p. 234.
64 Cantor, "Critique of Shapin's Interpretation," p. 255.
65 Cf. G. A. Clark, "NAGPRA, Science, and the Demon-Haunted World," *Skeptical Inquirer*, May/June 1999, vol. 23, no. 3, pp. 44-48.
   ("NAGPRA" is the Native American Graves Protection and Repatriation Act recently enacted into law in the United States.)
66 W. V. O. Quine, "On Empirically Equivalent Systems of the World," *Erkenntnis*, 9 (1975), pp. 327-28.
67 C. S. Peirce, "The Fixation of Belief," in his *Collected Papers*, 6 vols. (Cambridge, Mass.: Harvard University Press, 1931-35), vol. IV, pp. 223-47, at p. 244.
68 *Ibid.*
69 Roger Trigg, *Reason and Commitment* (London: Cambridge University Press, 1973), pp. 135-136.
70 B. Latour and S. Woolgar, *Laboratory Life: The Social Construction of Scientific Facts* (London: Sage, 1979), p. 69.
71 *Ibid.*, pp. 69-70.
72 *Ibid.*, p. 189.
73 *Ibid.*, p. 198.
74 H. M. Collins, "The Replication of Experiments in Physics," in B. Barnes and D. Edge, eds., *Science in Context* (London: The Open University Press, 1982), pp. 94-116, at p. 94.
75 *Ibid.*, p. 95.
76 Latour and Woolgar argue the same case but with a different example. For discussions of this example, of a piece with the following discussion of Collins, see J. R. Brown, *The Rational and the Social*, (London: Routledge, 1989), Ch. 4, and I. Hacking, "The Participant Irrealist at Large in the Laboratory," *British Journal of the Philosophy of Science*, 39 (1988), pp. 277-294.
77 Collins, p. 111.
78 Latour and Woolgar, *Laboratory Life*, p. 176.
79 Latour and Woolgar, *Laboratory Life*, p. 156ff.
80 Cf. J. R. Brown, *The Rational and the Social*, pp. 84ff.
81 Cf. Hacking's remark that "There is no call for irrealism on the basis of the observations of Latour and Woolgar. Their microsociology is used, in this realism, not to explain the construction of a scientific fact, bot to

explain the selection of fact ... to enter our stream of knowledge" ("The Participant Irrealist at Large in the Laboratory," pp. 292-293).

82 B. Barnes, *Interests and the Growth of Knowledge* (London: Routledge, 1977), p. 35.

Barnes' proposal would not work if he were, as Brown (*The Rational and the Social*, p. 16ff) suggests, a relativist. But as we have seen and contrary to what Brown claims, Barnes' sociology of knowledge does not commit him to relativism.

83 S. Harding, "After the Neutrality Ideal: Science, Politics, and 'Strong Objectivity'," *Social Research*, 59 (1992), pp. 567-587, at p. 578.

84 *Ibid.*, pp. 578-579.

85 *Ibid.*, p. 579.

86 T. Haskell, "Objectivity Is Not Neutralilty," *History and Theory*, 29 (1990), p. 132. The reference is to T. Nagel, *The View from Nowhere* (Oxford: Oxford University Press, 1986), pp. 4-6.

87 Harding, "After the Neutrality Ideal," p. 580.

88 Harding, "After the Neutrality Ideal," p. 581.

89 Harding, "After the Neutrality Ideal," p. 581.

90 Harding, "After the Neutrality Ideal," p. 583.

INDEX